SEROTONIN

SEROTONIN

BY

S. GARATTINI

Director, Institute of Pharmacological Research "Mario Negri", Milan; Professor of Chemotherapy and Pharmacology, University of Milan (Italy)

AND

L. VALZELLI

Psychopharmacology Section, Institute of Pharmacological Research "Mario Negri", Milan; Professor of Pharmacology, University of Milan (Italy)

WITH A FOREWORD BY

I. H. PAGE

Director, Research Division, Cleveland Clinic, Cleveland, Ohio (U.S.A.)

ELSEVIER PUBLISHING COMPANY Amsterdam · London · New York 1965

ELSEVIER PUBLISHING COMPANY
335 JAN VAN GALENSTRAAT, P.O. BOX 211, AMSTERDAM

AMERICAN ELSEVIER PUBLISHING COMPANY, INC.
52 VANDERBILT AVENUE, NEW YORK, N.Y. 10017

ELSEVIER PUBLISHING COMPANY LIMITED
RIPPLESIDE COMMERCIAL ESTATE, BARKING, ESSEX

LIBRARY OF CONGRESS CATALOG CARD NUMBER 63-16081

WITH 28 ILLUSTRATIONS AND 35 TABLES

PRINTED IN THE NETHERLANDS

FOREWORD

It has been only about fifteen years since serotonin was discovered and synthesized. But modern science is so alert and skillful that in that short period an almost unbelievable amount of information about it has accumulated. All sorts of revealing, and at times bizarre, facts and notions have appeared. Defense mechanism, vasoactive substance, neurotransmitter, psychotropic drug, hormone of gastrointestinal mobility, participant in thrombus formation and anaphylaxis are only some of the proposed functions that have confused and edified people. But then, people are seldom aware of the economy with which the body uses substances. Besides, it is just possible that all of these functions are subserved, one compound serving many, and widely diverse, functions. It seems clear in the case of serotonin that its action in the brain, for instance, is quite different from its action in the rest of the body.

Serotonin made the clinicians aware of the fact that the carcinoid tumor did, after all, have signs and symptoms. Before serotonin, it was considered to have neither. So often, the discovery of a substance has led to recognition of clinical conditions which were always there but which we did not see.

All of this is to point out the magnitude of the task Professors Garattini and Valzelli have undertaken. This is a book which documents what they have to say — and that is a lot. There are also several thousand references. I can testify to the fact that they have done their work well, which means they have done the rest of us an important service. May I, as one of their readers, acknowledge our debt?

I am sure my Italian colleagues must be as baffled as I am by the rapidity with which such diverse knowledge as that about serotonin accumulates. People are most extraordinarily ingenious; what they will discover! Serotonin in walnuts and bananas and in jelly-fish and in carcinoids and in brain leaves me with a sense of bewilderment, and fear that some will ask me what serotonin does. If they do, I will recommend that they read Garattini and Valzelli's book. They will be as likely to find the answer there as anywhere I know.

Serotonin is a product of international research. It is fitting that a book about it should be written by Italians because an Italian, Erspamer, made the first serotonin-containing extracts. Purification, identification and synthesis then occurred in the United States but with no idea that it was the then unknown substance contained in these tissue extracts. Study of serotonin has occupied thousands of scientists the world over which illustrates perfectly the unifying and cohesive power of research itself.

No physiological substance has been discovered, of which I am aware, that is believed to have such diverse actions in the body. I suspect that just as with other humoral agents such as catecholamines, serotonin will be found to function in the body at specific sites and in concentrations far lower than those used for pharmacological experiment. One of the great weaknesses of current pharmacology is the use of highly artificial conditions for the conduct of experiments. Anesthesia, trauma and high concentrations of test substances bear little resemblance to the conditions for the action of humoral substances during normal functions. Just as the tissue morphologist has gone to the "molecular level", so must the pharmacologist begin thinking in terms of the molecular equilibria which constitute the functional chemical fabric of the living body.

My congratulations are due to my Italian colleagues for the skill with which they have helped to build the large body of knowledge concerning the functions of the now large group of tryptamines which began so inauspiciously with serotonin.

The Cleveland Clinic Foundation,   IRVINE H. PAGE, M.D.
Cleveland, Ohio (U.S.A.), April 1965

ACKNOWLEDGEMENTS

The Authors gratefully acknowledge the help of Dr. J. Shewell, London, and Dr. D. B. Cater, Cambridge, for correcting the English text.

Milan, April 1965   S. GARATTINI

                                                                             L. VALZELLI

CONTENTS

Chapter 1

CHEMICAL PROPERTIES

STRUCTURE AND SALTS

Serotonin, 5-hydroxytryptamine, is usually extracted or prepared as the creatinine sulphate salt[48,52]. The base has the following structural formula:

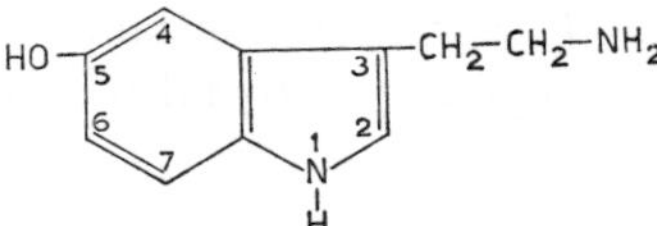

Fig 1. 5-Hydroxytryptamine.

The molecular weight of 5-hydroxytryptamine ($C_{10}H_{12}N_2O$) is 176.2 and that of the creatinine sulphate salt* ($C_{14}H_{23}N_5O_7S$) is 405.4. 5-Hydroxytryptamine also exists, with different degrees of stability, as the salt of other acids. The best known of these are the picrate, the hydrochloride and the salicylate. 5-Hydroxytryptamine, which does not possess any asymmetric carbon atoms and is therefore optically inactive, crystallizes out as small whitish lamellae. Each salt has a characteristic melting point. The hydrochloride melts[18,33] at 167–168°C; the picrate** (mol. wt. 423.3), according to various authors, melts with decomposition[1,24,33] between 185° and 194°C and the bisoxalate has a melting point of 202–204°C. The creatinine sulphate[52,53,57], whether extracted or synthesized, melts at 212–216°C.

SYNTHESIS

5-Hydroxytryptamine, originally obtained by extraction[8,22,50], is now synthesized by different methods. Only processes of historical interest will be considered here.

The first synthesis was described by Hamlin and Fischer[33] in which 5-benzyloxyindole[19] is converted successively to 5-benzyloxygramine, 5-benzyloxyindolyl-3-acetamide, 5-benzyloxytryptamine and by a catalytic debenzylation to one of the salts of 5-hydroxytryptamine. The synthesis described by Speeter *et al.*[57] uses as intermediates 5-benzyloxyindole and 5-benzyloxyindole-acetonitrile, while Asero *et al.*[1] starting from *m*-cresol obtained first 5-methoxyindole, then 5-methoxyindolyl-β-acetonitrile, 5-methoxytryptamine and finally 5-hydroxytryptamine picrate. Other

---

* 1 mg of 5-hydroxytryptamine creatinine sulphate corresponds to 0.43 mg of the base.
** 1 mg of 5-hydroxytryptamine picrate corresponds to 0.41 mg of the base.

types of synthesis have been described by Justoni and Pessina[44] and Noland and Hovden[46].

SOLUBILITY

The solubility of serotonin* in water is relatively high, and increases with increasing temperature. As much as 100 mg of serotonin may be dissolved in 1 ml of boiling water while about 20 mg/ml may be dissolved at 27°C. Serotonin is also quite soluble in glacial acetic acid but only slightly so in methanol and 95% ethanol, and is insoluble in absolute ethyl alcohol, acetone, pyridine, chloroform, ethyl ether, ethyl acetate or benzene. Since serotonin is only sparingly soluble in lipids, it has a very low oil:water partition coefficient: 0.00035 in buffered solution at pH 7.4 (ref. 11) or 0.055 (ref. 63), while the coefficient for tryptamine is 0.92 (ref. 63).

Dissociation constants are as follows: pK_{a1} (creatinine) 4.9, pK_{a2} (terminal amino group) 10.0 and pK_{a3} (ring hydroxyl group) 11.1 (ref. 63).

STABILITY

The stability of serotonin varies with different conditions. At low pH serotonin is very stable. For instance, Erspamer[20], some years ago, reported a loss of only 45% after boiling extracted serotonin with 0.1 N hydrochloric acid for 1 h. Under nitrogen, extracted serotonin at pH 1.5 resists a temperature of 97°C for 6 h without any loss of activity, while boiling for 4 h at pH 0.5 results in a 40% loss of activity[51]. On the other hand it is easily oxidized in air, developing a brown-violet colour, particularly at alkaline pH. In 0.1 N HCl, pure solutions of serotonin (0.5 μg/ml) are stable for several weeks at a temperature[62] of 5°C. Maximum stability is at pH 5–6 when serotonin resists boiling for some hours even in the presence of atmospheric oxygen[22].

At an almost neutral pH the stability of serotonin decreases, although it is still good. At pH 6.8–7.8 serotonin is stable at room temperature for more than 6 h and at boiling point for about 5 min. On the other hand at alkaline pH 5-hydroxytryptamine is stable at boiling temperature for only a few minutes (2 min at pH 9.6 and 1 min at pH 11, ref. 43)**. At room temperature serotonin is stable, even in strongly alkaline pH, for more than 2 h***. Its stability is also maintained in the presence of fairly strong reagents such as p-chloromercuribenzoate, potassium persulphate, iodine, potassium permanganate, sodium bisulphite, potassium ferricyanide and ceric sulphate.

The concentrations and conditions in which these substances do not inactivate

---

* Except where otherwise stated, the word "serotonin" is used to refer to the creatinine sulphate salt of 5-hydroxytryptamine, the only salt used in practice.

** In alkaline solution serotonin is split into creatinine sulphuric acid and 5-hydroxytryptamine. Also its methyl derivative is unstable[21] at alkaline pH.

*** A method of destroying norepinephrine and epinephrine in the presence of serotonin is based on this observation: the first two substances are completely inactivated after 30 min at pH 11 at room temperature. The significance of this observation will be considered when dealing with the methods of extraction and dosage of serotonin.

serotonin are described by Rapport *et al.*[51],*. Activated charcoal (Norit) and resins such as Amberlite IR 100 H or 4 B do not destroy serotonin at a temperature of 25°C for some minutes. Other resins such as Zeo-Karb 226 (H form) are, on the other hand, capable of adsorbing serotonin which can be eluted with acid ethanol[54]. Amberlite X E-64 also adsorbs serotonin[4]. If serotonin is adsorbed by deactivated charcoal elution can be carried out with phenol and water[15]. In contrast to the creatinine sulphate salt, 5-hydroxytryptamine hydrochloride is unstable at room temperature, being photosensitive and hygroscopic.

IDENTIFICATION

Several investigations have been made to establish the absorption spectrum of serotonin. In aqueous solution adjusted to pH 3.5 there is maximum absorption at 275 mμ with a subsidiary peak at 295 mμ and a minimum at 250 mμ, as shown in Fig. 2 (refs. 33, 48, 52, 53, 57, 65).

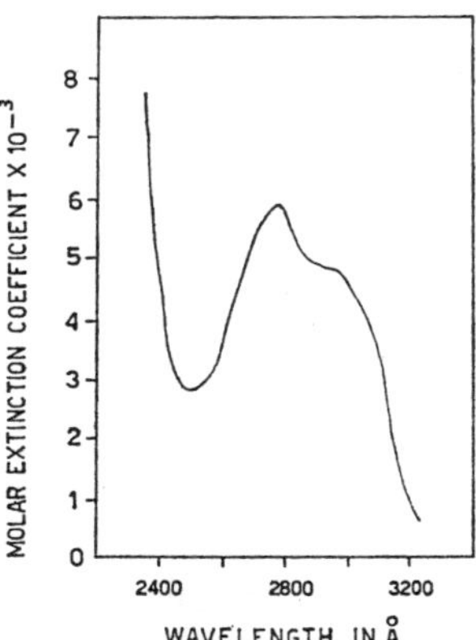

Fig. 2. Ultraviolet absorption spectrum of serotonin in water at pH 3.5.

Changing to an alkaline pH (11.6) results in a shift of the subsidiary peak[33] from 295 to 322 mμ. The bisoxalate salt has absorption points at 222, 276 and 298 mμ. The salt of 5-hydroxytryptamine with picric acid has a very similar spectrum giving at pH 5.3 a maximum at 275 mμ, a subsidiary peak at 293 mμ and a minimum around 247 mμ. Here again, alteration in pH affects the subsidiary peak, displacing it to about 332 mμ (refs. 23, 24). 5-Hydroxytryptophan has a very similar spectrum characterized by a maximum at 298 mμ, a minimum at 278 mμ and a subsidiary peak at 291 mμ which also shifts to 324 mμ at pH 11 (ref. 18). The infrared spectrum of serotonin, demonstrating the identity of the extracted and synthetic compounds, is described by Speeter *et al.*[57] and Stoll *et al.*[59] (see Fig. 3).

---

* Like serotonin, its precursor 5-hydroxytryptophan is very stable particularly in a nitrogen atmosphere at any pH[32]. However, 5-hydroxytryptophan, like histidine, is fairly sensitive to ultraviolet radiation, being converted to 5-hydroxytryptamine (10-min exposure causes a change[17] of 1 °/oo).

References p. 7

Another chemical property of serotonin is its ability, under certain conditions, to fluoresce under Wood's light. As will be seen later, this property is utilized in a method of assaying serotonin. There are three conditions where fluorescence can be obtained from serotonin:

(*a*) a yellow-green fluorescence which appears slowly after strong alkalinization[64];

(*b*) a yellow-gold fluorescence which can be obtained by treating serotonin with a solution made up of 9 parts of potassium bichromate (0.1 %) and 1 part of formaldehyde (37–41 %) and heating to 100–110°C for 5 min (ref. 55);

(*c*) a blue-green fluorescence which is developed after heating with a solution of acetone containing 0.2 % ninhydrin and 10 % acetic acid[40].

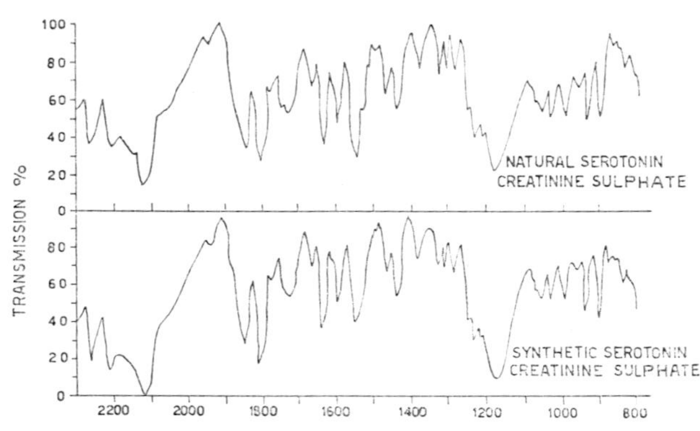

Fig. 3. Infrared absorption spectra of natural and synthetic 5-hydroxytryptamine creatinine sulphate complex. (Reported by Speeter *et al.*[57])

This last reaction, which is sensitive to 0.02 μg, seems to be highly specific for tryptamine derivatives which are not substituted in position 2, in the indole nitrogen, or which do not have long side chains, and may be due to the formation of β-carbolinic compounds[39].

The fluorescence emitted by serotonin can be selected by using a special apparatus called a "spectrophotofluorimeter" which can emit a light beam of a known wavelength and analyse the wavelength of the fluorescence developed[7]. In this way it has been observed that indoles activated at 275 mμ fluoresce at 360 mμ and 5-hydroxyindoles activated at 295 mμ fluoresce at 330 mμ. Variations of pH between 2 and 11 do not much affect this phenomenon, but at pH values below 2 (obtained with 3 N HCl)[61], 5-hydroxyindoles activated at 295 mμ fluoresce at 550 mμ, at a wavelength markedly different from those of the indoles not substituted in position 5 (ref. 60).

This phenomenon, which is probably due to the smaller amount of energy absorbed by the activation of 5-hydroxyindoles compared with that of the indoles, and to a more intense fluorescence[58], recently gave rise to a fairly specific reaction for detecting 5-hydroxytryptamine after chromatographic separation[13].

By using a special apparatus, it has been possible to observe the phosphorescence of serotonin. The sensitivity of this effect is 10 times greater than that of fluorescence[29].

With suitable reagents, serotonin gives rise to coloured substances which have been of some importance, especially in the past, in purification procedures. The following reactions, described by Rapport[51] and Erspamer[22,25] and also recently summarized by Cartier and Moreau[12], belong to this category:

(a) Hopkins-Cole's reaction[5] giving a blue-violet colour (1 mole of serotonin gives a colour equivalent to 0.8 moles of tryptophan);

(b) Ehrlich's reaction[31] with p-dimethylaminobenzaldehyde giving a blue-violet colour to begin with, which changes to blue-grey (1 mole of serotonin = 1.10 moles of tryptophan). Recently a dimethylaminocinnamaldehyde reaction, producing a blue colour, has been shown to be 10 times more sensitive than the Ehrlich reaction[35];

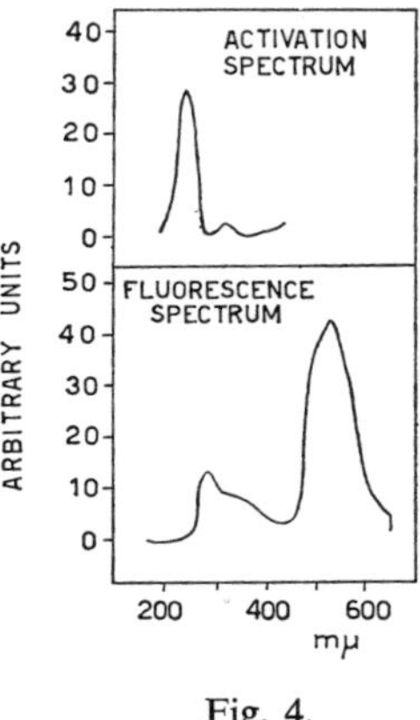

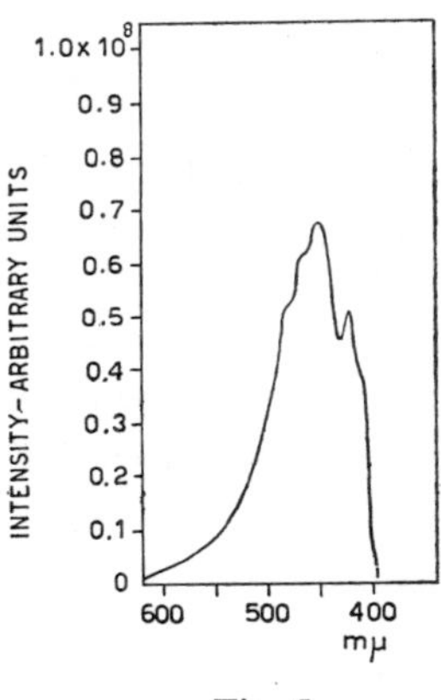

Fig. 4. Activation and fluorescence spectra of serotonin.

Fig. 5. Phosphorescence spectrum of serotonin; 0.1 mg/ml in methanol and ethanol 9:1. (Modified from Freed and Salmre[29].)

Fig. 4.   Fig. 5.

(c) Folin-Ciocalteu's reaction[28,34], characteristic for phenols, gives a blue colour[26] (1 mole of serotonin = 2.4 moles of tryptophan);

(d) Barac's reaction[3] with 2,6-quinonechloroimide[2], similar to Gibbs' reaction[30], reaching a sensitivity of 5 μg/ml;

(e) coupling reactions with diazonium salts[25–27] which can be carried out by different methods, using the diazonium salt of p-nitroaniline or sulphanilic acid, or Heinrich and Schuler's reagent (4-nitro-2-chloro-1-diazobenzene-α-naphthalene sulphuric acid)[36]. All these reactions with diazonium salts give rise to a red colour in an acid medium and a violet colour in an alkaline one.

In addition, serotonin reduces ammoniacal silver nitrate, reacts with gold chloride[26,52], ferrous ions[45], and shows a tendency to form complexes[49]. On the other hand, other reactions as Sakaguchi's[41], Bratton and Marshall's[10,56] and Pauly's diazoreaction for imidazoles[42], according to Rapport[5], are negative[26,52]. These reactions, which were mostly carried out on extracted material, have not always been repeated on synthetic serotonin. They are of value only if considered together[22] and particularly if carried out on material previously purified by paper chromatography. In this connection the R_F of serotonin in the presence of various solvents is recorded (see Table 1).

TABLE 1

Solvents	R_F of serotonin (a)	Reference
n-Butanol, acetic acid, H_2O (4 : 1 : 5)(b)	0.42(+)	16
	0.43	37
	0.44	54
	0.38	55
	0.27($'$)	16
n-Butanol, acetic acid, H_2O (12 : 3 : 5)	0.57(+)	38
n-Butanol saturated with HCl	0.18	25
	0.18	54
	0.15	65
	0.15($'$)	24
n-Butanol, methylamine (3 : 1)	0.56	54
n-Butanol, 25% methylamine in H_2O (8 : 3)	0.51($'$)	24
	0.55	25
n-Butanol saturated in H_2O	0.07($'$)	24
n-Butanol saturated with NH_4OH 3%	0.55–0.68	26
Chloroform, ethanol, H_2O (8 : 5 : 2)	0.43	9
Isopropanol, NH_3, H_2O (20 : 2 : 1)	0.47(+)	38
KCl 20% (c)	0.39(+)	16
	0.39($'$)	16
NaCl 8%	0.34(+)	14
NaCl 8%, acetic acid (100 : 1)	0.37(+)	14
n-Pentanol, pyridine, H_2O (2 : 2 : 1)	0.43	52
Propanol, H_2O (3 : 1)	0.48	54
Propanol, NH_3, H_2O (16 : 1 : 3)	0.47	54

(a) All data refer to 5-hydroxytryptamine creatinine sulphate except for those marked ($'$) or (+) which relate to the hydrochloride and the base respectively.
(b) according to ref. 47.
(c) according to ref. 6.

Serotonin can also be separated by paper electrophoresis in acetate buffer at pH 5.1; 200 V; 0.28 mA/cm for 4 h, according to Bracco et $al.$[9], under which conditions it moves to the negative pole.

REFERENCES

1 ASERO, B., COLÒ, V., ERSPAMER, V. AND VERCELLONE, A., *Ann. Chem., Liebigs*, 576 (1952) 69.
2 BARAC, G., *Arch. intern. pharmacodyn.*, 56 (1937) 427.
3 BARAC, G., *Arch. intern. physiol.*, 60 (1952) 185.
4 BERTLER, A., *Acta Physiol. Scand.*, 51 (1961) 75.
5 BLOCK, R. J. AND BOLLING, D., *The Aminoacid Composition of Proteins and Foods*, C. C. Thomas, Springfield, 1945, p. 91.
6 BOSCOTT, R. J., *Chem. Ind. (London)*, (1952) 472.
7 BOWMAN, R. L., CAULFIELD, P. A. AND UDENFRIEND, S., *Science*, 122 (1955) 32.
8 BRACCO, M. AND CURTI, P. C., *Experientia*, 10 (1954) 71.
9 BRACCO, M., CURTI, P. C. AND GIULIANO, V., *Experientia*, 12 (1956) 31.
10 BRATTON, A. C. AND MARSHALL, E. K., JR., *J. Biol. Chem.*, 128 (1939) 537.
11 BRODIE, B. B., personal communication.
12 CARTIER, P. AND MOREAU, J., *Semaine hôp.*, 34 (1958) 670.
13 CROSTI, P. F. AND BIANCHI, P., *Giorn. biochimica*, 7 (1958) 97.
14 CURZON, G., *Lancet*, 269 (1955) 1361.
15 DALGLIESH, C. E., *J. Clin. Pathol.*, 8 (1955) 73.
16 DALGLIESH, C. E., *Biochem. J.*, 64 (1956) 481.
17 DOEPFNER, W. AND CERLETTI, A., *Experientia*, 14 (1958) 376.
18 EK, A. AND WITKOP, B., *J. Am. Chem. Soc.*, 75 (1953) 500.
19 EK, A. AND WITKOP, B., *J. Am. Chem. Soc.*, 76 (1954) 5579.
20 ERSPAMER, V., *Experientia*, 2 (1946) 369.
21 ERSPAMER, V., *Arch. sci. biol. (Napoli)*, 31 (1946) 62.
22 ERSPAMER, V., *Rend. sci. Farmitalia*, 1 (1954) 129.
23 ERSPAMER, V. AND ASERO, B., *Nature*, 169 (1952) 800.
24 ERSPAMER, V. AND ASERO, B., *J. Biol. Chem.*, 200 (1953) 311.
25 ERSPAMER, V. AND BORETTI, G., *Experientia*, 6 (1950) 348.
26 ERSPAMER, V. AND BORETTI, G., *Arch. intern. pharmacodyn.*, 88 (1951) 296.
27 ERSPAMER, V. AND VIALLI, M., *Ricerca sci.*, 22 (1952) 1420.
28 FOLIN, O. AND CIOCALTEAU, V., *J. Biol. Chem.*, 73 (1927) 627.
29 FREED, S. AND SALMRE, W., *Science*, 128 (1958) 1341.
30 GIBBS, H. D., *Chem. Revs.*, 3 (1926) 291.
31 GRAHAM, C. E., SMITH, E. P., HIER, S. W. AND KLEIN, D., *J. Biol. Chem.*, 168 (1947) 711.
32 HADGRAFT, J. W., PRICE, S. A. AND WEST, G. B., *J. Pharm. and Pharmacol.*, 10 (1958) 87T.
33 HAMLIN, K. E. AND FISHER, F. E., *J. Am. Chem. Soc.*, 72 (1951) 5007.
34 HANIK, P. B. AND BERGEIM, O., *Practical Physiological Chemistry*, 11th ed., Blackiston, Philadelphia, 1937, p. 755.
35 HARLEY-MASONO, J. AND ARCHER, A. A. P. G., *Biochem. J.*, 69 (1958) 60P.
36 HEINRICH, P. AND SCHULER, W., *Helv. Chim. Acta*, 51 (1948) 320.
37 IORIO, M. A., GIORA, F. AND MARINI-BETTÒLO, G. B., *J. Chromatog.*, 5 (1961) 82.
38 JEPSON, J. B., *Lancet*, 269 (1955) 1009.
39 JEPSON, J. B. in G. P. LEWIS (Ed.), *5-Hydroxytryptamine*, Pergamon, London, 1958, p. 37.
40 JEPSON, J. B. AND STEVENS, B. J., *Nature*, 172 (1953) 772.
41 JORPES, E., *Biochem. J.*, 26 (1932) 1505.
42 JORPES, E., *Biochem. J.*, 26 (1932) 1507.
43 JOYCE, D., *Nature*, 182 (1958) 463.
44 JUSTONI, R. AND PESSINA, R., *Farmaco (Pavia)*, 10 (1955) 356.
45 LILLIE, R. D., *J. Histochem. and Cytochem.*, 9 (1961) 44.
46 NOLAND, W. AND HOVDEN, R., *J. Org. Chem.*, 24 (1959) 895.
47 PARTRIDGE, S. M., *Nature*, 158 (1946) 270.
48 RAPPORT, M. M., *J. Biol. Chem.*, 180 (1949) 961.
49 RAPPORT, M. M., *Ann. N. Y. Acad. Sci.*, 66 (1957) 665.
50 RAPPORT, M. M., GREEN, A. A. AND PAGE, I. H., *J. Biol. Chem.*, 174 (1948) 735.
51 RAPPORT, M. M., GREEN, A. A. AND PAGE, I. H., *J. Biol. Chem.*, 176 (1948) 1237.
52 RAPPORT, M. M., GREEN, A. A. AND PAGE, I. H., *J. Biol. Chem.*, 176 (1948) 1243.
53 RAPPORT, M. M., GREEN, A. A. AND PAGE, I. H., *Science*, 108 (1948) 329.

54 RODNIGHT, R., *Biochem. J.*, 64 (1956) 621.
55 SHEPHERD, D. M., WEST, G. B. AND ERSPAMER, V., *Nature*, 172 (1953) 357.
56 SMITH, H. W., FINKELITEIN, N., ALIMINOSA, L., CRAWFORD, B. AND GRABER, M., *J. Clin. Invest.*, 24 (1945) 388.
57 SPEETER, M. E., HEINZELMANN, R. V. AND WEISBLAT, D. I., *J. Am. Chem. Soc.*, 73 (1951) 5514.
58 SPRINCE, H., ROWLEY, G. R. AND HAMESON, D., *Science*, 125 (1957) 442.
59 STOLL, A., TROXLER, F., PEYER, J. AND HOFMANN, A., *Helv. Chim. Acta*, 38 (1955) 1452.
60 UDENFRIEND, S., BOGDANSKI, D. F. AND WEISSBACH, H., *Science*, 122 (1955) 972.
61 UDENFRIEND, S., WEISSBACH, H. AND CLARK, C., *J. Biol. Chem.*, 215 (1955) 337.
62 VALZELLI, L., unpublished results.
63 VANE, J. R. *Brit. J. Pharmacol.*, 14 (1959) 87.
64 VIALLI, M. AND ERSPAMER, V., *Arch. sci. biol. (Napoli)*, 28 (1942) 101.
65 ZUCKER, H. B., FRIEDMAN, B. K. AND RAPPORT, M. M., *Proc. Soc. Exptl. Biol. Med.*, 85 (1954) 282.

Chapter 2

METHODS OF ESTIMATING THE SEROTONIN CONTENT OF THE TISSUES

Several methods for the extraction and determination of serotonin in tissues are described in the literature. The procedures considered more important will be discussed in greater detail, while those devoid of practical interest will be merely mentioned. The following three considerations apply to all methods we are going to discuss:

(*a*) The extraction procedure must be started immediately after killing the animal in order to prevent inactivation of serotonin by monoamine oxidase[105] and other enzymes. In some cases it may be worthwhile using liquid air to stop all processes of synthesis or destruction of serotonin[54]. A systematic study of the stability of serotonin in different organs has not yet been carried out. It is accepted that serotonin does not undergo any appreciable degradation when tissues are kept at a temperature below —10°C. At room temperature the percentage of serotonin destroyed varies within different tissues: 40% of the serotonin content of lungs is lost within 30 min whereas only 10% of the brain serotonin content will disappear under identical conditions[99]. Experiments with the cerebral cortex of rabbits (20 mg/ml) homogenized in phosphate buffer at pH 6.9 confirm the stability of brain serotonin. This preparation withstands 70°C for 30 min and 100°C for 10 min without an appreciable loss of serotonin content[27]. Other studies on the stability of brain serotonin after death have been recently reported[65a]. At 15°C the ox intestine loses 10, 20 and 70% of its serotonin content after 1, 2 and 9 days respectively[41]. When sheep serum is kept at 4°C for 5, 10 and 15 days the serotonin decreases by 25, 50 and 75% respectively[41]. On the other hand whole blood, kept at 5°C, loses more than 50% of its serotonin content in a few hours[15,*].

(*b*) The serotonin content of blood may be a source of error when the serotonin content of tissues is analysed. Therefore the analysis of tissues should be preceded by a perfusion of the afferent arteries with saline. Blood serotonin content varies in different species, and perfusion becomes essential when the serotonin content of blood is marked. For instance, perfusion is essential in the rabbit because the blood serotonin content is high (3.5 to 9.4 μg/ml)[11,109] but not probably in the rat since rat blood contains only 0.15 to 0.34 μg/ml of serotonin[11,61] (for details see Appendix I).

(*c*) The existence of both free and bound serotonin has been proposed; however, none of the analytical methods available enables us to differentiate between these two

---

* The fragility of the platelets and/or the presence of monoamine oxidase in the plasma of some animal species[10] or traces of haemoglobin[28] could be responsible for the different stability of serotonin in the sera of various animal species.

biological entities*. It should be stated at once that this differential analysis may not be always important. In fact in basal conditions the bound form represents the bulk of the serotonin content, for the amount of free amine in comparison is negligible. Drugs altering the serotonin content of tissues on injection may act by changing the ratio between the free and bound amine. Thus the selective analysis of bound and free serotonin in tissue may be helpful for studying the mechanisms of action of drugs.

We do not have data to decide whether or not part of the bound serotonin is held by specific chemical linkages. If this were the case then the solvent used for serotonin extraction might be important. For instance, as Correale[25] demonstrated, different concentrations of acetone yield different results in analysis of brain serotonin content. Furthermore, Costa[26] demonstrated that the amount of serotonin extracted by 95% acetone increased if the tissue had been previously digested by chymotrypsin. This would suggest that, in the brain, part of the chemically-bound serotonin is susceptible to the action of chymotrypsin. On a quantitative basis the amount of serotonin which is thought to be protected from monoamine oxidase (MAO) destruction is greater than that shown with the use of chymotrypsin to be chemically bound. Thus mechanisms other than chemical binding must be invoked to explain the storage of serotonin in tissues with MAO activity. In fact a great part of the serotonin contained in the duodenal mucosa of the dog was recovered by Baker[6] in the "large granule" fraction. These particles can be separated from the mitochondria by the gradient centrifugation technique, using hypertonic solutions of sucrose. In these experiments the serotonin was extracted from the various precipitates with acetone according to the method of Amin et al.[2] and assayed using the rat uterus technique. The amount of serotonin stored in these intracellular particles is five times greater than that present in the supernatant fluid. Serotonin is released from these amine-carrying granules by hypotonic solutions. In the brain of rats only 25 to 27% of the total serotonin is recovered in the supernatant fluid of homogenates[73] centrifuged at 40,000 g at 1°C. Less than 20% of the total 5-HT was recovered in the supernatant fluid[55] after centrifugation at 100,000 g for 2 h. Other authors[104] confirmed these results using similar techniques and isotonic sucrose solutions for homogenization. It may be suggested that free serotonin corresponds to the amine found in the supernatant, whereas the amine present in the precipitate is bound serotonin. However, further experiments are required to support this suggestion because serotonin may be released in the supernatant fluid during the necessary manipulations.

The specificity of the different methods of assay has been discussed by many authors. It is fair to conclude that both chemical and biological methods have some handicaps. Compounds with similar characteristics of absorption and fluorescence may be a source of error in chemical methods. The bioassay procedures are also frequently misleading because antagonists or activators of the biological response to serotonin may be present in the extracts analysed. Among the substances that may

---

* It must be admitted that this concept is similar to Erspamer's proposals concerning the "active" and "inactive" enteramine present in extracts of rabbit's stomach and ox spleen[38,39]. The inactive form may become active by simply heating at a pH greater than 6 (ref. 41).

interfere with the bioassay of serotonin are both epinephrine and norepinephrine. Unfortunately adrenolytic drugs cannot be used because the inhibition they produce is relatively nonspecific when compared with the effects of epinephrine and serotonin. It is more useful to attempt a chemical inactivation of the catecholamines. These chemical methods are based on the greater stability of serotonin compared with adrenaline in alkaline medium. In practice adrenaline and noradrenaline are inactivated[65] by bringing the pH of the extract to 11 for 30 to 60 min or by heating the samples at pH 7.8 for 5 min at 98°C. A biological method for enzymatic destruction of adrenaline with a polyphenyloxidase* extracted from a common fungus, *Psalliota campestris*, has been suggested by Garven[53]. The mushroom is ground and homogenized and the liquid is lyophilized in vials containing 0.5 ml of extract. This enzymatic preparation is added to the samples prior to serotonin analysis. Catecholamines are destroyed at pH 5–6 at room temperature within 40 min. The serotonin content of kidney and lung assayed using this method increases about 25%, that of pancreas 45% and finally that of liver[53] 60% compared to results obtained without enzymatic destruction of adrenaline. An improvement of serotonin bioassay may be obtained by adding large doses of serotonin to the samples[91a].

METHODS OF EXTRACTION FOR BIOASSAY

When the serotonin contained in a tissue is determined by bioassay the extraction of the amine is usually carried out with acetone[43,48]. This solvent was suggested first by Erspamer[36,37] and has since been used by Twarog and Page[96], Feldberg and Toh[45], and many others. A good account of this procedure is given in detail by Amin *et al.*[2]: the tissue is rapidly isolated and carefully homogenized in acetone (20 ml/g of tissue). The mixture, stirred for 1 h at room temperature, is filtered through Whatman No. 1 paper. The residue is re-extracted for 30 min at room temperature with an equal volume of 95% acetone (v/wt). This second mixture is also filtered and the residue on the filter paper is washed twice with 5 ml 95% acetone (v/wt). The bulk of the serotonin contained in the tissue is dissolved in 95% acetone and the insoluble fraction is discarded because it contains most of the substance P and other substances[72] active on biological tests. The filtrates are combined in a 100-ml flask and dried under vacuum keeping the external temperature at 35°C. After the acetone has been evaporated the residue is suspended in 1 ml of distilled water, which in turn is extracted twice with 10 ml of petroleum ether (distillation point 40–60°C). The petroleum ether removes acid-soluble lipids which may be present and which could interfere with the bioassay[101]. When the tissue is particularly rich in lipids the purification with petroleum ether should be repeated 2–4 times. The aqueous phase is finally dried under vacuum and the residue, suspended in 1 ml of physiological saline, is ready for quantitative analysis. If the samples are not analysed immediately they should be kept dry at −17°C (this temperature is not critical and may be as high as −5°C). The

---

* Tyrosinase may also be used for this purpose. This is an anaerobic oxidase containing copper, which transforms the catechols into *o*-quinones[67].

critical point is the reconstitution of the extract for the bioassay. In order to avoid loss, the physiological saline must be added immediately before the analysis.

The method proposed by West[108] is apparently similar to that described above. West proposes the use of 5 ml of 80% acetone/g of tissue, so the extraction is carried out with more diluted acetone. The acetone is evaporated in air at 30°C and the residue, dissolved in saline, is used for the bioassay. This simplified technique may not be sufficiently accurate. Diluted acetone takes up serotonin[24] and other biologically active agents, such as substance P (ref. 2); thus the bioassay loses specificity and becomes misleading. Substance P is an important source of error in analysis of brain serotonin content when rat uterus technique is used. Since the biological activity of substance P can be inactivated by chymotrypsin[81,82] Correale[25] proposed the use of this enzyme to purify the tissue extracts made with diluted acetone. He suggests that the enzymatic reaction is complete within 2 h at 38°C, pH 7.5–8.5 and claims that the recovery of serotonin from tissues digested with chymotrypsin before extraction with 80% acetone is higher than that obtained with concentrated acetone, as proposed by Amin et al.[2]. Correale reports that the results of brain serotonin analysis following inactivation of substance P by chymotrypsin are 20 to 40% smaller than the values found in tissues extracted with 80% acetone without such inactivation.

METHOD OF EXTRACTION FOR CHEMICAL ASSAY

The extraction procedures to be used when serotonin is determined by physico-chemical methods are entirely different from those required for bioassay procedures. A method suitable for the determination of brain serotonin content was first described by Udenfriend et al.[98]. The modification proposed by Bogdanski et al.[11] is described as follows. Brain is homogenized in 2 parts of 0.1 N HCl. One part of the homogenate, containing 0.5 to 5 μg of serotonin (for rat brain this is about 1 to 10 g of tissue) is transferred into a large centrifuge tube with a glass stopper. Five ml of borate buffer* at pH 10 (pH 9.8 according to Dalgliesh and Dutton[32]) are added to the mixture previously brought to pH 10 with anhydrous sodium carbonate. Distilled water is used to bring the final volume to 15 ml. Five g of sodium chloride and 20 ml of n-butanol** are added; the mixture, after being shaken for 15 min, is then centrifuged and the supernatant transferred into another tube. The aqueous phase is removed by aspiration and the butanol phase is washed twice by shaking it with an equal volume of borate buffer. Fifteen ml of the butanol phase are transferred into another tube containing 30 ml of heptane and 3 ml of 0.1 N HCl. This mixture is shaken and

---

* 94.2 g of boric acid are dissolved in 3000 ml of distilled water and 165 ml of 10 N sodium hydroxide. An excess of both sodium chloride and n-butanol is also added. After the mixture has been shaken for 30 min the excess of n-butanol is removed by aspiration and the pH is adjusted to about 10 with HCl.

** Butanol and heptane should be purified as follows: one volume of butanol (or heptane) is washed with equal volumes of first 0.1 N sodium hydroxide and then 0.1 N HCl, and finally washed twice with distilled water.

centrifuged to remove the supernatant. The serotonin content is then determined by one of the physico-chemical methods to be described below.

A microanalytical procedure for extraction of 5-HT from small samples of tissues has been reported by Kuntzman *et al.*[69]. Serotonin and histamine[17a,106] and serotonin and noradrenaline[35a,77a,88,95a] may be extracted simultaneously with minor modifications of the method described.

A possible method of comparing biological with chemical assays after the same type of extraction has been devised recently by Towne *et al.*[95]. A good recovery of serotonin from rat brain was found after using formic acid and purified acetone (redistilled from $KMnO_4$).

EXTRACTION OF 5-HT IN BLOOD

When whole platelets or serum[13,58,94] are to be analysed a different extraction procedure is recommended. The use of anticoagulants and siliconized tubes kept at a temperature below 5°C is essential[15]. Cargill *et al.*[21] suggest that the analysis of blood serotonin content should be performed on 1.5 ml of heparinized blood, haemolysed with an equal volume of water. Butanol buffer is then added and the mixture is shaken for 2 min. The butanol phase is not washed but extracted directly with heptane and then with 0.1 N HCl. The aqueous phase is washed twice with ethyl ether, which is then removed by bubbling air through the mixture for 2 to 3 min. The solution is neutralized and brought to a suitable volume with the same physiological nutrient fluid used for the bioassay. According to these authors this method is accurate for blood analysis of some species of animals (*e.g.* rat, cat and dog) and unsatisfactory for others (*e.g.* guinea pig, rabbit and man). Minor modifications of this procedure however have been adapted for the estimation of serotonin in human platelets[27a,29a,75]. Davis[34] recently summarized the methods available for the analysis of serotonin in human serum and suggested the use of ascorbic acid to prevent the destruction of serotonin. Weissbach *et al.*[106] have shown that butanol extraction was not the most accurate way to extract blood serotonin. They suggest the following procedure: 1 ml of blood is haemolysed with 6 ml of water and 1 ml of sodium chloride solution (6.3%). The proteins are then precipitated with 1 ml $ZnSO_4$ (10%) followed by 0.5 ml N NaOH. The mixture is then shaken for 20 min and centrifuged at 2500 g. One ml of the supernatant, which should be completely clear, plus 0.3 ml of 12 N HCl are read in a spectrophotofluorimeter. Although this procedure is simpler than Bogdanski's methods[11], it is only applicable to blood and tissues* containing more than 1 μg of serotonin/g.

A more complicated but accurate method for measuring 5-HT in human blood is described by Waalkes[102]. This procedure involves the extraction of serotonin from

---

* Tissues are homogenized in 2 volumes of 0.1 N HCl and one part is brought to 8 ml with saline. Precipitation of proteins and further manipulations are kept identical to those suggested for blood.

a basic solution (3.5 ml of 10% $ZnSO_4$ + 2.1 ml of N NaOH + 1.5 ml of saturated EDTA) into organic solvents (ethyl acetate or n-butanol) and purification by means of an ion exchange medium, cotton acid succinate[77]. The measurement of the serotonin concentration is then made with a spectrophotofluorimetric method.

Erspamer and Sala[43], also Twarog and Page[96], store the blood for 2–3 h at room temperature and then for 15–20 h in a refrigerator. Serum is separated by centrifugation and extracted with 4 volumes of acetone according to the usual method for bioassay. Kärki and Paasonen, however, report a very low recovery (35–45%) by using acetone extraction [66] and Bracco et al. have questioned the stability of serotonin in blood[15]. These authors suggest the following method[14] for the extraction of serotonin from platelets: heparinized blood is diluted with 3 volumes of a solution made up of 5 parts of 0.2 M NaH_2PO_4, 5 parts of 0.2 M Na_2HPO_4 and 2 parts of 4% sodium oxalate. Platelets are isolated by refrigerated centrifugation and washed twice with 3 volumes of physiological saline. 30–60 mg of fresh platelets are suspended in 5 volumes of ethyl alcohol, stored in the refrigerator for 12 h and finally shaken for 10 min prior to the last centrifugation. The clear supernatant is evaporated under vacuum and the residue redissolved in 0.5 ml of chloroform and ethyl alcohol. This sample is then chromatographed on paper. Zucker et al.[111] employ oxalated blood and centrifuge it immediately after withdrawal from the vein. Fifty mg of platelets suspended in 2 ml of water and 0.5 ml of acetone are kept overnight in the refrigerator. Thirty-eight ml of cold acetone are added and after 2 h of refrigeration the mixture is centrifuged at low temperature using 900 g for 30 min. The clear supernatant is removed and evaporated under vacuum. The residue is redissolved in 2 ml of water in order to precipitate the lipids and centrifuged for 12 min at 20,000 g. An aliquot of the supernatant is again dried under vacuum, redissolved in 0.2 ml of 90% acetone and placed on Whatman No. 1 filter paper for chromatographic analysis.

PAPER CHROMATOGRAPHIC METHODS

Paper chromatographic techniques are more appropriate for the identification than for the quantitative analysis of 5-HT contained in the tissues of laboratory animals[41]. Dalgliesh's method[31] is particularly suitable for biological experiments and lends itself to the quantitative analysis of tissues unusually rich in 5-HT. Dalgliesh[31] worked mainly with urine, purified by adsorbing the aromatic compounds onto deactivated carbon. The deactivation, performed either with stearic acid[93] or octadecylamine (laurylamine)[3,30] is essential in order to block active carbon receptors which would impair elution of the adsorbate by aqueous solutions of phenolic acids. Provided that certain proportions between the amount of laurylamine added and the weight of the carbon are respected[3], it is possible to get 80 to 90% recovery. The sample, eluted from the carbon can be analysed by two-dimensional paper chromatography using one phase with butanol, acetic acid and water (4 : 1 : 5) and an aqueous phase containing 20% KCl (ref. 3).

As shown in Table 2 this method makes possible the separation of 5-HT from its biological precursor, 5-hydroxytryptophan, and/or its degradation product, 5-hydroxyindoleacetic acid[78].

TABLE 2

R_F OF SEROTONIN, 5-HYDROXYTRYPTOPHAN AND 5-HYDROXYINDOLEACETIC ACID IN
TWO-DIMENSIONAL PAPER CHROMATOGRAPHY

Compound	Butanol–acetic acid–H_2O	20% KCl in H_2O
Serotonin	0.57	0.77
5-Hydroxytryptophan	0.22	0.69
5-Hydroxyindoleacetic acid	0.84	0.72

The R_F values of these compounds in a variety of solvents are also given by Carcasona *et al.*[20] and Awapara *et al.*[4].

The reader is referred to Chapter 1 for the R_F values of serotonin dissolved in several other media. The methods used for chromatography of indolic substances have been critically reviewed by Jepson[64]. He recommended the use of two-dimensional chromatography, combining a basic solvent with an acid or a neutral one. The indolic substances of tissue extract can be separated using isopropanol, ammonia and water (200, 10 and 20) or *n*-butanol, acetic acid and water (120, 30 and 50) or isopropanol, ammonia and water and 20% aqueous solution of KCl. The separation is greatly facilitated by the purification of the extract to be analysed with carbon (deactivated with octadecylamine) as described above. A "multiple dipping"[89] technique can be used for staining. The sequence is as follows: ninhydrin, pyridine, Ehrlich's reagent*, sulphanilic acid[64]. Any of the colour reagents described in Chapter 1 may also be used, as well as fluorescence which develops after intense acidification according to Crosti's technique[29],**. When the results are in doubt, the spots attributed to serotonin may be eluted before staining and assayed biologically. However, 5-HT identification can be achieved exclusively by means of chromatographic techniques when the inseparability of the known from the unknown has been proved by every possible solvent and staining technique. The most suitable reagent suggested for the elution of stained spots is made up with sulphanilic acid (0.5% in 2% HCl) and 0.5% sodium nitrate. This reagent is sprayed on the paper followed by 4% pyridine. The orange colour which develops is then eluted with an excess of 4% pyridine in an aqueous solution. The maximal sensitivity of this method varies between 2 and 5 μg. Some advantages over paper chromatography have been obtained by using cellulose phosphate cation exchange paper[85a].

---

* A useful modification is described by Sprince[90].
** Collier's method is recommended for analysis of stamens, leaves or vegetable poisons[23] usually rich in serotonin.

RESIN CHROMATOGRAPHY

The estimation of serotonin contained in urines might be carried out by partition chromatography, either on weak acid carboxylic resins, such as Zeo-Karb 226 (ref. 87) or on aluminum oxide columns prepared according to specific requirements[16].

Recently a number of methods using resins have been developed for the analysis of serotonin in tissues. Gluckman and Abrams[56] homogenized tissues with 0.5 M perchloric acid to extract serotonin. The supernatant fluid is stirred with Dowex-1-OH to remove perchloric acid. After neutralization the sample is passed through a resin, Amberlite X E-97-NH$_4^+$ washed with ammonium acetate buffer at pH 7. Serotonin is then eluted with 1 M formic acid and fluorimetrically evaluated. A similar procedure, using Amberlite X E-64, and 1.2 N hydrochloric acid for elution, was developed by Bertler[8] for the analysis of serotonin in brain[7,8]. For a separation of amines from their precursor amino acids Awapara[4] suggests the use of Amberlite CG 50 prepared according to Hirs *et al.*[60]. In the same paper Awapara[4] reports a separation of several amines, including serotonin, by column chromatography using cellulose for the stationary phase and a mixture of butanol, acetic acid, and water for the mobile phase. Finally the technique of gas chromatography offers a new tool as shown by Sweeley and Williams[92] for the quantitative evaluation of 5-hydroxyindoleacetic acid and by Fales and Pisano for serotonin[43a].

CHEMICAL METHODS

Lack of specificity has outdated the colorimetric method described by Udenfriend *et al.*[98]. This method was carried out on tissues extracted according to the method of Bogdanski *et al.*[11]. 1-Nitroso-2-naphthol 0.1 % in 95 % ethyl alcohol and 2.5 % nitrous acid (prepared immediately before use with 5 parts 2 N H$_2$SO$_4$ and 0.2 parts of 2.5 % sodium nitrate) are added to the extracts. The excess of nitrosonaphthol is taken up with ethylene dichloride and the intensity of the violet colour which develops* is measured in a photometer at 540 mμ. It is possible to carry out a serotonin determination without colour reaction by analysing the absorption characteristics of the extract in the ultraviolet (275 mμ) with a spectrophotometer.

The above methods usually give values almost double in comparison with those given by the fluorimetric assay which today is considered sufficiently specific and highly sensitive. As described in Chapter 1, fluorimetric assay has been made possible with the availability of the spectrophotofluorimeter[12] (see Fig. 6).

When serotonin activation is carried out at 295 mμ and when serotonin is present in 3 N HCl, the fluorescence is read[11,97] at 550 mμ. Spectrophotofluorimetry is carried out on tissues extracted according to Bogdanski's method[11]. The spectrophotofluorimetric method is so sensitive that under optimal conditions 0.08 μg of serotonin/g of organ may be detected. Serotonin may be easily differentiated by this technique from tryptamine[59], noradrenaline[88] and a number of other amines[35].

---

* The colour varies from brown to red according to the experimental conditions.

The main reasons for preferring spectrofluorimetric analysis of serotonin are: precision, reproducibility, sensitivity, relative specificity and simplicity of technical manipulations.

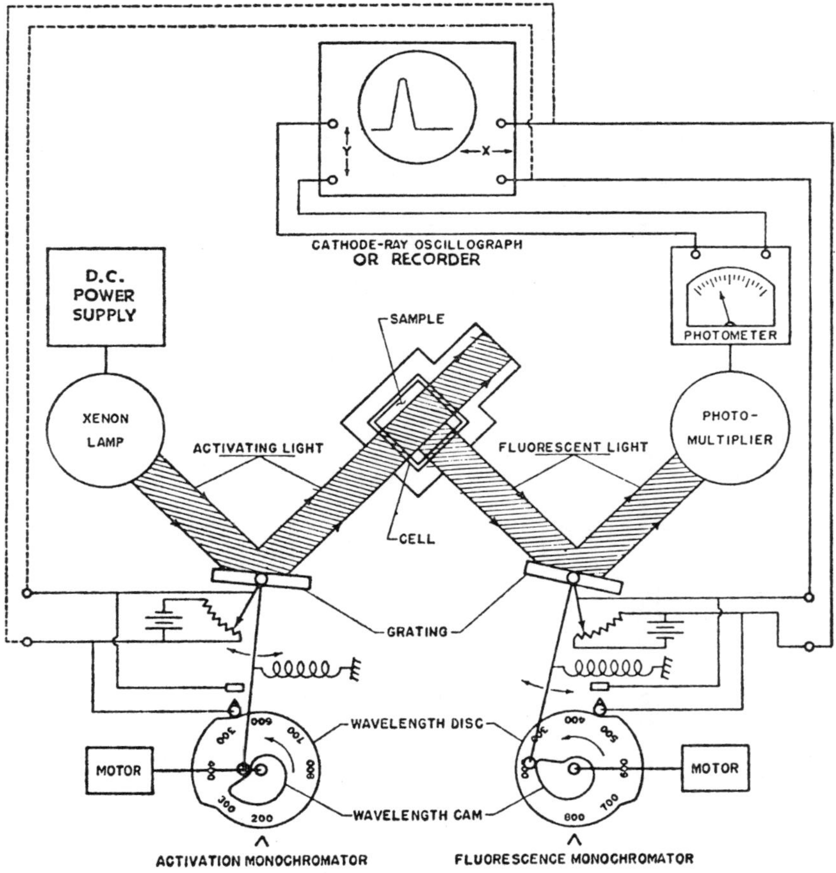

Fig. 6. Schematic of Aminco–Bowman Spectrophotofluorimeter. (From Aminco–Bowman instruction and service manual N. 768 B).

BIOLOGICAL METHODS

Rat uterus

This method was proposed by Erspamer[36,37,40,41] and is still being used with excellent results in many laboratories. Young rats, spayed not earlier than 2–3 months previously, are given subcutaneous injections on two successive days of 100–200 μg oestradiol dipropionate or 100 μg of diethylstilboestrol. The uterus is removed for bioassay 4 to 7 days after the 2nd injection.

TABLE 3

COMPOSITION OF SOME MEDIA USED FOR BIOASSAY OF SEROTONIN

Type of medium	Concentrations in g/l									Reference
	$NaCl$	KCl	$MgCl_2 \cdot 6H_2O$	$CaCl_2 \cdot 2H_2O$	$NaHCO_3$	Glucose	Dextrose	Buffer phosphate pH 7.4	Na_2SO_4	
Ringer, modified for perfusion of rabbit ear	8.2	0.84	0.06	0.04	0.2 0.5	1	—	0.01 M	—	80
Ringer, modified for rat isolated uterus	9	0.42	—	0.06	0.5	0.5	—	—	—	49
Ringer, modified for rat isolated colon	9	0.4	—	0.03	0.15	1	—	—	—	49
Locke, modified for rat isolated colon	9	0.42	—	0.06	0.5	—	1	—	—	63
Medium, for isolated heart of mollusc	28	0.65	10	1.1	0.2	—	—	—	4	48
Ringer–Locke, for rabbit intestine	9	0.42	—	0.24	0.5	—	1	—	—	17
Tyrode, for guinea pig ileum	8	0.2	0.1	0.2	1	1	—	—	—	19
Tyrode, modified for guinea pig ileum	8	0.2	0.01	0.24	1	1	—	—	—	62

The uterus is ready for the assay when it appears congested and swollen*. The preparation is then suspended in a nutrient bath containing small concentrations of calcium (see Table 3) and kept at 28–32°C.

Various investigators agree that ovariectomy of the animal is not essential for establishing optimal conditions for the bioassay[2,14,47,91,111]; good results can be obtained by using uteri from intact rats given diethylstilboestrol (10–100 μg/100 g body weight) or oestradiol (100 μg for 3 days before the experiments). In order to decrease the great sensitivity of this organ to acetylcholine, 0.1 to 1 μg/ml of atropine must be added to the fluid in the bath. These concentrations of atropine fail to modify the sensitivity of the uterus toward 5-hydroxytryptamine.

The reactivity of the uterus is within reasonable limits, proportional to the 5-HT concentrations and the variability of the response to the same doses of serotonin does not exceed 15%. Tachyphylaxis is not present and the response remains unaltered for at least 3–4 h. Under optimal conditions the 5-hydroxytryptamine threshold dose is 0.005 μg/ml and may often be as low as 0.001 μg/ml. Large doses of histamine have only a negligible inhibitory action on this test (10 μg of histamine inhibit 20% of the utero-stimulant action of 0.04 μg of serotonin)[41] while the antagonistic activity of greater importance is that of adrenaline, which even in concentrations as low as 0.1 to 0.2 μg produces a 50% inhibition of the effects of 0.2 μg of serotonin. In this preparation lysergic acid diethylamide (LSD) efficiently antagonizes serotonin[46,47].

Rat colon

This was proposed by Dalgliesh *et al.*[33] and Feldberg and Toh[45] in 1953. This bioassay utilizes the ascending colon, which next to the caecum is the most sensitive part of the rat intestine. The colon is suspended in an appropriate bath and perfused at 22–24°C with a special nutrient solution (see Table 3). The low temperature of the bath and the low content of calcium ions in the perfusion fluid reduce the spontaneous motility of the isolated colon without lessening the sensitivity of the preparation to serotonin. The rat colon[50] is sensitive to 0.001 μg/ml and often even to lower concentrations. The response is proportional to the dose used. Atropine in concentrations of 0.024 μg/ml inhibits the stimulating action of 0.1 μg/ml of acetylcholine without altering the activity of serotonin[51]. Histamine has no effect on this preparation, even in high concentrations[99] (10 μg/ml). In this test sympathomimetic substances, such as isopropyl-noradrenaline (a concentration of 0.01 μg inhibits 80% of the response to 0.1 μg/ml of serotonin), adrenaline and noradrenaline also inhibit the action of serotonin. This inhibition can be an important source of error when one assays heart, brain, etc.[22] and the other tissues which have great concentrations of these substances. Supek and Milkovic[91] carried out a parallel study of the sensitivity toward serotonin of the isolated uterus and colon. They reported that the sensitivity of both organs was similar but the colon was to be preferred by virtue of the greater regularity of the dose-response curve.

---

* It is not essential to use a uterus in oestrus. Rats weighing between 100–350 g may be used provided that the uteri are more than 3 mm in diameter[71].

Guinea pig ileum

The portion adjacent to the caecum, which is the most sensitive to the action of serotonin[47], is suspended in a 2-ml bath perfused with Tyrode solution (see Table 3)*. The bath is kept at 37°C (34°C according to Feldberg and Toh[45]). The intestine is hooked to the lever, so as to leave both ends open[1]. With this preparation 4–5 μg of serotonin in a 15-ml bath (0.27 to 0.33 μg/ml) does not cause tachyphylaxis. When the time interval between stimuli is approximately 3 min[86] even doses of 10 μg give reproducible results without tachyphylaxis. Both histamine and acetylcholine are notably active on this preparation. The antihistaminics fail to modify serotonin-induced contraction but atropine or small amounts of epinephrine (0.07 μg/ml) hinder the responses to serotonin. Tryptamine curtails serotonin-induced contractions while substance P may interfere by virtue of its ability to induce contractions[45]. More recent data suggest that this method of bioassay is not suitable for serotonin determinations when histamine[104a] and acetylcholine are also present unless the serotonin content is exceedingly large[19].

Fundus of the rat stomach

Up to date this is the latest biological test for the assay of serotonin[100]. The technical procedure is as follows: the stomach is isolated and the pyloric part is separated from the fundus. A strip of tissue is obtained from the fundus either by cutting it in a spiral or by opening the fundus along the straightest edge and cutting as indicated in Fig. 7.

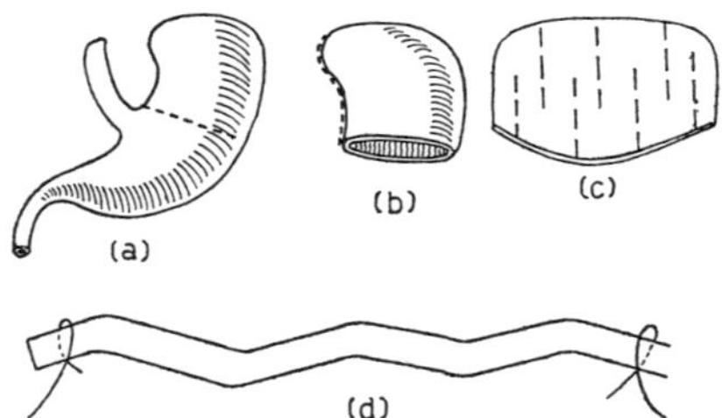

Fig. 7. Diagram showing the method of cutting the fundal strip from the rat stomach. (a) The whole stomach. (b) The fundus, left with a thin band of pyloric tissue and then cut along the dotted line. (c) The fundus opened into a sheet of tissue. Cut along the dotted lines. (d) The strip pulled out by cotton tied to each end, so that protrusions and bits of pyloric tissue can be cut off. (From Vane[100].)

The latter manipulation increases the sensitivity of the preparation. The stomach strip is suspended at 37°C in a suitable 5-ml bath containing the oxygenated nutrient fluid proposed for the rat uterus (Table 3). This solution is more suitable than Tyrode's, Krebs', Ringer's or the fluid proposed for the rat colon. The tissue begins to

---

* In this connection Innes et al.[62] using Tyrode with small concentrations of magnesium (0.01 μg/l, see Table 3) have been able to observe that varying the temperature alters the sensitivity toward agents inducing contractions. At optimal temperature (35°C), 5-hydroxytryptamine stimulates at doses of 0.025 to 0.1 μg/ml; at 25°C the response to 5-hydroxytryptamine is reduced, but not that to acetylcholine.

contract when the solution in the bath has not been renewed for more than 6–8 min. It is therefore necessary to perfuse the organ continuously at a speed of 1 ml/min. A frontal lever made of balsa wood should be preferred. The sensitivity of this preparation to serotonin can be as low as 50 picograms (0.00005 μg) though usually it is around 1 nanogram (0.001 μg)*. The doses of serotonin inducing contractions are 10 times greater than those of acetylcholine while several μg/ml of histamine are needed to make the rat stomach contract. It is possible to abolish the effects of acetylcholine by adding 10^{-7} g hyoscine bromide which does not interfere with the action of serotonin. The sympathomimetic amines inhibit the action of 5-hydroxytryptamine in this organ. Isopropyl-noradrenaline is the most active while adrenaline and noradrenaline are five times less active.

All the biological tests described are based on the fact that 5-hydroxytryptamine can produce contractions of mammalian smooth muscle. Many other organs such as the bladder, ureters, gallbladder and the entire stomach respond to serotonin. The following organs have been used in the past but never extensively: the isolated ears of guinea pig[54] and rabbit[47,70,83,84,111], strips[9,74,85] or rings of carotid artery[68,110]. Among invertebrates, the anterior retractor muscle of the byssus of *Mytilus edulis* is sensitive to serotonin and permits the study of the amount of serotonin present in serum diluted 1:10 even in the presence of high concentrations of histamine[9,18,74,85].

Mollusc heart

This technique has been found satisfactory by many investigators. Gaddum and Paasonen[48] and Welsh[107] have carried out an extensive study on the hearts of various species of molluscs. Only the methods used will be mentioned here:

Spisula (mactra) solida. If kept in aerated cool sea water the animal survives from 2 weeks to 3 months. The water should be changed twice a week (1 l/animal). The isolated heart is suspended in a 2-ml bath for isolated organs containing a special nutrient liquid (see Table 3). The heart starts to beat 5–60 minutes after dissection at a frequency of 12–25 beats/min. 5-Hydroxytryptamine, in concentrations of 0.1 to 0.5 μg/l increases the amplitude of the cardiac contraction; the maximal response is reached within 45–60 seconds and the cardiac activity returns to control levels immediately after washing. Furthermore there is no tachyphylaxis and the response to 5-hydroxytryptamine is proportional to a wide range of doses, the maximal effect being obtained with a concentration equal to 1 μg/ml.

As far as other substances are concerned, (*1*) LSD increases the amplitude and frequency of the heart beat and thus mimics the action of serotonin; (*2*) acetylcholine in concentrations between 1–100 μg/l usually inhibits the heart activity. Sometimes larger doses increase the frequency and cause a positive inotropic

---

* In place of measuring the contraction of the stomach strip preparation it has been proposed that the latent period before contraction, which is inversely proportional to serotonin concentration, should be measured[52].

effect until a diastolic block is produced. Benzoquinone chloride (6 mg/l)[79] is the only substance which effectively antagonizes the action of acetylcholine on the *Spisula* heart and it has the same effect on the heart of *Venus mercenaria*[74]. It is therefore necessary to add benzoquinone chloride (Mytolon) to the perfusing fluid when mollusc hearts are to be used for bioassay; (*3*) adrenaline and noradrenaline stimulate only when added in concentrations of 1000 μg/l or more; (*4*) histamine at the same concentration is without effect; (*5*) anterior pituitary hormone at 0.5 U/ml is also without effect.

The concentrations of the salts making up the nutrient liquid are of extreme importance. An increase in the amount of potassium chloride up to 0.9 mg/l causes a small increase in the amplitude of the response and in amounts of 2 mg/ml can cause a systolic blockade. An increase in calcium chloride up to 3 mg/ml* has a negative chronotropic effect while a diminution in the concentration of this salt causes increased frequency with a decreased amplitude of contractions.

Helix pomatia. This mollusc can survive for 5 months or more if kept simply in a beaker of water at room temperature. This preparation can be stimulated satisfactorily even with 1 μg/l of serotonin. The maximum response is reached after 20 sec. On this preparation LSD, ergometrine, dihydroergotamine and hydergine all increase the amplitude of the contractions but do not block the effect of serotonin. Adrenaline and histamine are without effect while acetylcholine inhibits the spontaneous activity of the heart when applied in concentrations of about 1 μg/l. Physostigmine, atropine, benzoquinone, decamethonium, hexamethonium and dibenamine are without effect even in concentrations of 10 mg/l.

The heart of *Venus mercenaria (Meretrix lusoria)* has been used extensively from the earliest experiments on serotonin[5,42]. For this purpose the heart is suspended at room temperature[96] in 10 ml of aerated artificial sea water. Sensitivity to serotonin varies from 0.0018–0.018 μg/ml and, like that to acetylcholine[76], bears an inverse relationship to the reduction of temperature[74,103]. A concentration of 6 μg/ml of Mytolon antagonizes the effects of acetylcholine[96].

Erspamer and Ghiretti[42] reported that some difficulties may be encountered using the heart of *Venus mercenaria* for the bioassay of serotonin in doses greater than 0.18 μg/ml. This heart is not sensitive to pitressin, angiotonin, histamine or adrenaline applied in concentrations up to 6 μg/ml, but it is very sensitive to tryptamine, *N*-methyl-5-hydroxytryptamine and to bufotenine.

The isolated heart of *Anodonta cygnea* L. is also suitable for the assay of serotonin (sensitivity 10^{-9} g/ml). Other known physiological substances are not very active on this preparation[44]. Embryos of some marine gastropods have been used to develop a new quantitative assay of serotonin reaching the sensitivity of 10^{-16} g/ml (ref. 68a).

CONCLUSIONS

This essentially fragmentary description of the many methods of pharmacological

---

* Sometimes lowering the calcium concentration changes an initially irregular rhythm into a regular one.

TABLE 4

EFFECT OF KNOWN PHYSIOLOGICAL SUBSTANCES ON BIOASSAY OF SEROTONIN

Substance	Rat isolated uterus	Rat isolated colon	Guinea pig ileum	Rat stomach strip	Venus mercenaria heart	Spisula solida heart
Adrenaline	—	—	—	—	+	+
Noradrenaline	—	—	—	—		+
Histamine	±	±	+	±		=
Acetylcholine	+	+	+	+	—	—
Substance P	±	—	=			=

— antagonism
+ effect similar to serotonin
± effect only at large concentrations unlikely to be present in the tissues of the common experimental animals
= no effect

assay requires a summary. The paper chromatographic methods are of no practical use in determining variations of serotonin content in the tissues of the common laboratory animals. However, chromatography is of value for the identification of serotonin. Quantitative analysis may also be performed should large amounts of material be available, or when the serotonin content is particularly great. Biological assay methods have the advantage of being both quantitative and very sensitive. However, the response obtained can be influenced by other biologically active substances normally present in the tissues. As can be seen from Table 4, adrenaline and noradrenaline are very potent antagonists of the action of serotonin in many biological systems. Catecholamines must be destroyed before carrying out a biological assay. This can be done by chemical or enzymological means. Finally by using specific inhibitors one must ascertain that the responses obtained are really due to serotonin. It is obvious when dealing with the tissues of animals which have been given drugs that may be taken up by the solvents used for serotonin extraction that the interference of such drugs with the action of serotonin must be excluded. Among the various biological preparations the isolated uterus of the spayed rat and the rat stomach strip appear to be of particular value, for they have been extensively utilized for the bioassay of serotonin and many of the possible sources of error have been carefully investigated.

The spectrophotofluorimetric method, after extraction with organic solvents or resins, is rapidly becoming the most widely used among the physicochemical methods by virtue of its convenience and high sensitivity. Even here, however, there may be interference from other substances, especially when studying the serotonin content of tissues after administration of drugs. The ideal experimental approach would be a constant combination of both biological and chemical methods. Should the results of both methods correspond, then one could feel quite certain that true variations in the serotonin content have been demonstrated.

References p. 24

REFERENCES

1   AMBACHE, N. AND ROCHA E SILVA, M., *Brit. J. Pharmacol.*, 6 (1951) 68.
2   AMIN, A. H., CRAWFORD, T. B. B. AND GADDUM, J. H., *J. Physiol. (London)*, 126 (1954) 569.
3   ASATOOR, A. AND DALGLIESH, C. E., *J. Chem. Soc.*, (1956) 2291.
4   AWAPARA, J., DAVIS, V. E. AND GRAHAM, O., *J. Chromatog.*, 3 (1960) 11.
5   BACQ, Z. M., FISCHER, P. AND GHIRETTI, F., *Arch. intern. physiol.*, 60 (1951) 165
6   BAKER, R. V., *J. Physiol. (London)*, 142 (1958) 563.
7   BERTLER, Å., *Acta Physiol. Scand.*, 51 (1961) 75.
8   BERTLER, Å. AND ROSENGREN, E., *Experientia*, 15 (1959) 382.
9   BIGELOW, F. S., *J. Lab. Clin. Med.*, 43 (1954) 759.
10 BLASCHKO, H., FRIEDMAN, P. J., HAWES, R. AND NILSSON, K., *J. Physiol. (London)*, 145 (1959) 384.
11 BOGDANSKI, D. F., PLETSCHER, A., BRODIE, B. B. AND UDENFRIEND, S., *J. Pharmacol. Exptl. Therap.*, 117 (1956) 82.
12 BOWMAN, R. L., CAUFIELD, P. A. AND UDENFRIEND, S., *Science*, 122 (1955) 32.
13 BRACCO, M., CURTI, P. C. AND GIULIANO, V., *Farmaco (Pavia), (Ed. sci.)*, 10 (1955) 595.
14 BRACCO, M., CURTI, P. C. AND GIULIANO, V., *Experientia*, 12 (1956) 31.
15 BRACCO, M., CURTI, P. C. AND GIULIANO, V., *Minerva med.*, 48 (1957) 204.
16 BUMPUS, F. M. AND PAGE, I. H., *J. Biol. Chem.*, 212 (1955) 111.
17 BURN, J. H., *Biological Standardization*, Oxford Medical Publ., 1950, p. 223.
17a CAIS, M., *Med. Exp.*, 5 (1961) 73.
18 CAMBRIDGE, G. W. AND HOLGATE, J. A., *J. Physiol. (London)*, 130 (1955) 22P.
19 CAMBRIDGE, G. W. AND HOLGATE J. A., *Brit. J. Pharmacol.*, 10 (1955) 326.
20 CARCASONA, A., UNTERHARNSCHEIDT, F. AND CERVOS-NAVARRO, J., *Klin. Wochschr.*, 37 (1959) 761.
21 CARGILL-THOMPSOHN, H. E. C., HARDWICK, D. C. AND WISEMAN, J. M., *J. Physiol. (London)*, 140 (1958) 10P.
22 CARLSSON, A., ROSENGREN, E., BERTLER, A. AND NILSSON, J. in S. GARATTINI AND V. GHETTI (Eds.), *Psychotropic Drugs*, Elsevier, Amsterdam, 1957, p. 363.
23 COLLIER, H. O. J., in G. P. LEWIS (Ed.), *5-Hydroxytryptamine*, Pergamon, London, 1958, p. 5.
24 CORREALE, P., *J. Neurochem.*, 1 (1956) 22.
25 CORREALE, P., *J. Neurochem.*, 2 (1958) 201.
26 COSTA, E., personal communication.
27 COSTA, E. AND APRISON, M. H., *Am. J. Physiol.*, 192 (1958) 95.
27a CRAWFORD, N. AND RUDD, B. T., *Clin. Chim. Acta*, 7 (1962) 114.
28 CROMARTIE, R. I. T. AND HARLEY-MASON, J., *Biochem. J.*, 66 (1957) 713.
29 CROSTI, P. F. AND BIANCHI, P., *Giorn. biochim.*, 7 (1958) 97.
29a CROSTI, P. F. AND LUCCHELLI, P. E., *J. Clin. Pathol.*, 15 (1962) 191.
30 DALGLIESH, C. E., *J. Clin. Pathol.*, 8 (1955) 73.
31 DALGLIESH, C. E., *Biochem. J.*, 64 (1956) 481.
32 DALGLIESH, C. E. AND DUTTON, R. W., *Cancer Research*, 11 (1957) 296.
33 DALGLIESH, C. E., TOH, C. C. AND WORK, T. S., *J. Physiol. (London)*, 120 (1953) 298.
34 DAVIS, R. B., *J. Lab. Clin. Med.*, 54 (1959) 344.
35 DUGGAN, D. E., BOWMAN, R. L., BRODIE, B. B. AND UDENFRIEND, S., *Arch. Biochem. Biophys.*, 68 (1957) 1.
35a EGGELS, P. H., HOEKSTRA, M. H. AND RÖLKENS, W. A., *Acta Physiol. Pharmacol. Neerl.*, 11 (1962) 300.
36 ERSPAMER, V., *Arch. exptl. Pathol. Pharmakol., Naunyn-Schmiedeberg's*, 196 (1940) 343.
37 ERSPAMER, V., *Arch. exptl. Pathol. Pharmakol., Naunyn-Schmiedeberg's*, 196 (1940) 391.
38 ERSPAMER, V., *Arch. exptl. Pathol. Pharmakol., Naunyn-Schmiedeberg's*, 200 (1942) 60.
39 ERSPAMER, V., *Arch. exptl. Pathol. Pharmakol., Naunyn-Schmiedeberg's*, 201 (1943) 377.
40 ERSPAMER, V., *Naturwissenschaften*, 40 (1953) 318.
41 ERSPAMER, V., *Rend. sci. Farmitalia*, 1 (1954) 1.
42 ERSPAMER, V. AND GHIRETTI, F., *J. Physiol. (London)*, 115 (1951) 470.
43 ERSPAMER, V. AND SALA, G., *Brit. J. Pharmacol.*, 9 (1956) 31.
43a FALES, H. M. AND PISANO, J. J., *Anal. Biochem.*, 3 (1962) 337.
44 FÄNGE, R., *Experientia*, 11 (1955) 156.

45   FELDBERG, W. AND TOH, C. C., *J. Physiol. (London)*, 119 (1953) 352.
46   GADDUM, J. H., *J. Physiol. (London)*, 121 (1953) 15P.
47   GADDUM, J. H. AND HAMEED, K. A., *Brit. J. Pharmacol.*, 9 (1954) 240.
48   GADDUM, J. H. AND PAASONEN, M. K., *Brit. J. Pharmacol.*, 10 (1955) 474.
49   GADDUM, J. H., PEART, W. S. AND VOGT, M., *J. Physiol. (London)*, 108 (1949) 467.
50   GARATTINI, S. AND VALZELLI, L., *Boll. soc. ital. biol. sper.*, 31 (1955) 1648.
51   GARATTINI, S. AND VALZELLI, L., *Boll. soc. ital. biol. sper.*, 32 (1956) 288.
52   GARCIA DE JALON, P. D., FERNANDEZ DE SAN JULIAN, C. AND GOMEZ ALONSO DE LA SIERRA, B.,
     Archivos inst. farmacol. exptl. (Madrid), 11 (1959) 71.
53   GARVEN, J. D., *Brit. J. Pharmacol.*, 11 (1956) 66.
54   GEY, K. F. AND PLETSCHER, A., *Experientia*, 15 (1960) 372.
55   GLUCKMAN, M., *Federation Proc.*, 19 (1960) 265.
56   GLUCKMAN, M. I. AND ABRAMS, R., *Federation Proc.*, 18 (1959) 395.
57   GRETTE, K., *Acta Pharmacol. Toxicol.*, 13 (1957) 177.
58   HARDISTY, R. M. AND STACEY, R. S., *J. Physiol. (London)*, 130 (1955) 711.
59   HESS, S. M. AND UDENFRIEND, S., *J. Pharmacol. Exptl. Therap.*, 127 (1959) 175.
60   HIRS, C. W. H., MOORE, S. AND STEIN, W. H., *J. Biol. Chem.*, 200 (1953) 493.
61   HUMPHRY, J. H. AND JAQUES, R., *J. Physiol. (London)*, 124 (1954) 305.
62   INNES, J. R., KOSTERLITZ, H. W. AND ROBINSON, J. A., *J. Physiol. (London)*, 137 (1957) 396.
63   JAQUES, R. AND SCHACHTER, M., *Brit. J. Pharmacol.*, 9 (1954) 49.
64   JEPSON, J. B., in I. SMITH (Ed.), *Chromatographic Techniques*, William Heinemann, London,
     1958, p. 114.
65   JOYCE, D., *Nature*, 182 (1958) 463.
65a JOYCE, D., *Brit. J. Pharmacol.*, 18 (1962) 370.
66   KÄRKI, N. T. AND PAASONEN M. K., *Acta Pharmacol. Toxicol.*, 16 (1959) 20.
67   KEILIN, D. AND MANN, T., *Proc. Roy. Soc., London, Series B*, 125 (1938) 187.
68   KIRPEKAR, S. M. AND LEWIS, J. J., *J. Pharm. and Pharmacol.*, 10 (1958) 255.
68a KOSHTOYANTS, KH. S., BUZNIKOV, G. A. AND MANUKHIN, B. N., *Comp. Biochem. Physiol.*,
     3 (1961) 20.
69   KUNTZMAN, R., SHORE, P. A., BOGDANSKI, D. AND BRODIE, B. B., *J. Neurochem.*, 6 (1961) 226.
70   KURIAKI, K. AND INOUE, T., *Compt. rend. soc. biol.*, 149 (1955) 2053.
71   LEITCH, J. L., MARYN, D. AND MOORE, J. B., *J. Am. Pharm. Assoc., Sci. Ed.*, 47 (1958) 535.
72   LEVY, J., *J. physiol. (Paris)*, 49 (1957) 881.
73   LIPTON, M. A. AND GORDON, P., *Abstract Fall Meeting Am. Soc. Pharmacol.*, Ann Arbor
     (Mich.), 1958, p. 22.
74   LUDUENA, F. P. AND BROWN, JR., T. G., *J. Pharmacol. Exptl. Therap.*, 105 (1952) 232.
75   MARSHALL, E. F., STIRLING, G. S., TAIT, A. C. AND TODRICK, A., *Brit. J. Pharmacol.*, 15 (1960)
     35.
76   McINTYRE, A. R., WILLIAMS, D. E. AND HUMDLER, F. L., *Federation Proc.*, 19 (1960) 283.
77   McINTYRE, F. C., ROTH, L. W. AND SHAW, J. L., *J. Biol. Chem.*, 170 (1947) 537.
77a MEAD, J. A. R. AND FINGER, K. F., *Biochem. Pharmacol.*, 6 (1961) 52.
78   MUSSINI, E. AND VALZELLI, L., *Atti soc. lombarda sci. med. e biol.*, 13 (1958) 301.
79   PAASONEN, M. K. AND VOGT, M., *J. Physiol. (London)*, 131 (1956) 617.
80   PAGE, J. H., *Methods in Med. Research*, i (1948) 253.
81   PERNOW, B., *Acta Physiol. Scand.*, 29 (1953) Suppl. 105.
82   PERNOW, B., *Acta Physiol. Scand.*, 34 (1955) 295.
83   RAPPORT, M. M., *J. Biol. Chem.*, 176 (1948) 1243.
84   RAPPORT, M. M., *Science*, 108 (1948) 329.
85   REID, G. AND RAND, M., *Nature*, 169 (1952) 801.
85a ROBERTS, M., *J. Pharm. and Pharmacol.*, 14 (1962) 746.
86   ROCHA E SILVA, M., VALLE, J. R. AND PICARELLI, Z. P., *Brit. J. Pharmacol.*, 8 (1953) 378.
87   RODNIGHT, R., *Biochem. J.*, 64 (1956) 621.
88   SHORE, P. A., *Pharmacol. Revs.*, 11 (1959) 276.
89   SMITH, I., *Chromatographic Techniques*, William Heinemann, London, 1958, p. 26.
90   SPRINCE, H., *J. Chromatog.*, 3 (1960) 97.
91   SUPEK, Z. AND MILKOVIC, S., *Experientia*, 12 (1956) 71.
91a SUPEK, Z. AND RANDIĆ, M., *Biochem. Pharmacol.*, 8 (1961) 79.

[92] SWEELEY, C. C. AND WILLIAMS, C. M., *Federation Proc.*, 20 (1961) 5.

[93] SYNGE, R. L. M. AND TISELIUS, A., *Acta Chem. Scand.*, 3 (1949) 231.

[94] TOH, C. C., *J. Physiol. (London)*, 126 (1954) 248.

[95] TOWNE, J. C., FENSTER, E. AND SCHAEFER, T., *Federation Proc.*, 20 (1961) 343.

[95a] TOWNE, J. C., FENSTER, E., SHERMAN, J. AND SCHAEFER, T., *Federation Proc.*, 21 (1962) 420.

[96] TWAROG, B. M. AND PAGE, J. H., *Am. J. Physiol.*, 175 (1953) 157.

[97] UDENFRIEND, S., BOGDANSKI, D. F. AND WEISSBACH, N., *Science*, 122 (1955) 972.

[98] UDENFRIEND, S., WEISSBACH, H. AND CLARK, C. T., *J. Biol. Chem.*, 215 (1955) 337.

[99] VALZELLI, L., unpublished results.

[100] VANE, J. R., *Brit. J. Pharmacol.*, 12 (1957) 344.

[101] VOGT, W., *Pharmacol. Revs.*, 10 (1958) 407.

[102] WAALKES, T. P., *J. Lab. Clin. Med.*, 53 (1959) 824.

[103] WAIT, R. B., *Biol. Bull.*, 85 (1943) 79.

[104] WALASZEK, R. J. AND ABOOD, L. G., *Federation Proc.*, 16 (1957) n. 579.

[104a] WARNATZ, H., SCHEIFFARTH, F., SCHMID, E. AND ZICHA, L., *Med. Exptl.*, 3 (1960) 101.

[105] WEISSBACH, H., BOGDANSKI, D. F., REDFIELD, B. AND UDENFRIEND, S., *Federation Proc.*, 16 (1957) 345.

[106] WEISSBACH, H., WAALKES, T. P. AND UDENFRIEND, S., *J. Biol. Chem.*, 230 (1958) 865.

[107] WELSH, J. H., *Nature*, 173 (1954) 955.

[108] WEST, G. B., *J. Pharm. and Pharmacol.*, 10 (1958) 92 T.

[109] WOOLLEY, D. W. AND EDELMAN, P. M., *Science*, 127 (1958) 281.

[110] WOOLLEY, D. W. AND SHAW, E., *J. Biol. Chem.*, 203 (1953) 69.

[111] ZUCKER, M. J., FRIEDMAN, B. K. AND RAPPORT, M. M., *Proc. Soc. Exptl. Biol. Med.*, 85 (1954) 282.

Chapter 3

BIOCHEMISTRY OF SEROTONIN

The metabolism of serotonin depends upon certain enzymatic systems and the availability of certain substrates. The major pathway begins from tryptophan (see Fig. 8 for the principal metabolic pathways of this amino acid) which is transformed into 5-hydroxytryptophan (5-HTP) through *tryptophan 5-hydroxylase* and then into 5-hydroxytryptamine by means of *5-hydroxytryptophan decarboxylase*. 5-Hydroxytryptamine is metabolized through *monoamine oxidase* (MAO) with the formation of 5-hydroxytryptophan acetaldehyde which is in its turn transformed into 5-hydroxyindoleacetic acid by *aldehyde dehydrogenase*.

In addition to this major pathway several other collateral metabolic steps have been discovered recently and these may become more important when they have been more fully investigated and understood.

TRYPTOPHAN 5-HYDROXYLASE

This is the enzyme transforming tryptophan into 5-hydroxytryptophan. The presence of this enzyme in mammalian tissues is still a matter for discussion. Udenfriend *et al.*[237] showed that [2-^{14}C]-*dl*-tryptophan is transformed into labelled 5-HTP by liver homogenates of rat and guinea pig. However, the same authors question the significance of this result because only a very small fraction of the original radioactivity was found in 5-HTP. Dalgliesh and Dutton could not confirm the presence of the step tryptophan–5-H TP*in vitro* by using an isolated perfused liver preparation of mouse on rat[78,79]. However, there is no doubt that an hydroxylation must occur *in vivo*, for the administration of the labelled tryptophan results in the presence of labelled 5-hydroxyindoleacetic acid in the urine[248]. This pathway which accounts for only 1% of the administered tryptophan in normal subjects, reaches about 60% in patients with carcinoid, a tumour of the intestinal argentaffin cells[248]. This finding suggested that the site where tryptophan is physiologically hydroxylated is represented by the argentaffin or enterochromaffin cells in the intestine[78,96,*]. An alternative hypothesis has also been considered, namely the possibility that tryptophan is first decarboxylated to yield tryptamine[260,270,338] and then tryptamine is hydroxylated and transformed into serotonin[150]. However, it has been shown that tryptamine cannot be considered a precursor of 5-hydroxytryptamine[239]. Tryptamine is in fact hydroxylated in the 6-position by a microsomal hepatic enzyme in the presence of a

---

* Although there is strong evidence in favour of the identity between the argentaffin reaction and the presence of serotonin, this view is not generally accepted[122,152,171].

system generating TPNH[153,153a]. A similar pathway is common also to α-methyl-tryptamine[226], *N,N*-diethyltryptamine[227] and skatol[148].

The venom-producing glands of the toad are rich in an enzyme system which can hydroxylate tryptophan in the 5-position[237,242]. A useful tool in studying tryptophan hydroxylase is represented by *Chromobacterium violaceum* producing a substance called violacein (see Fig. 9) which contains a 5-hydroxyindole[14]. According to Mitoma *et al.*[187] this bacterium is lacking in the principal metabolic pathways by which tryptophan is usually metabolized (tryptophan pyrrolase, kynurenine transaminase, tryptophanase, etc.) but contains an enzyme able to convert L-tryptophan into 5-hydroxytryptophan*.

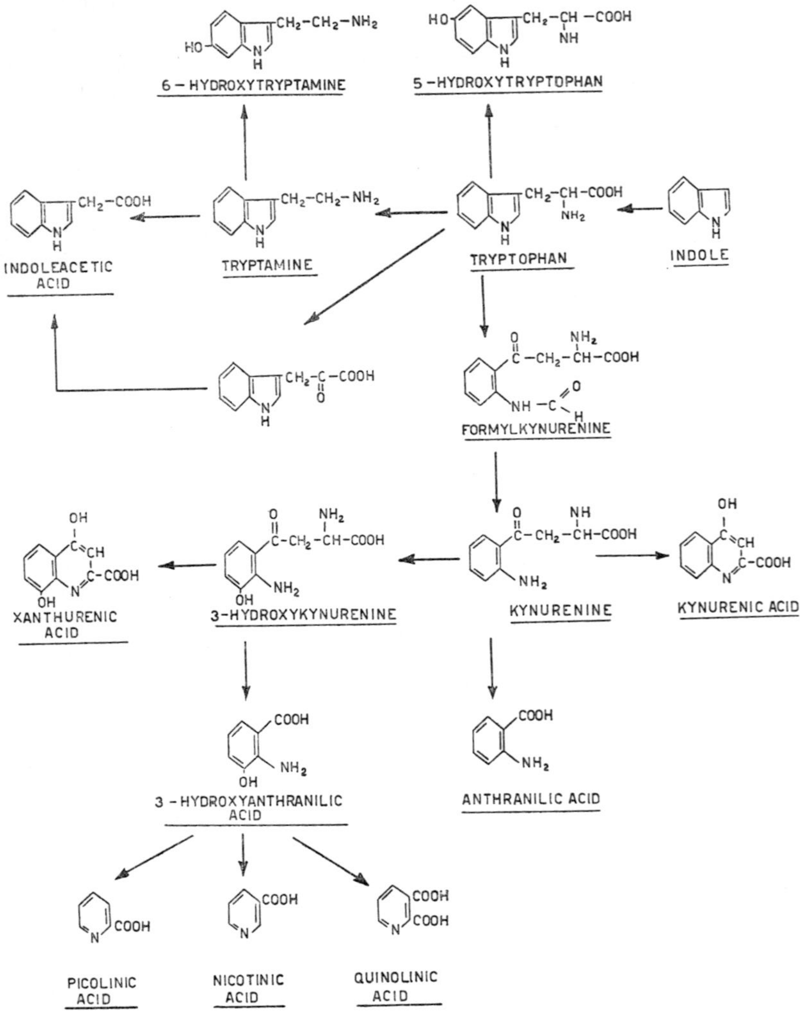

Fig. 8. Principal metabolic pathways of tryptophan.

* Other bacteria can hydroxylate indole compounds[125].

This enzyme which has been isolated and purified, shows an optimum of activity at pH 6, being almost inactive at pH 8. Beside 5-hydroxytryptophan the bacterium also produces N-acetyl-5-HTP, 5-HIA and 5-hydroxyindolepyruvic acid. It is now certain that the enzyme hydroxylating tryptophan is not phenylalanine hydroxylase[185] or the enzyme, present in microsomes of rabbit liver, which hydroxylates aromatic substances[186]. The enzyme isolated from *Chromobacterium violaceum* does not hydroxylate tryptamine, indoleacetic acid or kynurenine[188].

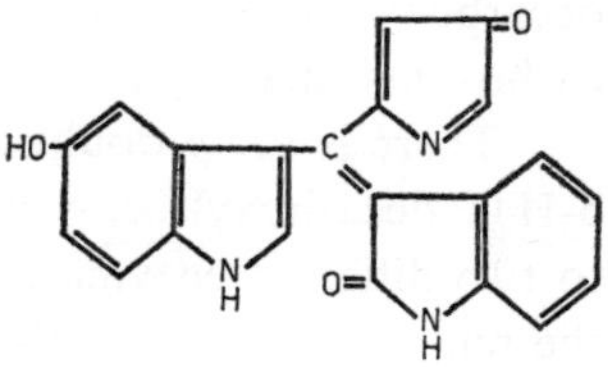

Fig. 9. Chemical structure of violacein.

It has also been shown that mouse can hydroxylate tryptophan yielding the formation of considerable amounts of serotonin[210,211].

Finally, Cooper reported the isolation and characterization of tryptophan 5-hydroxylase from guinea pig or rat intestinal mucosal cells and kidney[64,65]. The enzyme is confined in the particulate fraction although it has not yet been exactly localized. The reaction also proceeds anaerobically and requires the presence of cupric ions and ascorbic acid. The cupric ions cannot be replaced by Cu^+, Fe^{3+} or Mn^{2+}, but dehydroascorbic or D-ascorbic acid can replace the ascorbic acid.

The enzyme does not require sulphydryl groups, cytochrome c, DPN, DPNH, TPN or TPNH[64]. More recently, Freedland *et al.*[105a,105b] found a liver enzyme which hydroxylates tryptophan and which requires DPN and oxygen. This enzyme was shown to be identical with phenylalanine hydroxylase[203a–c].

Tryptophan 5-hydroxylase needs more investigation and this is most important because this enzyme is probably a limiting step in the overall metabolism of serotonin.

5-HYDROXYTRYPTOPHAN DECARBOXYLASE

This enzyme converts 5-hydroxytryptophan to 5-hydroxytryptamine. When it is highly purified it is quite unstable and sensitive to changes of pH and temperature[62]. The optimum range of activity[164a] is at pH 7.57–8.1. Contrary to what has been reported previously[62], the enzyme requires the presence of pyridoxal phosphate as cofactor. Pyridoxal phosphate may be replaced by pyridoxamine phosphate but not by pyridoxine or pyridoxal[48,*]. A diet poor in vitamin B_6 reduces considerably the levels of serotonin in several tissues[17,47,259,**]. 5-HTP decarboxylase is lowered but may be restored to normal levels by administering pyridoxine *in vivo* or pyridoxal phosphate *in vitro*[47,49]. The 5-HTP decarboxylase present in mouse mast-cell tumour also requires pyridoxal phosphate as coenzyme[132].

---

* The presence of phosphates is not required because β-glycerolphosphate, cocarboxylase, AMP, ATP or inhibitors of phosphate metabolism, like citrate and sodium fluoride do not substantially affect the 5-HTP decarboxylase[48].
** The decrease is not due to a decrease of food uptake[49].

Specificity

The enzyme decarboxylates the L-5-HTP, but not the D-isomer[106,106a]. Tryptophan and its derivatives hydroxylated in position 7 (refs. 102, 238), 6 and 2 (ref. 102) are not substrates for this enzyme. 4-Hydroxytryptophan is instead largely decarboxylated[99,102] and this raises the question of the meaning of the term 5-hydroxytryptophan decarboxylase. A number of other compounds tested for 5-hydroxytryptophan decarboxylase activity is reported in Table 5.

There is some doubt about the identity of 3,4-dioxyphenylalanine (DOPA) and 5-HTP decarboxylase. According to Clark *et al.*[62] the two activities should be referred to two different enzymes. The following reasons would support this conclusion: (*a*) the relative decarboxylation of 5-HTP or DOPA may change up to 20 times during the steps of the purification; (*b*) 5-HTP decarboxylase shows an optimum activity at pH 8.1 while DOPA decarboxylase has an optimum at pH 6.5; (*c*) pyridoxal phosphate may be easily removed from DOPA, but not from 5-HTP decarboxylase; (*d*) the inhibition exerted by hydroxylamine or semicarbazide may be prevented by pyridoxal phosphate for DOPA decarboxylase, but not for 5-HTP decarboxylase; (*e*) purified preparations of 5-HTP decarboxylase still act on DOPA, but 5-HTP is not a substrate of DOPA decarboxylase[137].

The kinetic of the two reactions is also different being of zero order for 5-HTP and of pseudo first order for DOPA decarboxylation[279]. This led Yuwiler *et al.*[279] to predict that a competitive inhibitor of the decarboxylase should be more active on DOPA than on 5-HTP.

On the other hand, there is evidence which suggests that the decarboxylation of DOPA and 5-HTP is carried out by the same enzyme. Several inhibitors of DOPA decarboxylase also block the decarboxylation of 5-HTP (refs. 84, 103, 137, 141, 220, 241, 274, 279–281). On the basis of studies carried out with inhibitors Yuwiler *et al.*[281] and Bertler and Rosengren[19] conclude that only one enzymatic site is involved in the decarboxylation of DOPA and 5-HTP. According to Werle and Aures[268] the differences observed between DOPA and 5-HTP decarboxylase would be related only to a different requirement of pyridoxal phosphate and not to a partial resolution of the apoenzyme. Further investigations are needed to elucidate this problem that may have physiological importance because the decarboxylation of DOPA and 5-HTP yields compounds biologically active.

It is however established that 5-HTP decarboxylase does not carry out the decarboxylation of histidine[46,121,154].

Distribution

5-HTP decarboxylase is widely distributed in nature. Mammalian kidney and liver are particularly rich in this enzyme which is present in the non-particulate fraction after ultracentrifugation[119]. Foetal liver[154] and kidney[149] of rat contain less 5-HTP decarboxylase activity than the respective adult tissues. Spleen and bone marrow do not decarboxylate 5-HTP[119]. Although platelets store large concentrations of serotonin, they do not contain this enzyme[62,244]. Careful studies on the distribution of 5-HTP decarboxylase in various areas of brain have been carried out by Gaddum and Giarman[108], Costa and Himwich[67] and Kuntzman *et al.*[164]. Their findings will be described in the section of the central nervous system. Pineal gland also shows 5-HTP decarboxylase activity[120,121]. Rat mast-cell[17a,164a] and mouse mast-cell tumours are rich in 5-HTP decarboxylase[131,132,210]. Finally the enzyme is present in certain organs of invertebrates[184,267] and in vegetables like banana[240].

Table 6 shows some data obtained by West[272] concerning the distribution of 5-HTP decarboxylase in several tissues of common laboratory animals.

Inhibitors

5-HTP decarboxylase is not inhibited by deoxypyridoxine[47], an antivitamin B_6 which inhibits other decarboxylases requiring pyridoxal phosphate as cofactor, like tyrosine decarboxylase[15].

Chelating agents like tetraethylthiuram disulphide, diphenylthiocarbazone, ethylenediaminetetraacetic acid (EDTA) are inhibitors of 5-HTP decarboxylation. The inhibition exerted by EDTA[47] is removed by Zn^{2+}, Mn^{2+} and Mg^{2+} (ref. 16) but not by iron. Diethyldithiocarbamate, a strong chelating agent for copper, does not inhibit 5-HTP decarboxylase. Some metals are inhibitors but only at high concentrations[48].

The inhibition exerted by hydroxylamine is removed by increasing the concentration of pyridoxal phosphate[48]. Inhibitors of the —SH groups [*p*-chloromercuribenzoate, *o*-iodosobenzoate, γ-(*p*-arsenioserphenyl)butyrate] are active on 5-HTP decarboxylase only at high concentrations and are not antagonized by BAL. This shows that —SH groups are not involved in the decarboxylation of 5-HTP[48].

Carbonyl reagents like semicarbazide[259], isoniazid[52] and some hydrazine-type monoamine oxidase inhibitors are inhibitors of the 5-HTP decarboxylation. Some tryptophan derivatives are also very active *in vitro* (example 2-phenyl-DL-tryptophan, tetrahydroharmane and *N*-methyltetrahydroharmane[106]).

In the nervous tissue, but not in other tissues, an excess of 5-HTP blocks its enzymatic decarboxylation[119].

Davison and Sandler[84] showed that phenylpyruvic, phenyllactic and phenylacetic acid (metabolites of phenylalanine in phenylketonuric patients) inhibit 5-HTP and DOPA decarboxylase.

It is interesting that 3,4-dioxyphenylalanine inhibits 5-HTP decarboxylase *per se*[19] and through the formation of its metabolite 3,4-dioxyphenylethylamine[279]. Also noradrenaline inhibits 5-HTP decarboxylase[50,51]. α-Methyl-*m*-tyrosine[141] and α-methyl-DOPA inhibit the decarboxylases of several amino acids including 5-HTP (refs. 141, 217, 241, 274). α-Methyl-DOPA is known to inhibit DOPA decarboxylase[220]. This suggests that it might be useful to study the effect on 5-HTP decarboxylase of all the inhibitors of DOPA decarboxylase recently summarized in a careful review[61].

A list of inhibitors of 5-HTP decarboxylase is given in Table 5.

TABLE 5

RESULTS OBTAINED WITH SOME SUBSTANCES TESTED ON 5-HTP DECARBOXYLASE

Substrates	Non-substrates	Inhibitors	Non-inhibitors
3,4-dihydroxyphenylalanine (DOPA)[279],** L-*erythro*-3,4-dihydroxyphenylserine[279] L-*threo*-3,4-dihydroxyphenylserine[279]	2-carbethoxy-DL-tryptophan[106] 2-phenyl-DL-tryptophan[106]	adrenaline[279] 1-aminoguanidine sulphate[279] diphenylthiocarbazone[47]	1-benzyl-2,5-dimethyl-5-hydroxy-tryptamine[279] (BAS) 2-carbethoxy-DL-tryptophan[106]
α-methyl-DOPA[217,262],** 5-hydroxy-*N*-acetyltryptophan[102],*** 5-acetoxy-*N*-acetyltryptophan[102],***		3,4'-dihydroxy-3'-carboxy-chalcone[279],*a and derivatives[281] 2-(3,4-dihydroxyphenyl)-ethylamine (dopamine)[279]	deoxypyridoxine[47] 2,5-dihydroxy-DL-tryptophan[106] dihydroxyindole-3-DL-alanine[106] harmine[106]
4-hydroxytryptophan[102]		ethylenediaminetetraacetic acid[47] 3,4-dihydroxyphenylalanine[279]	2-hydroxy-5-benzylhydroxy-tryptamine[106]
		phenylacetic acid[84],*g L-phenylalanine[84] β-phenylethylhydrazine[89]	2-hydroxytryptamine[106] 5-hydroxytryptamine[279] 2-hydroxy-DL-tryptophan[106]
	2-hydroxy-DL-tryptophan[106]	phenyllactic acid[84],*f phenyipyruvic acid[84],*d hydroxylamine[47]	5-(3-indoleacryloyl)salicylic acid[279] *N*-methyltetrahydronorharmane[106]
	D-5-hydroxytryptophan[106] 7-hydroxytryptophan[238] α-iso-5-hydroxytryptophan[102,106]	1-(5-hydroxyindolyl-3)-2-(carboxy-4-hydroxy-4-hydroxy-benzoyl)ethylene[130,279],*c	
	DL-6-hydroxytryptophan[102] DL-2-hydroxytryptophan[102] 5-methyltryptophan[102]	*m*-hydroxypropadrine[279] 5-hydroxytryptophan[119] β-(3-indolyl)acrylic (acid)[279]	α-methyltryptamine methane-sulphonate[279] (5-hydroxyindol-3-yl)-buten-3-one[279]
	5-methoxytryptophan[102] tyrosine[238] tryptophan[102,238]	α-methyl-DOPA[274],*e,**	sodium fluoride[48] tetrahydronorharmane[106]
	5-benzylhydroxytryptophan[102] 5-nitrotryptophan[102]	noradrenaline[50,279] semicarbazide[47]	tyrosine[279] tryptophan[279]

Substrates	Non-substrates	Inhibitors	Non-inhibitors
	5-hydroxy-β-methyltryptophan[102] 5-hydroxy-N,N-dimethyltryptophan[102]	5-hydroxy-α-methyltryptamine[129] tetraethylthiuramdisulphide[47] sodium thiocyanate[47]	
	5-hydroxy-2-methyltryptophan[102] 5-benzyloxy-6-methoxytryptophan[102]	thiosemicarbazide[279] 3,4,4'-trihydroxy-3'-carboxy-chalcone[279],* b	
	4-methyltryptophan[102] 4-benzylhydroxytryptophan[102] 6-benzylhydroxytryptophan[102]		

* For these compounds an inhibition of DOPA decarboxylase is known.

a[137]   b[137]   c[280]   d[103,137]   e[137,220]   f[103]   g[103]

** This substrate is slowly decarboxylated.

*** These compounds are converted in 5-HTP before being decarboxylated[102].

TABLE 6

THE RELATIVE 5-HTP DECARBOXYLASE ACTIVITY OF VARIOUS ANIMAL TISSUES

Recorded as the 5-HT formed (μg/h/g tissue) from added 5-HTP and 5-HT content (μg/g) [272].

Tissue	Rat		Mouse		Guinea pig		Hamster		Rabbit		Dog		Cat	
	5-HT	5-HTP decarb.	5-HT	5-HTP decarb.	5-HT	5-HTP decarb.	5-HT	5-HTP decarb.	5-HT	5-HTP decarb.	5-HT	5-HTP decarb.	5-HT	5-HTP decarb.
Brain	0.2	32	0.3	9	0.3	13	0.2	20	0.3	5	0.2	2	0.3	2
Duodenum	1.2	2	1.2	2	5	391	0.9	9	3.3	13	3.7	2	0.9	23
Ileum	1.2	1	1	1	3.4	190	1.3	1	3.7	6	4.3	1	0.5	8
Kidney	0.1	188	0.1	187	0.1	171	0.1	62	0.1	56	0.1	47	0.1	24
Liver	0.1	124	0.7	22	0.1	156	0.2	62	0.6	94	0.5	12	0.6	55
Skin	1.3	1	0.4	1	0.1	1	0.1	1	0.1	1	0.1	1	0.1	1
Spleen	2.5	1	2.7	6	1.1	2	20.5	93	24.3	19	4.6	2	8.5	2
Stomach	1.4	1	1	2	1.4	64	1.2	45	4.9	34	5.2	3	0.5	25

MONOAMINE OXIDASE

Monoamine oxidase* is an enzyme able to transform 5-hydroxytryptamine into 5-hydroxyindoleacetaldehyde** which is, at its turn, converted to 5-hydroxyindoleacetic acid[22,107] by an aldehyde dehydrogenase[234,264]. A soluble preparation of MAO has been obtained by several methods[12,71,264]. Little is known about the structure and the requirements of this enzyme. The amine oxidase of pea-seedlings contains copper[174] and on the other hand it has been reported that monoamine oxidase may contain a flavoprotein[204]. The monoamine oxidase activity is dependent and proportional to the oxygen tension[194]. That monoamine oxidase is responsible for the inactivation of serotonin is known since the time when this amine was called "serum vasoconstrictor"[40].

Distribution

This enzyme is widely distributed in nature. Table 7 shows that a certain number of species and tissues are able to oxidize 5-HT with the formation of 5-hydroxyindoleacetic acid (5-HIA).

It would be too tedious to enumerate all the tissues showing monoamine oxidase activity, therefore only certain interesting aspects will be reported here.

The literature concerning the presence of monoamine oxidase in blood is extensive. Some observations led to the conclusion that MAO was present in blood (refs. 142, 160, 161, 228, 271) although the opposite was reported by Blaschko[23,24,28]. More recently it was observed that serotonin is oxidized by the serum of horse, dog, pig and sheep, but not by the serum of rabbit, cat and human[26]. The enzyme responsible also oxidizes benzylamine and histamine[159] and is inhibited by carbonyl reagents more than by monoamine oxidase inhibitors. Further experiments finally established that the oxidation of 5-HT in pig plasma is not carried out by an amine oxidase but by a copper-protein, caeruloplasmin[25,31].

Rabbit serum does not contain MAO activity which is instead present in erythrocytes and probably in leucocytes[255]. Platelets have no monoamine oxidase activity[245].

Some tissues of various animal species contain monoamine oxidase activity but do not oxidize serotonin. This has been observed in liver[34] and brain[257] of cat, and in rat intestine[81]. Also the mouse mast-cell tumour oxidizes dopamine, but not serotonin[130]. On the other hand a monoamine oxidase present in *Sarcina lutea* oxidizes

---

* The distinction between mono- and diamine oxidase is losing of its meaning because several exceptions have been observed concerning both the specificity of the substrate[104,105] and the selectivity of the inhibitor. New criteria for the classification of this group of enzymes have been recently proposed[26,209,289].

** This aldehyde may be trapped by semicarbazide and transformed into 2,4-dinitrophenylhydrazone. The colour developed in alkaline medium by this compound may be used for the estimation of MAO[127,128]. Other simple methods are now available for the evaluation of MAO[266,278], although only the determination of the NH_3 produced should be considered as a specific method[69].

serotonin, has no activity on histamine and cadaverine, but is not inhibited by iproniazid[163]. These data suggest that different types of monoamine oxidase may occur according to the tissue and the animal species considered.

The foetal tissues have usually only a weak monoamine oxidase activity[91,190,269]. Histochemical investigations, using 5-HT as substrate, showed that MAO in kidney is more concentrated in the cortical part and that the proximal convoluted tubules are very rich in this enzyme[27]. Monoamine oxidase is localized mainly in mitochondria (refs. 11, 68, 138, 216).

TABLE 7

ORGANS AND TISSUES CONTAINING A MONOAMINE OXIDASE ACTING ON SEROTONIN

Dog	:	brain[247,257]
Eusepia officinalis	:	(ref. 34)
Guinea pig	:	brain[27,81], liver[34,126], intestine[81,94], lung[191], kidney[34]
Man	:	liver[165], lung[83], kidney[165], salivary glands[224]
Carcinoid patients	:	liver[83,165], lung[83], kidney[165]
Loligo forbesii	:	(ref. 34)
Mouse	:	brain[81,257], liver[81]
Octopus vulgaris	:	hepato-pancreas[94], salivary glands[22]
Pig	:	liver[34], kidney[34,57], brain[57]
Pisum sativum	:	(ref. 63)
Rabbit	:	brain[191,257], erythrocytes and leucocytes[255], intestine[191], lung[191]
Rat	:	brain[81], liver[231], lung[81], kidney[231], hepatoma[156]
Sarcina lutea	:	(ref. 163)
Sheep	:	lung[23]

Specificity

Monoamine oxidase is not specific for the oxidation of serotonin because it attacks a large number of amines. Tables 8 and 9 report some comparative data on the relative activity of certain enzyme preparations for several amines including serotonin. Among the indole derivatives, 4-, 6- and 7-hydroxytryptamine are oxidized by MAO preparations from guinea pig[98]. 2-Hydroxytryptamine and 2-hydroxy-5-benzoyltryptamine are not substrates for MAO but they are inhibitors[162]. Bufotenidine, but not bufotenine, is a substrate[23]. Also the N-methyl-5-HT[126] and the N,N-dimethyltryptamine-N-oxide[218] are oxidized by MAO. Polyglycol ethers and particularly the nonylphenol derivative increase MAO activity[269].

Inhibitors

A large number of compounds has been investigated *in vitro* in these last twenty years. Most of the early studies were summarized by Blaschko[23] in 1952*. However,

---

* Amphetamine[173], p-tolylcholine ether[43,44,66,82,246] and procainamide[246] were considered monoamine oxidase inhibitors.

TABLE 8

EFFECT OF MAO ON SEVERAL AMINES[264]

Substrate	Oxygen uptake	
	Mitochondria	Purified enzyme
Serotonin	100	100
Tyramine	240	190
Tryptamine	100	123
Heptylamine	20	9
Noradrenaline	20	30
Benzylamine	0	0
Phenylethylamine	0	0
Histamine	0	0
Spermine	0	0

the "era" of monoamine oxidase inhibitors was inaugurated by the increasing knowledge of the serotonin metabolism and by the finding of Zeller concerning the inhibitory effect exerted on MAO by iproniazid (*N*-isopropyl-isonicotinylhydrazide)[290,291] later confirmed by other authors[157,158,253].

Since iproniazid, some thousand compounds have been tested; it is therefore impossible to summarize the relationships between chemical structure and inhibition of MAO, and so the results are still unpredictable. Certain chemical structures with a tentative interpretation have been reported in some review articles[112,113,198,293,296].

The basic structure of a large group of monoamine oxidase inhibitors consists of a hydrazine group in which an atom of hydrogen is substituted by various different radicals R:

$$R - NH - NH - R_1$$

Alkylidene hydrazines[60] and hydrazone derivatives[292] do not show inhibitory effect on MAO. In the isonicotinylhydrazide structure the substitution of a terminal hydro-

TABLE 9

EFFECT OF SEVERAL MAO PREPARATIONS ON SELECTED AMINES[179]

Tissue	Relative MAO activity for		
	tyramine	dopamine	serotonin
Intestinal mucosa	100	52	74
Liver	75	28	2
Kidney	50	21	13
Lung	12	8	4
Brain	11	4	5
Skeletal muscle	1	0	0.7

gen with alkyl groups leads to an increase of activity from methyl to butyl, but a
decrease if the lateral chain has a length up to heptyl[13]. Isoniazid is completely inactive[291] (see Fig. 10). Iso lateral chains are usually less active than *n*-alkyl groups.

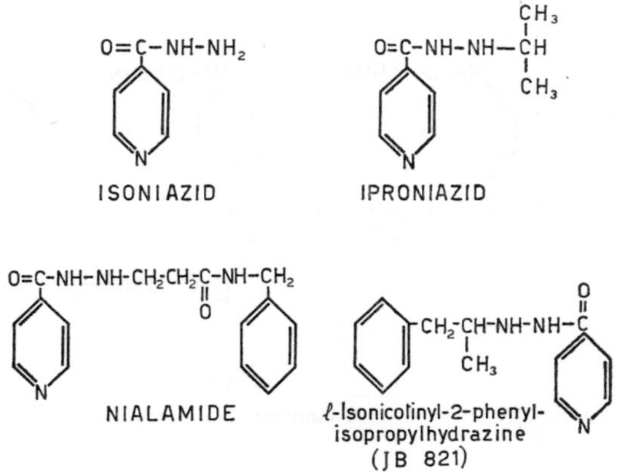

Fig. 10.

The hydrogenation of the pyridine group leads to a marked drop of activity[13]. The
hydrazide of nicotinic acid is not active[13], but the isopropylhydrazide still retains the
activity of iproniazid[54]. A derivative with good activity is nialamide[85] (*N*- [2(benzyl-

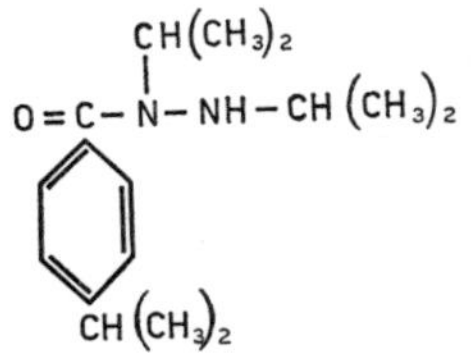

Fig. 11.

carbamyl)-ethylamino]-isonicotinamide). The isonicotinic acid may be replaced
by other groups. For instance other heterocyclic acids, aromatic acids[296] including
benzoic acid[110], retain strong activity when bound with alkylhydrazine or alkylaryl-
hydrazines.

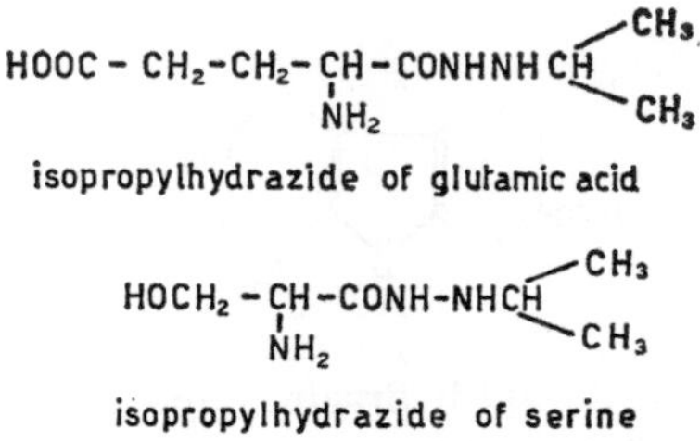

Fig. 12.

Sometimes activity may still be retained after both hydrogen atoms of the
β-hydrazine nitrogen but none of the hydrogen atoms on the α-nitrogen have been
substituted[292], *i.e.* the structure reported by Zeller *et al.*[296] (Fig. 11).

The cyclic groups may be replaced by amino acids[197,296] (see Fig. 12), fatty acids[296] or even an atom of hydrogen, obtaining compounds[80] with strong activity. Among the amino acids the D-form is usually markedly less active than the L-form[2,195].

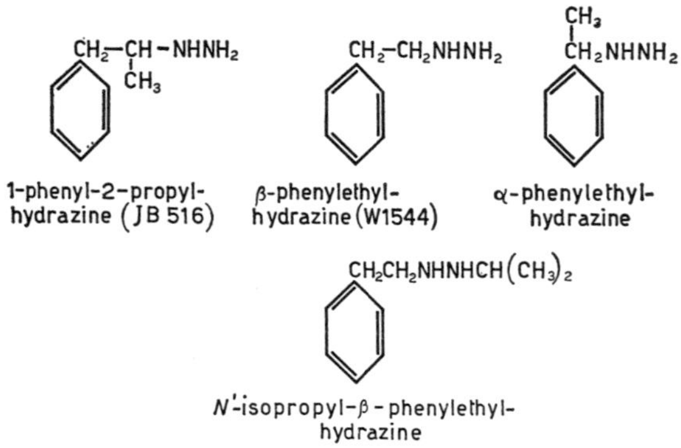

1-phenyl-2-propyl-
hydrazine (JB 516)

β-phenylethyl-
hydrazine (W1544)

α-phenylethyl-
hydrazine

N'-isopropyl-β-phenylethyl-
hydrazine

Fig. 13.

Also the substitutions in the NH_2 group of the amino acid structure result in a drop of activity[2]. Hydrazides of γ-aminobutyric acid and guanidobutyric acid are inactive[140].

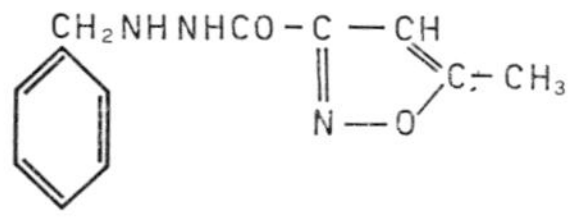

1-Benzyl-2-(5-methyl-3-isoxazolyl-carbonyl)-
hydrazine   Ro 5-0831/1

Fig. 14. Isocarboxazid.

Several arylalkylhydrazines are among the most active inhibitors available. Typical examples of this series are 1-phenyl-2-propylhydrazine (JB 516)[20,21,146,147] and α- or β-phenylethylhydrazine (W 1544)[81,87,112]. The D-form of 1-phenyl-2-propylhydrazine is four times more active than the L-form[18]. The respective N-alkyl derivatives retain strong inhibitory effects on monoamine oxidase[20,60] (Fig. 13).

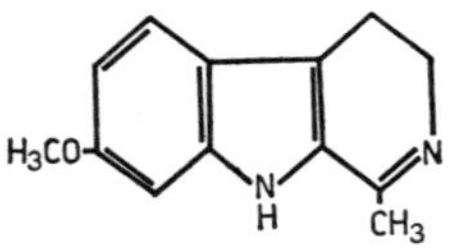

Fig. 15. Harmaline.

Benzyl hydrazines and their acyl derivatives are potent and long-lasting MAO inhibitors[113,201]. The substitution of a hydrogen in the 4-position of the benzene ring with a basic residue transforms long-acting into short-acting MAO inhibitors[114]. Also isocarboxazid, 1-benzyl-2-(5-methyl-3-isoxazolyl-carbonyl)hydrazine[201] (Fig. 14), is supposed to act through a release of benzylhydrazine[231].

Also phenylhydrazines (example *p*-methylphenylhydrazine[8]) may inhibit MAO activity.

The inhibition of the monoamine oxidase is however not necessarily confined to compounds containing a hydrazine group. In fact 2-hydroxytryptamine and other derivatives, tetrahydroharmane and *N*-methyltetrahydroharmane may inhibit MAO[106]. Harmaline is a very powerful short-acting inhibitor[196,245,249,252] (Fig. 15).

Very strong inhibitors of MAO have been recently found among amines: benzene-substituted α-phenylethylamines[180], 2-phenylcyclopropylamine (SKF 385)[173,208], 1-phenylcyclopropylamine[295], *N*-methyl-*N*-benzyl-2-propynylamine (A 19120) (refs. 221, 229, 230), α-methyl-[100,129] and α-ethyltryptamine[129] (Fig. 16).

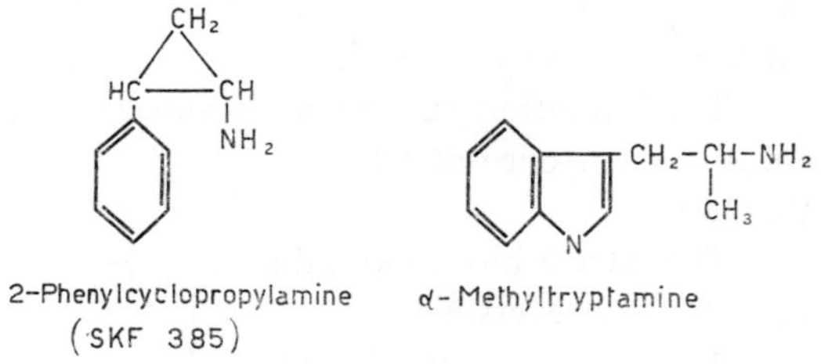

Fig. 16.

2-Phenylcyclopropylamine
(SKF 385)

α-Methyltryptamine

Finally, among others, are inhibitors of MAO some phenylhexamethylene-imines[136], pyridine aldoxime dodecyliodide[183], benactyzine[254], some alcohols[70] and hordenine[252].

It would be interesting to know if substances physiologically present in the body may influence the level of monoamine oxidase in tissues. At present there is only limited information on this subject. Cortisone is a weak inhibitor of MAO in female rats[214] but the enzyme is not influenced by adrenalectomy[111,214]. Thyroid hormones decrease MAO activity[297,298] (other observers have found the opposite effect[232,273]) while thyroidectomy may decrease[88] or increase it[223]. Riboflavin deficiency is without effect[277] contrary to what was previously found[139].

Mechanism of action of MAO inhibition

Despite the large number of compounds available, very little is known about the mechanism(s) by which they block the monoamine oxidase. It was interesting to see if different preparations of MAO from various organs showed a different sensitivity to MAO inhibitors. Usually homogenates are less sensitive to iproniazid than mito-chondria[53,123,288]. This suggested the possibility that in homogenates there is a soluble factor antagonizing the effects of iproniazid. Such a factor has been found in rat liver[123]. Mitochondrial MAO preparations coming from liver are more sensitive to iproniazid than that from brain[53]. However, liver and brain MAO of mouse are equally sensitive to iproniazid[81] as liver and brain MAO of rat to β-*p*-chlorophenyl-mercaptoethylhydrazine (S 231)[189]. Rat liver mitochondria are more sensitive to iproniazid than guinea pig liver mitochondria[287]. However, the differences are relatively small, because MAO activity of hog kidney mitochondria as well as that of brain or liver preparations from mouse, cat, dog, hog, rabbit and human are all inhibited between 90 and 100% by 10^{-4} M iproniazid[288]. *In vivo* the physiological disposition of a given MAO inhibitor may be relevant in explaining selective blockade at the level of some tissues (see Chapter 4).

For some MAO inhibitors it has been shown that they act competitively (*e.g.* iproniazid[80,82], 2-phenylcyclopropylamine[12], α-methyltryptamine[129]). This suggests that they are antagonized by increasing the concentration of the substrate[12,80,129,288].

Pre-incubation increases the inhibitory effect of iproniazid[82] and 1-phenyl-1,2-propylhydrazine[193], but not that of harmaline[193]. This fact suggested the possibility that iproniazid could be active through the formation of isopropylhydrazine, a powerful MAO inhibitor[80,82,284].

The inhibition exerted by iproniazid may be partially reversed by washing the enzyme in presence of the substrate[124] or by dialysis in presence of BAL or glutathione[80].

The MAO blockade induced by p-methylphenylhydrazine seems instead to be completely irreversible[8].

The studies on the MAO inhibitors have been useful in elucidating the "active centre" of the enzyme[55,208,283,285,286]. According to the latest views there would be a structure of this type: $-X-\ddot{Y}-$ where X is the acceptor for the α-carbon and $\ddot{Y}$ the acceptor for a proton derived from an α-hydrogen[294].

OTHER METABOLIC PATHWAYS

The three enzymes considered do not account for all the metabolic transformations to which serotonin is submitted.

Oxidation

Serotonin may be oxidized according to the following reaction[34] (Fig. 17) in the presence of caeruloplasmin, a copper-protein* deprived of monoamine oxidase activity[176,215].

According to Porter *et al.*[199] two oxidative reactions occur. First caeruloplasmin oxidizes serotonin to a p-quinonimine derivative and then this compound is transformed by non-enzymatic systems[41,76,235] to an adrenochrome-like structure and then to polymerized products[92,115,200,215]. The oxidation of 5-HT by caeruloplasmin has its optimum at pH 6.2 and is catalysed by Cu^{2+} (ref. 178). Copper itself and vanadium in the absence of caeruloplasmin can oxidize 5-HT[116,177,178]. Also

---

* Caeruloplasmin is probably a mixture of proteins[205] showing the same electrophoretic mobility as the α_2-globulin[250,251]. It has a molecular weight of about 151,000 and contains 0.34% of copper[42,143]. This protein present in serum, also known as "laccase", oxidizes p-phenylenediamine[144,145] and N,N-dimethyl-p-phenylenediamine[3].

AgNO$_3$ oxidizes 5-HT with the formation of a melanin-like pigment. The first intermediate compound of this reaction is a dimer of serotonin[92].

Fig. 17.

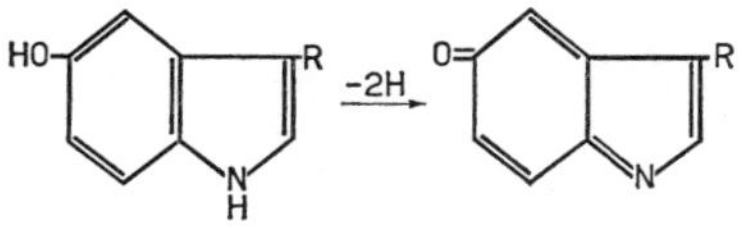

At 37°C, 400 μg of caeruloplasmin (about 1 μg of copper) destroy in 90 min 1 μg of serotonin[192].

Caeruloplasmin is present in serum of several animal species and shows a different sensitivity to the various substrates[109,178] (see Table 10). The serum of a patient with a carcinoid shows a high oxidase activity[178] while patients with Wilson disease have a low level[178,256,*].

TABLE 10

CAERULOPLASMIN ACTIVITY IN SERUM OF DIFFERENT ANIMAL SPECIES[109]

Species	Serum proteins g/100 ml ± S.E.	α$_2$-Globulins % of the serum proteins ±S.E	Ravin test O.D.* after 60 min (p-phenylenediamine)[202]		Akerfeldt test O.D. after 10 min (dimethyl-p-phenylenediamine)[3]	
			Average	Range	Average	Range
Man	7.5 ± 0.15	8.8 ± 0.52	0.42	(0.27–0.74)	0.63	(0.42–0.86)
Ox	7.1 ± 0.25	10.4 ± 0.54	0.11	(0.06–0.22)	1.06	(0.78–1.55)
Rat	6.4 ± 0.17	8.4 ± 0.37	0.58	(0.40–0.87)	0.01	n.s.**
Rabbit	6.3 ± 0.77	8.5 ± 2.04	0.27	(0.14–0.45)	1.43	(0.98–1.99)
Horse	5.8 ± 0.2	8.6 ± 0.87	0.15	(0.11–0.23)	1.20	(0.48–1.94)
Guinea pig	5.8 ± 0.25	17.7 ± 0.99	0.04	(0.02–0.05)	0.63	(0.41–0.85)
Chicken	4.1 ± 0.23	4.1 ± 0.40	0.01	n.s.**	0.01	n.s.**

* O.D. = optical density.
** n.s. = not significant.

Beside serotonin[1,25,31,75,166], other indoles are oxidized by caeruloplasmin, such as the 4-, 6- and 7-hydroxy derivatives[30]. Tryptophan, tryptamine, 5-HTP are not oxidized by caeruloplasmin[176,178,215]. 5-Hydroxyindoleacetic acid is instead slowly oxidized[74,75,178].

Noradrenaline, adrenaline and dopamine are also attacked by caeruloplasmin (refs. 178, 215).

Caeruloplasmin activity is inhibited by ascorbic acid[178], cyanide[75], isoniazid and iproniazid[135,282]. However, not all the MAO inhibitors are active against caerulo-

* Gill plates of *Mytilus edulis* contain an enzyme different from MAO able to oxidize serotonin and other indoles with the formation of brown pigments[29,32,33,170].

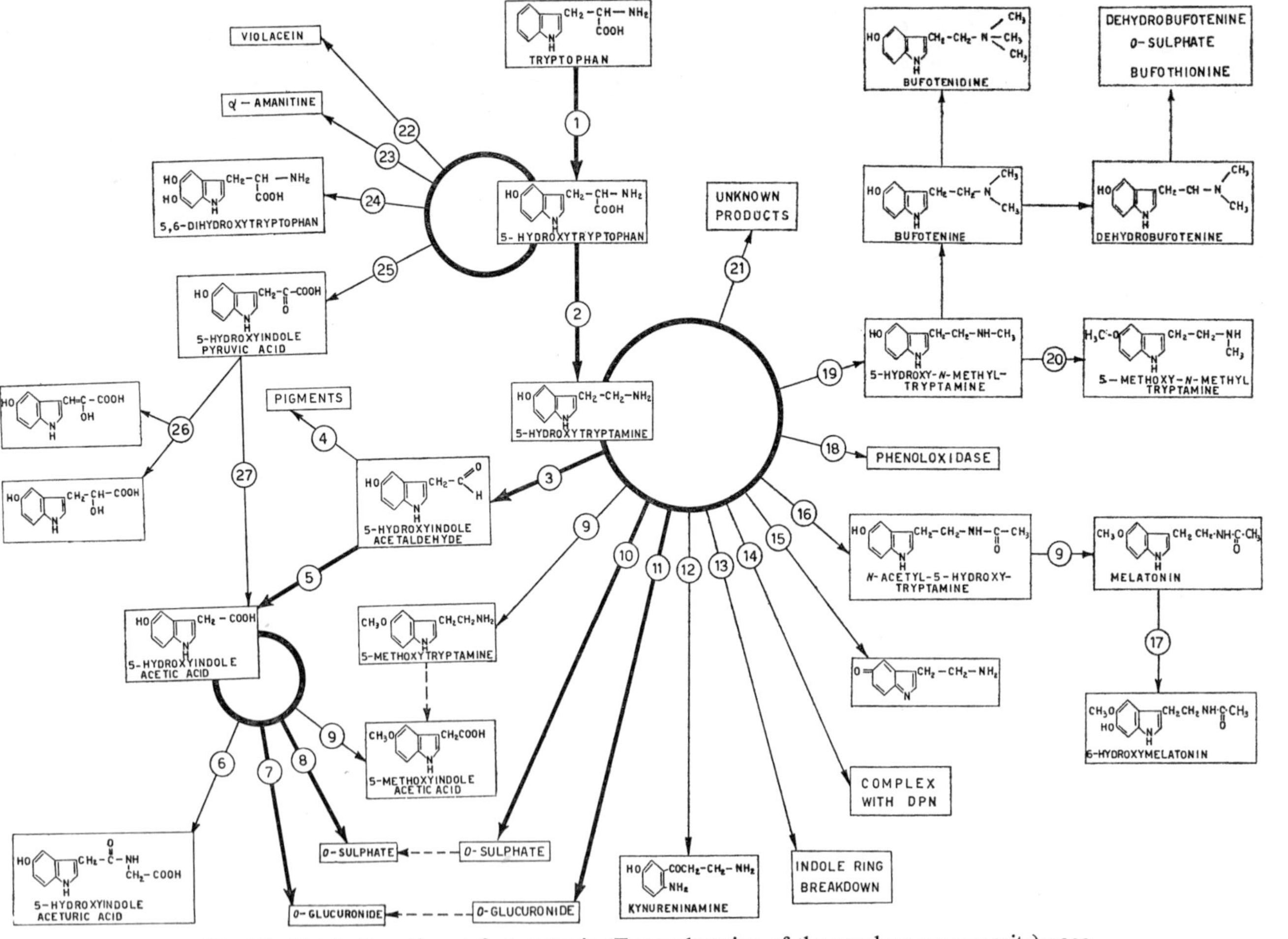

Fig. 18. Metabolic pathways for serotonin (For explanation of the numbers see opposite) page.

plasmin[118] (see Table 11). Serotonin, 5-HTP and 5-HIA may be inhibitors of the enzyme when N,N-dimethylphenylenediamine is used as a substrate[6,7].

TABLE 11

EFFECT OF MAO INHIBITORS ON CAERULOPLASMIN ACTIVITY[118]

Compound	Rat serum (substrate: p-phenylenediamine)		Rabbit serum (substrate: N,N-dimethyl-p-phenylenediamine)	
	Inhibitor		Inhibitor	
	0.1 μmole	1 μmole	3 μmoles	30 μmoles
Iproniazid	0	31	0	67
1-Phenyl-2-propylhydrazine (JB 516)	0	5	0	27
Phenylcyclopropylamine	0	0	0	0
Marplan (RO 5/0831)	0	21	0	—
Nialamide	10	41	—	—
Phenylhydrazine	25	100	100	100
Isoniazid	9	13	0	0

It is possible that caeruloplasmin may be important for the inactivation of serotonin in serum particularly after injections of 5-HT and in cases of increased caeruloplasmin activity, like in pregnancy, infections and neoplasms.

Similar to the oxidation induced by caeruloplasmin may be the transformation of serotonin obtained by Bracco *et al.*[37–39] with mammalian serum.

When serotonin is incubated with blood, erythrocytes or haemoglobin there is a loss of the fluorescence and the onset of a pink product[35,72]. This reaction, common also to 5-HTP and 5-HIA, but not to tryptophan or tryptamine, is inhibited by ascorbic acid, cyanide and carbon monoxide[35,172,206].

Cytochrome c and cytochrome oxidase can destroy serotonin, but the final metabolic products have not yet been identified[233,234]. In some tissues as heart and kidney, cytochrome oxidase seems to play an important part in the metabolism of 5-hydroxy-indoles[147a–c,256a].

Legend to Fig. 18.

1. Tryptophan-5-hydroxylase present in rat intestine[64], toad[237], and *Chromobacterium violaceum*[187]
2. 5-Hydroxytryptophan decarboxylase
3. Monoamine oxidase
4. Formation of a brown pigment[27,191]
5. Aldehyde dehydrogenase
6. Glycine conjugation[181]
7. Glucuronic conjugation of 5-HIA (ref. 261)
8. Sulphonic conjugation of 5-HIA (ref. 219)
9. Hydroxyindole-*O*-methyltransferase (refs. 9, 258)
10. Sulphonic conjugation of 5-HT (ref. 58)
11. Glucuronic conjugation of 5-HT (refs. 181, 265)
12. Under action of ultraviolet light[134]
13. Reaction suggested[95,191]
14. Nonenzymatic complex with DPN (refs. 4, 203)
15. Caeruloplasmin
16. *N*-Acetylation[181,263a]
17. Hydroxylation of melatonin[162,169]
18. Phenoloxidase present in Harding-Passey melanoma (ref. 175)
19. *N*-Methylation present in invertebrates[93,97,151,236] and in rabbit lung [8a]. *N,N*-Dimethyltryptamine is present in seeds of *Piptadenia peregrina*[225]
20. Methylation occurring in *Phalaris aurundinacea*
21. Oxidation by haemoglobin[172]
22. Violacein
23. Produced by *Amanita phalloides*[275]
24. Dihydroxytryptophan and dihydroxytryptamine[56]
25. Transamination[260]
26. Methylation occurring in *Phalaris aurundinacea*
27. Decarboxylation[260]
   For other references see text.

The possibility of the formation of a 5,6-dihydroxytryptamine has been considered[77]. An o-dihydroxytryptamine has been found in urines of carcinoid patients[212], rats and rabbits[182]. On the other hand, the 5-methoxy-N-acetyltryptamine (melatonin[168]) is transformed into a 6-hydroxymelatonin[162,169].

Conjugation

The hydroxyl group of indoles may be conjugated with glucuronate or sulphate. The formation of 5-O-glucuronide-tryptamine has been reported[117,181,261] particularly when monoamine oxidase pathway is blocked[155,261]. Also 4-hydroxytryptamine[101] and 5-hydroxyindoleacetic acid[261] are conjugated with glucuronic acid. o-Aminobenzoate inhibits 5-HT-glucuronic conjugation[86].

5-O-sulphate-tryptamine is formed *in vitro* by homogenates of rat liver[58] in the presence of Mg^{2+}, ATP and SO_4^{2-} (ref. 59) and it is hydrolysed by sulphatase of *Charonia lampas* (*Tritonalia sauliae*)[155a]. Iproniazid and probably also other MAO inhibitors increase the excretion of 5-O-sulphate-tryptamine[59]. Also 5-HIA is excreted as a sulphonic ester[219] in carcinoid patients and rats[73].

O-Methylation

It is possible that the hydroxyl group of serotonin is methylated in the same way as that described for catecholamines[276]. This pathway has been shown in pineal gland[10]. The enzyme called hydroxyindole-O-methyltransferase is different from catechol-O-methyltransferase and is highly localized in pineal gland, being practically absent in other tissues[9,258]. The enzyme requires δ-adenosylmethionine and hydroxylates N-acetylserotonin to form 5-methoxy-N-acetyltryptamine (melatonin[263], 5-HIA and 5-HT[10]).

The 5-methoxyindoleacetic acid may be formed directly or by the action of monoamine oxidase on 5-methoxytryptamine[167,169].

N-Methylation and N-acetylation

N-Methylation occurs mainly in invertebrates[97,151,236] and probably in *Amanita phalloides*[275]. In mammals this pathway is not common[243] although Bumpus and Page[45] found traces of N-methylserotonin in human urine. Furthermore, Axelrod isolated a serotonin N-methylating enzyme from rabbit lung[8a]. N-Acetylation has been shown in pineal glands[169,263] and N-acetylserotonin has been found in urine[181]. The enzyme acetylating serotonin has been recently studied[263a].

Transaminase

5-Hydroxytryptophan is not subjected to the action of tryptophan pyrrolase[90] with production of 5-hydroxykynurenine[36] but is transaminated[260].

The enzyme responsible is 5-hydroxytryptophan-α-oxoglutarate transaminase[207] which yields the formation of 5-hydroxyindolepyruvic acid. 5-Hydroxyindolepyruvic acid is a substrate of the 2-indolylpyruvic tautomerase[222].

The transaminase is thermolabile, has an optimum at pH 8.2 and is adaptive[207].

Other reactions

Serotonin and other indoles can form complexes with DPN[4,203] and *N*-methyl-nicotinamide[5].

Ultraviolet light transforms serotonin into kynureninamine[134], similar to the reaction tryptophan–kynurenine[133].

A map of the above-mentioned and other metabolic pathways for serotonin is shown in Fig. 18.

REFERENCES

1   ABOOD, L., *Conference of the Brain Research Foundation, Chicago, Ill.*, Jan. 12, 1957.
2   AESCHLIMANN, J. A., *Ann. N.Y. Acad. Sci.*, 80 (1959) 564.
3   AKERFELDT, S., *Science*, 127 (1957) 117.
4   ALIVISATOS, S. G. A., MOURKIDES, G. A. AND JIBRIL, A., *Nature*, 186 (1960) 718.
5   ALIVISATOS, S. G. A., MOURKIDES, G. A. AND JIBRIL, A., *Federation Proc.*, 19 (1960) 308.
6   APRISON, M. H., HANSON, K. M. AND AUSTIN, D. C., *Federation Proc.*, 17 (1958) 15.
7   APRISON, M. H., HANSON, K. M. AND AUSTIN, D. C., *J. Nervous Mental Diseases*, 128 (1959) 249.
8   ARAI, A., *Japan. J. Pharmacol.*, 9 (1960) 159.
8a AXELROD, J., *J. Pharmacol.*, 138 (1962) 28.
9   AXELROD, J. AND WEISSBACH, H., *Science*, 131 (1960) 1312.
10 AXELROD, J. AND WEISSBACH, H., *J. Biol. Chem.*, 236 (1961) 211.
11 AZIOKA, J. AND TANIMUKAI, H., *J. Neurochem.*, 1 (1957) 311.
12 BARBATO, L., *Federation Proc.*, 20 (1961) 237d.
13 BARSKY, J., PACHA, W. L., SARKAR, S. AND ZELLER, E. A., *J. Biol. Chem.*, 234 (1959) 389.
14 BEER, R. J. S., JENNINGS, B. E. AND ROBERTSON, A., *J. Chem. Soc.*, (1954) 2679.
15 BEILER, J. M. AND MARTIN, G. J., *J. Biol. Chem.*, 169 (1947) 345.
16 BEILER, J. M. AND MARTIN, G. M., *Federation Proc.*, 13 (1954) 180.
17 BEILER, J. M. AND MARTIN, G. M., *J. Biol. Chem.*, 211 (1954) 39.
17a BENDITT, E. P., in G. P. LEWIS (Ed.), *5-Hydroxytryptamine*, Pergamon, London, 1958, p. 32.
18 BERNSTEIN, J., LOSE, K. A., SMITH, C. I. AND RUBIN, B., *J. Am. Chem. Soc.*, 81 (1959) 4433.
19 BERTLER, A. AND ROSENGREN, E., *Experientia*, 15 (1959) 382.
20 BIEL, J. R., NUHFER, P. A. AND CONWAY, A. C., *Ann. N.Y. Acad. Sci.*, 80 (1959) 568.
21 BIEL, J. H., DRUKKER, A. E., SHORE, P. A., SPECTOR, S. AND BRODIE, B. B., *J. Am. Chem. Soc.*, 80 (1958) 1519.
22 BLASCHKO, H., *Biochem. J.*, 50 (1952) 52P.
23 BLASCHKO, H., *Pharmacol. Revs.*, 4 (1952) 415.
24 BLASCHKO, H., *Brit. Med. Bull.*, 9 (1953) 146.
25 BLASCHKO, H., *Farmaco (Pavia), Ed. sci.*, 15 (1960) 532.
26 BLASCHKO, H., FRIEDMAN, P. J., HAWES, R. AND NILSSON, K., *J. Physiol. (London)*, 145 (1959) 384.
27 BLASCHKO, H. AND HELLMANN, K., *J. Physiol. (London)*, 122 (1953) 419.
28 BLASCHKO, H. AND HOPE, D. B., *J. Physiol. (London)*, 129 (1955) 11.
29 BLASCHKO, H. AND LEVINE, W. G., *Biochem. Pharmacol.*, 3 (1960) 168.
30 BLASCHKO, H. AND LEVINE, W. G., *Brit. J. Pharmacol.*, 15 (1960) 625.
31 BLASCHKO, H. AND LEVINE, W. G., *J. Physiol. (London)*, 154 (1960) 599.
32 BLASCHKO, H. AND MILTON, A. S., *J. Physiol. (London)*, 148 (1959) 54P.
33 BLASCHKO, H. AND MILTON, A. S., *Brit. J. Pharmacol.*, 15 (1960) 42.
34 BLASCHKO, H. AND PHILPOT, F. J., *J. Physiol. (London)*, 122 (1953) 403.
35 BLUM, J. J. AND LING, N. S., *Biochem. J.*, 73 (1959) 530.
36 BOYLAND, E., SIMS, P. AND WILLIAMS, D. C., *Biochem. J.*, 62 (1956) 546.
37 BRACCO, M. AND CURTI, P. C., *Boll. soc. ital. ematol.*, 6 (1958) 31.
38 BRACCO, M., CURTI, P. C., GIULIANO, V. AND PESSINA, G., *Boll. soc. ital. ematol.*, 6 (1958) 26.
39 BRACCO, M., CURTI, P. C. AND PESSINA, G., *Boll. soc. ital. ematol.*, 6 (1958) 27.
40 BRADLEY, T. R., BUTTERWORTH, R. P., REID, G. AND TRAUTNER, E. M., *Nature*, 166 (1950) 911.
41 BRODIE, B. B., AXELROD, J., SHORE, P. A. AND UDENFRIEND, S., *J. Biol. Chem.*, 208 (1954) 741
42 BROMAN, L., *Nature*, 181 (1958) 1655.
43 BROWN, B. G. AND HEY, P., *J. Physiol. (London)*, 118 (1952) 15P.
44 BROWN, B. G. AND HEY, P., *Brit. J. Pharmacol.*, 11 (1956) 58.
45 BUMPUS, P. M. AND PAGE, I. H., *J. Biol. Chem.*, 212 (1955) 11.
46 BURKHALTER, A., *Federation Proc.*, 20 (1961) 257.
47 BUXTON, J. AND SINCLAIR, H. M., *Biochem. J.*, 62 (1956) 27P.
48 BUZARD, J. A. AND NYTCH, P. D., *J. Biol. Chem.*, 227 (1957) 225.
49 BUZARD, J. A. AND NYTCH, P. D., *J. Biol. Chem.*, 229 (1957) 409.
50 BUZARD, J. A. AND NYTCH, P. D., *Federation Proc.*, 17 (1958) 780.
51 BUZARD, J. A. AND NYTCH, P. D., *J. Biol. Chem.*, 234 (1959) 884.

52   CANAL, N., MAFFEI-FACCIOLI, A. AND MANZINI, B., *Boll. soc. ital. biol. sper.*, 34 (1958) 1525.
53   CANAL, N. AND MAFFEI-FACCIOLI, A., in P. B. BRADLEY, P. DENIKER AND C. RADOUCO-THOMAS (Eds.), *Neuropsychopharmacology*, Elsevier, Amsterdam, 1959, p. 288.
54   CANAL, N., MAFFEI-FACCIOLI, A. AND MANZINI, E., *Boll. soc. ital. biol. sper.*, 34 (1958) 1525.
55   CARBON, J. A., BURKARD, W. P. AND ZELLER, E. A., *Helv. Chim. Acta*, 41 (1958) 1883.
56   CARLISLE, D. B., *Biochem. J.*, 63 (1956) 32P.
57   CASTELLI, B. AND DE GIUSEPPE, L., *Farmaco (Pavia)*, *Ed. sci.*, 10 (1955) 1011.
58   CHADWICK, B. T. AND WILKINSON, J. H., *Biochem. J.*, 68 (1958) 1P.
59   CHADWICK, B. T. AND WILKINSON, J. H., *Biochem. J.*, 76 (1960) 102.
60   CHESSIN, M., DUBNICK, B., LEESON, G. AND SCOTT, C. C., *Ann. N.Y. Acad. Sci.*, 80 (1959) 597.
61   CLARK, W. G., *Symposium on Catecholamines*, suppl. to *Pharmacol. Revs.*, 11 (1959) 330.
62   CLARK, C. T., WEISSBACH, H. AND UDENFRIEND, S., *J. Biol. Chem.*, 210 (1954) 139.
63   CLARCKE, A. J. AND MANN, P. J. G., *Biochem. J.*, 65 (1957) 763.
64   COOPER, J. R., *Ann. N.Y. Acad. Sci.*, 92 (1961) 208.
65   COOPER, J. R. AND MELCER, I., *Pharmacologist*, 2 (1960) 67.
66   CORNE, S. J. AND GRAHAM, J. D. P., *J. Physiol. (London)*, 135 (1957) 339.
67   COSTA, E. AND HIMWICH, H. E., in F. BRÜCKE (Ed.), *Biochemistry of the Central Nervous System*, Pergamon, London, 1959, p. 282.
68   COTZIAS, G. C. AND DOLE, V. P., *Proc. Soc. Exptl. Biol. Med.*, 78 (1951) 157.
69   COTZIAS, G. C. AND GREENOUGH, J. J., *Nature*, 183 (1959) 1732.
70   COTZIAS, G. C. AND GREENOUGH, J. J., *Nature*, 185 (1960) 384.
71   COTZIAS, G. C., SELIN, I. AND GREENHOUGH, J. J., *Science*, 120 (1954) 144.
72   CROMARTIE, R. I. T. AND HARLEY-MASON, J., *Biochem. J.*, 66 (1957) 713.
73   CURZON, G., *Arch. Biochem. Biophys.*, 66 (1957) 497.
74   CURZON, G., *Proc. Roy. Soc. Med.*, 52 (1959) 64.
75   CURZON, G. AND VALLET, L., *Biochem. J.*, 74 (1960) 279.
76   DALGLIESH, C. E., *Arch. Biochem. Biophys.*, 58 (1955) 214.
77   DALGLIESH, C. E., in G. P. LEWIS (Ed.), *5-Hydroxytryptamine*, Pergamon, London, 1957, p. 58.
78   DALGLIESH, C. E. AND DUTTON, R. W., *Brit. J. Cancer*, 11 (1957) 296.
79   DALGLIESH, C. E. AND DUTTON, R. W., *Biochem. J.*, 65 (1957) 21P.
80   DAVISON, A. N., *Biochem. J.*, 67 (1957) 316.
81   DAVISON, A. N., *Bull. soc. chim. biol.*, 40 (1958) 1737.
82   DAVISON, A. N., *Arch. Biochem. Biophys.*, 77 (1958) 368.
83   DAVISON, A. N. AND SANDLER, M., *Clin. Chim. Acta*, 1 (1956) 450.
84   DAVISON, A. N. AND SANDLER, M., *Nature*, 181 (1958) 186.
85   DELAHUNT, C. B. AND PEPIN, J., *Federation Proc.*, 18 (1959) 1867.
86   DONALDSON, R. M., ARABEJETY, J. AND GRAY, S. J., *J. Clin. Invest.*, 38 (1959) 933.
87   DUBNICK, B., LEESON, G. A., CHESSIN, M. AND SCOTT, C., *Abstracts Fall Meeting Am. Soc. Pharmacol.*, Ann Arbor, Mich., 1958, p. 10.
88   DUBNICK, B., LEESON, G. A. AND LEVERETT, R., *Pharmacologist*, 2 (1960) 67.
89   DUBNICK, B., LEESON, G. A. AND PHILLIPS, G. E., *Federation Proc.*, 18 (1959) 861.
90   EK, A. AND WITKOP, B., *J. Am. Chem. Soc.*, 76 (1954) 5579.
91   EPPS, H. M. R., *Biochem. J.*, 39 (1945) 37.
92   ERIKSEN, N., MARTIN, G. M. AND BENDITT, E. P., *J. Biol. Chem.*, 235 (1960) 1662.
93   ERSPAMER, V., *Rend. sci. Farmitalia*, 1 (1954) 98.
94   ERSPAMER, V., *J. Physiol. (London)*, 127 (1955) 118.
95   ERSPAMER, V., *J. Physiol. (London)*, 133 (1956) 1.
96   ERSPAMER, V., in S. GARATTINI AND V. GHETTI (Eds.), *Psychotropic Drugs*, Elsevier, Amsterdam, 1957, p. 414.
97   ERSPAMER, V., *Farmaco (Pavia)*, *Ed. sci.*, 13 (1958) 842.
98   ERSPAMER, V., FERRINI, R. AND GLÄSSER, A., *J. Pharm. and Pharmacol.*, 12 (1960) 761.
99   ERSPAMER, V. AND GLÄSSER, A., *Ateneo parmense*, 31 (1960) 511.
100 FERRINI, R., GLÄSSER, A. AND MANTEGAZZINI, P., *Ateneo parmense*, 31 (1960) 515.
101 ERSPAMER, V., GLÄSSER, A., NOBILI, B. M. AND PASINI, C., *Experientia*, 16 (1960) 506.
102 ERSPAMER, V., GLÄSSER, A., PASINI, C. AND STOPPANI, G., *Nature*, 189 (1961) 483.
103 FELLMAN, J. H., *Proc. Soc. Exptl. Biol. Med.*, 93 (1956) 413.
104 FOUTS, J. R., *Federation Proc.*, 13 (1954) 210.

[105] FOUTS, J. R., BLANKSMA, L. A., CARBON, J. A. AND ZELLER, E. A., *J. Biol. Chem.*, 147 (1943) 415.
[105a] FREEDLAND, R. A., WADZINKI, I. M. AND WAISMAN, H. A., *Biochem. Biophys. Res. Comm.*, 5 (1961) 94.
[105b] FREEDLAND, R. A., WADZINKI, I. M. AND WAISMAN, H. A., *Biochem. Biophys. Res. Comm.*, 6 (1961) 227.
[106] FRETER, K., WEISSBACH, H., REDFIELD, B. G., UDENFRIEND, S. AND WITKOP, B., *J. Am. Chem. Soc.*, 80 (1958) 983.
[106a] FRETER, K., WEISSBACH, H., UDENFRIEND, S. AND WITKOP, B., *Proc. Soc. Exptl. Biol. Med.*, 94 (1957) 725.
[107] FREYBURGER, W. A., GRAHAM, B. E., RAPPORT, M. M., SEAY, P. H., GOVIER, W. M., SWOAP, O. F. AND VAN DER BROOK, M. J., *J. Pharmacol.*, 105 (1952) 80.
[108] GADDUM, J. H. AND GIARMAN, N. J., *Brit. J. Pharmacol.*, 11 (1956) 88.
[109] GARATTINI, S., GIACHETTI, A. AND PIERI, L., *Arch. Biochem. Biophys.*, 91 (1960) 83.
[110] GARATTINI, S. AND JORI, A., to be published.
[111] GARATTINI, S., LAMESTA, L., MORTARI, A., PALMA, V. AND VALZELLI, L., *J. Pharm. and Pharmacol.*, 13 (1961) 385.
[112] GARATTINI, S. AND VALZELLI, L., in *Il sindromi depressive, Atti Symposium, Rapallo*, Minerva Medica, Torino, 1960, p. 7.
[113] GARDNER, T. S., WENIS, E. AND LEE, J., *J. Med. Pharm. Chem.*, 2 (1960) 133.
[114] GARDNER, T. S., WENIS, E. AND LEE, J., *J. Med. Pharm. Chem.*, 3 (1961) 241.
[115] GELLER, E. AND YUWILER, A., *Federation Proc.*, 18 (1959) 922.
[116] GELLER, E., EIDUSON, S. AND YUWILER, A., *J. Neurochem.*, 5 (1959) 73.
[117] GESSNER, P. K., KHAIRALLAH, P. A., MC ISAAC, W. M. AND PAGE, I. H., *J. Pharmacol. Exptl. Therap.*, 130 (1960) 126.
[118] GIACHETTI, A., PALMA, V. AND PIERI, L., to be published.
[119] GIARMAN, N. J., *Federation Proc.*, 15 (1956) 428.
[120] GIARMAN, N. J. AND DAY, M., *Biochem. Pharmacol.*, 1 (1958) 235.
[121] GIARMAN, N. J., DAY, M. AND PEPEU, G., *Federation Proc.*, 18 (1959) n. 1555.
[122] GLENNER, G. G. AND LILLIE, R. D., *J. Histochem. Cytochem.*, 5 (1957) 279.
[123] GLUCKMAN, M. I., DANIELS, A. AND MARAZZI, A. S., *J. Pharmacol. Exptl. Therap.*, 122 (1958) 25A.
[124] GLUCKMAN, M. I. AND MARAZZI, A. S., *Federation Proc.*, 17 (1958) 371.
[125] GODTFREDSEN, W. O., KORSBY, G., LORCK, H. AND VANGEDAL, S., *Experientia*, 14 (1958) 88.
[126] GOVIER, M., HOWES, B. G. AND GIBBENS, A. J., *Science*, 118 (1953) 596.
[127] GREEN, A. L. AND HAUGHTON, T. M., *Biochem. J.*, 76 (1960) 44P.
[128] GREEN, A. L. AND HAUGHTON, T. M., *Biochem. J.*, 78 (1961) 172.
[129] GREIG, M. E., WALK, R. A. AND GIBBONS, A. J., *J. Pharmacol. Exptl. Therap.*, 127 (1959) 110.
[130] HAGEN, P., *Federation Proc.*, 18 (1959) 1575.
[131] HAGEN, P. AND LEE, F. L., *J. Physiol. (London)*, 143 (1958) 7P.
[132] HAGEN, P., WEINER, N., ONO, S. AND LEE, F. L., *J. Pharmacol. Exptl. Therap.*, 130 (1960) 9.
[133] HAKIM, A. A., *Federation Proc.*, 18 (1959) 240.
[134] HAKIM, A. A. AND THIELE, K. A., *Biochem. Biophys. Research Commun.*, 2 (1960) 242.
[135] HANSON, K. M., AUSTIN, D. C. AND APRISON, M. H., *J. Appl. Physiol.*, 14 (1959) 363.
[136] HARRIS, E. S., ALBURN, H. E. AND SEIFTER, J., *Abstracts Fall Meeting Am. Soc. Pharmacol.*, Ann Arbor, Mich., 1958, p. 17.
[137] HARTMAN, W. J., AKAVIE, R. I. AND CLARK, W. G., *J. Biol. Chem.*, 216 (1955) 507.
[138] HAWKINS, J., *Biochem. J.*, 50 (1952) 577.
[139] HAWKINS, J., *Biochem. J.*, 51 (1952) 399.
[140] HEILBRONN, E., *Biochem. J.*, 76 (1960) 45P.
[141] HESS, S. M., CONNAMACHER, R. H. AND UDENFRIEND, S., *Federation Proc.*, 20 (1961) 344a.
[142] HIRSCH, J. C., *J. Exptl. Med.*, 97 (1953) 345.
[143] HOLMBERG, C. G. AND LAURELL, G. B., *Acta Chem. Scand.*, 1 (1947) 944.
[144] HOLMBERG, C. G. AND LAURELL, G. B., *Acta Chem. Scand.*, 2 (1948) 559.
[145] HOLMBERG, C. G. AND LAURELL, G. B., *Acta Chem. Scand.*, 5 (1951) 476 and 921.
[146] HORITA, A., *J. Pharmacol. Exptl. Therap.*, 122 (1958) 176.
[147] HORITA, A., *Ann. N.Y. Acad. Sci.*, 80 (1959) 590.
[147a] HORITA, A., *Biochem. Pharmacol.*, 11 (1962) 147.

147b HORITA, A., *Biochem. Pharmacol.*, 11 (1962) 672.
147c HORITA, A. AND WEBER, L. J., *Biochem. Pharmacol.*, 7 (1961) 47.
148 HORNING, E. C., SWEELEY, C. C., DALGLIESH, C. E. AND KELLY, W., *Biochim. Biophys. Acta*, 32 (1959) 566.
149 HUANG, I., TANNENBAUM, S. AND HSIA, D. Y. Y., *Nature*, 186 (1960) 717.
150 ICHIHARA, K., SAKAMOTO, A., INAMORI, K. AND SAKAMOTO, Y., *J. Biochem. (Tokyo)*, 44 (1957) 649.
151 ILUSEN, H. AND CHEN, K. K., *J. Biol. Chem.*, 116 (1936) 87.
152 JACOBSON, W., in G. E. W. WOLSTENHOLME AND M. P. CAMERON (Eds.), *Ciba Foundation Symposium on Chemistry and Biology of Pteridines*, 1954, p. 314.
153 JEPSON, J. B., UDENFRIEND, S. AND ZALTZMAN, P., *Federation Proc.*, 18 (1959) 1003.
153a JEPSON, J. B., ZALTZMAN, P. AND UDENFRIEND, S., *Biochim. Biophys. Acta*, 62 (1962) 91.
154 KAMESWARAN, L. AND WEST, G. B., *J. Pharm. and Pharmacol.*, 13 (1961) 191.
155 KEGLEVIĆ, D., SUPEK, Z., KVEDER, S., ISKRIĆ, S., KEČKEŠ, S. AND KISIĆ, A., *Biochem. J.*, 73 (1959) 53.
155a KISHIMOTO, Y., TAKAHASHI, N. AND EGAMI, F., *J. Biochem.*, 49 (1961) 436.
156 KIZER, D. E. AND CHAN, S. K., *Federation Proc.*, 19 (1960) 397.
157 KOELLE, G. B. AND VALK, A. DE T., JR., *Federation Proc.*, 13 (1954) 376.
158 KOELLE, G. B. AND VALK, A. DE T., JR., *J. Physiol. (London)*, 126 (1954) 434.
159 KOLN, B., *Zentr.-Veterinärmed.*, 3 (1956) 570, quoted from *Chem. Abstr.*, 51 (1957) 4520.
160 KOLB, E., *Zentr.-Veterinärmed.*, 4 (1957) 265.
161 KOLB, E., *Arch. exptl. Veterinärmed.*, 10 (1957) 874.
162 KOPIN, I. J., PARE, G. M. B., AXELROD, J. AND WEISSBACH, H., *Biochim. Biophys. Acta*, 40 (1960) 377.
163 KORZENOVSKY, M., WALTERS, G. P. AND HUGHES, M. S., *Federation Proc.*, 18 (1959) 1045.
164 KUNTZMAN, R., SHORE, P. A., BOGDANSKI, D. AND BRODIE, B. B., *J. Neurochem.*, 6 (1961) 226.
164a LAGUNOFF, D. AND BENDITT, E. P., *Am. J. Physiol.*, 196 (1959) 993.
165 LANGEMANN, H., in G. P. LEWIS (Ed.), *5-Hydroxytryptamine*, Pergamon, London, 1957, p. 151.
166 LEACH, B., *Conference of the Brain Research Foundation, Chicago, Ill.*, Jan. 12, 1957.
167 LERNER, A. B., CASE, J. D., BIEMANN, K., HEINZELMAN, R. V., SEMUSZKOVIC, J., ANTHONY, W. C. AND KRIVIS, A., *J. Am. Chem. Soc.*, 81 (1959) 5264.
168 LERNER, A. B., CASE, J. D. AND HEINZELMAN, P. V., *J. Am. Chem. Soc.*, 81 (1959) 6084.
169 LERNER, A. B., CASE, J. D. AND TAKAHASHI, Y., *J. Biol. Chem.*, 235 (1960) 1992.
170 LEVINE, W. G., *Biochem. J.*, 76 (1960) 43P.
171 LILLIE, R. D., *J. Histochem. Cytochem.*, 5 (1957) 188.
172 LING, N. S. AND BLUM, J. J., *Federation Proc.*, 17 (1958) 98.
173 MAASS, A. R. AND NIMMO, M. J., *Nature*, 184 (1959) 547.
174 MANN, P. J. G., *Biochem. J.*, 76 (1960) 44P.
175 MANSOUR, T. E., *Biochim. Biophys. Acta*, 30 (1958) 492.
176 MARTIN, G. M., BENDITT, E. P. AND ERIKSEN, N., *Federation Proc.*, 17 (1958) 447.
177 MARTIN, G. M., BENDITT, E. P. AND ERIKSEN, N., *Federation Proc.*, 18 (1959) 1937.
178 MARTIN, G. M., BENDITT, E. P. AND ERIKSEN, N., *Arch. Biochem. Biophys.*, 90 (1960) 208.
179 MCCAMAN, R. E., *Federation Proc.*, 20 (1961) 344c.
180 MCCOUBRY, A., *Biochem. Pharmacol.*, 2 (1959) 264.
181 MCISAAC, W. M. AND PAGE, I. H., *Science*, 128 (1958) 537.
182 MCISAAC, W. M. AND PAGE, I. H., *J. Biol. Chem.*, 234 (1959) 858.
182a MEIER, R. AND SCHULER, W., *Helv. Physiol. Pharm. Acta*, 15 (1957) 284.
183 MELTZER, H. Y., *Federation Proc.*, 20 (1961) 227.
184 MILTON, A. S. AND GOSSELIN, R. E., *Federation Proc.*, 19 (1960) 126.
185 MITOMA, C. AND LEEPER, L. C., *Federation Proc.*, 13 (1954) 266.
186 MITOMA, C., POSNER, H. S., ROITS, H. C. AND UDENFRIEND, S., *Arch. Biochem. Biophys.*, 61 (1956) 431.
187 MITOMA, C., WEISSBACH, H. AND UDENFRIEND, S., *Nature*, 175 (1955) 994.
188 MITOMA, C., WEISSBACH, H. AND UDENFRIEND, S., *Arch. Biochem. Biophys.*, 63 (1956) 122.
189 MØLLER NIELSON, I. AND HUUS, I., *Meeting British Scandinavian Pharmacological Societies, Copenhagen*, July 26–28, 1960.
190 NACHMIAS, V. T., *J. Neurochem.*, 6 (1960) 99.

191 NAKAI, N., *Nature*, 181 (1958) 1734.
192 NAKAJIMA, H. AND THUILLIER, J., *Compt. rend. soc. biol.*, 152 (1958) 270.
193 NICKERSON, M. AND PARMAR, S. S., *Federation Proc.*, 20 (1961) 165e.
194 NOVICK, W. J., *Federation Proc.*, 20 (1961) 237e.
195 PLETSCHER, A., in F. BRÜCKE (Ed.), *Biochemistry of the Central Nervous System*, Pergamon, London, 1959, p. 124.
196 PLETSCHER, A. AND BESENDORF, H., *Experientia*, 15 (1959) 25.
197 PLETSCHER, A. AND GEY, K. F., *Helv. Physiol. Pharmacol. Acta*, 16 (1958) C 26.
198 PLETSCHER, A., GEY, K. F. AND ZELLER, P., in E. JUCKER (Ed.), *Progress in Drug Research*, Birkhäuser, Basel, 1960, Vol. 2.
199 PORTER, C. C., TITUS, D. C., SANDERS, B. E. AND SMITH, E. V. C., *Science*, 126 (1957) 1014.
200 OUGH, C. E. M. AND QUASTEL, J. H., *Biochem. J.*, 31 (1937) 2306.
201 RANDALL, L. O. AND BAGDON, R. E., *Ann. N.Y. Acad. Sci.*, 80 (1959) 626.
202 RAVIN, H. A., *Lancet*, 270 (1956) 762.
203 REMILY, C. AND WOLFE, R. G., *Biochem. Biophys. Research Communs.*, 3 (1960) 457.
203a RENSON, J., GOODWIN, F., WEISSBACH, H. AND UDENFRIEND, S., *Biochem. Biophys. Research Communs.*, 6 (1961) 20.
203b RENSON, J., WEISSBACH, H. AND UDENFRIEND, S., *J. Biol. Chem.*, 237 (1962) 2261.
203c RENSON, J., WEISSBACH, H. AND UDENFRIEND, S., *Federation Proc.*, 21 (1962) 323.
204 RICHTER, D., *Biochem. J.*, 31 (1937) 2022.
205 RICHTERISCH, R., GAUTIER, E., STILLHART, H. AND ROSSI, E., *Helv. Paediat. Acta*, 15 (1960) 424.
206 RODNIGHT, R., *J. Physiol. (London)*, 141 (1958) 10P.
207 SANDLER, M., SPECTOR, R. G., RUTHVEN, C. R. J. AND DAVISON, A. N., *Biochem. J.*, 74 (1960) 42P.
208 SARKAR, S., BANERJEE, R., ISE, M. S. AND ZELLER, E. A., *Helv. Chim. Acta*, 43 (1960) 439.
209 SARKAR, S. AND ZELLER, E. A., *Federation Proc.*, 20 (1961) 238f.
210 SCHINDLER, P., *Biochem. Pharmacol.*, 1 (1958) 323.
211 SCHINDLER, R., DAY, M. AND FISCHER, G. A., *Cancer Research*, 19 (1959) 47.
212 SCHNOCKLOTH, R. E., MCISAAC, W. M. AND PAGE, I. H., *J. Am. Med. Assoc.*, 170 (1959) 1143.
213 SCHWARTZ, M. A., *Federation Proc.*, 20 (1961) 167e.
214 SCHWEPPE, J. S., ZELLER, E. A. AND HIGGINS, M., *Proc. Staff Meetings Mayo Clinic*, 26 (1951) 371.
215 SIVA SANKAR, D. V., *Federation Proc.*, 18 (1959) 441.
216 SJOERDSMA, A., SMITH, T. E., STEVENSON, T. D. AND UDENFRIEND, S., *Proc. Soc. Exptl. Biol. Med.*, 89 (1955) 36.
217 SMITH, S. E., *Brit. J. Pharmacol.*, 15 (1960) 319.
218 SMITH, T. E., RENSON, J. F., WEISSBACH, H. AND UDENFRIEND, S., *Federation Proc.*, 20 (1961) 238a.
219 SNOW, P. J. D., LENNARD-JONES, J. E., GURZON, G. AND STACEY, R. S., *Lancet*, 269 (1955) 1004.
220 SOURKES, T. L., *Arch. Biochem. Biophys.*, 51 (1954) 444.
221 SPECTOR, S., KUNTZMAN, R., HIRSCH, C. AND BRODIE, B. B., *Federation Proc.*, 19 (1960) 279.
222 SPENCER, R. P. AND ZAMCHECK, N., *Biochem. Biophys. Research Communs.*, 3 (1960) 386.
223 SPINKS, A. AND BURN, J. H., *Brit. J. Pharmacol.*, 7 (1952) 93.
224 STRÖMBLAD, B. C. R., *J. Physiol. (London)*, 147 (1959) 639.
225 STROMBERG, V. L., *J. Am. Chem. Soc.*, 76 (1954) 1707.
226 SZARA, S., *Experientia*, 17 (1961) 76.
227 SZARA, S., HEARST, E. AND PUTNEY, F., *Federation Proc.*, 19 (1960) 23.
228 TABOR, C. W., TABOR, H. AND ROSENTHAL, S. M., *J. Biol. Chem.*, 208 (1954) 645.
229 TAYLOR, J. D., WYKES, A. A., GLADISH, Y. C. AND MARTIN, W. B., *Nature*, 187 (1960) 941.
230 TAYLOR, J. D., WYKES, A. A., GLADISH, Y. C., MARTIN, W. B. AND EVERETT, G. M., *Federation Proc.*, 19 (1960) 278.
231 TITUS, E. AND UDENFRIEND, S., *Federation Proc.*, 13 (1954) 411.
232 TRENDELENBURG, U., *Brit. J. Pharmacol.*, 8 (1953) 454.
233 UDENFRIEND, S., in G. P. LEWIS (Ed.), *5-Hydroxytryptamine*, Pergamon, London, 1958, p. 43.
234 UDENFRIEND, S., in F. BRÜCKE (Ed.), *Biochemistry of the Central Nervous System*, Pergamon, 1959, p. 301.
235 UDENFRIEND, S., CLARK, C. T., AXELROD, J. AND BRODIE, B. B., *J. Biol. Chem.*, 208 (1954) 731.

236 UDENFRIEND, S., CLARK, C. T. AND TITUS, E., *Experientia*, 8 (1952) 379.
237 UDENFRIEND, S., CLARK, C. T. AND TITUS, E., *Federation Proc.*, 12 (1953) 282.
238 UDENFRIEND, S., CLARK, C. T. AND TITUS, E., *J. Am. Chem. Soc.*, 75 (1953) 501.
239 UDENFRIEND, S., GREVELING, C. R., POSNER, H., REFIELD, B. G., DALY, J. AND WITKOP, B., *Arch. Biochem. Biophys.*, 83 (1959) 501.
240 UDENFRIEND, S., LOVENBERG, W. AND SJOERDSMA, A., *Arch. Biochem. Biophys.*, 85 (1959) 487.
241 UDENFRIEND, S., LOVENBERG, W. M. AND WEISSBACH, H., *Federation Proc.*, 19 (1960) 7.
242 UDENFRIEND, S. AND TITUS, E., in W. D. McELROY AND H. B. GLASS (Eds.), *Amino Acid Metabolism*, Johns Hopkins, Baltimore, 1955, p. 954.
243 UDENFRIEND, S., TITUS, E., WEISSBACH, H. AND PETERSON, R. E., *J. Biol. Chem.*, 219 (1956) 335.
244 UDENFRIEND, S. AND WEISSBACH, H., *Federation Proc.*, 13 (1954) 412.
245 UDENFRIEND, S. AND WEISSBACH, H., *Proc. Soc. Exptl. Biol. Med.*, 97 (1958) 748.
246 UDENFRIEND, S., WEISSBACH, H. AND BOGDANSKI, D. F., *J. Pharmacol. Exptl. Therap.*, 120 (1957) 255.
247 UDENFRIEND, S., WEISSBACH, H. AND BOGDANSKI, D. F., *Ann. N.Y. Acad. Sci.*, 66 (1957) 602.
248 UDENFRIEND, S., WEISSBACH, H. AND SJOERDSMA, A., *Science*, 123 (1956) 669.
249 UDENFRIEND, S., WITKOP, B., REDFIELD, B. G. AND WEISSBACH, H., *Biochem. Pharmacol.*, 1 (1958) 160.
250 URIEL, J., *Bull. soc. chim. biol.*, 39 (1957) Suppl. 1, 105.
251 URIEL, J., *Nature*, 181 (1958) 999.
252 VINCENT, D. AND SEGOUZAC, G., *Thérapie*, 14 (1960) 914.
253 VIOLLIER, G., QUIRING, E. AND STAUB, H., *Helv. Chim. Acta*, 36 (1953) 724.
254 VITEK, V. AND RYSANEK, K., *Nature*, 186 (1960) 244.
255 WAALKES, T. P. AND COBURN, H., *Proc. Soc. Exptl. Biol. Med.*, 99 (1958) 742.
256 WALSHE, J. M., *Ann. Internal Med.*, 51 (1959) 1110.
256a WEBER, L. J. AND HORITA, A., *Life Sciences*, 1 (1963) 44.
257 WEINER, N., *Federation Proc.*, 18 (1959) 1808.
258 WEISSBACH, H. AND AXELROD, J., *Federation Proc.*, 19 (1960) 50.
259 WEISSBACH, H., BOGDANSKI, D. G., REDFIELD, B. G. AND UDENFRIEND, S., *J. Biol. Chem.*, 227 (1957) 617.
260 WEISSBACH, H., KING, W., SJOERDSMA, A. AND UDENFRIEND, S., *J. Biol. Chem.*, 234 (1959) 81.
261 WEISSBACH, H., LOVENBERG, W., REDFIELD, B. G. AND UDENFRIEND, S., *J. Pharmacol. Exptl. Therap.*, 131 (1961) 26.
262 WEISSBACH, H., LOVENBERG, W. AND UDENFRIEND, S., *Biochem. Biophys. Research Communs.*, 3 (1960) 225.
263 WEISSBACH, H., REDFIELD, B. G. AND AXELROD, J., *Biochim. Biophys. Acta*, 43 (1960) 352.
263a WEISSBACH, H., REDFIELD, B. G. AND AXELROD, J., *Biochim. Biophys. Acta*, 54 (1961) 190.
264 WEISSBACH, H., REDFIELD, B. G. AND UDENFRIEND, S., *J. Biol. Chem.*, 229 (1957) 953.
265 WEISSBACH, H., REDFIELD, B. G. AND UDENFRIEND, S., *J. Biol. Chem.*, 234 (1959) 81.
266 WEISSBACH, H., SMITH, T. E., DALY, J. W., WITKOP, B. AND UDENFRIEND, S., *J. Biol. Chem.*, 235 (1960) 1160.
267 WELSH, J. H. AND MOORHEAD, M., *Gumna J. Med. Sci. (Japan)*, 8 (1959) 211.
268 WERLE, E. AND AURES, D., *Z. physiol. Chem.*, Hoppe-Seyler's, 216 (1959) 1.
269 WERLE, E. AND HENNING, E., *Z. Vitamin- Hormon- u. Fermentforsch.*, 11 (1960) 159.
270 WERLE, E. AND MONNIKEN, G., *Biochem. Z.*, 291 (1937) 325.
271 WERLE, E. AND ROSWER, F., *Biochem. Z.*, 322 (1952) 320.
272 WEST, G. B., *J. Pharm. and Pharmacol.*, 10 (1958) Suppl. 92 T.
273 WESTERMANN, E., *Arch. exptl. Pathol. Pharmakol.*, Naunyn-Schmiedeberg's, 228 (1956) 159.
274 WESTERMANN, E., BALZER, H. AND KNELL, J., *Arch. exptl. Pathol. Pharmakol.*, Naunyn-Schmiedeberg's, 234 (1958) 194.
275 WIELAND, T., MOTZEL, W. AND MERZ, H., *Ann. Chem.*, Liebigs, 581 (1953) 10.
276 WILKINSON, S., *J. Chem. Soc.*, (1958) 2079.
277 WISEMAN, M. H. AND SOURKES, T. L., *Biochem. J.*, 78 (1961) 123.
278 WYKES, A. A., GLADISH, Y. C. AND TAYLOR, J. D., *Federation Proc.*, 18 (1959) 1826.
279 YUWILER, A., GELLER, E. AND EIDUSON, S., *Arch. Biochem. Biophys.*, 80 (1959) 162.
280 YUWILER, A., GELLER, E. AND EIDUSON, S., *Federation Proc.*, 18 (1959) 2383.
281 YUWILER, A., GELLER, E. AND EIDUSON, S., *Arch. Biochem. Biophys.*, 89 (1960) 143.

282   ZARAFONETIS, C. J. D. AND KALAS, J. P., *Am. J. Med. Sci.*, 239 (1960) 203.
283   ZELLER, E. A., *Cornell University Symposium papers on Biochemistry and Nutrition*, 1956.
284   ZELLER, E. A., *J. Clin. Exptl. Psychopathol. & Quart. Rev. Psychiat. Neurol.*, 19 (1958) 27.
285   ZELLER, E. A., *Pharmacol. Revs.*, 11 (1959) 387.
286   ZELLER, E. A., *Experientia*, 16 (1961) 399.
287   ZELLER, E. A. AND BARSKY, J., *Proc. Soc. Exptl. Biol. Med.*, 81 (1952) 459.
288   ZELLER, E. A., BARSKY, J. AND BERMAN, E. R., *J. Biol. Chem.*, 214 (1955) 267.
289   ZELLER, E. A., BARSKY, J., BERMAN, E. R., CHERKAS, M. S. AND FOUTS, J. R., *J. Pharmacol. Exptl. Therap.*, 124 (1958) 282.
290   ZELLER, E. A., BARSKY, J., BERMAN, E. R. AND FOUTS, J. R., *J. Lab. Clin. Med.*, 40 (1952) 965.
291   ZELLER, E. A., BARSKY, J., FOUTS, J. R., KIRCHHEIMER, W. F. AND VAN ORDEN, L. S., *Experientia*, 8 (1952) 349.
292   ZELLER, E. A., BARSKY, J., FOUTS, J. R. AND LAZANAS, J. C., *Biochem. J.*, 60 (1955) V.
293   ZELLER, E. A., BLANKSMA, L. A., BURKARD, N. P., PACHA, W. L. AND LAZANAS, J. C., *Ann. N.Y. Acad. Sci.*, 80 (1959) 583.
294   ZELLER, E. A., PACHA, W. L. AND SARKAR, S., *Biochem. J.*, 76 (1960) 45P.
295   ZELLER, E. A. AND SARKAR, S., *Federation Proc.*, 19 (1960) 21.
296   ZELLER, P., PLETSCHER, A., GAY, F. K., GUTMAN, H., HEGEDÜS, B. AND STRAUB, O., *Ann. N.Y. Acad. Sci.*, 80 (1959) 555.
297   ZILE, M. H., *Endocrinology*, 66 (1960) 311.
298   ZILE, M. H. AND LARDY, H. P., *Arch. Biochem. Biophys.*, 82 (1959) 411.

Chapter 4

METABOLISM OF SEROTONIN *IN VIVO*

Having surveyed the enzymes synthesizing and metabolizing 5-HT the metabolism of the amine *in vivo* will now be discussed. There is a static aspect represented by the distribution in nature, the level in the tissues and the localization at subcellular level and there is also a dynamic aspect to be considered, *i.e.* the turnover of 5-HT in the body. The data obtained with the help of drugs releasing 5-HT or blocking its destruction are very interesting in this connection. In this chapter only the general lines of the problem will be discussed. The details concerning peculiar aspects of 5-HT metabolism in a certain tissue and particularly for brain, will be described in other chapters.

DISTRIBUTION OF 5-HT IN NATURE

5-HT is widely distributed in both the vegetal and animal kingdoms. The biological importance in plants of an analogue of serotonin, the auxin 3-indolylacetic acid, has been known for a long time. Recently it has also been found that serotonin is present in several edible vegetables (see Table 12). This is quite interesting because the presence of 5-HT has been rather closely associated with poisonous plants (cowhage[51], stinging nettle[76,80] and *Paneolus campanulatus*[349]). The pathways of the 5-HT metabolism in plants is under investigation by several groups. It has been reported that in banana the presence of 5-HT is not associated with the presence of tryptamine, 5-HTP or 5-HIA[389]. Also 5-HTP decarboxylase seems to be absent[353]. Serotonin and tryptamine in edible fruits are not apparently associated, for instance the blue plum[353] contains tryptamine but not serotonin, while the reverse is shown by avocado[353], bananas[389] and plantains[145]. The presence of 5-HT in plants is well established in several families[60a]. It will be of interest to know if the processes of serotonin metabolism in plants are comparable to those already known in animal tissues.

In the animal kingdom there is a large number of data in the literature about the presence and the amount of 5-HT in the different tissues. A note of caution should be given for some of these data, where obtained with non-specific methods, like some of the bioassay procedures without suitable controls. The different methods used for extraction and evaluation, as well as the different strains, food and experimental conditions used, may account for the differences, and these are sometimes really striking. Most of the data available are reported in a set of tables (see Appendix I). Extensive data on the distribution of 5-HT in invertebrates have been reported by several authors[38a,214a,240–242,265,347,365,383–387].

TABLE 12

SEROTONIN CONTENT IN VEGETABLES AND FRUITS

Serotonin is present in	*References*	*Serotonin is absent in*	*References*
Avocado	353	Apple	389
Banana	315a, 353, 371	Black currant	389
Blue-red plum	353	Blue plum	353
Brinjal	315a	Cherry	389
Egg-plant	353	Fig	389
Passion fruit	145	Gooseberry	389
Pineapple (Australian)	60	Lemon	389
Plantain (matoke banana)	145, 239	Mango	145
Red plum	353	Orange	353, 389
Tomato	353, 389–391	Pineapple (Afr.)	145, 145a
Papaw	145	Potato	315a, 353, 389
		Prune	389
		Raspberry	389
		Rhubarb	389
		Spinach	353
		Strawberry	389

Since it would be tedious to enumerate all the species and tissues in which serotonin has been found, only special aspects will be discussed hereafter. There is no doubt that the most important localization of 5-HT in mammals (at least from a quantitative viewpoint) is represented by the *gastrointestinal tract*[118]. 5-HT is believed to be linked with the so-called enterochromaffin cell system[23,23a,92,*]. These cells are characterized by granules showing chromaffin and argentaffin reactions, combining with diazonium salts and showing fluorescence after fixation with formaldehyde[**]. That in enterochromaffin cells there is biosynthesis of 5-HT from tryptophan is now well established by the extensive investigation of carcinoidosis (see p. 92). In this tumour the neoplastic proliferation of the argentaffin cells is associated with an enormous increase of the production of 5-HT and occasionally also 5-HTP. [2-^{14}C]-Tryptophan administered by oral route to these patients is followed by the elimination of labelled 5-HTP, 5-HT and 5-HIA[109,325]. These findings strongly suggest that in the normal intestinal tissue 5-HT is synthesized and stored in the enterochromaffin cells[93,125].

In fact it has been recently reported that the enzyme hydroxylating tryptophan in 5-position can be isolated from intestine[81].

Against the objections[161,231–233] concerning the identity of enterochromaffin granules with 5-HT are the reports that after formaldehyde treatment, 5-HT resembles the enterochromaffin granules in many histochemical preparations[23,23a,130,198,***]. How-

---

*   A classification of this system can be found in a review by Erspamer[116].
**   Also mast cells in precancerous mouse skin are fluorescent after treatment with formaldehyde[284].
***   Other chromaffin granules contain dopamine[33,131].

ever, other indolealkylamines may show the same reaction[94,198]. Collier[79] in a recent review of this problem, quotes that 5-HT was not detected in the intestinal mucosa of *teleostei*, in which cells are present with granules indistinguishable from those of the enterochromaffin cells. Furthermore, Adam observed that in the venom gland of the scorpion 5-HT is present in large quantities although there are no cells resembling the mammalian enterochromaffin cells[1,2]. Jacobson[195] believes that not all the 5-HT in the intestine is stored in the argentaffin cells. In fact some samples of carcinoid tissue, consisting only of argyrophil cells (and not showing fluorescence) contained high amounts of 5-HT (from 0.33 to 1.1 mg/g). Furthermore it has been reported that patients with atypical carcinoids excreted 5-HTP in their urine without showing at biopsy typical argentaffin cells[291,326,*]. However, apart from these objections, there is considerable agreement with the hypothesis that 5-HT is synthesized by the enterochromaffin cells in intestine.

The different parts of the intestine do not contain the same concentration of 5-HT. This difference is not predictable, but varies according to animal species[114,382]. For instance, according to Erspamer[116] the mouse contains 3.1 μg of 5-HT/g of large intestine and 1.6 μg/g of small intestine. In rats, 8–15 % of the total gastrointestinal 5-HT is contained in the stomach, 30–40 % in the small intestine and the remaining 45–60 % in the large intestine[116].

Other data from West[388] show that the distribution of 5-HT in stomach or duodenum is not uniform. In the cat, for instance, stomach contains 0.5 μg/g 5-HT and duodenum 0.9 μg/g, while in the guinea pig these values are respectively 1.4 and 5 μg/g. These differences in the presence of 5-HT at the various levels of the gastrointestinal tract have been found also in man[116]. However, these data, when obtained with bioassay, might not be representative of the 5-HT content, because in the intestine there is another unknown biologically active principle which is only related to 5-HT[94,132].

A careful study of Feldberg and Toh[132] shows that 5-HT is mainly present in the mucosa of the gastrointestinal tract in dog and rabbit; the submucosa and the *muscularis* layers contain only small amounts of 5-HT. Again the mucosa of the gastrointestinal tract contains different concentrations of 5-HT according to the area: in dog only 0.6 μg/g are present in the mucosa of the oesophagus, but 10 μg/g in the mucosa of the pyloric region[132].

In *liver*, the concentrations of 5-HT are generally low and they do not exceed the value of 1 μg/g, except in the chicken where a value of 1.6 μg/g has been reported[377].

In *lung* the values range from 0.06–0.6 μg/g in the guinea pig to more than 2 μg/g in mouse and rabbit (see Appendix I).

In *spleen* 5-HT is present in high concentrations up to 20 μg/g as in the case of the hamster[262]. The levels in spleen depend probably upon the number of platelets

---

* A possible explanation could be that 5-HT must be concentrated in a small volume to give positive argentaffin reactions. Thus a large quantity of 5-HT spread in a cell might not be enough to show the enterochromaffin granules[25].

TABLE 13

SOME RICH SOURCES OF 5-HT IN NATURE (> 100 μg/g or ml fresh material)

Source	Content	References
Rat, mast cells	800 μg/g	124
Eledone moschata, post-salivary gland	760 μg/g	116, 118
Bombinator pachypus, skin	700–1000 μg/g	116, 118
Human carcinoid tumour	580–2500 μg/g	226
Calliactis parasitica, coelenteric tissue	550 μg/g	240
Rabbit platelets	500 μg/g	193
Octopus vulgaris, posterior salivary gland	420–510 μg/g	116, 118
Spiny dogfish (*Squalus acanthias*), ejaculate	437 μg/ml	234
Discoglossus pictus, skin	410 μg/g	116, 118
Goose, platelets	350 μg/g	55
Wasp	300 μg/g sting apparatus	79, 196
Stinging nettle	200 μg/g sting fluid	79, 80
Walnut	170–340 μg/g	214b
Cowhage	150 μg/g	79
Banana, peel	150 μg/g	353
Vejovis sp., sting segment	138 μg/g	387
Mouse, neoplastic mast cells	130 μg/g	323

present[53] although in diseased spleens from human beings and rats, this correlation was not observed[245a].

It is still to be established if 5-HT is present in *plasma* in a free form because it is difficult to avoid the disruption of platelets during the manipulation of the blood. Moreover, in serum of some species there is an amine oxidase destroying 5-HT[369]. Values of 0.002–0.019 g/ml are reported in the literature as the physiological leak-out of platelets in plasma[66,85b,233b,284a]. The content of 5-HT in the blood depends on the number of platelets available[283,357,406], although the values of 5-HT for a given number of platelets vary according to the species[55,193,408]. For instance, 10^9 platelets contain 0.4 μg of 5-HT in rat and 7.5 μg in rabbit[193]. No difference has been observed between the content of 5-HT in arterial or venous blood[343]. Other cells in blood, like polymorphonuclear leucocytes or erythrocytes, apparently do not contain 5-HT. Red cells of man[68,244,331], but not of dogs[194] may adsorb 5-HT. One case with a high number of eosinophils showed an increase of 5-HT[246]. The binding of 5-HT to serum proteins is a matter of discussion[89a,214c].

Mast cells, normal or neoplastic, are rich in 5-HT only in rats and mice (refs. 73, 85, 100, 101, 110, 148, 164, 212, 220, 300, 301). Mice with mast-cell tumours show an increased concentration of tissue 5-HT[110] and urinary 5-HIA[162,323], but 5-HT could not be detected in mast cells (normal or neoplastic) in man[266] and dog[323]. However, 5-HT is present in mast cells occurring in human carcinoid tumours[113a].

The difference also explains why the values of 5-HT in the *skin* of rat and mouse are higher than in any other species[323,*].

---

* The skin of *amphibia* is also very rich in 5-HT[116,147b].

In *kidney*, the 5-HT concentration is rather low but without any striking differences among the animal species[348].

The data concerning *brain* will be discussed in the chapter dealing with the 5-HT in the central nervous system.

In the *body fluids* 5-HT is not particularly represented. Platelet-poor plasma[329], urine[227,329] and cerebrospinal fluid[4,8,12,329] contain an amount of 5-HT below 0.1 μg/ml. In urine[227], but not in cerebrospinal fluid[324,325,329] of carcinoid patients, there is an increase of 5-HT. This finding is explained by the difficult penetration of 5-HT through the blood-brain barrier[84,207,352,397].

The *ejaculate* contains very high values of 5-HT in the spiny dog-fish[233a,234]. In human semen, 5-HT is contained in high concentration according to Katsh[206] but only in negligible amounts according to Eliasson[112a] and Mann *et al.*[234a]. Similarly no 5-HT was found in semen of bull, ram, boar and dog[234a].

It might be interesting to remember that 5-HT is not present in adrenal gland, skeletal muscle, placenta (*post partum*)[217], sciatic[153], vagus and femoral nerves, and diaphragm[348].

The discussion about the levels of 5-HT in tissues has been limited, because in our opinion the value given may not be indicative of the importance of 5-HT for that particular tissue. The values measured are the result of two processes: synthesis–destruction and storage–release and they do not indicate the turnover of 5-HT. If they are high, they only suggest a predominance of synthesis over destruction, or storage over release, and huge amounts of 5-HT could indicate that in certain tissues, 5-HT is only accumulated without having any function.

For these reasons it is always difficult to suggest any correlation or dependence between 5-HT present in two different tissues. Thus a relationship between gastro-intestinal tract and blood does not seem apparent[118] because the total 5-HT in blood may represent a small fraction of the total gastrointestinal 5-HT (examples dog and guinea pig) or a very high fraction (examples rat and rabbit). In the case of the cat the total 5-HT present in serum is more than 3 times higher than the total 5-HT present in the gastrointestinal tract[116]. Although it is accepted that platelets absorb 5-HT from the intestine, this process is not so simple as to allow one to predict the values of 5-HT in the serum from the intestinal concentration. Similar conclusions seem to apply also to other suggested correlations like spleen, lung or liver and intestine. The relations between blood and spleen are supported by the importance of the spleen for the destruction of platelets. However, the ratio between spleen and serum 5-HT may vary from less than 1 (frog) to more than 10 (cat)[127]. The size of the spleen in relation to the body weight might play a role in explaining these differences. Also enzymatic differences might be involved although it is accepted that spleen is poor in both 5-HTP decarboxylase[154] and monoamine oxidase[185]. The possibility that platelets absorb serotonin at the sites where they are made seems to be unlikely, because bone marrow does not contain 5-HTP decarboxylase[154]. The fact that the portal blood contains more 5-HT than the blood of other areas[343], and that hepatic vein blood contains more 5-HT than vena cava blood[129] suggest the possibility that all the 5-HT in the body might be made by the intestine and released in the circulating platelets. This hypothesis is mainly based upon the difficulty of demonstrating the occurrence of tryptophan hydroxylase in other tissues. In fact, this enzyme has been found only in the intestine, and, to a small extent, in the kidney[81] (see Chapter 3). In rats and mice, however, intestine is not the only source of 5-HTP, because a complete synthesis of 5-HT from tryptophan has been shown in neoplastic mast cells (refs. 168, 169, 221, 253, 300, 301). It would not be surprising if in the future other cells, in other animal species, will be recognized as able to hydroxylate tryptophan in the 5-position.

Assuming that intestine is responsible for 5-HTP synthesis, there is another aspect to be discussed, *i.e.*, the occurrence of a transport mechanism able to carry 5-HTP in all the tissues where it can be decarboxylated. A rough calculation, based upon the rate of metabolism of 5-HT in lung, kidney, liver*, brain and skin would indicate that the requirement of 5-HTP would be quite large. To date, 5-HTP has not been detected in normal intestine, probably because it is rapidly decarboxylated. Another possibility is that platelets are the carrier for serotonin in all the tissues, which will take up 5-HT from platelets, after the release of the amine, through the "Nephrosine"-type factors[344]. This is, however, unlikely because monoamine oxidases are widely spread in all tissues, and because in this instance the occurrence of 5-HTP decarboxylase would be without meaning. If, on the other hand, it is assumed that other tissues might synthesize 5-HTP, then the origin of 5-HT in platelets could be related to exchange with cells different from the enterochromaffin ones.

Recent experiments show that the removal of large intestine does not substantially decrease the concentration of serotonin in the blood and tissues of rats, but reduces the excretion of urinary 5-HIA[28,30]. Similar results were obtained also in dog[129,286] and patients[31,176,286]. The complete removal of the gastrointestinal tract lowers after 3 days the level of 5-HT in serum, spleen, lung, but not in skin and brain, and reduces the amount of 5-HIA eliminated in urine[30] by more than 90%. These results, although obtained in conditions of severe impairment of the physiological functions and of acute alimentary deficiency, suggest that a limited biosynthesis occurs also in tissues other than the intestine. The gastrointestinal tract seems responsible for the 5-HT levels in serum and of the urinary 5-HIA. Alternative explanations are however to be considered. For instance, a drop of serum 5-HT was observed in rats after removal of the adrenals[245]. On the other hand, tryptophan deficiency causes a 50% decrease of 5-HIA in urine within 3 days[346].

5-HT METABOLISM, THE EFFECT OF PHYSIOLOGICAL AGE

The 5-HT content of the whole foetus increases from undetectable amounts to 0.2 μg/g at the end of pregnancy[106]. The presence of 5-HT in the gastrointestinal tract during the embrionic development is strictly related to the development of the entero-chromaffin system[116]. In several organs, including brain[204], there is an increase of

---

* The capacity to take up 5-HTP has been shown for liver and kidney, but not for spleen[299].

serotonin in relation to the age[29,246]. An organ which shows the presence of 5-HT very early in development is the spleen[118]. There is a parallel between the activity of 5-HTP decarboxylase[189], the levels of 5-HT in tissues and the excretion of 5-HIA[306]. Also monoamine oxidase activity, very low in the foetus, increases with the age[309]. In goat the blood of the foetus contains twice the concentration of 5-HT compared with the blood of the mother[264].

During pregnancy, particularly in the first and last months, an increase of 5-HT metabolism has been reported[217,269]. However there are no changes in the urinary excretion of 5-HIA in pregnant women[163] and rats[10], while the excretion of histamine is remarkably increased. An increased excretion of urinary 5-HIA was observed *post partum*[34a]. Placenta does not contain high concentrations of 5-HT[217]. It has been suggested that serotonin might have a role in the growth processes[219].

Food intake may have importance in determing the 5-HT metabolism. A diet poor in tryptophan decreases the concentration of 5-HT in the intestine and brain of several animal species[112,398] and the urinary excretion of 5-HIA[346]. A diet rich in this amino acid increases intestinal 5-HT in mice[333] but not in rats[335]. In mice, there is an increase in spleen and platelets, but not in brain 5-HT[335], with 1% tryptophan in the diet.

An increase of intestinal 5-HT was observed after a diet rich in meat[333,335], without any change in brain, spleen and platelets 5-HT[335] or in the excretion of urinary 5-HIA[335]. Also a diet rich in liver, but not in soya bean or calcium caseinate, increases intestinal 5-HT[335].

It has been suggested that the increase of intestinal 5-HT after meat diet or antibiotic treatment[333] is in relation with an increase of the anaerobic bacterial flora, increasing the hydroxylation of tryptophan[335].

Increased elimination of 5-HIA in urine has been observed after ingestion of bananas[9,20,72,85a,89,228,281,*], walnuts[214b] and pineapples[60], but not after oranges or grapefruit[20]. However, banana feeding does not change tissue 5-HT[155,247] although it slightly increases intestinal 5-HT[247]. Similarly, oral administration of 5-HT increases only urinary excretion but not tissue 5-HT[89].

EXCRETION AND TURNOVER

Very low levels of 5-HT are normally found in urine (see Table 14), but many breakdown products of 5-HT metabolism are present. An exact knowledge of the urinary metabolites (5-HIA and others) might be very helpful in calculating the amount of serotonin metabolized in the body.

5-HIA is eliminated by glomerular filtration and tubular secretion[103,104,147]. Its elimination is subject to diurnal variations[200] but according to other authors no difference has been found in the excretion during the day or night[223].

---

* Also the excretion of total, but not free[238], catecholamines is increased after banana feeding[88] because banana is rich in noradrenaline and dopamine[371].

TABLE 14

EXCRETION OF 5-HT AND 5-HIA IN URINE OF SEVERAL ANIMAL SPECIES

(Numbers in brackets indicate references)

Species	Normal values of urinary excretion				% of administered 5-HT excreted in urine			
	5-HT		*5-HIA*		*5-HT*		*5-HIA*	
Dog	0.8–0.36 µg/ml	(348)	0.36–1.6 mg/h	(315)	1	(348)	20–30	(356)
			2 mg/day	(341)			40	(341)
			1.4–2.7 mg/day	(355)			25	(119)
			2.5–4 µg/ml	(119)				
			3.1–7.3 µg/ml	(147)				
Guinea pig			0.3 µg/ml	(119)			1	(119)
							4	(250)
Hog			1.5 µg/ml	(119)				
Horse			0.3 µg/ml	(119)				
Man	40–150 µg/day	(285)	6.06 ± 2.19 mg/day	(5–7)	n.d.	(107)	30–45	(35)
	100 µg/day	(375)	3–8 mg/day	(322)			25	(64)
	0.014–0.72 µg/ml	(227)	2–10 mg/day	(375)			20	(119)
			2.2–12.9 mg/day	(65)			59	(107)
			2–3.8 µg/ml	(119)			83–105	(89)
			0.7–13.2 mg/day	(134)			37	(140, 307)
			3.6 ± 1 mg/day	(215)			39	(259)
			203 µg/h	(135)			25	(334)
			4.4–8.2 mg/day	(355)				
			4.20 ± 1.2 mg/day	(105)				
			15.2 mg/day	(281)				

TABLE 14 (*continued*)

Species	Normal values of urinary excretion				% of administered 5-HT excreted in urine		
	5-HT		5-HIA		5-HT	5-HIA	
Kid			1.25 μg/ml	(119)			
Lamb			2.25 μg/ml	(119)			
Mouse						31	(255)*
Mouse (bearing mast cell tumour)			70.5 μg/day	(162, 323)			
Ox			0.3 μg/ml	(119)			
Rabbit			0.3 μg/ml	(119)		1	(119)
			0.39–0.65 mg/day	(355)		4	(250)
Rat	0.08 μg/h	(3)	6 μg/h	(3)**	0.7–1.7 (3)	25	(207)
			1.34 μg/ml	(117)		47–67	(3)**
			81.4 μg/kg/24 h	(117)		77	(95)
			8.9 μg/100 g b.w.	(338)		40	(141)
			1.25–1.5 μg/ml	(119)		38–33	(119)
			7.2–12.2 μg/24 h	(10)			
			18–20 μg/24 h	(113)			
Toad			0.27 μg/ml	(119)			

* Recovered in the whole mouse 2 h after administration of 2 mg 5-HT/mouse.
** Hydrated rats.

n.d. = not detectable.

Again many species differences have been observed. In the urine of some animals, like guinea pig, rabbit and horse, 5-HIA is present in small amounts (less than 0.3 μg/ml). In other species (dog, rat, human) 5-HIA is present in concentrations ranging from 1.5 to 4 μg/ml, although there is a remarkable variability (see Table 14). Is this representing a qualitative difference in the metabolism of 5-HT or is it only indicative of a different turnover? The latter possibility might be ruled out by experiments showing that the administration of 5-HT in guinea pig and rabbit resulted in a poor excretion of 5-HIA in urine[249,250], while a high proportion of injected 5-HIA is excreted[249]. It seems that in rabbits the major metabolite of 5-HT is 5-hydroxyindole-aceturic acid[244]. To understand the significance of 5-HIA in urine, the metabolic paths described in Chapter 3 must be borne in mind. There are alternative metabolic routes, besides monoamine oxidase, which are able to transform 5-HT. 5-HIA is considered the major product of 5-HT metabolism, but other substances found in the urine suggest that the problem is not a simple one. These substances include N-methyl-5-HT[61], N-acetyl-5-HT[243,244], O-sulphate[75,91,329] and O-glucuronide conjugates[2a,243,381,*]; 5-O-sulphate-indoleacetic acid[74], 5-hydroxytryptophol[217a] and 5-hydroxyindoleaceturic acid[243,244]. This latter is not present in rat urine[15]. In addition, there are a number of unidentified metabolites[61,207,326,341]. That monoamine oxidase is not the sole route for degradation of exogenous serotonin is shown by the slightly increased toxicity of the latter during a block of MAO[37] and by the similarity in disappearance of serotonin in normal or MAO-blocked mice or rats[361]. In the rat, the endogenous excretion of 5-HIA is around 150 μg/kg/day (see Table 14) but this value should be doubled because the administration of 5-HT results in a recovery of less than 50% (ref. 107) as urinary 5-HIA**. It may thus be assumed that a rat excretes around 400 μg/kg/day of 5-HT metabolites in urine. Since it is estimated that the rat contains about 300 μg of 5-HT/kg[30], it can be seen that the total 5-HT content is replenished in about 18 hours. In man, the metabolism of 5-HT is represented by 3–8 mg/day excreted as 5-HIA (see Table 14). This value should be at least doubled to obtain the actual 5-HT metabolism, because when 5-HT is administered, there is only a 20 to 59% recovery from the urine in the form of 5-HIA***. In man it has been calculated that about 10–20 mg of 5-HT/day are metabolized (about 2% of the tryptophan intake[326]).

But not all the tissues participate to the same degree in the turnover of 5-HT, expressed by 5-HIA and other 5-HT metabolites present in urine †.

Two methods have been used for calculating turnover of 5-HT in tissue: radio-

---

* Recently it has been shown that also 4-HTP *in vivo* is decarboxylated to 4-HT and then transformed into 4-HIA or 4-O-glucuronide-HT[128].
** The lower recovery obtained by administering 5-HT might be related with the antidiuretic effect of 5-HT which tends to retard the excretion of the 5-HT metabolites. The administration of 5-HTP results in an increased excretion of 5-HIA amounting to about 80% of that expected[34].
*** As already stated several times, these values may go up to 1 g/day in carcinoid patients. Also the level of 5-HIA in blood is increased in carcinoid (4 μg/ml) compared with normal subjects (0.2–0.4 μg/ml)[322].
† Other routes of excretion are not quantitatively important[208]. After intraperitoneal administration of serotonin only 0.5–5% is excreted in the gut content[244].

active-labelled precursors of serotonin[358] and fast-acting monoamine oxidase (MAO) inhibitors[358].

By administering [^{14}C]tryptophan and [^{14}C]5-HTP to rabbits, it has been established that the half-life of 5-HT in the stomach is about 10 h, in intestine about 17 h, and in platelets more than 24 h (ref. 358) (5–6 days according to Heyssel[184a]). The half-life of gastrointestinal serotonin is 7–12 h in man and 6–8 h in dogs[129]. In carcinoid tissue the turnover was much slower: about 5 days[324]. In the cat lung preparation the added 5-HT has a half-life of 4–20 min in the perfusion blood[149]. Since the amount of labelled serotonin in brain was too small, an estimation has been made with powerful monoamine oxidase inhibitors, like harmaline[255,363]. The half-life of 5-HT in brain was found to be around 10–30 min[358]. A shorter time has been suggested[58] because alternative pathways to MAO may occur even in brain, and because it will certainly take some time before harmaline can reach and act on the MAO. The calculation of 5-HT turnover with MAO inhibitors is not applicable to other tissues because, in contrast to brain, other metabolic pathways are quite important. Assuming that the reported figures are correct, one may calculate that in rabbit the brain will contribute to the pool of 5-HT metabolites 1 μg/kg body weight every ∼20 min, making a total of 72 μg per 24 h. This seems quite a large amount considering that the serum (platelets) will contribute to the pool only about 60–100 μg/day and the gastrointestinal tract about 320 μg/day. Unfortunately, the amount of 5-HT metabolites excreted per day in the rabbit is not known, and thus it is impossible to calculate the percentage contribution for the tissues considered. A similar calculation made for rat, assuming that the turnover of 5-HT in the different tissues is similar to that in rabbit tissues, will give the following values per kg of body weight and per day: brain ∼150 μg, serum ∼30 μg, spleen ∼10 μg and gastrointestinal tract ∼110 μg. These amounts make a total of about 300 μg/kg/day to which should be added the quantity of 5-HT present in skin and other tissues. This value is comparable to the 400 μg/kg/day calculated to be the daily excretion of 5-HT metabolites (see p. 62).

Studies performed in rats show that the recovery of 5-HIA in urine is different if 5-HT or 5-HIA are given. Urinary 5-HIA represents 38% of the 5-HT injected, but 76% of the 5-HIA that is given[107]. An inhibitor of glucuronide formation, sodium o-aminobenzoate[14,302,394] increases in rats the excretion of 5-HIA after 5-HT but not after 5-HIA injection[107]. The major site of glucuronide conjugation is presumably the liver. These data indirectly suggest that serotonin, but not 5-HIA is conjugated[107]. In fact, the decrease of 5-HIA excretion after treatment with monoamine oxidase inhibitors is associated with an increase in 5-HT-O-glucuronide[208,381] and 5-HT-O-sulphate[75]. If this is correct, it should be expected that the 5-HIA present in urine is mainly derived from the tissues metabolizing 5-HT preferentially through MAO.

PATHOLOGICAL CONDITIONS AND DRUG TREATMENTS

Although many of these data will be discussed elsewhere, it seems useful to summarize here some of the findings.

References p. 73

As already mentioned, there are two main conditions accompanied by an increased production and metabolism of serotonin: a tumour of the mast cells in mice and a tumour (carcinoid) of the enterochromaffin (argentaffin) cells in man. Both conditions are characterized by a large concentration of serotonin in the tumour cells, an increase of blood and tissue serotonin and an increase of the urinary excretion of 5-HIA and other indole substances. Both types of tumour cells have the capacity to synthesize serotonin from tryptophan. The metabolism of serotonin in the mast cells has already been discussed and the metabolism of serotonin in carcinoids will be evaluated, in relation to the pathological implication, in the chapter devoted to the gastrointestinal tract. Other pathological conditions with abnormalities in the metabolism of serotonin are not so well known.

The abnormal metabolism of 5-HT present in carcinoids is not common to other types of cancer[163], except for patients with laryngeal or bronchial tumour[52] or argentaffinoma of the lung[201], in which 5-HT and 5-HIA are eliminated in concentrations smaller than in carcinoids, but higher than in normal subjects.

In phenylketonuria, there is a low excretion of 5-HIA in urine[139,260,*] as well as a low level of 5-HT in blood[260]. This finding has been related to an inhibition exerted by phenylpyruvic, phenylacetic and phenyllactic acids, present in these patients, on decarboxylase enzymes[138,171] including the 5-HTP decarboxylase[99]. The removal of phenylalanine from the diet decreases the formation of phenylpyruvic[395] acid and at the same time increases the excretion of 5-HIA in phenylketonuric subjects[18,26,261]. Furthermore, the administration of 5-HTP increases the excretion of urinary 5-HIA (refs. 18, 261) to a lesser extent than in normal subjects, suggesting an impairment of the 5-HTP decarboxylase activity[261]. An alternative explanation would be a competition of the phenylalanine (increased in phenylketonuria) for the hydroxylase system which catalyses the conversion of tryptophan to 5-HTP. However, in a carcinoid patient, the administration of phenylalanine failed to decrease the excretion of 5-HIA, while phenylacetic acid was effective in reducing the concentration of urinary 5-HIA (ref. 290). In disagreement with this finding are recent data showing that *o*-hydroxyphenylacetic, phenylpyruvic and phenylacetic acids, at lower dosage than in the previous investigation[290], increase the excretion of 5-HIA in normal subjects[339].

A block in 5-HIA excretion has been described in patients with Hartnup disease[22]. In this pathological condition there is also an increased excretion of indoleacetic acid[22] and a complex alteration in the metabolism of tryptophan[197] related to abnormal activity of the intestinal microorganisms[308].

Patients with liver diseases excrete normal amounts of 5-HIA[108,111,317,**] and convert 5-HTP to 5-HIA[41]. However, this conversion is increased in cirrhotic patients,

---

* This effect is accompanied by an increased excretion of indoleacetic and indolelactic acids in urine[11].

** Other authors, however, reported an increased excretion of 5-HIA in patients with hepatic diseases[222,263]. This increase is evident also after administration of 5-HT and is probably connected with a decreased formation of the glucuronide conjugate of 5-HT[222]. Vitamin E treatment restores the normal excretion of 5-HIA[222,263].

suggesting a decrease of the conjugation of 5-HT, which would make available more substrate for the monoamine oxidase activity[107]. In this connection it may be recalled that rats with hepatic lesions produced by carbon tetrachloride show an increase of 5-HIA excretion[143].

Variable data have been reported in hypertensive patients: tendency to a low[40,368] or high excretion of 5-HIA[181,348] or normal levels[102].

In symptomatic nontropical sprue there is an increase of 5-HIA in urine, returning to normal values after elimination of gluten and gliadin from the diet[177,178,215].

Elevation of excretion of 5-HIA in diseases of the intestinal tract[204] has been reported. Other data show that the conversion of 5-HT to 5-HIA is substantially unchanged in patients with regional enteritis and ulcerative colitis[111]. Decrease of 5-HT in platelets has been observed in cases of thrombocytopenia[19,36,87,118,175,404,407], in leukaemic subjects[175,277,*] and in patients with inflammatory or rheumatoid processes[213,304]. Collagenous vascular diseases[181] and surgical stress[334] are other conditions with decreased excretion of 5-HIA in urine. Subnormal values of blood serotonin were also found in patients after resection of the stomach[303]. Human pathological spleens from autopsy or operations show a normal value of 5-HT[115]. As already reported, 5-HT is not detectable in cerebrospinal fluid of normal subjects[4,8,12,329] or carcinoid patients[324,325,329]. However, 5-HT content in cerebrospinal fluid is higher in children during tuberculous meningitis[4] or when there is a high concentration of proteins[12]. Also head injuries increase the amount of 5-HIA in cerebrospinal fluid of man, dogs and cats[289].

No change in the 5-HIA elimination has finally been reported in a number of other diseases[163,181] or after surgical operations[147a]. The significance of alterations in the excretion of 5-HIA is discussed by Rosenberg and Davis[285a]. Heavy smokers show an increased elimination of 5-HIA in urines[299a].

The effect of drugs on serotonin metabolism has been carefully studied, measuring the level of serotonin in various tissues or the excretion of 5-HIA. A summary of the effect of various drugs, alone or in combination, is reported in Appendices II and III. Also in the other chapters, the effect of drugs on serotonin metabolism is mentioned when relevant. Two groups of drugs are particularly interesting: the "releasers" of serotonin and the "inhibitors" of monoamine oxidase. It must be underlined however that, at present, there are no drugs interfering selectively with the metabolism of serotonin. Reserpine and its analogues or derivatives release not only serotonin, but also noradrenaline[32,57,70], dopamine[69] and histamine[63,370,372]. Iproniazid and its congeners increase not only serotonin, but also noradrenaline[310], dopamine[69,160] and their *O*-methylated derivatives[13,159,350], tryptamine[18a,161a,183,252,320], *o*- and *m*-hydroxyphenylethylamine[354], phenylethylamine[199], octopamine (*p*-hydroxyphenylethanolamine)[202], tyramine[319a] and various other amines[62,96].

---

* The low concentration of platelet 5-HT in patients with blood disorders appears in relation with a decreased uptake of 5-HT[175]. In a patient with congenital afibrinogenaemia, the content of 5-HT in the whole blood was particularly high[173].

References p. 73

The nature of the release exerted by reserpine is discussed in the paragraph devoted to the problem of storage of serotonin. When the dose of reserpine is high

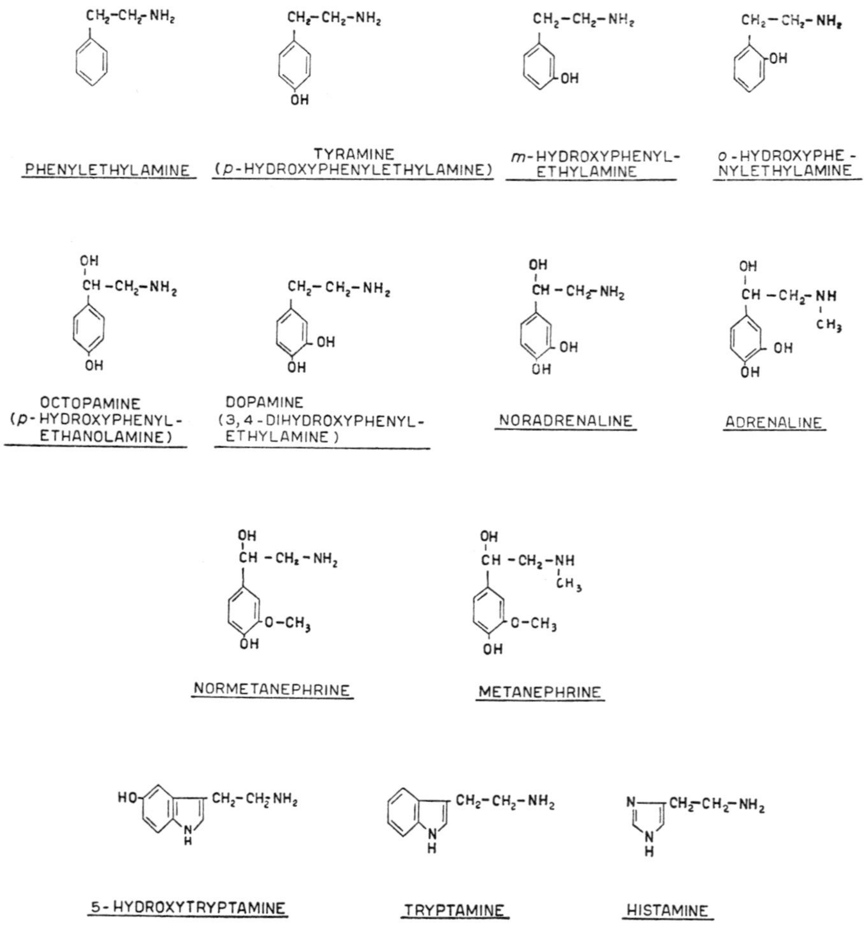

Fig. 19. Structures of amines.

enough there is, in the first hours after administration, an increase in the excretion of the 5-HIA in urine[123,314,342]. This effect has been observed in several animal species (rat[123], dog[126,170], rabbit[144], and man [364]). After this release continuing reserpine treatment, the excretion of 5-HIA becomes normal[126,268] although the levels of serotonin in tissues are very low[122]. This means that the biosynthesis of serotonin is not impaired by the reserpine treatment[59,292,*]. The effect on tissue serotonin exerted by reserpine is related to the animal species and to the tissue considered.

---

* However, West found after a subacute treatment with reserpine a decrease of 5-HTP decarboxylase[388].

Except for guinea pig[370] the serotonin of mammalian intestine[122,180] is quite resistant to the effect of reserpine, while in the chicken it is easily released[82,121,189a]. In rats and other species, including man[166,172], platelets[179,248,313,376] are very sensitive

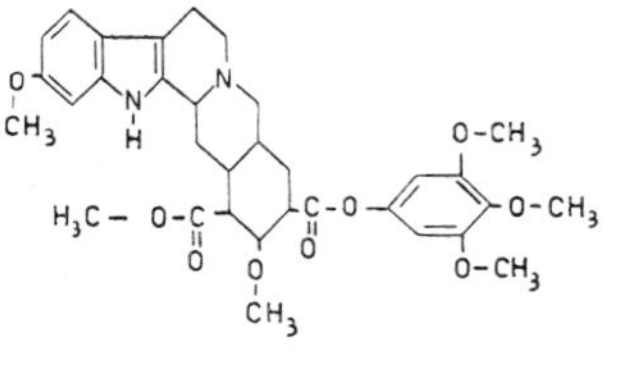

RESERPINE

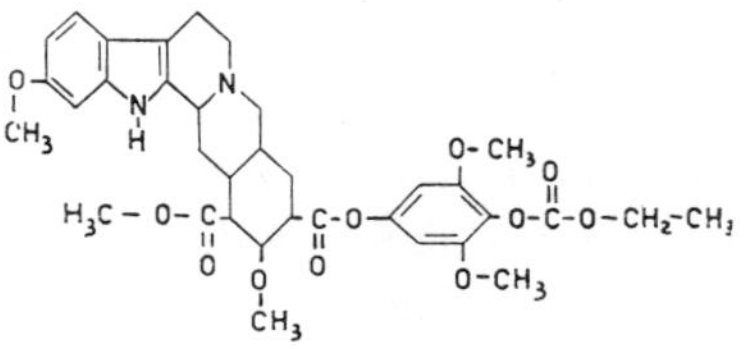

SU 3118
CARBETHOXYSYRINGOYL METHYLRESERPATE

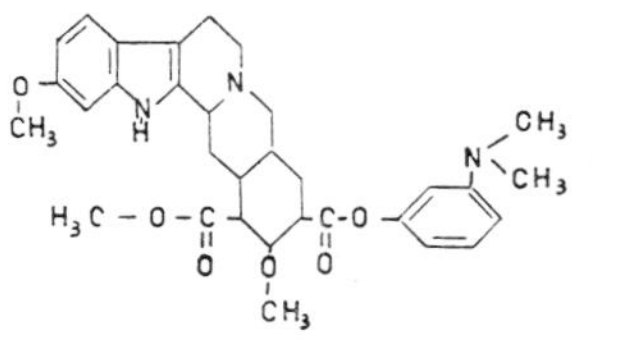

SU 5171
METHYL -18-*O*-(3-*N,N*-DIMETHYLAMINO-BENZOYL) RESERPATE

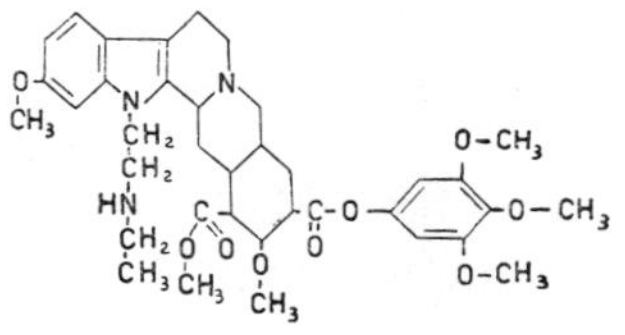

DL 152
DIETHYLAMINOETHYL -1-RESERPINE

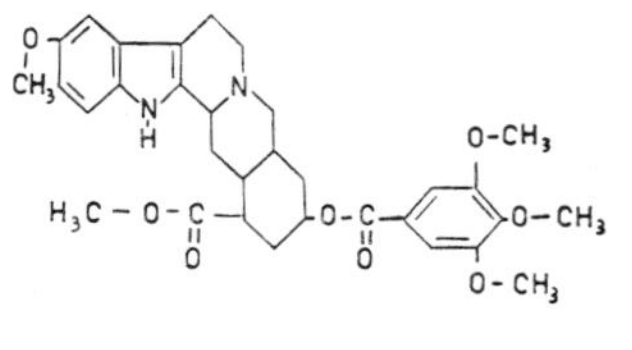

10-METHOXYDESERPIDINE

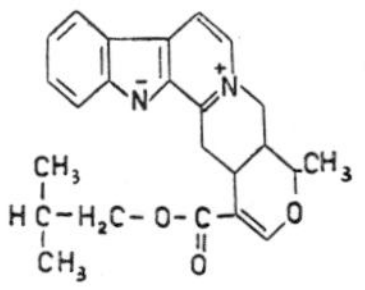

Ph 458
ISOBUTYLSERPENTINATE

Fig. 20. Structures of reserpine and analogues.

to reserpine, as is the spleen[123], but lung, kidney, carcinoid tissue[324], mast cells[73] and thyroid[256] are rather resistant. When reserpine decreases intestinal serotonin, a depletion of enterochromaffin cells has also been observed[25,236]. The effect of reserpine on brain serotonin will be discussed in Chapter 10. Reserpine releases 5-HT also in amphibians and invertebrates[229,230,267]. Among reserpine derivatives, syrosyngopine releases serotonin in spleen at doses without activity on other tissues[151]. Similarly 10-methoxydeserpidine, isobutylserpentinate, DL 152 (see formulae) are inactive on brain and only weakly active on the concentration of serotonin in other tissues[150]. In agreement with these data is the observation that 10-methoxydeserpidine does not increase the urinary elimination of 5-HIA in dogs[218]. Dimethylaminobenzoyl methylreserpate (SU 5171) behaves like reserpine on tissue serotonin[56,151,152]. The effect of many reserpine derivatives on platelet serotonin has been reported[71]. A series of derivatives of reserpine, benzoquinolizines, release 5-HT from different tissues[282,305] and increase the excretion of 5-HIA[272,274] also in man. A short-acting reserpine analogue is the methylether of methylreserpate[90]. Vincamine is another alkaloid able to increase the urinary excretion of 5-HIA[232a]. Also segontin, N-(1-phenylpropyl-2)-1,1-diphenylpropylamine-3, increases the excretion of urinary 5-HIA[214e].

A treatment able to reduce the elimination of 5-HIA in urine is represented by the class of drugs known as "monoamine oxidase inhibitors". This reduced excretion of 5-HIA[81a,319,338] is accompanied by an increase of serotonin in tissues and platelets (refs. 24, 270, 271, 276, 321, 359, 363) (see Appendix II). The values of intestinal 5-HT are, however, not particularly influenced[271]. As already pointed out, this is not a selective increase of 5-HT because other amines are increased as well. There are no marked differences in the effect of MAO inhibitors on tissues of different animal species[400].

The type of physiological disposition of a certain monoamine oxidase inhibitor will determine the degree of inhibition of this enzyme in various tissues. Iproniazid (refs. 67, 97, 184) and N-benzyl-N-methyl-2-propynylamine[340] inhibit liver MAO more rapidly than brain MAO, but phenelzine[187], phenylisopropylhydrazine[158,186,188] and phenylcyclopropylamine[297] are more selective for brain MAO particularly when administered by parenteral route[187].

Monoamine oxidase inhibitors might be divided, according to their effect, into long-acting (iproniazid[184,312], phenelzine[77], phenylisopropylhydrazine[186], nialamide[287]) and short-acting (harmaline[255,363], phenylcyclopropylamine[83,165], α-methyl-[142,167], and β-methyltryptamine[167], and α-ethyltryptamine[166a]).

Short-acting MAO inhibitors[275] or amphetamine[254] antagonize the effect of long-acting MAO inhibitors.

Benactyzine at high doses lowers the excretion of 5-HIA[367] probably by an inhibition of monoamine oxidase[366].

An analogue of serotonin (1-benzyl-2-methyl-5-methoxytryptamine or BAS) displaces serotonin from platelets[396] but does not change the tissue serotonin[330] or inhibit monoamine oxidase[137,330,399] as previously reported[133,136].

Vitamin B$_6$ deficiency induces in the chicken a decrease of 5-HT in tissue[377] but does not change 5-HIA excretion in rat urine[141].

Oral administration of 3-indoleacetic acid[288] and dimethyltryptamine[336] increases the excretion of 5-HIA; there is, however, the possibility that this is a derivative hydroxylated in position "6" (ref. 337).

Also a loading with tryptophan increases the excretion of 5-HIA in urine[224,362], particularly in carcinoid patients[362]. However, other authors did not find any change in 5-HIA excretion after tryptophan administration[134]. As will be discussed elsewhere, the administration of 5-HTP results in an increase of 5-HT also in tissues, such as liver, kidney, heart and uterus, which normally contain no measurable amount of the substance[95,351,360]. After administering serotonin by a parenteral route, there is an increase in platelet and tissue 5-HT[120,207,378], but not in intestinal 5-HT[143]. The recovery of 5-HT as 5-HIA in urine is reported for various animal species in Table 14. The subcutaneous injection of 5-HT induces a more lasting variation of 5-HT metabolism than the intraperitoneal administration[120]. 5-HT is absorbed after inhalation by guinea pig but not by rabbit[235].

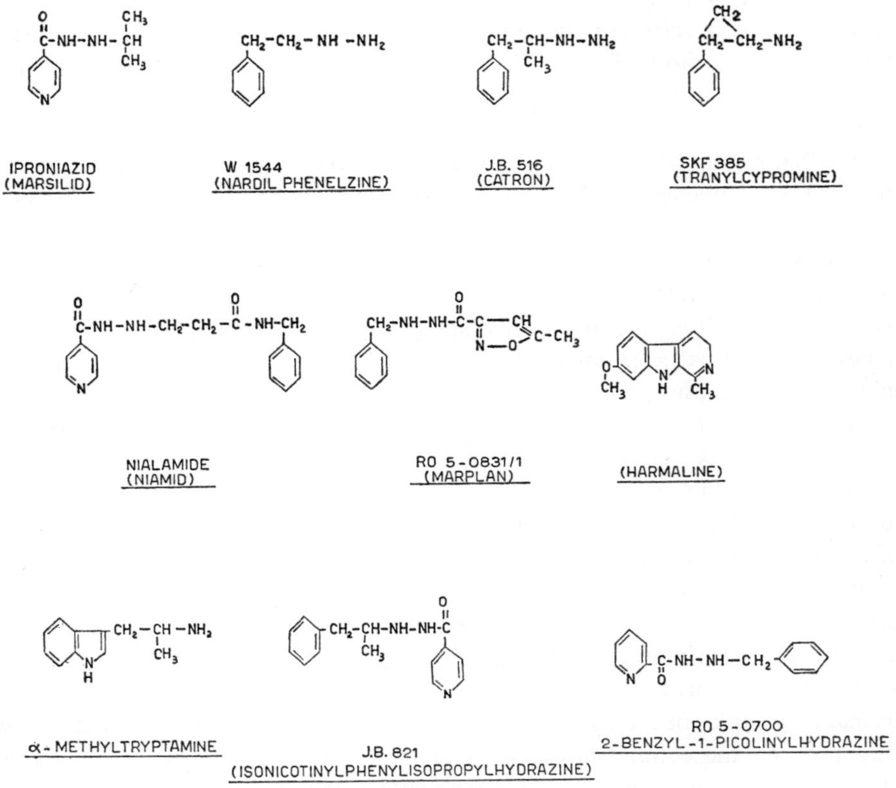

Fig. 21. Structures of MAO inhibitors.

Treatment with α-methyl-DOPA, an inhibitor of decarboxylases[327,328,354], decreases 5-HT in tissues[182,278,328] and excretion of 5-HIA in urine and increases excretion of 5-HTP in carcinoids[251,318]. Similar results were obtained with α-methyl-m-tyrosine[216,278]. This decrease is not specific for 5-HT metabolism, but common to all the amines formed by decarboxylation[251,318]. The block of decarboxylase is however associated with a depletion of the amines[182,278] and particularly of noradrenaline[181a,181b].

The effect of chlorpromazine in decreasing excretion of 5-HIA in carcinoid patients[78,146,223,293] or in animals[162] is related only to a chemical interference with the reaction of identification[316]. Other authors, however, did not find any effect of chlorpromazine on the excretion of 5-HIA[27,298]. Chlorpromazine interferes also with the decarboxylation[153a] of 5-HTP.

STORAGE AND UPTAKE

Serotonin is normally stored intracellularly in a pharmacologically inactive form (form A and form I already suggested several years ago by Erspamer[116]).

In the first investigations carried out with encephalic tissue no specific linkage was found between serotonin and any of the cell constituents[39,191] but later research established that serotonin is stored in the intestinal mucosa of dog[17] and rat[38] mainly in "large granules" sedimenting with mitochondria by ultracentrifugation over 1.6 *M* sucrose[16,157].

The "large granules" are, however, distinct from mitochondria, both morphologically[21] and biochemically[17], but analogous to the chromaffin granules of the adrenal medulla and the histamine-carrying granules of mast cells[17,225]. The concentration of serotonin in the granules is according to Baker[17], about 9 μg/mg of protein nitrogen. The granules carrying serotonin in brain are probably different from the "large granules" of the intestinal mucosa and from the granules storing acetylcholine[392]. Guinea pig brain particles have a diameter[393] of 0.02–0.3 μ and contain with serotonin also an acid phosphatase but not β-glucuronidase[392].

A problem still unsolved is the way in which serotonin is bound in its stored form. Serotonin, as well as histamine, is associated with heparin in the granules of mouse mast cell tumour[21,148,225]. Between serotonin and heparin there is a certain proportionality[164], and this suggests the possibility that a complex may be formed (refs. 209–212). However, there is some disagreement on this last point[296].

Born *et al.* suggested a possible complex between 5-HT and ATP[43,49,50]. In the fraction of the dog small intestine (mucous membrane) rich in 5-HT there is a molar ratio of 3.1 $\pm$ 0.6 between 5-HT and ATP[280]. This ratio does not change during repeated passage through a density gradient[279] and is similar to that found in the chromaffin granules of the adrenal medulla[280]. Also in platelets there is a clear correlation between the concentration of 5-HT and ATP[42,43,48] similar to what was shown for adrenaline[47]. In normal human platelets the ratio between ATP and 5-HT is about 12.8 (ref. 50). However, the occurrence of the complex ATP–5-HT has neither been detected chemically nor found in disrupted platelets[190].

Other experiments show that in rabbit platelets 5-HT is not linked to any of the 8 protein fractions present in platelets[294].

For the elucidation of the nature of the storage of 5-HT, the studies on release and uptake of the amine are important.

A release of 5-HT from platelets or granules occurs in the following conditions: during clotting[43,401,402] quite independently of fibrin formation[173], in the presence of hypotonic solution[16,373,393] or mild detergents[373], after repeated freezing and thawing (ref. 373), or mechanical agitation[393]. The release of 5-HT during early inflammatory processes or anaphylactic reactions is discussed in Chapter 6.

It is interesting because of the possible physiological implications, that alkaline extracts of mammalian tissues are capable of releasing 5-HT from platelets[344] and mast cells[156]. The active principle, called "nephrosin", is present particularly in kidney and releases not only serotonin, but also histamine from platelets[344,345]. Nephrosin is also present in other tissues (in liver and lung but not in spleen, skin and heart[344]) although in a lower concentration than in kidney. The 5-HT release induced by nephrosin is not mediated by thromboplastin or thrombin and does not require the presence of Ca^{2+}. However EDTA, oxalate and citrate inhibit the release[344].

A release of 5-HT has been described in platelets after treatment with thrombin[403] and in granules in the presence of venom lecithinase[393].

The drug more widely used for studies on the release of 5-HT is certainly reserpine which shows its effect both *in vivo* and *in vitro*[71,179,311]. Low temperature prevents the effect of reserpine[44,71,311]. Only Rauwolfia alkaloids with reserpine-like properties cause the release of 5-HT from platelets[71,203,257]. Short-acting reserpine congeners such as the benzoquinolizine derivatives (tetrabenazine)[258], and α-methyl amino acids are releasers of the amines[182,278].

It has been calculated that 1 molecule of reserpine releases about 100 molecules of serotonin[71].

This fact leads to the hypothesis that the effect of reserpine is not a simple chemical displacement of serotonin from the binding sites, but a more complex activity dealing with the mechanism able to maintain serotonin concentrated in platelets. It has been shown that platelets can extract serotonin from the medium[142a,405] even though the outside concentration is far lower than the inside[46,174,194,*]. The existence of a concentrating mechanism (active transport) is supported by the following data: (*a*) 5-HT may be accumulated in platelets against a concentration gradient of 1000:1 (refs. 46, 190). The concentration of 5-HT in platelets does not saturate, in normal conditions, all the binding sites[120]. In fact an increase of 5-HT in platelets may be obtained by giving 5-HTP[360], 5-HT[54,120], monoamine oxidase inhibitors[86,273,312], or spontaneously in carcinoid patients[329]. The degree of 5-HT accumulation in platelets depends on the animal species, and in human platelets may reach a value of 1 mg of $5\text{-HT}/10^8$ platelets[46]. The time required for uptake is, however, relatively slow in comparison with other substances as is shown by these figures: the uptake of 5-HT is of the order of 0.04 μmole/sec/cm^2 of platelet surface, for K^+ it is 0.68 and for glycine 16 (ref. 46); (*b*) at low temperature (0–2°) the uptake of 5-HT is markedly decreased (refs. 44, 378); however, the concentration of 5-HT in platelets still exceeds that in plasma[44]; (*c*) the uptake requires inorganic ions. It has been established that the presence of K^+ increases the uptake of serotonin by platelets[45,379,380]. NH_4^+, Li^+ and Mg^{2+} are inactive in this respect[379]. On the other hand chelating agents tend to decrease the 5-HT uptake[214]; (*d*) the uptake is sensitive to metabolic inhibitors. On this point the results are often contradictory. Sano *et al.*[294,295] reported that glucose and ATP increase the uptake of 5-HT by platelets, but these results were not confirmed by other experiments[190,379]. However, the presence of phosphates stimulates 5-HT uptake when the platelets are suspended in saline[378–380]. In some research it is not always clearly

---

* This property does not seem to be shared by the "granules" rich in serotonin[16]. Also, the platelets cannot take up serotonin in the newborn infant[246].

realized that the concentration of 5-HT in the medium is a critical factor in determining whether one is studying mainly passive diffusion or active transport.

The uptake is inhibited by large concentrations of cyanide[46,294], sodium acetate[46], malonate[294], and fluoride[295,379,380]. 2,4-Dinitrophenol inhibits uptake in large doses[45,294], and at pH 5.7 (ref. 379), but is practically inactive at a pH of 7.4 when the active transport is counteracted by the passive diffusion of 5-HT[379,380]. These results, with the observation that the uptake of 5-HT by platelets occurs also in anaerobiosis[190], indicate that "the oxidative formation of ATP is not a requirement for the uptake of serotonin *in vitro*" (ref. 190).

Reserpine is a strong inhibitor of the uptake of serotonin by platelets[50,172,190,192], and by brain mitochondria[373,374], but this effect is decreased at pH 8.0 when a diffusion-type mechanism appears to predominate[380].

Other compounds inhibit the uptake of 5-HT by platelets by a mechanism not yet elucidated: imipramine[237,*], tryptamine[332], tyramine[332], dopamine[332], ouabain[380], digitoxin[380] and strophanthin K [380]. These compounds differ from reserpine because they share with the alkaloid the capacity to inhibit uptake, but not the property to release serotonin.

---

* Imipramine decreases the uptake of 5-HT also from neoplastic murine mast cells[100].

REFERENCES

1 ADAM, K. R. AND WEISS, C., *J. Exptl. Biol.*, 35 (1958) 39.
2 ADAM, K. R. AND WEISS, C., *Nature*, 183 (1959) 1398.
2a AIRAKSINEN, M. M., *Biochem. Pharmacol.*, 8 (1961) 245.
3 AIRAKSINEN, M. M., PAASONEN, M. K. AND PELTOLA, P., *Ann. Med. Exptl. et Biol. Fenniae (Helsinki)*, 38 (1960) 257.
4 AKCASU, A., AKCASU, M. AND TUMAY, S. B., *Nature*, 187 (1960) 324.
5 AMERIO, A. AND MICELLI, O., *Boll. soc. ital. biol. sper.*, 34 (1958) 1265.
6 AMERIO, A. AND MICELLI, O., *Boll. soc. ital. biol. sper.*, 34 (1958) 1268.
7 AMERIO, A. AND MICELLI, O., *Minerva nefrol.*, 7 (1959) 1.
8 AMIN, A. H., CRAWFORD, T. B. B. AND GADDUM, J. R., *J. Physiol. (London)*, 126 (1954) 596.
9 ANDERSON, J. A., ZIEGLER, M. R. AND DOEDEN, D., *Science*, 127 (1958) 236.
10 ANGERVALL, L., EUERBÄCK, L. AND WESTLING, H., *Experientia*, 16 (1960) 209.
11 ARMSTRONG, M. D. AND ROBINSON, K. S., *Arch. Biochem. Biophys.*, 52 (1954) 287.
12 ASHCROFT, G. W. AND SHARMAN, D. F., *Nature*, 186 (1960) 1050.
13 AXELROD, J., *Science*, 127 (1958) 754.
14 AXELROD, J., INSCOE, J. K. AND TOMKINS, G. M., *J. Biol. Chem.*, 232 (1958) 835.
15 BAGLIONI, C., FASELLA, P., TURANO, C. AND SILIPRANDI, N., *Giorn. biochim.*, 9 (1960) 8.
16 BAKER, R. V., *J. Physiol. (London)*, 142 (1958) 563.
17 BAKER, R. V., *J. Physiol. (London)*, 145 (1959) 473.
18 BALDRIDGE, R. C., BOROFSKY, L., BAIRD, H. III, REICHLE, F. AND BULLOCK, D., *Proc. Soc. Exptl. Biol. Med.*, 100 (1959) 529.
18a BALDRIDGE, T. E., MILLER, L. V., HAVERBACK, B. J. AND BRUNJES, S., *New Engl. J. Med.*, 267 (1962) 421.
19 BALLERINI, G., BERETTA, P. AND LA PAGLIA, S., *Ann. univ. Ferrara*, 2 (1958) 115.
20 BARBEAU, A. AND WILKOFF, L. J., *Can. Med. Assoc. J.*, 80 (1959) 717.
21 BARNETT, R. J., HAGEN, P. AND LEE, F. L., *Biochem. J.*, 69 (1958) 36P.
22 BARON, D. N., DENT, C. E., HARRIS, H., HART, E. W. AND JEPSON, J. B., *Lancet*, 271 (1956) 421.
23 BARTER, R. AND EVERSON-PEARSE, A. G., *Nature*, 172 (1953) 810.
23a BARTER, R. AND EVERSON-PEARSE, A. G., *J. Pathol. Bacteriol.*, 69 (1955) 25.
24 BARTLET, A. L., *Brit. J. Pharmacol.*, 15 (1960) 140.
25 BENDITT, E. P. AND WONG, R. L., *J. Exptl. Med.*, 105 (1957) 509.
26 BERENDES, H., ANDERSON, J. A., PRIGGIE, B., RUTTENBERG, D. AND ZIEGLER, M. R., *Staff Meeting Rept., Univ. Minnesota, Med. Bull.*, 29 (1958) 498.
27 BERGER, F. N., CAMPBELL, G. L., HENDLEY, C. D., LUDWIG, B. J. AND LYNES, T. E., *Ann. N.Y. Acad. Sci.*, 66 (1957) 686.
28 BERTACCINI, G., *Naturwissenschaften*, 45 (1958) 548.
29 BERTACCINI, G., *Ricerca sci.*, 28 (1958) 2261.
30 BERTACCINI, G., *J. Physiol. (London)*, 153 (1960) 239.
31 BERTACCINI, G. AND CHIEPPA, S., *Lancet*, 278 (1960) 881.
32 BERTLER, A., CARLSSON, A. AND ROSENGREN, E., *Naturwissenschaften*, 43 (1956) 521.
33 BERTLER, A., FALCK, B., HILLARP, N. A., ROSENGREN, E. AND TERP, A., *Acta Physiol. Scand.* 47 (1959) 251.
34 BESSMAN, S. P., MERLIS, J. K. AND BORGES, F., *Proc. Soc. Exptl. Biol. Med.*, 95 (1957) 502.
34a BETTENDORF, G., SCHANDER, U. AND SCHMERMUND, H. J., *Klin. Wochschr.*, 37 (1959) 936.
35 BIANCHI, P. A., CROSTI, P. F. AND LUCCHELLI, P. E., *Progr. med.*, 16 (1960) 174.
36 BIGELOW, F. S., *J. Clin. Invest.*, 32 (1953) 555.
37 BILLA, B. AND VALZELLI, L., *Boll. soc. ital. biol. sper.*, 34 (1958) 1404.
38 BLASCHKO, H., in M. HARRINGTON (Ed.), *Hypotensive Drugs*, Pergamon, New York and London, 1956.
38a BOGDANSKI, D. F., BONOMI, L. AND BRODIE, B. B., *Life Sciences*, 1 (1963) 80.
39 BOGDANSKI, D. F., WEISSBACH, H. AND UDENFRIEND, S., *J. Neurochem.*, 1 (1957) 272.
40 BORGES, F. J. AND BESSMAN, S. P., *Proc. Soc. Exptl. Biol. Med.*, 93 (1956) 513.
41 BORGES, F. J., MERLIS, J. K. AND BESSMAN, S. P., *J. Clin. Invest.*, 38 (1959) 715.
42 BORN, G. V. R., *Biochem. J.*, 62 (1956) 33P.
43 BORN, G. V. R., *J. Physiol. (London)*, 133 (1956) 61P.

44 BORN, G. V. R. AND BRICKNELL, J., *J. Physiol. (London)*, 147 (1959) 153.
45 BORN, G. V. R. AND GILLSON, R. E., *J. Physiol. (London)*, 141 (1958) 39P.
46 BORN, G. V. R. AND GILLSON, R. E., *J. Physiol. (London)*, 146 (1959) 472.
47 BORN, G. V. R. AND HORNYKIEWICZ, O., *J. Physiol. (London)*, 136 (1957) 30P.
48 BORN, G. V. R., HORNYKIEWICZ, O. AND STAFFORD, A., *Brit. J. Pharmacol.*, 13 (1958) 411.
49 BORN, G. V. R., INGRAM, G. I. C. AND STACEY, R. S., *J. Physiol. (London)*, 135 (1957) 63P.
50 BORN, G. V. R., INGRAM, G. I. C. AND STACEY, R. S., *Brit. J. Pharmacol.*, 13 (1958) 62.
51 BOWDEN, K., BROWN, B. G. AND BATTY, J. E., *Nature*, 174 (1954) 925.
52 BOYLAND, E., GASSON, J. E. AND WILLIAMS, D. C., *Lancet*, 271 (1956) 975.
53 BRACCO, M. AND CURTI, P. C., *Recenti progr. in med.*, 25 (1958) 425.
54 BRACCO, M., CURTI, P. C. AND GIULIANO, V., *Farmaco (Pavia) (Ed. sci.)*, 11 (1956) 177.
55 BRACCO, M., CURTI, P. C. AND GIULIANO, V., *Experientia*, 12 (1956) 31.
56 BRODIE, B. B., FINGER, K. F., ORLANS, F. B., QUINN, G. P. AND SULSER, F., *J. Pharmacol. Exptl. Therap.*, 129 (1960) 250.
57 BRODIE, B. B., OLIN, J. S., KUNTZMAN, R. G. AND SHORE, P. A., *Science*, 125 (1957) 1293.
58 BRODIE, B. B., SPECTOR, S., KUNTZMAN, R. G. AND SHORE, P. A., *Naturwissenschaften*, 45 (1958) 243.
59 BRODIE, B. B., TOMICH, E. G., KUNTZMAN, R. G. AND SHORE, P. A., *J. Pharmacol. Exptl. Therap.*, 119 (1957) 461.
60 BRUCE, D. W., *Nature*, 188 (1960) 147.
60a BULARD, C. AND LÉOPOLD, C. A., *Compt. rend.*, 247 (1958) 1382.
61 BUMPUS, F. M. AND PAGE, I. H., *J. Biol. Chem.*, 212 (1955) 111.
62 BURKARD, W. P., GEY, K. F. AND PLETSCHER, A., *Biochem. Pharmacol.*, 3 (1960) 249.
63 BURKHALTER, A., COHN, W. H. AND SHORE, P. A., *Biochem. Pharmacol.*, 3 (1960) 328.
64 BUSCAINO, G. A. AND STEFANACHI, L., *A.M.A. Arch. Neurol. Psychiat.*, 80 (1958) 78.
65 CALI, G., DE BONIS, P. AND PIETROPAOLO, C., *Progr. med.*, 15 (1959) 340.
66 CAMBRIDGE, G. W. AND HOLGATE, J. A., *J. Physiol. (London)*, 142 (1958) 53P.
67 CANAL, N. AND MAFFEI-FACCIOLI, A., in P. B. BRADLEY, P. DENIKER AND C. RADOUCO-THOMAS (Eds.), *Neuropsychopharmacology*, Elsevier, Amsterdam, 1959, p. 288.
68 CARGILL-THOMPSON, H. E. C., HARDWICK, D. C. AND WISEMAN, J. M., *J. Physiol. (London)*, 140 (1958) 10P.
69 CARLSSON, A., LINDQVIST, M., MAGNUSSON, T. AND WALDECK, B., *Science*, 127 (1958) 471.
70 CARLSSON, A., ROSENGREN, E., BERTLER, A. AND NILSSON, J., in S. GARATTINI AND V. GHETTI (Eds.), *Psychotropic Drugs*, Elsevier, Amsterdam, 1957, p. 363.
71 CARLSSON, A., SHORE, P. A. AND BRODIE, B. B., *J. Pharmacol. (London)*, 120 (1957) 334.
72 CARTIER, P., MOREAU, J. AND GEFFREY, Y., *Compt. rend. soc. biol.*, 152 (1958) 902.
73 CASS, R., MARSHALL, P. B. AND RILEY, J. P., *J. Physiol. (London)*, 141 (1958) 510.
74 CHADWICK, B. T. AND WILKINSON, J. H., *Biochem. J.*, 68 (1958) 1P.
75 CHADWICK, B. T. AND WILKINSON, J. H., *Biochem. J.*, 70 (1960) 102.
76 CHESHER, G. B. AND COLLIER, O. J., *J. Physiol. (London)*, 130 (1955) 41P.
77 CHESSIN, M., DUBNICK, B., KRAMER, E. R. AND SCOTT, C. C., *Federation Proc.*, 15 (1956) 409.
78 COLE, J. W. AND BERTINO, G. G., *Proc. Soc. Exptl. Biol. Med.*, 93 (1956) 100.
79 COLLIER, H. O. J., in G. P. LEWIS (Ed.), *5-Hydroxytryptamine*, Pergamon, London, 1957, p. 5.
80 COLLIER, H. O. J. AND CHESHER, G. B., *Brit. J. Pharmacol.*, 11 (1956) 186.
81 COOPER, J. R. AND MELCER, I., *Pharmacologist*, 2 (1960) 67.
81a CORNE, J. A. AND GRAHAM, J. D. P., *J. Physiol. (London)*, 135 (1957) 339.
82 CORREALE, P. AND CORTESE, I., *Boll. soc. ital. biol. sper.*, 32 (1956) 525.
83 COSTA, E., *Pharmacologist*, 1 (1959) 82.
84 COSTA, E. AND APRISON, M. H., *Am. J. Physiol.*, 192 (1958) 95.
85 COUPLAND, R. E. AND RILEY, J. F., *Nature*, 187 (1960) 1128.
85a CRAWFORD, M. A., *Lancet*, i (1962) 352.
85b CRAWFORD, N., *Clin. Chim. Acta*, 8 (1963) 39.
86 CROSTI, P. F. AND BIANCHI, P. A., *Atti XXI Congr. Soc. ital. cardiologia*, II (1959) 139.
87 CROSTI, P. F., BIANCHI, P. A. AND LUCCHELLI, P. E., *Haematologica Latina*, 3 (1960) 161.
88 CROUT, J. R. AND SJOERDSMA, A., *J. Pharm. Pharmacol.*, 11 (1959) 190.
89 CROUT, J. R. AND SJOERDSMA, A., *New Engl. J. Med.*, 261 (1959) 23.

89a CRUCHAND, A., GIRARD, J. P., DOMINÉ, E. AND MICHELI, H., *Intern. Arch. Allergy Appl. Immunol.*, 19 (1961) 65.

90 CUENCA, E., KUNTZMAN, E., COSTA, E. AND BRODIE, B. B., *Federation Proc.*, 20 (1961) 322a.

91 CURZON, G., *Arch. Biochem. Biophys.*, 66 (1957) 497.

92 DALGLIESH, C. E. AND DUTTON, R. W., *Biochem. J.*, 65 (1957) 21P.

93 DALGLIESH, C. E. AND DUTTON, R. W., *Brit. J. Cancer*, 11 (1957) 296.

94 DALGLIESH, C. E., TOH, C. C. AND WORK, T. S., *J. Physiol. (London)*, 120 (1953) 298.

95 DAVIDSON, J., SJOERDSMA, A., LOOMIS, L. N. AND UDENFRIEND, S., *J. Clin. Invest.*, 36 (1957) 1594,

96 DAVIS, E. J. AND DO ROPP, R. S., *Biochem. Biophys. Research Communs.*, 2 (1960) 361.

97 DAVISON, A. N., *Bull. soc. chim. biol.*, 40 (1958) 1737.

98 DAVISON, A. N., *Arch. Biochem. Biophys.*, 77 (1958) 368.

99 DAVISON, A. N. AND SANDLER, M., *Nature*, 181 (1958) 186.

100 DAY, S. M. AND GREEN, J. P., *Pharmacologist*, 2 (1960) 68.

101 DAY, S. M. AND GREEN, J. P., *Federation Proc.*, 19 (1960) 133.

102 DE BONIS, P., PIETROPAOLO, C. AND CALÌ, G., *Ricerca sci.*, 29 (1959) 1184.

103 DESPOPOULOS, A., *Federation Proc.*, 15 (1956) 48.

104 DESPOPOULOS, A. AND WEISSBACH, H., *Am. J. Physiol.*, 189 (1957) 548.

105 DICK, M. AND TODRICK, A., *Biochem. J.*, 63 (1956) 17P.

106 DIXON, J. B., *J. Physiol. (London)*, 147 (1959) 144.

107 DONALDSON, R. M., ARABEHETY, J. AND GRAY, S. J., *J. Clin. Invest.*, 38 (1959) 933.

108 DONALDSON, R. M. AND GRAY, S. J., *Gastroenterology*, 36 (1959) 7.

109 DONALDSON, R. M., GRAY, S. J. AND LETSON, V. G., *Lancet*, ii (1959) 1002.

110 DONALDSON JR., R. M., MALKIEL, S. AND GRAY, S. J., *Proc. Soc. Exptl. Biol. Med.*, 103 (1960) 261.

111 DONALDSON, R. M., ZAIDMAN, I. AND GRAY, S. J., *Gastroenterology*, 38 (1960) 937.

112 EBER, O. AND LEMBECK, F., *Pflügers Arch. ges. Physiol.*, 265 (1958) 563.

112a ELIASSON, R., *J. Urol.*, 86 (1961) 676.

113 ENERBÄCK, L., *Endocrinology*, 67 (1960) 717.

113a ENERBÄCK, L. *Nature*, 197 (1963) 610.

114 ERSPAMER, V., *Arch. exptl. Pathol. Pharmakol., Naunyn-Schmiedeberg's*, 196 (1940) 343.

115 ERSPAMER, V., *Virchows Arch. pathol. Anat. u. Physiol.*, 310 (1943) 59.

116 ERSPAMER, V., *Rend. sci. Farmitalia*, 1 (1954) 1.

117 ERSPAMER, V., *Experientia*, 10 (1954) 471.

118 ERSPAMER, V., *Pharmacol. Revs.*, 6 (1954) 425.

119 ERSPAMER, V., *J. Physiol. (London)*, 127 (1955) 118.

120 ERSPAMER, V., *J. Physiol. (London)*, 133 (1956) 1.

121 ERSPAMER, V., *Naturwissenschaften*, 43 (1956) 61.

122 ERSPAMER, V., *Lancet*, 270 (1956) 511.

123 ERSPAMER, V., *Experientia*, 12 (1956) 63.

124 ERSPAMER, V., *Z. Vitamin- Hormon- u. Fermentforsch.*, 9 (1957) 74.

125 ERSPAMER, V., *XXI Congresso intern. cienc. fisiol. Buenos Ayres*, 9–15 August, 1959, p. 216.

126 ERSPAMER, V. AND CICERI, C., *Experientia*, 13 (1957) 87.

127 ERSPAMER, V. AND FAUSTINI, F., *Naturwissenschaften*, 40 (1953) 317.

128 ERSPAMER, V., GLÄSSER, A., NOBILI, B. M. AND PASINI, C., *Experientia*, 16 (1960) 506.

129 ERSPAMER, V. AND TESTINI, A., *J. Pharm. Pharmacol.*, 11 (1959) 618.

130 EVERSON-PEARSE, A. G., *Riv. istochim.*, 2 (1956) 103.

131 FALCK, B., HILLARP, N. A. AND TORP, A., *J. Histochem. Cytochem.*, 7 (1959) 323.

132 FELDBERG, W. AND TOH, C. C., *J. Physiol. (London)*, 119 (1953) 352.

133 FELDSTEIN, A., FREEMAN, H., HOPE, J. M., DILNER, I. M. AND HOAGLAND, H., *Am. J. Psychiat.*, 116 (1959) 219.

134 FELDSTEIN, A., HOAGLAND, H. AND FREEMAN, H., *Science*, 128 (1958) 358.

135 FELDSTEIN, A., HOAGLAND, H. AND FREEMAN, H., *J. Nervous Mental Disease*, 129 (1959) 62.

136 FELDSTEIN, A., HOAGLAND, H. AND FREEMAN, H., *Science*, 130 (1959) 500.

137 FELDSTEIN, A., HOAGLAND, H. AND FREEMAN, H., *Science*, 132 (1960) 1659.

138 FELLMAN, J. H., *Proc. Soc. Exptl. Biol. Med.*, 93 (1956) 413.

139 FERRARI, V., CAMPAGNARI, F. AND GUIDA, A., *Minerva med. terap.*, 46 (1955) 119.

140 FERRARI, V. AND CASTELLI, B., *Rass. fisiopatol. clin. e terap.*, 26 (1954) 689.

141   FERRARI, V., GINOULHIAC, E. AND TENCONI, L. T., *Boll. soc. ital. biol. sper.*, 33 (1957) 1418.
142   FERRARI, R., GLASSER, A. AND MANTEGAZZINI, P., *Ateneo parmense*, 31 (1960) 515.
142a FICHERA, C., *Boll. soc. ital. biol. sper.*, 38 (1962) 1283.
143   FIORE-DONATI, L., CHIECO-BIANCHI, L. AND BERTACCINI, G., *Arch. intern. pharmacodyn.*, 123 (1959) 115.
144   FISCHER, P. AND LECOMTE, J., *Compt. rend. soc. biol.*, 150 (1956) 1026.
145   FOY, J. M. AND PARRATT, J. R., *J. Pharm. Pharmacol.*, 12 (1960) 360.
145a FOY, J. M. AND PARRATT, J. R., *J. Pharm. Pharmacol.*, 13 (1961) 382.
146   FRASER, H. P., EISEMMAN, A. J. AND BROOKS, J. W., *Federation Proc.*, 16 (1957) 298.
147   FRICK, M. H., *Nature*, 187 (1960) 609.
147a FRICK, M. H. AND VIRKKULA, L., *Ann. Med. Exptl. Biol. Fenniae (Helsinki)*, 39 (1961) 101.
147b FRYDMAN, B. AND DENLOFEN, V., *Experientia*, 17 (1961) 545.
148   FURTH, J., HAGEN, P. AND HIRSCH, E. I., *Proc. Soc. Exptl. Biol. Med.*, 95 (1957) 824.
149   GADDUM, J. H., HEBB, C. O., SILVER, A. AND SWAN, A. A., *Quart. J. Exptl. Physiol.*, 38 (1953) 255.
150   GARATTINI, S., LAMESTA, L., MORTARI, A. AND VALZELLI, L., *J. Pharm. Pharmacol.*, 13 (1961) 548.
151   GARATTINI, S., KATO, R. AND VALZELLI, L., *Experientia*, 16 (1960) 120.
152   GARATTINI, S., MORTARI, A., VALSECCHI, A. AND VALZELLI, L., *Nature*, 183 (1959) 1273.
153   GARVEN, J. D., *Brit. J. Pharmacol.*, 11 (1956) 66.
153a GEY, K. F., BURKARD, W. P. AND PLETSCHER, A., *Biochem. Pharmacol.*, 8 (1961) 383.
154   GIARMAN, N. J., *Federation Proc.*, 15 (1956) 428/1393.
155   GIARMAN, N. J., FREEDMAN, D. X. AND PICARD, AMI L., *Nature*, 186 (1960) 480.
156   GIARMAN, N. J., POTTER, L. T. AND DAY, M., *Experientia*, 16 (1960) 492.
157   GIARMAN, N. J. AND SCHANBERG, S., *Biochem. Pharmacol.*, 1 (1958) 301.
158   GOGERTY, J. M. AND HORITA, A., *J. Pharmacol. Exptl. Therap.*, 129 (1960) 357.
159   GOLDSTEIN, M., FRIEDHOFF, A. J. AND SIMMONS, C., in P. B. BRADLEY, P. DENIKER AND C. RADOUCO-THOMAS (Eds.), *Neuropsychopharmacology*, Elsevier, Amsterdam, 1959, p. 572.
160   GOLDSTEIN, M., FRIEDHOFF, A. J., SIMMONS, C. AND PROCHOROFF, N. N., *Experientia*, 15 (1959) 254.
161   GOMORI, G., *Arch. Pathol.*, 45 (1948) 4.
161a GOTTFRIES, C. G. AND MAGNUSSON, T., *Acta Psychiat. Scand.*, 38 (1962) 223.
162   GOTTLIEB, L. S., HAZEL, M., BROITMAN, S. AND ZAMCHECK, M., *Federation Proc.*, 19 (1960) 181.
163   GRANOWITZ, E. AND PLETSCHER, A., *Helv. Med. Acta*, 24 (1957) 21.
164   GREEN, J. P. AND DAY, M., *Biochem. Pharmacol.*, 3 (1960) 190.
165   GREEN, H. AND ERICKSON, R. W., *J. Pharmacol. Exptl. Therap.*, 129 (1960) 237.
166   GREEN, J. P., PAASONEN, M. K. AND GIARMAN, N. J., *Proc. Soc. Exptl. Biol. Med.*, 94 (1957) 428.
166a GREIG, M. E. AND GIBBSON, A. J., *Arch. intern. pharmacodyn.*, 136 (1962) 147.
167   GREIG, M. E., WALK, R. A. AND GIBBSON, A. J., *J. Pharmacol. Exptl. Therap.*, 127 (1959) 110.
168   HAGEN, P. AND LEE, F. L., *J. Physiol. (London)*, 143 (1958) 7P.
169   HAGEN, P., WEINER, N., ONO, S. AND LEE, F. L., *J. Pharmacol. Exptl. Therap.*, 130 (1960) 9.
170   HANS, M. J. AND VAN DEN DRIESSCHE, R., *Compt. rend. soc. biol.*, 152 (1958) 1273.
171   HANSON, A., *Naturwissenschaften*, 45 (1958) 423.
172   HARDISTY, R. M., INGRAM, G. I. C. AND STACEY, R. S., *Experientia*, 12 (1956) 424.
173   HARDISTY, R. M. AND PINNINGER, J. L., *Brit. J. Haematol.*, 2 (1956) 139.
174   HARDISTY, R. M. AND STACEY, R. S., *J. Physiol. (London)*, 130 (1955) 711.
175   HARDISTY, R. M. AND STACEY, R. S., *Brit. J. Haematol.*, 3 (1957) 292.
176   HAVERBACK, B. J., *Clin. Research*, 6 (1958) 57.
177   HAVERBACK, B. J. AND DAVIDSON, J. D., *Gastroenterology*, 35 (1958) 570.
178   HAVERBACK, B. J., DYCE, B. AND THOMAS, H. V., *New Engl. J. Med.*, 262 (1960) 754.
179   HAVERBACK, B. J., DUTCHER, T. F., SHORE, P. A., TOMICH, E. G., TERRY, L. L. AND BRODIE, B. B., *New Engl. J. Med.*, 256 (1957) 343.
180   HAVERBACK, B. J., SHORE, P. A., TOMICH, E. G. AND BRODIE, B. B., *Federation Proc.*, 15 (1956) 434.
181   HAVERBACK, B. J., SJOERDSMA, A. AND TERRY, L. L., *New Engl. J. Med.*, 255 (1956) 270.
181a HESS, S. M., CONNAMACHER, R. H., OZAKI, M. AND UDENFRIEND, S., *J. Pharmacol. Exptl. Therap.*, 134 (1961) 129.
181b HESS, S. M., CONNAMACHER, R. H. AND UDENFRIEND, S., *Federation Proc.*, 20 (1961) 344a.

182 HESS, S. M., OZAKI, M. AND UDENFRIEND, S., *Pharmacologist*, 2 (1960) 81.
183 HESS, S. M., REDFIELD, P. G. AND UDENFRIEND, S., *J. Pharmacol. Exptl. Therap.*, 127 (1959) 178.
184 HESS, S. M., WEISSBACH, H., REDFIELD, B. G. AND UDENFRIEND, S., *J. Pharmacol. Exptl. Therap.*, 124 (1958) 189.
184a HEYSSEL, R. M., *J. Clin. Invest.*, 40 (1961) 2134.
185 HOPE, D. B. AND SMITH, A. D., *Biochem. J.*, 74 (1960) 101.
186 HORITA, A., *J. Pharmacol. Exptl. Therap.*, 122 (1958) 176.
187 HORITA, A., *Federation Proc.*, 19 (1960) 279.
188 HORITA, A., MCGRATH, W. R. AND ELTHERINGTON, L. G., *Federation Proc.*, 18 (1959) 403.
189 HUANG, I., TANNENBAUM, S. AND HSIA, D. Y. Y., *Nature*, 186 (1960) 717.
189a HUBER, W. G. AND LINK, R. P., *Proc. Soc. Exptl. Biol. Med.*, 109 (1962) 118.
190 HUGHES, F. B. AND BRODIE, B. B., *J. Pharmacol. Exptl. Therap.*, 127 (1959) 96.
191 HUGHES, F. B., KUNTZMAN, R. AND SHORE, P. A., *Federation Proc.*, 16 (1957) 308.
192 HUGHES, F. B., SHORE, P. A. AND BRODIE, B. B., *Experientia*, 14 (1958) 178.
193 HUMPHREY, J. H. AND JAQUES, R., *J. Physiol. (London)*, 124 (1954) 305.
194 HUMPHREY, J. H. AND TOH, C. C., *J. Physiol. (London)*, 124 (1954) 300.
195 JACOBSON, W., in G. P. LEWIS (Ed.), *5-Hydroxytryptamine*, Pergamon, London, 1957, p. 38.
196 JAQUES, R. AND SCHACHTER, M., *Brit. J. Pharmacol.*, 9 (1954) 53.
197 JEPSON, J. B., *Biochem. J.*, 64 (1956) 14.
198 JEPSON, J. B., in G. P. LEWIS (Ed.), *5-Hydroxytryptamine*, Pergamon, London, 1957, p. 37.
199 JEPSON, J. B., LOVENBERG, W., SJOERDSMA, A., OATES, J., ZALTZMAN, P. AND UDENFRIEND, S., *Pharmacologist*, 1 (1959) 74.
200 JOHNSEN, H. A., SMITH, R. E. AND SIMON, W., *Clin. Research*, 6 (1958) 268.
201 JOSEPH, M. AND TAYLOR, R. R., *Brit. Med. J.*, 20 (1960) 568.
202 KAKIMOTO, Y. AND ARMSTRONG, M. D., *Federation Proc.*, 19 (1960) 295.
203 KÄRKI, N. T. AND PAASONEN, M. K., *Nature*, 185 (1960) 109.
204 KÄRKI, N. T., PAASONEN, M. K. AND PELTOLA, P., *Ann. Med. Exptl. et Biol. Fenniae (Helsinki)*, 38 (1960) 231.
205 KATO, R., *J. Neurochem.*, 5 (1960) 202.
206 KATSH, S., *J. Urol.*, 81 (1959) 570.
207 KEGLEVIC-BROVET, D., SUPEK, Z., KVEDER, S., ISKRIC, S. AND KECKES, S., *Proc. 2nd U.N. Intern. Conf. Peaceful Uses of Atomic Energy, Geneve*, 15 (1958) 481.
208 KEGLEVIC, D., SUPEK, Z., KVEDER, S., ISKRIC, S., KECKES, S. AND KISIC, A., *Biochem. J.*, 73 (1959) 53.
209 KELLER, R., *Experientia*, 13 (1957) 112.
210 KELLER, R., *Experientia*, 14 (1958) 181.
211 KELLER, R., *Arzneimittel-Forsch.*, 8 (1958) 390.
212 KELLER, R., *Schweiz. med. Wochschr.*, 90 (1960) 503.
213 KERBY, G. P. AND TAYLOR, S. M., *Proc. Soc. Exptl. Biol. Med.*, 102 (1959) 45.
214 KERBY, G. P. AND TAYLOR, S. M., *J. Clin. Invest.*, 40 (1961) 44.
214a KERKUT, G. A. AND COTTRELL, G. A., *Comp. Biochem. Physiol.*, 8 (1963) 53.
214b KIRBERGER, E., *Deut. med. Wochschr.*, 87 (1962) 929.
214c KERP, L. AND KASEMIR, H., *Arch. exptl. Pathol. Pharmakol., Naunyn-Schmiedeberg's*, 243 (1962) 187.
214d KIRBERGER, E. AND BRAUN, L., *Biochim. Biophys. Acta*, 49 (1961) 391.
214e KIRBERGER, E. AND MOHS, U., *Med. Klin. (Munich)*, 56 (1961) 2211.
215 KOWLESSAR, O. D., WILLIAMS, R. C., LAW, D. H. AND SLEISENGER, M. H., *New Engl. J. Med.*, 259 (1958) 340.
216 KUNTZMAN, E., COSTA, E., GESSA, G. L., HIRSCH, C. AND BRODIE, B. B., *Federation Proc.*, 20 (1961) 308b.
217 KURIAKI, K. AND INOUE, T., *Compt. rend. soc. biol.*, 150 (1956) 1835.
217a KVEDER, S., ISKRIĆ, S. AND KEGLEVIĆ, D., *Biochem. J.*, 85 (1962) 447.
218 LA BARRE, J. AND HANS, M. J., *Compt. rend. soc. biol.*, 154 (1960) 474.
219 LABORIT, H., NIASSAUT, P., BROUSOLLE, B. AND JOUANY, J. M., *Presse méd.*, 67 (1959) 927.
220 LAGUNOFF, D. AND BENDITT, E. P., *Am. J. Physiol.*, 196 (1959) 993.
221 LAGUNOFF, D., LAM, K. B., ROEPER, E. AND BENDITT, E. P., *Federation Proc.*, 16 (1957) 363.
222 LAMIERI, C., *Boll. soc. ital. biol. sper.*, 36 (1960) 1353.

223 LANGEMANN, H. AND GOERRE, J., *Schweiz. med. Wochschr.*, 87 (1957) 607.
224 LAUER, J. W., INSKIP, W. M., BERNSHON, J. AND ZELLER, E. A., *A.M.A. Arch. Neurol. Psychiat.*, 80 (1958) 122.
225 LEE, F. L., HAGEN, P. AND FURTH, J., *Federation Proc.*, 17 (1958) 387.
226 LEMBECK, F., *Nature*, 172 (1953) 910.
227 LEMBECK, F. AND NEUHOLD, K., *Arch. exptl. Pathol. Pharmakol., Naunyn-Schmiedeberg's*, 226 (1955) 456.
228 LEWIS, C. E., *Proc. Soc. Exptl. Biol. Med.*, 99 (1958) 523.
229 LIEBECQ-HUTTER, S. AND BACQ, Z. M., *Experientia*, 14 (1958) 180.
230 LIEBECQ-HUTTER, S., RENSON, J. F. AND FISHER, P., *Arch. intern. physiol. biochim.*, 67 (1959) 144.
231 LILLIE, R. D., *J. Histochem. Cytochem.*, 4 (1956) 120.
232 LILLIE, R. D., BURTNER, H. J. AND GRECO HEUSON, J. P., *J. Histochem. Cytochem.*, 1 (1953) 154.
232a LINÉT, A., KREJČÍ, I. AND HÁVA, M., *Acta Biol. Med. Ger.*, 9 (1962) 158.
233 LISON, L., *Histochimie et Cytochimie Animales*, Gauthier-Villars, Paris, 1953.
233a MAGALINI, S. I. AND STEFANINI, M., *Proc. Soc. Exptl. Biol. Med.*, 92 (1956) 788.
233b MANN, T., *Biol. Bull.*, 119 (1960) 354.
234 MANN, T., *Nature*, 188 (1960) 941.
234a MANN, T., SEAMARK, R. F. AND SHARMAN, D. F., *Brit. J. Pharmacol.*, 17 (1961) 208.
235 MARIANI, L., VALSECCHI, A. AND ZUANAZZI, G. F., *Boll. soc. ital. biol. sper.*, 134 (1958) 1409.
236 MARKS, B. H., SERGEN, R. W. AND HAYES, E. R., *Proc. Soc. Exptl. Biol. Med.*, 97 (1958) 413.
237 MARSHALL, E. F., STIRLING, G. S., TAIT, A. C. AND TODRICK, A., *Brit. J. Pharmacol.*, 15 (1960) 35.
238 MARSHALL, P. B., *J. Pharm. Pharmacol.*, 10 (1958) 781.
239 MARSHALL, P. B., *J. Pharm. Pharmacol.*, 11 (1959) 639.
240 MATHIAS, A. P., ROSS, D. M. AND SCHACHTER, M., *Nature*, 180 (1957) 658.
241 MATHIAS, A. P., ROSS, D. M. AND SCHACHTER, M., *J. Physiol. (London)*, 142 (1958) 56P.
242 MATHIAS, A. P., ROSS, D. M. AND SCHACHTER, M., *J. Physiol. (London)*, 151 (1960) 296.
243 McISAAC, W. M. AND PAGE, I. H., *Science*, 128 (1958) 537.
244 McISAAC, W. M. AND PAGE, I. H., *J. Biol. Chem.*, 234 (1959) 858.
245 MEDAKOVIC, M. AND SPUZIC, I., *Nature*, 183 (1959) 1685.
245a MELLINKOFF, S., CRADDOCK, C., FRANKLAND, M., KENDRICKS, F. AND GREIPEL, M., *Am. J. Digest. Diseases*, 7 (1962) 347.
246 MITCHELL, R. G. AND CASS, R., *J. Clin. Invest.*, 38 (1959) 595.
247 MUSSINI, E. AND VALZELLI, L., *Atti soc. lombarda sci. med. e biol.*, 13 (1958) 310.
248 NAESS, K. AND SCHANCHE, S., *Nature*, 177 (1956) 1130.
249 NAKAI, K., *Nature*, 181 (1958) 1734.
250 NAKAI, K., *Jap. J. Pharmacol.*, 8 (1958) 63.
251 OATES, J. A., GILLESPIE, L., UDENFRIEND, S. AND SJOERDSMA, A., *Science*, 131 (1960) 1890.
252 OATES, J. A. AND ZALTZMAN, P. Z., *Ann. N.Y. Acad. Sci.*, 80 (1959) 977.
253 ONE, S. AND HAGEN, P., *Nature*, 184 (1959) Suppl. 15, 1143.
254 ORNESI, A., *Thesis*, Università di Milano, 1960.
255 OZAKI, M., WEISSBACH, H., OZAKI, A., WITKOP, B. AND UDENFRIEND, S., *J. Med. Pharm. Chem.*, 2 (1960) 591.
256 PAASONEN, M. K., *Experientia*, 14 (1958) 95.
257 PAASONEN, M. K. AND KÄRKI, N. T., *Brit. J. Pharmacol.*, 14 (1959) 164.
258 PAASONEN, M. K. AND PLETSCHER, A., *Experientia*, 15 (1959) 477.
259 PAGE, I. H., CORCORAN, A. C., UDENFRIEND, S., SJOERDSMA, A. AND WEISSBACH, H., *Lancet*, 268 (1955) 198.
260 PARE, C. M. B., SANDLER, M. AND STACEY, R. S., *Lancet*, 272 (1957) 551.
261 PARE, C. M. B., SANDLER, M. AND STACEY, R. S., *Lancet*, 275 (1958) 1099.
262 PARRATT, J. R. AND WEST, G. B., *J. Physiol. (London)*, 137 (1957) 617.
263 PASSERI, M., *Boll. soc. ital. biol. sper.*, 36 (1960) 1355.
264 PEPEU, G. AND GIARMAN, N. J., *Pharmacologist*, 2 (1960) 91.
265 PHILLIPS, J. H., *Nature*, 178 (1956) 932.
266 PHILLIPS, J. H., BURCH, G. E. AND HIBBS, R. G., *Circulation Research*, 8 (1960) 692.
267 PICCINELLI, D., *Arch. intern. pharmacodyn.*, 117 (1958) 452.
268 PIETROPAOLO, C., DE BONIS, P. AND CALÌ, G., *Ricerca sci.*, 29 (1959) 1223.
269 PIGEAUD, H., NELKEN, S. AND BÉTHOUX, R., *Presse méd.*, 68 (1960) 170.

270 PLETSCHER, A., *Experientia*, 12 (1956) 479.
271 PLETSCHER, A., *Helv. Physiol. Pharmacol. Acta*, 14 (1956) C76.
272 PLETSCHER, A., *Science*, 126 (1957) 507.
273 PLETSCHER, A. AND BERNSTEIN, A., *Nature*, 181 (1958) 1133.
274 PLETSCHER, A., BESENDORF, H. AND BÄCHTOLD, H. P., *Arch. exptl. Pathol. Pharmakol., Naunyn-Schmiedeberg's*, 232 (1958) 499.
275 PLETSCHER, A. AND BESENDORF, H., *Experientia*, 15 (1959) 25.
276 PLETSCHER, A. AND GEY, K. F., *Science*, 128 (1958) 900.
277 POLLI, E. E., BIANCHI, P. A. AND CROSTI, P. F., *Haematol. Latina (Milan)*, 1 (1958) 250.
278 PORTER, C. C., TOTARO, J. A. AND LEIBY, C. M., *Pharmacologist*, 2 (1960) 81.
279 PRUSOFF, W. H., *Brit. J. Pharmacol.*, 15 (1960) 520.
280 PRUSOFF, W. H., *Biochem. J.*, 76 (1960) 65P.
281 PUENTE-DUANY, G. A., RIEMER, W. E. AND MIALE, J. B., *Proc. Soc. Exptl. Biol. Med.*, 98 (1958) 499.
282 QUINN, G. P., SHORE, P. A. AND BRODIE, B. B., *J. Pharmacol. Exptl. Therap.*, 127 (1959) 103.
283 RAND, M. AND REID, G., *Nature*, 168 (1951) 385.
284 RILEY, J. F., *Experientia*, 14 (1958) 141.
284a ROBERTSON, J. I. S. AND ANDREWS, T. M., *Lancet*, i (1961) 578.
285 RODNIGHT, R., *Biochem. J.*, 64 (1956) 621.
285a ROSENBERG, J. C. AND DAVIS, R. B., *J. Surg. Res.*, 2 (1962) 337.
286 ROSENBERG, J. C., DAVIS, R., MORAN, W. H. AND ZIMMERMANN, B., *Federation Proc.*, 18 (1959) 503.
287 ROWE, R. P., BLOOM, B. M., P'AN, S. Y. AND FINGER, K. F., *Proc. Soc. Exptl. Biol. Med.*, 101 (1959) 832.
288 RYSANEK, K. AND VITEK, V., *Experientia*, 15 (1959) 217.
289 SACHS, E., *J. Neurosurg.*, 14 (1957) 22.
290 SANDLER, M. AND CLOSE, H. G., *Lancet*, 277 (1959) 316.
291 SANDLER, M. AND SNOW, P. J. D., *Lancet*, 274 (1958) 137.
292 SANDLER, M. AND WEST, G. B., *J. Physiol. (London)*, 140 (1958) 9P.
293 SANO, I., KAKIMOTO, Y., OKAMOTO, T., NAKAJIMA, H. AND KUDO, Y., *Schweiz. med. Wochschr.*, 87 (1957) 214.
294 SANO, I., KAKIMOTO, Y. AND TANIGUKI, K., *Am. J. Physiol.*, 195 (1958) 495.
295 SANO, I., KAKIMOTO, Y., TANIGUKI, K. AND TAKESADA, M., *Am. J. Physiol.*, 197 (1959) 81.
296 SANYAL, R. K. AND WEST, G. B., *J. Pharm. Pharmacol.*, 11 (1959) 548.
297 SARKAR, S., BANERJEE, R., ISE, M. S. AND ZELLER, E. A., *Helv. Chim. Acta*, 43 (1960) 439.
298 SAUER, W. G., DEARING, W. H., FLOCK, E. V., WAUGH, J. M., DOCKERTY, M. B. AND ROTH, G. M., *Gastroenterology*, 34 (1958) 216.
299 SCHANBERG, S. AND GIARMAN, N. J., *Biochim. Biophys. Acta*, 41 (1960) 556.
299a SCHIEVELBEIN, H., SURBERG, U. AND WERLE, E., *Klin. Wochschr.*, 40 (1962) 52.
300 SCHINDLER, R., *Biochem. Pharmacol.*, 1 (1958) 323.
301 SCHINDLER, R., DAY, S. M. AND FISCHER, G. A., *Cancer Research*, 19 (1959) 47.
302 SCHMID, R., AXELROD, J., HAMMAKER, L. AND SWARM, R. L., *J. Clin. Invest.*, 37 (1958) 1123.
303 SCHMID, E., KINZELMEIER, H. AND SENG, I., *Experientia*, 15 (1959) 230.
304 SCHMID, E., SCHEIFFARTH, F. AND ZICHA, L., *Lancet*, 277 (1959) 354.
305 SCHWARTZ, D. E., PLETSCHER, A., GEY, K. F. AND RIEDER, J., *Helv. Physiol. Acta*, 18 (1960) 10.
306 SERRA, U., *Folia Endocrinol. (Pisa)*, 11 (1958) 410.
307 SERRA, U., *Folia Endocrinol. (Pisa)*, 12 (1959) 219.
308 SHAW, K. N. F., REDLICH, D., WRIGHT, S. W. AND JEPSON, J. B., *Federation Proc.*, 19 (1960) 194.
309 SCHIMIZU, N. AND MORIKAWA, N., *Nature*, 184 (1959) 650.
310 SHORE, P. A. AND BRODIE, B. B., *Science*, 127 (1958) 704.
311 SHORE, P. A., CARLSSON, A. AND BRODIE, B. B., *Federation Proc.*, 15 (1956) 483.
312 SHORE, P. A., GILLESPIE, L., SPECTOR, S. AND PROCKOP, D., *Naturwissenschaften*, 45 (1958) 340.
314 SHORE, P. A., SILVER, S. L. AND BRODIE, B. B., *Science*, 122 (1955) 284.
315 SIGEL, B., POTASH, I. M. AND WOLCOTT, M. W., *Proc. Soc. Exptl. Biol. Med.*, 105 (1960) 237.
315a SINHA, S. N., SANYAL, R. K. AND SINHA, Y. K., *Indian J. Med. Research*, 49 (1961) 681.
316 SJOERDSMA, A., personal communication.
317 SJOERDSMA, A., *New Engl. J. Med.*, 261 (1959) 231.

318 SJOERDSMA, A., *Ann. N.Y. Acad. Sci.*, 88 (1960) 933.

319 SJOERDSMA, A., GILLESPIE, L. A. AND UDENFRIEND, S., *Lancet*, 275 (1958) 159.

319a SJOERDSMA, A., LOVENBERG, W., OATES, J. A., CROUT, J. R. AND UDENFRIEND, S., *Science*, 130 (1959) 225.

320 SJOERDSMA, A., OATES, J. A., ZALTZMAN, P. AND UDENFRIEND, S., *J. Pharmacol. Exptl. Therap.*, 126 (1959) 217.

321 SJOERDSMA, A., SMITH, T. E., STEVENSON, T. D. AND UDENFRIEND, S., *Proc. Soc. Exptl. Biol. Med.*, 89 (1955) 36.

322 SJOERDSMA, A. AND UDENFRIEND, S., *J. Clin. Invest.*, 34 (1955) 914.

323 SJOERDSMA, A., WAALKES, T. P. AND WEISSBACH, H., *Science*, 125 (1957) 1202.

324 SJOERDSMA, A., WEISSBACH, H., TERRY, L. L. AND UDENFRIEND, S., *Am. J. Med.*, 23 (1957) 5.

325 SJOERDSMA, A., WEISSBACH, H. AND UDENFRIEND, S., *Am. J. Med.*, 20 (1956) 520.

326 SMITH, A. N., NYHUS, L. M., DALGLIESH, C. E., DUTTON, R. W., LENNOX, B. AND MACFARLANE, P. S., *Scot. Med. J.*, 2 (1957) 24.

327 SMITH, S. E., *J. Physiol. (London)*, 148 (1959) 18P.

328 SMITH, S. E., *Brit. J. Pharmacol.*, 15 (1960) 319.

329 SNOW, P. J. D., LENNARD-JONES, J. E., CURZON, G. AND STACEY, R. S., *Lancet*, 269 (1955) 1004.

330 SPECTOR, S., SHORE, P. A. AND BRODIE, B. B., *Science*, 132 (1960) 735.

331 STACEY, R. S., *J. Physiol. (London)*, 132 (1956) 39P.

332 STACEY, R. S., personal communication, *Joint Meeting of British and Scandinavian Pharmacological Soc., Copenhagen*, July 1960.

333 STACEY, R. S. AND SULLIVAN, T. J., *J. Physiol. (London)*, 137 (1957) 63P.

334 SULLENBERGER, J. W., ANLYAN, W. G. AND WEAVER, W. T., *Surgery*, 46 (1959) 22.

335 SULLIVAN, T. J., *Brit. J. Pharmacol.*, 15 (1960) 513.

336 SZARA, S., *Experientia*, 12 (1956) 441.

337 SZARA, S. AND AXELROD, J., *Experientia*, 15 (1959) 216.

338 TABACHNICK, I. I. A. AND RUBIN, A. A., *Proc. Soc. Exptl. Biol. Med.*, 101 (1959) 435.

339 TASHIAN, R. E., *Proc. Soc. Exptl. Biol. Med.*, 103 (1960) 407.

340 TAYLOR, J. D., WYKES, A. A., GLADISH, Y. C. AND MARTIN, W. B., *Nature*, 187 (1960) 941.

341 TITAS, E. AND UDENFRIEND, S., *Federation Proc.*, 13 (1954) 411.

342 TODRICK, A., DICK, M. AND TAIT, A. C., *Brit. Med. J.*, 1 (1958) 496.

343 TOH, C. C., *J. Physiol. (London)*, 126 (1954) 248.

344 TOH, C. C., *J. Physiol. (London)*, 133 (1956) 402.

345 TOH, C. C., *J. Physiol. (London)*, 138 (1957) 488.

346 TOWSEND, E. E., KATIS, J. G. AND SOURKES, T. L., *Federation Proc.*, 17 (1958) 324.

347 TWAROG, B. M., *J. Cellular Comp. Physiol.*, 44 (1954) 141.

348 TWAROG, B. M. AND PAGE, I. H., *Am. J. Physiol.*, 175 (1953) 157.

349 TYLER JR., V. E., *Science*, 128 (1958) 718.

350 UDENFRIEND, S., in *Biochemistry of the Central Nervous System*, Pergamon, London, 1959, p. 301.

351 UDENFRIEND, S., BOGDANSKI, D. F. AND WEISSBACH, H., *Federation Proc.*, 15 (1956) 493.

352 UDENFRIEND, S., BOGDANSKI, D. F. AND WEISSBACH, H., *Ann. N.Y. Acad. Sci.*, 66 (1957) 602.

353 UDENFRIEND, S., LOVENBERG, W. AND SJOERDSMA, A., *Arch. Biochem. Biophys.*, 85 (1959) 487.

354 UDENFRIEND, S., LOVENBERG, W. M. AND WEISSBACH, H., *Federation Proc.*, 19 (1960) 7.

355 UDENFRIEND, S., TITUS, E. AND WEISSBACH, H., *J. Biol. Chem.*, 216 (1955) 499.

356 UDENFRIEND, S., TITUS, E., WEISSBACH, H. AND PETERSON, R. E., *J. Biol. Chem.*, 219 (1956) 335.

357 UDENFRIEND, S. AND WEISSBACH, H., *Federation Proc.*, 13 (1954) 412.

358 UDENFRIEND, S. AND WEISSBACH, H., *Proc. Soc. Exptl. Biol. Med.*, 97 (1958) 748.

359 UDENFRIEND, S., WEISSBACH, H. AND BOGDANSKI, D. F., *Ann. N.Y. Acad. Sci.*, 66 (1957) 602.

360 UDENFRIEND, S., WEISSBACH, H. AND BOGDANSKI, D. F., *J. Biol. Chem.*, 224 (1957) 803.

361 UDENFRIEND, S., WEISSBACH, H. AND BOGDANSKI, D. F., *J. Pharmacol. Exptl. Therap.*, 120 (1957) 255.

362 UDENFRIEND, S., WEISSBACH, H. AND SJOERDSMA, A., *Science*, 123 (1956) 669.

363 UDENFRIEND, S., WITKOP, B., REDFIEND, B. C. AND WEISSBACH, H., *Biochem. Pharmacol.*, 1 (1958) 160.

364 VALCOURT, A. J., *Federation Proc.*, 16 (1957) 130.

365 VIALLI, M. AND CASATI, C., *Rend. ist. Lomb. sci. lettere*, B 92 (1958) 329.

366 VITEK, V. AND RYSAMEK, K., *Nature*, 186 (1960) 244.

367   VOJTECHOVSKY, M., VITEK, V., RYSANEK, K. AND BULTASOVA, H., *Experientia*, 14 (1958) 422.
368   WAALKES, T. P., *J. Lab. Clin. Med.*, 53 (1959) 824.
369   WAALKES, T. P. AND COBURN, H., *Proc. Soc. Exptl. Biol. Med.*, 99 (1958) 742.
370   WAALKES, T. P., COBURN, H. AND TERRY, L. T., *J. Allergy*, 30 (1959) 408.
371   WAALKES, T. P., SJOERDSMA, A., CROVELING, C. R., WEISSBACH, H. AND UDENFRIEND, S., *Science*, 127 (1958) 648.
372   WAALKES, T. P. AND WEISSBACH, H., *Proc. Soc. Exptl. Biol. Med.*, 93 (1956) 394.
373   WALASZEK, E. J. AND ABOOD, L. G., *Federation Proc.*, 16 (1957) 133.
374   WALASZEK, E. J. AND ABOOD, L. G., *Proc. Soc. Exptl. Biol. Med.*, 101 (1959) 37.
375   WALDENSTRÖM, J., PERNOW, B. AND SILVER, H., *Acta Med. Scand.*, 156 (1956) 73.
376   WEINER, M. AND UDENFRIEND, S., *Circulation*, 15 (1957) 353.
377   WEISSBACH, H., BOGDANSKI, D. F., REDFIELD, B. G. AND UDENFRIEND, S., *J. Biol. Chem.*, 227 (1957) 617.
378   WEISSBACH, H., BOGDANSKI, D. F. AND UDENFRIEND, S., *Arch. Biochem. Biophys.*, 73 (1958) 492.
379   WEISSBACH, H. AND REDFIELD, B. G., *J. Biol. Chem.*, 235 (1960) 3287.
380   WEISSBACH, H., REDFIELD, B. G. AND TITUS, E., *Nature*, 185 (1960) 99.
381   WEISSBACH, H., REDFIELD, B. G. AND UDENFRIEND, S., *Federation. Proc*, 17 (1958) 418.
382   WEISSBACH, H., WAALKES, T. P. AND UDENFRIEND, S., *J. Biol. Chem.*, 230 (1958) 865.
383   WELSH, J. H., *Federation Proc.*, 13 (1954) 162.
384   WELSH, J. H., *Nature*, 186 (1960) 811.
384a WELSH, J. H., *Am. Zoologist*, 1 (1961) 267.
385   WELSH, J. H. AND MOORHEAD, M., *Gunma J. Med. Sci.*, 8 (1959) 211.
386   WELSH, J. H. AND MOORHEAD, M., *Science*, 129 (1959) 1491.
387   WELSH, J. H. AND MOORHEAD, M., *J. Neurochem.*, 6 (1960) 146.
388   WEST, G. B., *J. Pharm. Pharmacol.*, 10 (1958) 92T.
389   WEST, G. B., *J. Pharm. Pharmacol.*, 10 (1958) 589.
390   WEST, G. B., *J. Pharm. Pharmacol.*, 11 (1959) 275T.
391   WEST, G. B., *J. Pharm. Pharmacol.*, 11 (1959) 319.
392   WHITTAKER, V. P., *Biochem. Pharmacol.*, 1 (1959) 319.
393   WHITTAKER, V. P., *Biochem. J.*, 73 (1959) 37P.
394   WILLIAMS, R. T., *Detoxication Mechanisms*, Chapman and Hall, London, 1959, p.447.
395   WOOLF, L. I., GRIFFITHS, R., MONCRIEFF, A., COATES, S. AND DILLISTONE, F., *Arch. Disease Childhood*, 33 (1958) 31.
396   WOOLLEY, D. W. AND EDELMAN, P. M., *Science*, 127 (1958) 281.
397   WOOLLEY, D. W. AND SHAW, E., *Proc. Natl. Acad. Sci. U.S.*, 40 (1954) 228.
398   ZBINDEN, G., PLETSCHER, A. AND STUDER, A., *Z. ges. exptl. Med.*, 129 (1957) 615.
399   ZELLER, E. A., *Science*, 132 (1960) 1659.
400   ZELLER, E. A., BARSKY, J. AND BERMAN, E. R., *J. Biol. Chem.*, 214 (1955) 267.
401   ZUCKER, M. B., *Am. J. Physiol.*, 142 (1944) 12.
402   ZUCKER, M. B. AND BORRELLI, J., *J. Appl. Physiol.*, 7 (1955) 425.
403   ZUCKER, M. B. AND BORRELLI, J., *J. Appl. Physiol.*, 7 (1955) 432.
404   ZUCKER, M. B. AND BORRELLI, J., *Am. J. Clin. Pathol.*, 26 (1956) 13.
405   ZUCKER, M. B. AND BORRELLI, J., *Am. J. Physiol.*, 186 (1956) 105.
406   ZUCKER, M. B., FRIEDMAN, B. K. AND RAPPORT, M. M., *Proc. Soc. Exptl. Biol. Med.*, 85 (1954) 282.
407   ZUCKER, M. B., LEWIS, J. H. AND BORRELLI, J., *Proc. 6th Intern. Congress Haematology, Boston*, 1956, Grune and Stratton, New York, 1958, p. 540.
408   ZUCKER, M. B. AND RAPPORT, M. M., *Federation Proc.*, 13 (1954) 170.

Chapter 5

SEROTONIN AND GASTROINTESTINAL TRACT *

SPECIAL METABOLIC ASPECTS

It has already been described in Chapter 4 that the gastrointestinal tract is very rich in serotonin. The importance of the enterochromaffin cells for the production and storage of 5-HT has also been discussed. The distribution is not uniform because different concentrations of 5-HT may be found at the different levels of the gastro-intestinal tract. The 5-HT is practically concentrated in the mucosal layer, stored in the so-called "large granules" which are distinguishable from the mitochondria. The serotonin in the gastrointestinal tract is less sensitive to the action of precursors and releasers of 5-HT as well as to the monoamine oxidase inhibitors.

The administration of 5-HT or 5-HTP results in an increase of the tissue serotonin, but the increase at the gastrointestinal level is very weak. Reserpine and its derivatives are more active in releasing 5-HT from platelets and spleen than from the gastrointestinal tract. Regional differences after 5-HT treatment have been observed; for instance, 5-HT of the prepyloric area but not that of the stomach fundus is released by reserpine treatment. Several monoamine oxidase inhibitors fail to increase gastro-intestinal 5-HT both by acute or chronic administration.

EFFECT OF SEROTONIN AND RESERPINE ON GASTRIC EROSIONS AND HAEMORRHAGE

Rats and mice receiving 5-HT in large doses (30–50 mg/kg) develop in a rather short time gastric erosions and haemorrhages[31,50,123,262,263]. This effect may also be produced by using the serotonin precursor, 5-hydroxytryptophan (300 mg/kg)[117,263,**] and a serotonin releaser, such as reserpine (5–10 mg/kg)[25,30,31]. Both serotonin[97] and reserpine[6] show increased ulcerogenic activity in rats submitted to adrenalectomy. Ulcer induced by prolonged treatment with 5-HT in adrenalectomized rats is not prevented by cortisol or DOCA[34]. A careful description of these ulcers in comparison with other types of experimentally induced gastric erosions has been reported by Radouco-Thomas *et al.*[211]. Carcinoid patients with hyperserotoninaemia have a high incidence of gastric ulcers[49a,180]. A number of compounds antagonize the ulcer produced by 5-HT or its precursor. LSD 25 and chlorpromazine[262,263] prevent the effect of

---

* Reviews on this subject have been written by Warner[258a] and Haverback and Wirtschafter[121a].
** Tryptophan and tryptamine are inactive[263].

both, while atropine inhibits the ulcer induced by 5-HTP[117] but not by 5-HT[262]. The reason for this discrepancy is not yet known. The importance of the parasympathetic system in this kind of reaction is further suggested by the failure to produce ulcers by 5-HT treatment in vagotomized mice[31]. Unexpectedly a monoamine oxidase inhibitor, iproniazid, may also prevent the effect of 5-HT[31]. However, since iproniazid was given a few minutes before 5-HT, it has still to be established whether this protection is related to the block of MAO or to a direct antiulcer effect of iproniazid[278]. Reserpine too can induce gastric ulcers, as has been shown in mice[31], rats with or without pyloric ligation[25,30,146,157a,169,170,268], rabbits[268] and man[8,72,128,208,259]. Before the onset of the ulcer, reserpine induces histological changes in the gastric mucosa (ref. 106, 231) which heal in about 5 days after a single injection[268]. As for 5-HT, the gastric haemorrhage induced by reserpine is antagonized by vagotomy[31] and iproniazid[30,31,36,148,170,278]. On the other hand, LSD 25 and BOL 148 are almost ineffective[31,147]. The reserpine-induced ulcer is increased by gonadotrophin[267] and is healed by atropine[31], hexamethonium[31], parasympathicolytics[146], thenalidine[191] and chlorobenzoxamine[171,210,211]. Imipramine is slightly active in reducing the ulcers induced by 5-HT[262] and prevents the ulcerogenic effect of reserpine[36,69] and that of restraint[37].

These findings suggest the possibility that the reserpine effect can be explained in terms of a serotonin release[205].

A definite relationship between serotonin release in the gastrointestinal tract and the onset of gastric erosions might be suggested by the fact that syrosingopine (Su 3118)

TABLE 15

EFFECT OF SOME DRUGS OR EXPERIMENTAL CONDITIONS ON THE ULCER INDUCED BY SEROTONIN AND RESERPINE IN RATS OR MICE

(Numbers in brackets indicate references)

| | *Serotonin* | *Reserpine* | |
| | *Gastric ulcer* | *Gastric ulcer* | *Gastric hyperacidity* |
Compound or condition			
Adrenalectomy	E (97)	E (6)	
Amphetamine	N (262)	P (31)	
Atropine	N (262)	P (31)	P (150)
Benadryl	P (147)	P (147)	
BOL	P (147)	N (31)	
Chlorpromazine	P (262)		
Hexamethonium		P (31)	
Imipramine	P (262)	P (69)	
Iproniazid	P (31)	P (31, 244a, 278)	N (36a)
LSD 25	P (147, 262)	N (31, 147)	N (96)
UML	P (147)	N (147)	
Vagotomy	P (31)	P (31)	P (31)

P = protection; N = no effect; E = enhancement.

and dimethylaminobenzoylreserpate (Su 5171), two reserpine derivatives, have different ulcerogenic activities[31,134] as well as different properties in releasing intestinal 5-HT[99]. Rauwolfia extracts deprived of reserpine[149,151,155], and reserpiline[152,156], raumitorine[157] and tetrabenazine[31,278] do not have any ulcerogenic effect. However, other data suggest that the effects of reserpine on the gastric functions are not related to 5-HT only. The effect of reserpine also seems to be connected with a parasympathetic predominance or to a histamine-like effect because the ulcerogenic properties are associated with increased gastric secretion and acidity. These activities observed in man[8,22,55,121,144,145,214,215,225,229,*], dogs[154,207,223] cats[96] and rats[150] are blocked by anticholinergics such as oxyphenonium[12,14,35,145,176,207], or by hexamethonium[145,176], but not by pyribenzamine[14]. Atropine also exerts a protective effect in rats[150] and dogs[154], but not in man[22,23,54,145,229]. Furthermore, the effect of reserpine is present also in vagotomized and vagosympathectomized patients[215] or dogs[223].

Reserpine and serotonin induce a marked fall of gastric acidity when administered in the lateral ventricle in cats. This effect differs from the hyperacidity elicited by reserpine given intravenously, because it is prevented by LSD[96].

These data do not prove beyond doubt the hypothesis that reserpine acts by stimulation of the parasympathetic[12,16,175,183,230]. It has been suggested that reserpine could act through a release of neurohumoral substance[223].

On the other hand the gastric ulcer produced by 5-HT or 5-HTP is not associated with an increase of gastric acidity and volume**. There is, instead, in man[224,226], dogs[29,117,118,120,227], and rats[117,193,234] a certain decrease of gastric secretion and acidity after 5-HT or 5-HTP treatment***. Iproniazid also decreases gastric acidity[235] and does not affect the hypersecretion induced by reserpine[170b]. The inhibitory effect of 5-HT is particularly evident when gastric secretion is artificially stimulated by histamine. In this condition 5-HT or 5-HTP are "antihistaminics" in several animal species[29,66,219,239]. However, this antagonism between histamine and 5-HT was not observed by White et al.[260]. The antagonistic effect on gastric acidity does not apply to the production of ulcer, because 5-HT and histamine are synergists in this respect[132]. Serotonin inhibits also the hyperacidity elicited by an antiserotonin agent (UML) in patients with peptic ulcer[213a].

The inhibitory effect of 5-HT on gastric acidity might be partially explained by the increased secretion of gastric mucus[29,238,261] whose buffering effect is well known. It is possible that ulcer production by reserpine is mediated through both serotonin and histamine, but the hyperacidity is due to histamine and parasympathetic predominance. It is in fact known that reserpine releases not only serotonin but also histamine[257].

Another problem to be solved is whether the gastric ulcers induced in rats by various "amine liberators" are mediated by serotonin. Polymyxin B, Compound 48/80,

---

*   Syrosingopine is less effective than reserpine in increasing gastric acidity[7].

**   Pepsin excretion was reduced in rats[235] by 5-HTP but increased in dogs by 5-HT or phenylisopropylhydrazine[260].

*** Other authors reported slight increases of the gastric acidity after 5-HT [48].

dextran and ovomucoid elicit acute haemorrhagic ulcers restricted to the glandular portion of the stomach in rats[87,131,132,193,233]. The ulcers induced by polymyxin B are antagonized by antihistaminics[87,188,193] and particularly by cyproheptadine[188], but not by atropine[188,193] or LSD[87]. However, histamine is not ulcerogenic in rats although it considerably increases the number of ulcers produced by serotonin[132].

Other data suggest that the ulcers induced by prolonged administration of corticosteroids may be related to the metabolism of 5-HT in the gastrointestinal tract[244].

In patients with achlorhydria, ulcer, or increased gastric acidity, the excretion of 5-HIA in the urine is normal[39]. Patients with gastritis show instead a decreased elimination[226a]. Certain types of irritation of the gastrointestinal tract increase the content of 5-HT (*e.g.* alcohol[277] and intestinal obstruction[248]). An atypical case of carcinoid secreting 5-hydroxytryptophan was associated with multiple peptic ulcers[49a].

The introduction of a hyperosmolar meal into the upper small intestine induces a complex of symptoms called the "dumping syndrome". These symptoms are relieved by antiserotonin agents[133a,200a] and are probably in connection with a release of serotonin in the intestine[70a,200a].

EFFECT OF SEROTONIN ON THE SMOOTH MUSCLE OF THE GASTRO-INTESTINAL TRACT

The stimulation induced by 5-HT on the smooth muscle of the gastrointestinal tract has been used to measure the concentration of 5-HT in biological fluids (see Chapter 2). However such stimulation occurs at different concentrations of 5-HT according to the animal species and, in the same species, according to the part of the gastrointestinal tract tested. The *in vitro* sensitivity of various gastrointestinal preparations to 5-HT is shown in Table 16.

An important question concerns the specificity of the action of serotonin on smooth muscle. Is this a direct effect or can it be explained by the action of other known chemical mediators? Gaddum and his group concluded, after a series of investigations, that the effect of serotonin is specific and not related to the effect of acetylcholine or histamine. The term "tryptamine receptors"* was proposed to classify the receptors of the smooth muscle sensitive to serotonin. The concept of the specific effect of serotonin is supported by results obtained studying the effect of drugs acting on known chemical mediators toward serotonin-induced stimulation of the intestine. Although there is a considerable lack of agreement in the results obtained studying the antagonists of 5-HT, it may be stated that the intestinal stimulation induced by 5-HT is not related only to an effect on acetylcholine or histamine. In some intestinal preparations, as rat duodenum[173], rat ileum[173], rat colon[62,84,101,247], rabbit ileum[216] and frog rectum[256], 5-HT maintains its stimulating effect even in

---

* Tryptamine is several times less active than 5-HT in stimulating intestinal motility[90]. It was suggested by studying antagonists that tryptamine and 5-HT receptors may not be necessarily identical[10]. Recently, Vane suggested that amphetamine-like agents would act by stimulating the tryptamine receptors (rat stomach strip)[253].

TABLE 16

SENSITIVITY OF VARIOUS GASTROINTESTINAL PREPARATIONS TO 5-HT

Species	Type of preparation	Range of the concentrations required (g/ml)		References
Brown trout	intestine or stomach	$1 \cdot 10^{-9}$		47
Cat	ileum	$5 \cdot 10^{-4}$	$1 \cdot 10^{-5}$	74, 76
Frog	rectum	$1 \cdot 10^{-8}$		77, 256
Guinea pig	colon	$1 \cdot 10^{-7}$		74
	ileum	$2.5 \cdot 10^{-6}$	$1 \cdot 10^{-9}$	88, 93, 173, 218
	taenia coli	$2 \cdot 10^{-9}$	$5 \cdot 10^{-8}$	40
Helix pomatia	stomach	$5 . 10^{-13}$		113a
Mouse	jejunum	$5 \cdot 10^{-8}$	$3 \cdot 10^{-9}$	213
Rabbit	colon	$1\text{–}5 \cdot 10^{-9}$		74, 236
	ileum	$2 \cdot 10^{-7}$	$1 \cdot 10^{-8}$	74, 76
Rat	colon	$1 \cdot 10^{-8}$	$2 \cdot 10^{-8}$	84
	duodenum	$1 \cdot 10^{-5}$	$5 \cdot 10^{-8}$	75, 172, 173
	ileum	$2 \cdot 10^{-6}$	$2 \cdot 10^{-7}$	173
	stomach *in toto*	$1 \cdot 10^{-6}$	$1 \cdot 10^{-8}$	51
	stomach (fundus strip)	$2 \cdot 10^{-10}$	$5 \cdot 10^{-11}$	60a, 251
Tortoise	jejunum	$1 \cdot 10^{-6}$		247

the presence of large concentrations of atropine. In the rat's stomach fundus strip, 5-HT is about 10 times more active than acetylcholine. Furthermore hyoscine hydrobromide (scopolamine) at the concentration of 10^{-7} g inhibits acetylcholine, leaving 5-HT completely unaffected[251]. In other preparations, as in the case of guinea pig ileum, contradictory results have been reported. Rocha e Silva *et al.*[218], finding a complete antagonism between atropine and 5-HT, suggested that 5-HT was stimulating the post-ganglionic cholinergic fibres of the intramural nervous system. According to other authors, only a partial antagonism was observed between 5-HT and atropine on the guinea pig ileum[49,173,216,236]. However, when a comparison was made between the inhibitory effects exerted by atropine on 5-HT and on acetylcholine, it was concluded that 5-HT may elicit contraction of the guinea pig ileum even when the preparation is made insensitive to acetylcholine* (refs. 41, 42, 49, 85, 90, 93, 142, 212). These data contradict the hypothesis that 5-HT acts through acetylcholine release, although they cannot exclude a cholinergic component in the intestinal stimulation induced by 5-HT. On the other hand, anticholinesterase drugs potentiate or increase the effect of 5-HT on guinea pig[2] and rabbit[2,217] intestine, but not on rat ileum[173]. Also the response to acetylcholine is potentiated in the presence of serotonin[71].

---

* On the isolated rat uterus, the antiserotonin effect of atropine is dependent on the pH of the medium[163].

More experimental support is required to exclude the possibility that 5-HT acts on the intestine by releasing histamine. Several antihistaminics show also antiserotonin effect, however, they are usually more active on histamine than on serotonin[212]. Diphenhydramine and mepyramine do not inhibit the action of 5-HT on isolated guinea pig ileum at doses inhibiting histamine activity[84,88,90,93,*]. Hyoscine hydrobromide is without effect on 5-HT at doses very active in inhibiting histamine on the rat's stomach[251]. Also, *in vivo*, several antihistaminics (pyrilamine, diphenhydramine, tripelennamine, chlorpheniramine and phenindamine) fail to prevent the effect of 5-HT on intestinal motility in humans[125] or in dogs[120].

RECEPTORS M AND D

A further elaboration of the "tryptamine receptor" concept has been developed by Gaddum who distinguished two types of receptors: "the M receptors blocked with morphine and the D receptors blocked with dibenzyline" (benzylphenoxyisopropyl-chloroethylamine).

The M receptors, although not accurately localized, are supposed to be in the nervous tissue (sympathetic ganglia[28] or post-ganglionic nerve fibres in the guinea pig's intestine), whereas D receptors are situated in plain muscle fibres[92]. The M receptors are distinct from the receptors for nicotine and $BaCl_2$ because they are not blocked by hexamethonium or large doses of nicotine[92,216,218]. The two types of receptors are present in the ileum of guinea pig[11,92]. On this preparation morphine, or its congeners[184,185], or dibenzyline only partially antagonize the effect of 5-HT, even when their concentration is considerably increased[92,186]. This finding offers an interesting tool for studying the types of receptors present in the smooth muscles from different organs and species and for analysing the action of drugs antagonizing or mimicking 5-HT. For instance, rat uterus[92] is similar to rat stomach[251] and rat colon[184], and contains almost entirely D receptors, because it is rather insensitive to the inhibiting effect of morphine. According to this hypothesis it was suggested that the diethylamide of lysergic acid and its congeners, dihydroergotamine and 5-benzoyl-gramine inhibit the effect of 5-HT by blocking the D receptors, whereas atropine and cocaine are inhibitors of the M receptors[92,95]. Also inhibitors of morphine, like nalorphine and levallorphan would antagonize the effect of 5-HT on the guinea pig's ileum by acting on M receptors[142]. LSD would be only a weak inhibitor of the effect of 5-HT on the guinea pig intestine because in this tissue the D receptors are less important than the M receptors[93]. Similarly, an excess of 5-HT makes the tissue insensitive to low concentrations of 5-HT in the case of the guinea pig ileum (M receptors) but not of rat uterus (D receptors)[93]. It is worth mentioning that isolated organs desensitized to the effect of 5-HT by large doses of this amine are also insensitive to tryptamine, but not to acetylcholine, histamine or substance P[40,90,93]. On

---

* The indazole analogue (3-β-aminoethyl-5-hydroxyindazole) of 5-HT also stimulates guinea pig ileum and is partially blocked by atropine, but not by antihistaminics[1].

the other hand, isolated organs desensitized to nicotine are still responsive to 5-HT[201].

Other authors approaching this problem from a different angle believe that 5-HT stimulates organs containing smooth muscle by a dual action (one direct on the smooth muscle fibres and another indirect on the nervous tissue)[2,21,129,142]. The concept of the two receptors, which recalls the mechanism of action of acetylcholine, does not explain all the data available. For instance, Lewis questions the use of morphine as a selective inhibitor of one type of 5-HT effect, showing that morphine antagonizes not only drugs acting on the nervous structures but also drugs stimulating smooth muscle[174].

SEROTONIN ANTAGONISTS

Many compounds show the capacity to antagonize the effect of 5-HT on the isolated intestine *in vitro*. A list of these drugs is reported at the end of this book (see Appendix IV) and a review on this subject has been published recently[130]. The antagonists of the 5-HT effect on intestine and on other smooth muscles, belong to different chemical and pharmacological classes.

Several lysergic acid derivatives show important anti-5-HT effects[220]. Among the most studied and the most powerful are LSD[40,51,59,91,111,173], its bromo derivative BOL[9,40,92,111,112,247], and 1-methyllysergic acid (butanolamide). LSD is more powerful as a 5-HT antagonist on duodenum and uterus than on intestine[91,101]. It is interesting that on rat's isolated uterus very low concentrations of LSD potentiate the contraction induced by 5-HT[59,68]. This effect is not known for the intestine. Several antihistaminics show antagonism also against the effect of 5-HT on isolated intestine preparations[212]. Promethazine (phenergan)[51,101,173], homochlorcyclizine (*N-p*-chloro-benzhydryl-*N'*-methylhomopiperazine) and its congeners[139], and cyproheptadine (1-methyl-4-(5-dibenzo[*a,e*]cycloheptatrienylidene)-piperidine)[243] are among the most active although they show also a considerable antihistaminic effect.

Chlorpromazine[24,51,59,101,114] and several other phenothiazines[115,182,220a] antagonize 5-HT at doses not active against acetylcholine[101,182] but usually already active against histamine[182]. Diethazine (diparcol), a phenothiazine derivative showing anti-parkinson activity, antagonizes the effect of 5-HT on isolated rat colon[101]. Other compounds, structurally related to phenothiazine, such as imipramine[69] and amitriptyline[242] are also antagonists of 5-HT.

Among indole derivatives an anti-5-HT effect has been reported for: 5-halogen-tryptamines[209], 2-methyl-3-ethyl-5-dimethylaminoindole (medmain)[76,275], 5-benzoyl-gramine[92], 6-methylgramine, and others[10,94]. BAS or 1-benzyl-2-methyl-3-(2-amino-ethyl)-5-methoxyindole also shows an antiserotonin effect[9] but only at rather high doses and the effect is easily reversible[189]. Sympathicomimetic agents are very active in inhibiting 5-HT[51,251]. Among several which have been studied, isopropylnor-adrenaline was found to be particularly effective[51,102] (see Table 17).

Also adrenolytic agents antagonize the 5-HT effect on intestine. Dihydroergot-amine is rather specific[40,86,92], dibenamine is more active as an antiserotonin than as

TABLE 17

ANTAGONISM BETWEEN SYMPATHOMIMETIC AGENTS AND SEROTONIN ON THE ISOLATED COLON OF RAT

Compound	Concentration ($\mu g/ml$)	% Inhibition after successive stimulation with 5-HT (0.1 $\mu g/ml$)			
		1	2	3	4
L-Adrenaline	0.01	83	0		
	0.1	100	3	0	
	0.5	100	90	53	18
L-Noradrenaline	0.01	90	0		
	0.1	100	10	0	
	0.5	100	100	50	0
L-Isopropylnoradrenaline	0.001	80	0		
	0.01	100	65	0	
	0.1	100	80	48	0

an antiacetylcholine[75,86,140], but piperoxane is almost inactive[86] showing that adrenolytic and antiserotonin properties are not necessarily related.

Substance P may be considered an antiserotonin agent since it inhibits the stimulating effect of 5-HT[21] and the depressant effect on the peristaltic reflex exerted by 5-HT at large concentrations[19]. On the other hand, 5-HT prevents the effect of low but not of high concentrations of substance P on the intestine[20].

An excess of potassium tends to antagonize the activity of 5-HT on smooth muscles[80].

Among other agents showing an antiserotonin effect it is worth mentioning botulinum toxin[2], cocaine[92,140], heparin[138,240], local anaesthetics[236] and dock leaf extract[58]. It is obvious that spasmolytic agents may be effective in antagonizing the contraction induced by 5-HT on the intestine.

Reserpine in large concentrations depresses the effect of 5-HT on the guinea pig's ileum or other isolated organs, but this action is not specific since other stimulating agents are also inhibited[53,101,108,109]. This effect is reversed by citrate[109]. On the isolated rabbit duodenum, reserpine shows a stimulating effect followed by depression. The stimulation is, however, probably not related to 5-HT release because it is not inhibited by BOL[110].

Monoamine oxidase inhibitors do not increase the effect of 5-HT on the rat's stomach although they potentiate the effect of tryptamine. It is supposed that this may be related to the fact that 5-HT does not diffuse inside the cells and therefore does not come in contact with the enzyme[252]. Direct experiments performed with labelled serotonin on guinea pig taenia coli show that the amine is taken up very slowly[37a].

STIMULATION OF INTESTINAL MOTILITY *in vivo*

5-HT also stimulates the intestine *in vivo* in several animal species including man[56,125], dogs[33,88,120], cats[33,166], rabbits[88,105], guinea pigs[74], rats[74,75,187], mice[38,74], pigeons[73,153], frogs[74] and fishes[79]. The 5-HT must be injected, it has no effect when given by mouth[65,119] except in mice[57]. Usually the stimulation is short-lasting[56], tends to decrease with repeated administrations (tachyphylaxis)[65,104,105] and is followed by a period of inactivity[125,*].

The effect of serotonin is quantitatively different according to the species considered. The stimulation of motility also depends upon the part of the intestine observed. For instance after administration of 5-HT (0.5 mg i.v.) in man, Debray and Besançon[65] reported 67% of positive (increased motility) responses in the duodenum, 63% in jejunum, 47% in small intestine, 15% in ileum and no responses in descending colon. The pattern of the motor response elicited by serotonin is similar to the spontaneous one[65a]. The motility of the oesophagus is also not affected by 5-HT[226]. As a result of the intestinal stimulation there is an increase in the weight of faeces eliminated[38,74]. In dogs, serotonin induces stimulation of the intestine even if the intestinal mucosa is denervated[261]. Also the precursor of 5-HT, 5-hydroxytryptophan, increases intestinal motility in several animal species including man (dog[33], cat[33], rat[33], mouse (refs. 38, 272), and man[64,119,227]). Vagotomy inhibits the peristalsis but not the contraction induced by 5-HT[104,105,120] or 5-HTP[33]. In rabbits, atropine prevents the effect of 5-HT[104,105]. In dogs, atropine does not affect[119,120,179] or partially antagonizes[261] the increased intestinal motility induced by perfusion of 5-HT, but prevents the activity of 5-HTP[119]. In human beings atropine[125] abolishes the intestinal stimulation induced by 5-HT. Methantheline[56,104,125,233], dibenamine[75], morphine[187] and BAS[125] prevent the stimulation induced by 5-HT *in vivo*. On the other hand, antihistaminics[120,125], prostigmine[104,105] and hexamethonium[56,120,125] are almost completely inactive.

LSD[104,105,120], BOL[119] and UML[65] in doses which are relatively low in comparison with the concentration of 5-HT, do not antagonize the effect of 5-HT on the intestinal motility. However, LSD was found active in preventing the increased motility of intestinal villi induced by 5-HT in dogs[179]. This test shows that chlorpromazine also has antiserotonin properties[179], although by itself it depresses the motility of the intestinal villi[103]. Atropine is inactive in this respect[179].

A simple test to evaluate the intestinal effect of serotonin was suggested by Woolley who showed that 5-HTP, administered to mice at a dose of 1 mg i.p., induces profound diarrhoea[272]. This effect is also induced, with a lesser intensity, by 4-hydroxytryptophan[78]. The diarrhoea induced by 5-HTP in mice is prevented by a number of antiserotonin agents including BAS[272,276], BAB[276], chlorpromazine[60], imipramine[60], homochlorcyclizine[139], several indole derivatives[273] and nicotinamide[105]. The test

---

* A decrease of the intestinal motility of the stomach and large intestine after 5-HT administration in man has been reported[225,226].

could probably be made more specific by comparison with the effect of the various drugs on the diarrhoea induced by carbaminoylcholine or other intestinal stimulants. The effect of some drugs on this experimental condition is reported in Table 18 (ref. 97).

TABLE 18

EFFECT OF SELECTED DRUGS ON DIARRHOEA INDUCED BY DL-5-HYDROXYTRYP-TOPHAN OR CARBACHOL

Compound	Diarrhoea induced by	
	5-HTP	Carbachol
Atropine	+	+++
BAB	++	—
BAS	++	+
BOL 148	++	—
Chlorisondamine	+++	+
Chlorpromazine	+++	+++
Chlorprothixene	+++	+++
Cyproheptadine	+	++
Dibenamine	+	+
Hexamethonium	++	+
LSD 25	—	—
Papaverine	+	+
Thenalidine	—	—
Tripolidine	—	—
UML 491	—	—

+++ = very active
++ = active
+ = slightly active
— = not active

Other techniques have been used to show the effect of serotonin or its precursor on the intestine. For instance, it has been reported that 5-HT (10 mg/kg s.c.) delays the transit of a charcoal meal through the intestine of mouse[38].

This, however, reflects an action at the level of the stomach because 5-HT does not influence the progress of the charcoal meal when the latter is given at the duodenal level. Furthermore, in the same experiment 5-HTP considerably increased the weight of the faeces excreted[38]. Propulsive motility of the intestine, visualized by the transit of phenol red, was decreased in rats receiving 0.1–1 mg/kg (s.c.) and 10 mg/kg (oral) of 5-HT or 0.4–40 mg/kg (s.c.) of 5-HTP[177]. 5-HT tends to increase the time taken for the stomach to empty after phenol red[177] or carmine meal[241]. α-Methyl-DOPA, a powerful inhibitor of amino acid decarboxylases, increases the time taken to empty the stomach as well as slowing the transit of the carmine meal through the intestine[241]. These results are not unequivocal in suggesting that 5-HT is a mediator of the intestinal motility.

References p. 96

Other data relevant to the topic under discussion come from experiments carried out with reserpine. As has already been mentioned, the picture obtained with reserpine is complicated by the fact that this alkaloid releases not only 5-HT, but also histamine and catecholamines* which have profound activity on the intestinal motility. One of the typical effects of reserpine in animals is an increased excretion of faeces which may become of a diarrhoeic type even in human beings[264-266]. Diarrhoea is observed also in animals after a single treatment with large doses of reserpine and it is particularly marked in adrenalectomized rats[6]. It has been shown that the motility of the intestinal tract is increased after reserpine[52,73,89,153,206,214] and tetrabenazine[170a]. According to Bein, this effect of reserpine is due to the lack of catecholamine (inhibition of sympathetic system) inducing a preponderance of parasympathetic activity[15,16]. In dogs, atropine blocks the increased gastric motility induced by reserpine[52] but not the increased motility of the intestinal villi[178]. The possibility that the increased intestinal motility induced by reserpine is due to the release of 5-HT (refs. 25a, 26, 27, 89, 136, 222, 255) from the intestine has also to be considered. While studying some reserpine derivatives it was observed that compounds inactive in decreasing intestinal 5-HT were also unable to induce diarrhoea in rats[100]. However, the question is still open and the answer must wait until there is more experimental evidence.

A considerable dissociation between central effects and gastrointestinal stimulation has been achieved recently with a new reserpine derivative, methyl-18-epireserpate methyl ether (Su 9064)[13].

It may also be of some interest to recall that guanethidine, a new antihypertensive agent inducing diarrhoea as a side-effect in patients[160], lowers 5-HT in the intestine of mice and rabbits[17].

Physiological amounts of perfused hydrochloric acid release serotonin from the upper intestine of rats[213b].

INTESTINAL MOTILITY IN CARCINOID PATIENTS

Carcinoid patients show a very high level of serotonin and serotonin synthesis and thus they provide a unique tool for research in this field[107,158,164,165,249]. The well-known fact that carcinoid patients have a high incidence of borborygmus of the bowel and diarrhoea[67,197,237a,245] has been associated with this high availability of serotonin. Thorson, in an exhaustive review, considered that intestinal disturbances were very common in carcinoid patients[245,**].

Lembeck, in a review of cases, found that 43 out of 51 carcinoid patients have diarrhoea[166] accompanied by a high level of serum 5-HT[113,199,245] and a high excretion of 5-HIA[61,81,116,124,133,181,195,237,246] or 5-HTP[190,191a] in urine (see also

---

* In the intestine of some animal species (ox, sheep, dog) dopamine accounts for more than 90% of the total catecholamines[232].
** Other article reviews on this subject have been prepared by Holtz[126], Valzelli[250], Olson and Gray[192], Hedinger[122a], and Christodoulopoulos and Klotz[53a].

Chapter 4). The diarrhoea is of an intermittent nature but it has not yet been established if it is connected with an increased release of serotonin. This would however be proved by the use of agents, like histamine[63], noradrenaline[3,198] and fatty meals[32], which are able to induce diarrhoea and to increase the level of 5-HT in the plasma of carcinoid patients. On the other hand it has been established that 5-HT releases histamine[82,83] and that carcinoid patients excrete high amounts of histamine in their urine[49a,199,200,258]. Because of the small number of carcinoid patients it has not yet been possible to assess the therapeutical value of the antiserotonin agents for relief of diarrhoea in this disease. It has been reported that chlorpromazine, but not BAS or the bromo derivative of lysergic acid, has some value[227a,228]. A long treatment with UML (1-methyllysergic acid butanolamide) was found to relieve the diarrhoea in one patient without affecting the excretion of the urinary 5-HIA[159] but other authors did not confirm this finding[189a,227a]. It may be worth mentioning that iproniazid exacerbates diarrhoea in carcinoids when 5-HIA excretion was decreased[135]. No clear results have been obtained with α-methyl-DOPA[190a,191a].

Phenylacetic acid, an inhibitor of 5-HTP decarboxylase, does not affect the intestinal symptoms of carcinoid patients[221].

Other types of diarrhoeic patients do not show any change in the excretion of urinary 5-HIA[143].

PHYSIOLOGICAL ROLE OF SEROTONIN IN INTESTINAL PERISTALSIS

The great interest shown by various authors in the effects of serotonin in gastroenterology is based upon the idea that 5-HT may play a role in the physiological functions of the gastrointestinal tract. Only in few instances, however, has this been proved by direct evidence. The outstanding work of the Bülbring group devoted to this approach will now be briefly summarized.

In contrast to other authors, who showed that 5-HT was able to inhibit the peristaltic reflex when applied to the serosal surface[141,167], Bülbring studied the effect of 5-HT on the mucosal surface because she considered that 5-HT is concentrated in the intestinal mucosa[84] and that the sensory receptors initiating the reflex are also situated in the mucosa. With a suitable apparatus[44,46] it was possible to study the peristaltic reflex in isolated loops of guinea pig and rabbit jejunum by measuring the pressure threshold and the fluid transport after application of 5-HT outside or inside the intestine. 5-HT (10^{-9}–10^{-5} g/ml) added to the fluid passing through the lumen of the intestine "stimulates peristalsis, lowers the threshold of intraluminar pressure required to elicit the reflex, increases the frequency of contractions and increases the volume of fluid transported"[46]. Also, Lembeck observed an increase of the peristaltic waves when serotonin was added inside the lumen of guinea pig ileum or ascending colon[167,*]. These effects are explained by a stimulation of the sensory receptors similar to the action of 5-HT in other parts of the body[4,5,70] and on the vagal gastric and

---

* 5-HT restores weak peristaltic activity to guinea pig ileum after cooling[18].

intestinal receptors[196]. Other sensory stimulants, phenyldiguanide, veratrine, vera-tridine and protoveratrine, facilitate the peristaltic reflex like 5-HT[41].

The lysergic acid derivatives (LSD and BOL) applied inside the intestinal loop exert an opposite effect to that of 5-HT.

Furthermore during peristalsis a release of 5-HT in the fluid was observed, proportional to the intensity of the intraluminar pressure. Similar results were obtained by Lee[161,162] using the colon of rabbits and guinea pigs. The release of 5-HT tends to decrease with the number of induced peristaltic waves, but this effect may be prevented by using iproniazid and 5-HTP alone or combined with pyridoxal phosphate. It is interesting that the release of 5-HT still occurs in relation with the rise of the intraluminal pressure* even when the peristaltic reflex is abolished by the presence of procaine or hexamethonium[45,46]. The release of 5-HT after a rise of in-traluminal pressure was also confirmed *in vivo* in guinea pigs[42], rats[56a] and rabbits[89,137]. In rabbits the release of 5-HT into the lumen progressively decreases in the more distal portions of the gut[122]. 5-HTP stimulates peristalsis *in vivo*, but reserpine did not abolish the peristaltic reflex in spite of the extremely low mucosal 5-HT content and the very small amount of 5-HT released into the lumen. However, the biosynthesis of 5-HT was still active providing that 5-HTP was given[43].

These observations led to the conclusion that 5-HT is involved in the physio-logical mechanisms of the peristaltic reflex, probably because of its capacity to stimu-late the sensory receptors of the mucosa when it is released by changes of pressure in the lumen of the gut.

INTERACTIONS BETWEEN SEROTONIN AND IONS

Very little is known about the intrinsic mechanism by which 5-HT is able to stimulate the intestinal motility and more generally the smooth muscles. It has been reported that there is a decrease of glycogen and phosphorylase activity in the smooth muscle after 5-HT[168]. Interactions between serotonin and ions have also been considered. Serotonin affects efflux of sodium and potassium from red beet roots[204], erythrocytes[202] and frog skin[203]. Substitution of Na^+ with Li^+ decreases the smooth muscle responsiveness to 5-HT[127]. Woolley observed in the isolated uterus of rat that the chelation of calcium by means of EDTA considerably decreased the effect of serotonin. When EDTA was neutralized by calcium ions, 5-HT again exerted its stimulating properties[269]. A similar observation was made in strips of isolated rat stomach[143a]. The contraction induced by calcium was prevented by an antiserotonin like BAS[271] In further experiments, a substance was extracted which made serotonin soluble in a fat solvent (benzene–butanol) in the presence of calcium[138a,270]. This substance, which is thought to be present in the serotonin receptors, is present in various parts of the gastrointestinal tract[270]. From these findings it was concluded

---

* Intestinal occlusion and neostigmine do not affect the intestinal 5-HT or the excretion of urinary 5-HIA[137,194,235a]. The type of intestinal flora may change the 5-HT levels in the in-testinal wall[14a].

that calcium was essential to bring the serotonin, which is insoluble in fat, into contact with the receptors on the smooth muscle fibres. On the other hand, serotonin increases the transport of radioactive calcium through the cell membrane, simulated by a model system[274]. An increased absorption of labelled calcium by 5-HT was also shown in the isolated everted intestinal tract and in intact rats when calcium was given by the oral route[98,254].

REFERENCES

1   AINSWORTH, C., *J. Am. Chem. Soc.*, 79 (1957) 5245.
2   AMBACHE, N., in G. P. LEWIS (Ed.), *5-Hydroxytryptamine*, Pergamon, London, 1958, p. 203.
3   ANDREWS, T. M., PEART, W. S. AND ROBERTSON, J. I. S., *J. Physiol. (London)*, 155 (1960) 8–9 P.
4   ARMSTRONG, D., DRY, R. L. M., KEELE, C. A. AND MARKHAM, J. W., *J. Physiol. (London)*, 117 (1952) 70 P.
5   ARMSTRONG, D., DRY, R. L. M., KEELE, C. A. AND MARKHAM, J. W., *J. Physiol. (London)*, 120 (1953) 326.
6   ASHWIN, J. G., *Proc. Soc. Exptl. Biol. Med.*, 104 (1960) 188.
7   BACHRACH, W. H., *Proc. Soc. Exptl. Biol. Med.*, 99 (1958) 295.
8   BACHRACH, W. H., *Am. J. Digest. Diseases*, 4 (1959) 117.
9   BARLOW, R. B. AND KHAN, I., *Brit. J. Pharmacol.*, 14 (1959) 99.
10 BARLOW, R. B. AND KHAN, I., *Brit. J. Pharmacol.*, 14 (1959) 265.
11 BARLOW, R. B. AND KHAN, I., *Brit. J. Pharmacol.*, 14 (1959) 553.
12 BARRETT, W. E., PLUMMER, A. J., EARL, A. E. AND ROGIE, B., *J. Pharmacol. Exptl. Therap.*, 113 (1955) 3.
13 BARRETT, W. E., RUTLEDGE, R. AND PLUMMER, A. J., *Federation Proc.*, 20 (1961) 393.
14 BARRETT, W. E., RUTLEDGE, R. AND ROGIE, B., *Federation Proc.*, 13 (1954) 334.
14a BEAVER, M. H. AND WOSTMANN, B. S., *Brit. J. Pharmacol.*, 19 (1962) 385.
15 BEIN, H. J., *Experientia*, 9 (1953) 107.
16 BEIN, H. J., *Ann. N.Y. Acad. Sci.*, 61 (1955) 4.
17 BEIN, H. J., in J. R. VANE, G. E. W. WOLSTENHOLME AND M. O'CONNOR (Eds.), *Adrenergic Mechanisms*, Churchill, London, 1960, p. 162.
18 BELESLIN, D. AND VARAGIC, V., *Brit. J. Pharmacol.*, 13 (1958) 266.
19 BELESLIN, D. AND VARAGIC, V., *Brit. J. Pharmacol.*, 13 (1958) 321.
20 BELESLIN, D. AND VARAGIC, V., *J. Pharm. Pharmacol.*, 11 (1959) 99.
21 BELESLIN, D. AND VARAGIC, V., *Arch. intern. pharmacodyn.*, 126 (1960) 321.
22 BEMAN, F. M., KNOLL, H. C. AND DEHAR, C. J., *Clin. Research Proc.*, 83 (1955) 1007.
23 BEMAN, F. M., KNOLL, H. C. AND DEHAR, C. J., *Clin. Research Proc.*, 83 (1955) 131.
24 BENDITT, E. P. AND ROWLEY, D. A., *Science*, 123 (1956) 24.
25 BENDITT, E. P. AND WONG, R. L., *Am. J. Pathol.*, 32 (1956) 638.
25a BENDITT, E. P. AND WONG, R. L., *J. Exptl. Med.*, 105 (1957) 409.
26 BERTELLI, A., GALLI, G. AND GENOVESE, E., *Boll. soc. ital. biol. sper.*, 32 (1956) 822.
27 BERTLER, Å., *Acta Physiol. Scand.*, 51 (1961) 75.
28 BINDLER, E. H. AND GYERMEK, L., *Federation Proc.*, 20 (1961) 319.
29 BLACK, J. W., FISHER, S. W. AND SMITH, A. N., *J. Physiol. (London)*, 141 (1958) 27.
30 BLACKMAN, J. G., CAMPION, D. S. AND FASTIER, F. N., *Proc. Univ. Otago Med. School*, 36 (1958) 1.
31 BLACKMAN, J. G., CAMPION, D. S. AND FASTIER, F. N., *Brit. J. Pharmacol.*, 14 (1959) 112.
32 BLEEHEN, N. N., *Lancet*, 269 (1955) 1362.
33 BOGDANSKI, D. F., WEISSBACH, H. AND UDENFRIEND, S., *J. Pharmacol. Exptl. Therap.*, 122 (1958) 182.
34 BOIS, P. AND SELYE, H., *Rev. can. biol.*, 15 (1956) 238.
35 BOM, F., FRIDERICHSEN, T. AND JENSEN, A. R., *Ugeskrift Laeger*, 118 (1956) 329.
36 BONFILS, S., DUBRASQUET, M. AND LAMBLING, A., *Thérapie*, 15 (1960) 1096.
36a BONFILS, S., DUBRASQUET, M. AND LAMBLING, A., *Gastroenterologia*, 98 (1962) 217.
37 BONFILS, S., DUBRASQUET, M., ORY-LAVALLÉE, A. AND LAMBLING, A., *Compt. rend. soc. biol.*, 154 (1960) 924.
37a BORN, G. V. R., *J. Physiol. (London)*, 161 (1962) 160.
38 BRITTAIN, R. T. AND COLLIER, H. O. J., *J. Physiol. (London)*, 141 (1958) 14 P.
39 BRUMMER, P. AND JAUROLA, A., *Acta Med. Scand.*, 162 (1958) 397.
40 BÜLBRING, E. AND BURNSTOCK, G., *Brit. J. Pharmacol.*, 15 (1960) 611.
41 BÜLBRING, E. AND CREMA, A., *Brit. J. Pharmacol.*, 13 (1958) 444.
42 BÜLBRING, E. AND CREMA, A., *J. Physiol. (London)*, 146 (1959) 18.
43 BÜLBRING, E. AND CREMA, A., *J. Physiol. (London)*, 146 (1959) 29.
44 BÜLBRING, E., CREMA, A. AND SAXBY, O. B., *Brit. J. Pharmacol.*, 13 (1958) 440.

45 BÜLBRING, E. AND LIN, R. C. Y., *J. Physiol. (London)*, 138 (1957) 12 P.
46 BÜLBRING, E. AND LIN, R. C. Y., *J. Physiol. (London)*, 140 (1958) 381.
47 BURNSTOCK, G., *Brit. J. Pharmacol.*, 13 (1958) 216.
48 CALI, G. AND CORDOVA, C., *Progr. med. (Napoli)*, 12 (1956) 752.
49 CAMBRIDGE, G. W. AND HOLGATE, J. A., *Brit. J. Pharmacol.*, 10 (1955) 326.
49a CAMPBELL, A. C. P., GOWENLOCK, A. H., PLATT, D. S. AND SNOW, P. J. D., *Gut*, 3 (1962) 61.
50 CARRETI, D. AND MISSERE, G., *Chir. patol. sper.*, 3 (1955) 323.
51 CASENTINI, S. AND GALLI, G., *Boll. soc. ital. biol. sper.*, 32 (1956) 1640.
52 CASTIAU, J., *Arch. intern. pharmacodyn.*, 114 (1958) 478.
53 CASTIAU, J., *Compt. rend. soc. biol.*, 152 (1958) 1268.
53a CHRISTODOULOPOULOS, J. B. AND KLOTZ, A. P., *Gastroenterology*, 40 (1961) 429.
54 CLARK, M. L. AND SCHNEIDER, E. M., *Clin. Research Proc.*, 3 (1955) 206.
55 CLARK, M. L. AND SCHNEIDER, E. M., *Gastroenterology*, 29 (1955) 877.
56 CLIFTON, J. A., ATKINSON, M., HENDRIX, T. R. AND INGELFINGER, F. J., *J. Lab. Clin. Med.*, 48 (1956) 796.
56a COLE, J. W., SCHNEIDER, J. AND MCKALEN, A., *Proc. Soc. Exptl. Biol. Med.*, 107 (1961) 58.
57 COLLIER, H. O. J., in G. P. LEWIS (Ed.), *5-Hydroxytryptamine*, Pergamon, London, 1958, p. 5.
58 COLLIER, H. O. J., in G. P. LEWIS (Ed.), *5-Hydroxytryptamine*, Pergamon, London, 1958, p. 204.
59 COSTA, E., *Proc. Soc. Exptl. Biol. Med.*, 91 (1956) 39.
60 COSTA, E., GARATTINI, S. AND VALZELLI, L., *Experientia*, 16 (1960) 461.
60a CUGURRA, F. AND PARODI, S., *Arch. intern. pharmacodyn.*, 142 (1963) 52.
61 CURZON, G., *Lancet*, 269 (1955) 1361.
62 DALGLIESH, C. E., TOH, C. C. AND WORK, T. S., *J. Physiol. (London)*, 120 (1953) 298.
63 DAUGHERTY, G. W., MANGER, W. M., ROTH, G. M., FLOCK, E. V., CHILDS, D. S. AND WAUGH, J. M., *Proc. Mayo Clin.*, 30 (1955) 595.
64 DAVIDSON, J. D., SJOERDSMA, A., LOOMIS, L. N. AND UDENFRIEND, S., *Clin. Research Proc.*, 5 (1957) 304.
65 DEBRAY, C. AND BESANÇON, F., *Semaine hôp.*, 36 (1960) 3241.
65a DEBRAY, C. AND BESANÇON, F., *J. physiol. (Paris)*, 53 (1961) 525.
66 DE CORRAL SALETA, J., *Proc. 20th Intern. Physiol. Congress, Bruxelles*, 1956, p. 225.
67 DE GENNES, L. AND DE FOSSEY, M., *Presse méd.*, 64 (1956) 1066.
68 DELAY, J. AND THUILLIER, J., *Compt. rend. soc. biol.*, 150 (1956) 1335.
69 DOMENJOZ, R. AND THEOBALD, W., *Arch. intern. pharmacodyn.*, 20 (1959) 450.
70 DOUGLAS, W. W. AND TOH, C. C., *J. Physiol. (London)*, 120 (1953) 311.
70a DRAPANAS, T., MCDONALD, J. C. AND STEWART, J. D., *Ann. Surg.*, 156 (1962) 528.
71 DRASKOCI, M. AND HARANATH, P. S. R. K., *Brit. J. Pharmacol.*, 14 (1959) 2.
72 DUNCAN, D. A. AND FLEESON, W., *J. Am. Med. Assoc.*, 170 (1959) 1661.
73 EARL, A. E., *J. Pharmacol. Exptl. Therap.*, 113 (1955) 17.
74 ERSPAMER, V., *Ricerca sci.*, 22 (1952) 694.
75 ERSPAMER, V., *Arch. intern. pharmacodyn.*, 93 (1953) 293.
76 ERSPAMER, V., *Rend. sci. Farmitalia*, (1954) 1.
77 ERSPAMER, V., *Pharm. Revs.*, 6 (1954) 425.
78 ERSPAMER, V., GLÄSSER, A. AND MANTEGAZZINI, P., *Experientia*, 16 (1960) 505.
79 EULER, U. S. VON, AND OSTLUND, E., *Acta Physiol. Scand.*, 38 (1957) 364.
80 EVANS, D. H. L., SCHILD, H. O. AND THESLEFF, S., *J. Physiol. (London)*, 143 (1958) 474.
81 FEIN, S. B. AND KNUDTSON, K. P., *Cancer*, 9 (1956) 148.
82 FELDBERG, W. AND SMITH, A. M., *J. Physiol. (London)*, 122 (1953) 409.
83 FELDBERG, W. AND SMITH, A. M., *Brit. J. Pharmacol.*, 8 (1953) 406.
84 FELDBERG, W. AND TOH, C. C., *J. Physiol. (London)*, 119 (1953) 352.
85 FERGUSON, J., *Federation Proc.*, 16 (1957) No. 160.
86 FINGL, E. AND GADDUM, J. H., *Federation Proc.*, 12 (1953) No. 1057.
87 FRANCO-BROWDER, S., MASSON, G. M. C. AND CORCORAN, A. C., *J. Allergy*, 30 (1959) 1.
88 FREYBURGER, W. A., GRAHAM, B. E., RAPPORT, M. M., SEAY, P. H., GOVIER, W. M., SWOAP, O. F. AND VANDER BROOK, M. J., *J. Pharmacol. Exptl. Therap.*, 105 (1952) 80.
89 FUKADA, C., *Folia Pharmacol. Japon.*, 55 (1959) 34.
90 GADDUM, J. H., *J. Physiol. (London)*, 119 (1953) 363.
91 GADDUM, J. H., *J. Physiol. (London)*, 121 (1953) 15 P.

92   GADDUM, J. H., in G. P. LEWIS (Ed.), *5-Hydroxytryptamine*, Pergamon, London, 1958, p. 195.
93   GADDUM, J. H. AND HAMEED, K. A., *Brit. J. Pharmacol.*, 9 (1954) 240.
94   GADDUM, J. H., HAMEED, K. A., HATHWAY, D. E. AND STEPHENS, F. F., *Quart. J. Exptl. Physiol.*, 40 (1955) 49.
95   GADDUM, J. H. AND PICARELLI, P. Z., *Brit. J. Pharmacol.*, 12 (1957) 323.
96   GAITONDE, B. B., SATOSKAR, R. S. AND MANDREKAR, S. S., *Arch. intern. pharmacodyn.*, 127 (1960) 118.
97   GARATTINI, S., GIACHETTI, A. AND PALMA, V., to be published.
98   GARATTINI, S., GROSSI, E., PAOLETTI, P., PAOLETTI, R. AND POGGI, M., *Nature*, 191 (1961) 185.
99   GARATTINI, S., KATO, R. AND VALZELLI, L., *Experientia*, 16 (1960) 120.
100 GARATTINI, S., LAMESTA, L., MORTARI, A., PALMA, V. AND VALZELLI, L., *J. Pharm. Pharmacol.*, 13 (1961) 385.
101 GARATTINI, S. AND VALZELLI, L., *Boll. soc. ital. biol. sper.*, 31 (1955) 1648.
102 GARATTINI, S. AND VALZELLI, L., *Boll. soc. ital. biol. sper.*, 32 (1956) 288.
103 GATI, T., LUDANY, G. AND SANTHA, A., *Arch. intern. pharmacodyn.*, 113 (1957) 390.
104 GEORGES, G. AND HEROLD, M., *Compt. rend. soc. biol.*, 151 (1957) 77.
105 GEORGES, G. AND HEROLD, M., *Thérapie*, 13 (1958) 56.
106 GÉRARD, A., *Rev. belge pathol. méd. exptl.*, 27 (1960) 133.
107 GIARMAN, N. J., GREEN, V. S., GREEN, J. P. AND PAASONEN, M. K., *Proc. Soc. Exptl. Biol. Med.*, 94 (1957) 761.
108 GILLIS, C. N. AND LEWIS, J. J., *Nature*, 178 (1956) 859.
109 GILLIS, C. N. AND LEWIS, J. J., *J. Pharm. Pharmacol.*, 8 (1956) 606.
110 GILLIS, C. N. AND LEWIS, J. J., *Nature*, 179 (1957) 820.
111 GINZEL, K. H., in S. GARATTINI AND V. GHETTI (Eds.), *Psychotropic Drugs*, Elsevier, Amsterdam, 1957, p. 48.
112 GINZEL, K. H., *J. Physiol. (London)*, 137 (1957) 62 P.
113 GOBLE, A. J., HAY, D. R. AND SANDLER, M., *Lancet*, 269 (1955) 1016.
113a GRYGLEWSKI, R. AND SUPNIEWSKI, J., *Bull. acad. polon. sci.*, 11 (1963) 53.
114 GYERMEK, L., *Lancet*, 269 (1955) 724.
115 GYERMEK, L., LAZAR, I. AND CSAK, A. Z., *Arch. intern. pharmacodyn.*, 107 (1956) 62.
116 HAÜSON, A. AND SERIN, F., *Lancet*, 269 (1955) 1359.
117 HAVERBACK, B. J. AND BOGDANSKI, D. F., *Proc. Soc. Exptl. Biol. Med.*, 95 (1957) 392.
118 HAVERBACK, B. J., BOGDANSKI, D. F. AND HOGBEN, C. A. M., *Gastroenterology*, 34 (1958) 188.
119 HAVERBACK, B. J. AND DAVIDSON, J. D., *Gastroenterology*, 35 (1958) 570.
120 HAVERBACK, B. J., HOGBEN, C. A. M., MORAN, N. AND TERRY, L., *Gastroenterology*, 32 (1957) 1058.
121 HAVERBACK, B. J., STEVENSON, T. D., SJOERDSMA, A. AND TERRY, L., *Am. J. Med. Sci.*, 230 (1955) 601.
121a HAVERBACK, B. J. AND WIRTSCHAFTER, S. K., *Advan. Pharmacol.*, 1 (1962) 309.
122 HAYAMA, T., *Federation Proc.*, 19 (1960) 188.
122a HEDINGER, C., *Schweiz. med. Wochschr.*, 89 (1959) 1362.
123 HEDINGER, C. AND VERAGUTH, F., *Schweiz. med. Wochschr.*, 87 (1957) 1175.
124 HEILMEYER, L. AND CLOTTEN, R., *Deut. med. Wochschr.*, 83 (1958) 338.
125 HENDRIX, T. R., ATKINSON, M., CLIFTON, J. A. AND INGELFINGER, F. J., *Am. J. Med.*, 23 (1957) 886.
126 HOLTZ, P., *Deut. med. Wochschr.*, 83 (1958) 681.
127 HORN, L., KROHN, H. F. AND ZWEIFACH, B. W., *Federation Proc.*, 20 (1961) 301.
128 HUSSAR, A. E. AND BRUNO, E., *Gastroenterology*, 31 (1956) 500.
129 INNES, I. R., KOSTERLITZ, H. W. AND ROBINSON, J. A., *J. Physiol. (London)*, 137 (1957) 396.
130 JACOB, J., *Les Antagonistes de la Sérotonine. Actualités Pharmacologiques*, Masson, Paris, 1960, p. 131.
131 JASMIN, G., *Rev. can. biol.*, 15 (1956) 107.
132 JASMIN, G. AND BOIS, P., *Rev. can. biol.*, 18 (1959) 367.
133 JEPSON, J. E., *Lancet*, 269 (1955) 1009.
133a JOHNSON, L. P., SLOOP, R. D. AND JESSEPH, J. E., *J. Am. Med. Assoc.*, 180 (1962) 493.
134 JORI, A. AND LEONARDI, A., *Boll. soc. ital. biol. sper.*, 35 (1959) 860.
135 KABAKOW, B., WEINSTEIN, J. B. AND ROSS, G., *Federation Proc.*, 17 (1958) 382.

136 KÄRKI, N. T. AND PAASONEN, M. K., *J. Neurochem.*, 3 (1959) 352.
137 KÄRKI, N. T., PAASONEN, M. K. AND PELTOLA, P., *Ann. Med. Exptl. et Biol. Fenniae (Helsinki)*, 38 (1960) 231.
138 KELLER, R., *Experientia*, 13 (1957) 112.
138a KERBI, G. P. AND TAYLOR, S. M., *Proc. Soc. Exptl. Biol. Med.*, 107 (1961) 936.
139 KIMURA, E. T., YOUNG, P. R. AND RICHARDS, R. K., *J. Allergy*, 31 (1960) 237.
140 KOSTERLITZ, H. W. AND ROBINSON, J. A., *J. Physiol. (London)*, 129 (1955) 18 P.
141 KOSTERLITZ, H. W. AND ROBINSON, J. A., *J. Physiol. (London)*, 136 (1957) 249.
142 KOSTERLITZ, H. W. AND ROBINSON, J. A., *Brit. J. Pharmacol.*, 13 (1958) 296.
143 KOWLESSAR, O. D., WILLIAMS, R. C., LAW, D. H. AND SLEISINGER, M. H., *New Engl. J. Med.*, 259 (1958) 340.
143a KOZÁK, J. AND LANG, N., *Nature*, 197 (1963) 293.
144 KROGSGAARD, A. R., *Nord. Med.*, 54 (1955) 1851.
145 KROGSGAARD, A. R., *Acta Med. Scand.*, 158 (1957) 1.
146 LA BARRE, J., *Arch. intern. pharmacodyn.*, 114 (1958) 483.
147 LA BARRE, J., *Compt. rend. soc. biol.*, 153 (1959) 364.
148 LA BARRE, J., *Compt. rend. soc. biol.*, 154 (1960) 444.
149 LA BARRE, J. AND CASTIAU, J., *Compt. rend. soc. biol.*, 151 (1957) 2222.
150 LA BARRE, J. AND DESMAREZ, J. J., *Compt. rend. soc. biol.*, 151 (1957) 1451.
151 LA BARRE, J. AND GILLO, L., *Compt. rend. soc. biol.*, 151 (1957) 1978.
152 LA BARRE, J. AND GILLO, L., *Compt. rend. soc. biol.*, 152 (1958) 530.
153 LA BARRE, J. AND HANS, M. J., *Arch. intern. pharmacodyn.*, 116 (1958) 255.
154 LA BARRE, J. AND LIEBER, C. S., *Compt. rend. soc. biol.*, 151 (1957) 1449.
155 LA BARRE, J. AND LIEBER, C. S., *Compt. rend. soc. biol.*, 151 (1957) 1979.
156 LA BARRE, J., LIEBER, C. S. AND CASTIAU, J., *Compt. rend. soc. biol.*, 152 (1958) 532.
157 LA BARRE, J., LIEBER, C. S. AND CASTIAU, J., *Compt. rend. soc. biol.*, 152 (1958) 1270.
157a LAMBERT, R., *Compt. rend. soc. biol.*, 156 (1962) 81.
158 LANGEMANN, H. AND KAKI, J., *Klin. Wochschr.*, 34 (1956) 237.
159 LANZ, R., *Schweiz. med. Wochschr.*, 90 (1960) 1046.
160 LAURENCE, D. R. AND ROSENHEIM, M. L., in J. R. VANE, G. E. W. WOLSTENHOLME AND M. O'CONNOR (Eds.), *Adrenergic Mechanisms*, Churchill, London, 1960, p. 201.
161 LEE, C. Y., *J. Physiol. (London)*, 148 (1959) 63 P.
162 LEE, C. Y., *J. Physiol. (London)*, 152 (1960) 405.
163 LEITCH, J. L., MARYN, D., DEBLEY, V. G. AND HALEY, T. J., *J. Pharmacol. Exptl. Therap.*, 120 (1957) 408.
164 LEMBECK, F., *Nature*, 172 (1953) 910.
165 LEMBECK, F., *Arch. exptl. Pathol. Pharmakol., Naunyn-Schmiedeberg's*, 221 (1954) 50.
166 LEMBECK, F., in G. P. LEWIS (Ed.), *5-Hydroxytryptamine*, Pergamon, London, 1958, p. 147.
167 LEMBECK, F., *Arch. ges. Physiol., Pflüger's*, 265 (1958) 567.
168 LEONARD, S. L. AND DAY, H. T., *Proc. Soc. Exptl. Biol. Med.*, 104 (1960) 338.
169 LEUSEN, J. AND LACROIX, E., *Arch. intern. pharmacodyn.*, 121 (1959) 114.
170 LEUSEN, J., LACROIX, E. AND DEMEESTER, G., *Experientia*, 15 (1959) 73.
170a LEUSEN, I., LACROIX, E. AND DEMEESTER, G., *Arch. intern. pharmacodyn.*, 119 (1959) 225.
170b LEUSEN, I., THOENEN, H. AND LACROIX, E., *Arch. intern. pharmacodyn.*, 141 (1963) 190.
171 LEVIS, S. AND BEERSAERTS, J., *Arch. intern. pharmacodyn.*, 126 (1960) 359.
172 LEVY, J. AND MICHEL-BER, E., *Compt. rend.*, 242 (1956) 3007.
173 LEVY, J. AND MICHEL-BER, E., *J. physiol. (Paris)*, 48 (1956) 1051.
174 LEWIS, G. P., *Brit. J. Pharmacol.*, 15 (1960) 425.
175 LIEBOWITZ, D. AND CARBONE, J. V., *New Engl. J. Med.*, 257 (1957) 227.
176 LIEBOWITZ, D. AND CARBONE, J. V., *Proc. Soc. Exptl. Biol. Med.*, 100 (1959) 27.
177 LISH, P. M., CLARK, B. B. AND ROBBINS, S. I., *Am. J. Physiol.*, 197 (1959) 22.
178 LUDANY, G., GATI, T. AND HIDEG, J., *Arch. intern. pharmacodyn.*, 114 (1958) 227.
179 LUDANY, G., GATI, T., SZABO, ST. AND HIDEG, J., *Arch. intern. pharmacodyn.*, 118 (1959) 62.
180 MAC DONALD, R. A., *Am. J. Med.*, 21 (1956) 867.
181 MACFARLANE, P. S., DALGLIESH, C. E., DUTTON, R. W., LENNOX, B., NYHUS, L. M. AND SMITH, A. N., *Scot. Med. J.*, 1 (1956) 148.
182 MARIANI, L. AND VILLANI, R., *Boll. soc. ital. biol. sper.*, 35 (1959) 1598.

183 McQUEEN, E. G., DOYLE, A. E. AND SMIRK, F. H., *Circulation*, 11 (1955) 161.
184 MEDAKOVIC, M., *Acta Med. Jugoslav.*, 11 (1957) 186.
185 MEDAKOVIC, M., *Arch. intern. pharmacodyn.*, 114 (1958) 201.
186 MEDAKOVIC, M., *Acta Med. Jugoslav.*, 12 (1958) 176.
187 MEDAKOVIC, M., *Acta Med. Jugoslav.*, 12 (1958) 283.
188 MORENO, O. M. AND BRODIE, D. A., *Federation Proc.*, 20 (1961) 252.
189 MURELLI, B., VALSECCHI, A. AND VALZELLI, L., *Boll. soc. ital. biol. sper.*, 33 (1957) 859.
189a MUSTALA, O. O. AND AIRAKSINEN, M. M., *Acta Med. Scand.*, 171 (1962) 483.
190 NELSON, S. D., *Ulster Med. J.*, 26 (1957) 186.
190a NICHOLSON, G. I. AND IRVINE, R. O. H., *Brit. Med. J.*, 2 (1962) 961.
191 NIKODIJEVIC, B. AND VANOV, S., *Experientia*, 16 (1960) 464.
191a OATES, J. A. AND SJOERDSMA, A., *Am. J. Med.*, 32 (1962) 333.
192 OLSON, T. E. AND GRAY, S. J., *Am. J. Gastroenterol.*, 29 (1958) 280.
193 OWENS, F. J. AND MASSON, G. M. C., *J. Lab. Clin. Med.*, 52 (1958) 932.
194 PAASONEN, M. K., PELTOLA, P. AND VANHAKARTANO, P. A., *Ann. Med. Exptl. et Biol. Fenniae (Helsinki)*, 38 (1960) 220.
195 PAGE, I. H., CORCORAN, A. C., UDENFRIEND, S., SJOERDSMA, A. AND WEISSBACH, H., *Lancet*, 268 (1955) 198.
196 PAINTAL, A. S., *J. Physiol. (London)*, 126 (1954) 271.
197 PEART, W. S., ANDREWS, T. M. AND ROBERTSON, J. I. S., *Lancet*, 280 (1961) 577.
198 PEART, W. S., ROBERTSON, J. I. S. AND ANDREWS, T. M., *Lancet*, 276 (1959) 715.
199 PERNOW, B. AND WALDENSTRÖM, J., *Lancet*, 267 (1954) 951.
200 PERNOW, B. AND WALDENSTRÖM, J., *Am. J. Med.*, 23 (1957) 16.
200a PESKIN, G. W. AND MILLER, L. D., *Arch. Surg.*, 85 (1962) 701.
201 PICK, E. P., *Arch. intern. pharmacodyn.*, 126 (1960) 374.
202 PICKLES, V. R., *J. Physiol. (London)*, 134 (1956) 474.
203 PICKLES, V. R., *J. Physiol. (London)*, 138 (1957) 495.
204 PICKLES, V. R. AND SUTCLIFFE, J. F., *Biochim. Biophys. Acta*, 17 (1955) 244.
205 PLETSCHER, A., SHORE, P. A. AND BRODIE, B. B., *Science*, 122 (1955) 130.
206 PLUMMER, A. J., BARRETT, W. E. AND RUTLEDGE, R., *Am. J. Digest. Diseases*, 22 (1955) 337.
207 PLUMMER, A. J., EARL, A. E., SCHNEIDER, J. A., TRAPOLD, J. AND BARRETT, W., *Ann. N.Y. Acad. Sci.*, 59 (1954) 8.
208 PUTZ, F. AND SPIES, R., *Wien. med. Wochschr.*, 104 (1954) 202.
209 QUADBECK, G. AND RÖHM, E., *Z. physiol. Chem., Hoppe-Seyler's*, 297 (1954) 229.
210 RADOUCO-THOMAS, C., *Arch. exptl. Pathol. Pharmakol., Naunyn-Schmiedeberg's*, 238 (1960) 132.
211 RADOUCO-THOMAS, C., LATASTE-DOROLLE, C., ROGG-EFFRON, C., VOLUTER, G., MEYER, M., CHUAMONTET, J. M. AND LARNE, D., *Arzneimittel-Forsch.*, 10 (1960) 588.
212 RAPPORT, M. M. AND KOELLE, G. B., *Arch. intern. pharmacodyn.*, 92 (1953) 464.
213 REID, G. AND RAND, M., *Nature*, 169 (1952) 801.
213a RESNICK, R. H.,, ADELARDI, C. F. AND GRAY, S. J., *Gastroenterology*, 42 (1962) 22.
213b RESNICK, R. H. AND GRAY, S. J., *J. Lab. Clin. Med.*, 59 (1962) 462.
214 RIDER, J. A., *Proc. Soc. Exptl. Biol. Med.*, 90 (1955) 736.
215 RIDER, J. A., MOELLER, H. C. AND GIBBS, J. O., *Gastroenterology*, 33 (1957) 737.
216 ROBERTSON, P. A., *J. Physiol. (London)*, 121 (1953) 54P.
217 ROBERTSON, P. A., *J. Physiol. (London)*, 125 (1954) 37P.
218 ROCHA E SILVA, M., VALLE, J. R. AND PICARELLI, Z. P., *Brit. J. Pharmacol.*, 8 (1953) 378.
219 ROSA, L., CENACCHI, G. C. AND TOSCHI, G. P., *Bologna med.*, 4 (1958) 73.
220 ROTHLIN, E., *Ann. N.Y. Acad. Sci.*, 66 (1957) 668.
220a RUMMEL, W., *Med. Exptl.*, 4 (1961) 126.
221 SANDLER, M., DAVIES, A. AND RIMINGTON, C., *Lancet*, 2 (1959) 318.
222 SANDLER, M. AND WEST, G. B., *J. Physiol. (London)*, 140 (1958) 9P.
223 SCHAPIRO, H. AND WOODWARD, E. R., *Federation Proc.*, 20 (1961) 248.
224 SCHMID, E. AND KINZLMEIER, H., *Arch. exptl. Pathol. Pharmakol., Naunyn-Schmiedeberg's*, 236 (1959) 51.
225 SCHMID, E. AND KINZLMEIER, H., *Gastroenterologia*, 91 (1959) 248.
226 SCHMID, E. AND KINZLMEIER, H., *Gastroenterologia*, 91 (1959) 254.

226a SCHMID, E., ROSENBUSCH, U., HEINKEL, K., SCHWEMMLE, K. AND SCHÖN, H., *Gastroenterologia*, 96 (1961) 275.

227 SCHMID, V., ZICHA, L. AND SCHEIFFARTH, F., *Med. Exptl.*, 2 (1960) 266.

227a SCHNECKLOTH, R. E., McISAAC, W. M. AND PAGE, I. H., *J. Am. Med. Assoc.*, 170 (1959) 1143.

228 SCHNECKLOTH, R., PAGE, I. H., DEL GRECO, F. AND CORCORAN, A. C., *Circulation*, 16 (1957) 523.

229 SCHNEIDER, E. M. AND CLARK, M. L., *Am. J. Digest. Diseases*, 1 (1956) 22.

230 SCHNEIDER, J. A., *Am. J. Physiol.*, 181 (1955) 64.

231 SCHOFIELD, G., *Proc. Univ. Otago Med. School*, 36 (1958) 2.

232 SCHÜMANN, H. J., in J. R. VANE, G. E. W. WOLSTENHOLME AND M. O'CONNOR (Eds.), *Adrenergic Mechanisms*, Churchill, London, 1960, p. 6.

233 SELYE, H., JEAN, P. AND CANTIN, M., *Proc. Soc. Exptl. Biol. Med.*, 103 (1960) 444.

234 SHAY, H. S., SUN, D. C. H. AND GRUENSTEIN, M., *Federation Proc.*, 16 (1957) 118.

235 SHAY, H. S., SUN, D. C. H. AND GRUENSTEIN, M., *Federation Proc.*, 17 (1958) 580.

235a SIGEL, B., POTASH, I. M. AND WOLCOTT, M. W., *Proc. Soc. Exptl. Biol. Med.*, 105 (1960) 237.

236 SINHA, Y. K. AND WEST, G. B., *J. Pharm. Pharmacol.*, 5 (1953) 370.

237 SJOERDSMA, A., WEISSBACH, H. AND UDENFRIEND, S., *J. Am. Med. Assoc.*, 159 (1955) 397.

237a SMITH, A. N., *Proc. Royal Soc. Med.*, 52 (1959) 24.

238 SMITH, A. N., in G. P. LEWIS (Ed.), *5-Hydroxytryptamine*, Pergamon, London, 1958, p. 183.

239 SMITH, A. N., BLACK, J. W. AND FISHER, E. W., *Nature*, 180 (1957) 1127.

240 SMITH, G. AND SMITH, A. N., *Surg. Gynecol. Obstet.*, 101 (1955) 691.

241 SMITH, S. E., *Brit. J. Pharmacol.*, 15 (1960) 319.

242 STONE, C. A., personal communication.

243 STONE, C. A., WENGER, H. C., LUDDEN, C. T., STAVORSKI, J. M. AND ROSS, C. A., *J. Pharmacol. Exptl. Therap.*, 131 (1961) 73.

244 TELFORD, J. M. AND WEST, G. B., *Brit. J. Pharmacol.*, 15 (1960) 532.

244a THOENEN, H., LACROIX, E. AND LEUSEN, I., *Gastroenterologia*, 98 (1962) 225.

245 THORSON, A. H., *Acta Med. Scand.*, (1958) Suppl. 334.

246 THORSON, A. H., BIORK, G., BJOKMAN, G. AND WALDENSTROM, J., *Am. Health J.*, 47 (1954) 795.

247 TOH, C. C. AND MOHIUDDIN, A., *Brit. J. Pharmacol.*, 13 (1958) 113.

248 TYCE, G. M., STODIE, G. H. C. AND BOLLMAN, J. L., *Federation Proc.*, 20 (1961) 252.

249 UDENFRIEND, S., WEISSBACH, H. AND SJOERDSMA, A., *Science*, 123 (1956) 669.

250 VALZELLI, L., *Rec. progr. med.*, 29 (1960) 426.

251 VANE, J. R., *Brit. J. Pharmacol.*, 12 (1957) 344.

252 VANE, J. R., *Brit. J. Pharmacol.*, 14 (1959) 87.

253 VANE, J. R., in J. R. VANE, G. E. W. WOLSTENHOLME AND M. O'CONNOR (Eds.), *Adrenergic Mechanisms*, Churchill, London, 1960, p. 357.

254 VERTUA, R. AND POGGI, M., *Ricerca sci.*, 30 (1960) 1600.

255 VIALLI, M. AND QUARONI, E., *Rivista istochim. norm. patol.*, 2 (1956) 111.

256 VOGT, W., *Arch. exptl. Pathol. Pharmakol., Naunyn-Schmiedeberg's*, 222 (1954) 427.

257 WAALKES, T. P. AND WEISSBACH, H., *Proc. Soc. Exptl. Biol. Med.*, 93 (1956) 394.

258 WALDENSTRÖM, J., PERNOW, B. AND SILVER, H., *Acta Med. Scand.*, 156 (1956) 73.

258a WARNER, R. R. P., *J. Mt. Sinai Hosp., N.Y.*, 26 (1959) 450.

259 WEST, W. O., *Ann. Internal Med.*, 48 (1958) 1033.

260 WHITE, T. T., MACALEXANDER, R. A. AND MAGEE, D. F., *Surg. Gynecol. Obstet.*, 109 (1959) 168.

261 WHITE, T. T. AND MAGEE, D. F., *Gastroenterology*, 35 (1958) 289.

262 WILHELMI, G., *Helv. Physiol. Acta*, 15 (1957) C 83.

263 WILHELMI, G. AND SCHINDLER, W., *Arch. exptl. Pathol. Pharmakol., Naunyn-Schmiedeberg's*, 236 (1959) 49.

264 WILKINS, R. W., *Ann. N.Y. Acad. Sci.*, 59 (1954) 36.

265 WILKINS, R. W. AND JUDSON, W. E., *New Engl. J. Med.*, 248 (1953) 48.

266 WINSOR, T., *Ann. N.Y. Acad. Sci.*, 59 (1954) 61.

267 WIRTHEIMER, C., *Compt. rend. soc. biol.*, 153 (1959) 1284.

268 WONG, R. L. AND BENDITT, E. P., *Am. J. Pathol.*, 33 (1956) 623.

269 WOOLLEY, D. W., *Proc. Natl. Acad. Sci. U.S.*, 44 (1958) 197.

270 WOOLLEY, D. W., *Proc. Natl. Acad. Sci. U.S.*, 44 (1958) 1202.

271 WOOLLEY, D. W., *Science*, 128 (1958) 1277.

272 WOOLLEY, D. W., *Proc. Soc. Exptl. Biol. Med.*, 98 (1958) 367.

273   WOOLLEY, D. W., *Biochem. Pharmacol.*, 3 (1959) 51.
274   WOOLLEY, D. W. AND CAMPBELL, N. K., *Biochim. Biophys. Acta*, 40 (1960) 543.
275   WOOLLEY, D. W. AND SHAW, E., *J. Biol. Chem.*, 203 (1953) 69.
276   WOOLLEY, D. W., VAN WINKLE, E. AND SHAW, E., *Proc. Natl. Acad. Sci. U.S.*, 43 (1957) 128.
277   ZBINDEN, G. AND PLETSCHER, A., *Schweiz. Z. Pathol. u. Bakteriol.*, 21 (1958) 1137.
278   ZBINDEN, G., PLETSCHER, A. AND STUDER, A., *Schweiz. med. Wochschr.*, 89 (1959) 289.

Chapter 6

IMPORTANCE OF SEROTONIN
IN ANAPHYLACTIC, ALLERGIC AND
INFLAMMATORY REACTIONS

In this chapter will be discussed problems ranging from the bronchospasm induced by serotonin to the release of serotonin after antigen–antibody reaction, and to the possibility that serotonin has a role in the symptomatology induced by anaphylactic, anaphylactoid and inflammatory reactions*.

This is a field which is rapidly developing and therefore no definite conclusions can be drawn. Moreover, in these reactions there are striking differences according to the animal species used. Many of the problems here discussed for serotonin are similar to those studied some years ago in relation with the suggested importance of histamine in allergic reactions.

There are two general approaches used by the various investigators to study the possible importance of serotonin in anaphylaxis and inflammation: firstly, the measurement of the levels of serotonin in various tissues during or after such reactions, and secondly, the administration of drugs interfering with serotonin metabolism (reserpine and monoamine oxidase inhibitors) or inhibiting the pharmacological effects of serotonin (lysergic acid derivatives). Since these approaches are indirect, a word of caution must be given on the validity of the conclusion reached, whether positive or negative.

BRONCHOSPASM BY SEROTONIN IN GUINEA PIGS

Serotonin injected at a dose of 50–500 μg/kg (refs. 180, 287) or given by aerosol at a concentration of 0.5–1 % induces bronchospasm in the guinea pig[184]. The gross symptomatology is similar to that elicited by histamine, acetylcholine or anaphylactic shock. According to Herxheimer[183,188] the dyspnoea, cyanosis and bronchospasm occurring after 30 sec to 1 min of exposure to a 1 % serotonin aerosol correspond to the effects obtained with 0.5 % histamine or 0.25 % acetylcholine aerosol. After prolonged exposure to serotonin aerosol the animals become tolerant[185].

It is likely that the major component of the bronchospasm is due to a direct effect of serotonin on the bronchial smooth muscles[43,121]. Resection of the spinal cord does not prevent the bronchospasm produced by serotonin in guinea pigs[244].

Autoradiographic experiments show that [3-^{14}C]serotonin is selectively fixed in the muscular layer of the bronchi in guinea pigs[292].

---

* For reviews on these problems see refs. 98, 226, 255, 310, 416, 468, 469.

TABLE 19

ISOLATED BRONCHIAL MUSCLE

Minimum effective doses expressed as μg of the substances per ml in bath. From Brockle-hurst[43]

	Bronchioles			Trachea		
Animal species	Serotonin	Histamine	Acetyl-choline	Serotonin	Histamine	Acetyl-choline
Cat	0.01	2.0	0.1	0.02	>20	0.05
Rat	0.01	>5	0.04	0.1	>5	0.04
Dog	0.05	0.3	0.1	as bronchioles		
Guinea pig	0.4	0.4	0.4	as bronchioles		
Rabbit	> 8	0.5	0.2	>8	>5	0.4
Rh. monkey	>20	0.5	0.1	0.02	0.5	0.01
Man	>20	0.2	0.1			

Bold numerals denote relaxation after large doses. In the frog lung, 5-hydroxytryptamine induces relaxation ($5 \cdot 10^{-9}$ g/100 ml) while histamine and acetylcholine are constrictors[41,41a].

A detailed description of other effects of serotonin on the respiratory function in guinea pigs is given by Krueger and Smith[246]. According to these authors serotonin decreases the activity of the cilia (ciliary rate), contracts the posterior tracheal wall, increases the response of the tracheal mucosa to trauma, induces vasoconstriction in the tracheal wall and increases respiratory rate. These effects are shared by iproniazid and (+) air ions, while an opposite picture may be obtained by reserpine treatment or (—) air ions[246].

Table 19 shows the concentrations of serotonin, histamine and acetylcholine required to induce contraction of bronchioles and trachea *in vitro* in several animal species. Serotonin and histamine have about the same activity in the bronchial muscles of the guinea pig. The effect of serotonin *in vitro* is reduced[122] by a decrease of the pH.

In the isolated perfused lung the respiratory volume is reduced by serotonin (refs. 29, 32, 242, 429, 430). This effect is prevented or antagonized by antiserotonin agents like 2-bromodiethyllysergamide (BOL 148)[242], 1-benzyl-2-methyl-5-methoxy-tryptamine (BAS), 1-benzyl-2-methyl-5-methoxybufotenine (BAB)[429] and chlor-promazine[242].

The bronchospasm elicited *in vivo* by serotonin aerosol has been widely used as a test to evaluate antiserotonin agents*.

Diethylamide of lysergic acid (LSD 25)[20,184,187,188,197,209,244] and its derivatives BOL 148 (refs. 242, 407) and UML 491 (butanolamide of 1-methyllysergic acid)[407] are specific inhibitors of the serotonin bronchospasm and they are inactive in the bronchospasm induced by histamine[187,244,407] or bradykinin[66].

Also 1,4-dimethyl-7-isopropylazulene shows activity against serotonin without

---

* See a recent review on 5-hydroxytryptamine antagonists[160].

interfering with histamine-induced bronchospasm[138,426,427]. Serotonin itself given by subcutaneous route protects guinea pigs from the bronchospasm induced by serotonin aerosol[69]. This may be an aspect of the tachyphylaxis which occurs after repeated administration of serotonin[185]. This effect does not occur when histamine or acetylcholine are given by aerosol.

On the other hand, histamine or acetylcholine given by general route do not prevent the bronchospasm induced by aerosol of each one of the bronchoconstrictor agents[69].

There are various other pharmacological agents which can prevent the bronchospasm induced by serotonin, but they are not specific because they are effective also against histamine.

Many antihistamine drugs show a clear antiserotonin activity[20,197,286,401]. The most specific antihistamine agents are considered to be tripolidine[124], chlorpyribenzamine[401] and mepyramine[187].

Phenothiazine derivatives show a strong antiserotonin activity but they have also antihistamine[68,187,401] and antibradykinin activity[268].

Chlorpromazine has been studied most extensively[20,209,286,401]. It prevents the fixation of serotonin to the muscular layer of the bronchi[292] and other tissues[293] of guinea pigs.

The ratio between antihistamine and antiserotonin activity (bronchospasm) is about 3 for chlorpromazine and 0.1 for thiopropazate. Other phenothiazine compounds have approximately equal activity against histamine or serotonin bronchospasm[286].

Sympathomimetic agents, because of their bronchodilator action, are effective in reducing the bronchospasm by serotonin[29,117,209]. Table 20 shows the effect of a number of drugs on the bronchospasm induced by histamine and serotonin.

TABLE 20

EFFECT OF DRUGS ON BRONCHOSPASM INDUCED BY SEROTONIN, HISTAMINE OR ANAPHYLACTIC SHOCK IN GUINEA PIGS
(Numbers indicate references)

Compound	Serotonin bronchospasm	Histamine bronchospasm	Anaphylactic shock
Acetylcholine	— 69	— 69	
Acetylsalicylic acid	— 65	— 65	
Adrenaline	+ 180, 209 ± 187	+ 187	+ 187
Aminophenazone	— 480	+ 480	
Aminophylline	+ 187	+ 187, 352	+ 187
Aminopyrine	— 65	— 65	
Amphetamine	— 209		
Antazoline (Antistine)	+ 294, 401	+ 294, 401	+ 2, 11
Antergan (2339 R.P.)		+ 38, 165	+ 166

TABLE 20 (*continued*)

Compound	Serotonin bronchospasm	Histamine bronchospasm	Anaphylactic shock
Atropine	+ 188, 197, 209	+ 38, 187	+ 11, 185, 404 ± 183
Azamethonium	— 209		
1-Benzyl-5-methoxy-2-methyl- tryptamine (BAS)	(*)		— 326
Benzyloxy-5-tryptamine	— 69		
Benzyloxy-6-tryptamine	— 69		
BOL	+ 407	— 407	
Bufotenine	+ 69		
Caffeine	± 209		+ 186
Carisoprodol	+ 480	+ 480	
Chlorcyclizine	+ 230	+ 230	+ 230
Chlorisondamine	+ 209	— 124	
Chlorphenamine (Chlortrimeton)		+ 159	+ 11, 404
Chlorpromazine	+ 69, 187, 209, 401, 478	+ 68, 187, 401	+ 187, 286, 291 ± 175 — 186
Chlorpyramine (Synopen)	+ 401	+ 401	
Chlorpyribenzamine	+ 401	+ 401	
Cocaine	+ 187 — 124, 209	+ 187 — 124	+ 187
Cyproheptadine	+ 124	+ 428	
Dexametazone	+ 482	+ 482	
Dibenamine	+ 124 ± 209	— 124	
Dibucaine (Nupercaine)	— 209		
Dihydroergotamine	+ 188 ± 184		— 184
10-(3-Dimethylaminopropyl)-2- (trifluoromethyl)phenothiazine (Vesprin)	+ 294	+ 294	
1,4-Dimethyl-7-isopropylazulene (Guaiazulen)	+ 138	— 138	+ 138
Diphenhydramine (Benadryl)	+ 294 ± 197	+ 197, 294	
Ephedrine	+ 209	+ 38	
Ergotamine	+ 124 — 407	— 124	
γ-Globulin–histamine complex	+ 397	+ 397	
Hexamethonium	+ 187	+ 187	+ 187
Histamine	— 69	— 69	
Homochlorcyclizine (SA-97)	+ 230	+ 230	+ 230
Hydralazine	— 209		
DL-5-Hydroxyacetyltryptophan	— 69		
5-Hydroxyindoleacetic acid	± 69		
5-Hydroxytryptamine	+ 69	— 69	
DL-5-Hydroxytryptophan	— 69		
Imipramine (Tofranil)	+ 294	+ 294	
Isopropylnoradrenaline	+ 117, 209	+ 117	
Isothipendyl (Andantol)	+ 401	+ 401	

TABLE 20 (*continued*)

Compound	Serotonin bronchospasm	Histamine bronchospasm	Anaphylactic shock
Levopromazine	+ 69, 478		
LSD 25	+ 157, 184, 187, 188, 197, 209, 244, 407, 478	— 187, 244, 407	+ 342 — 175, 184, 187, 188, 326, 404
Meprobamate	— 478		— 285, 287
Mepyramine (Pyrilamine)	— 184, 185, 187, 188, 407 ± 197	+ 39, 183, 187, 407	+ 11, 166, 175, 187
Methamphetamine (Pervitin)	+ 209		
α-Methyl-DOPA	+ 470**		
Methylergobasine	+124	—124	
Methylphenenetidate	+ 209		
Morphine	—124	—124	
Nikethamide	— 209		
Noradrenaline	± 209	+ 38	
Oxyphenonium (Antrenyl)	+ 209		
Papaverine	+ 187	+ 187 ± 124	— 187
Perphenazine (Trilafon)	+ 294	+ 294	
Phenindamine (Thephorin)	+ 401	+ 401	
Phenoxybenzamine	—124	+124	
Phentolamine	+124	+124	
Phenylbutazone	— 65	— 65	
Prednisolone	+ 481	+ 481	
Procaine	— 124, 209***	+ 180 — 124	+ 440 — 11
Prochlorphenazine	+ 69		
Promazine	+ 294	+ 294	+ 287
Promethazine	+ 188, 294	+ 185, 294	+ 11, 166, 175
Propantheline	+ 187	+ 187	+ 186, 187
Prothipendyl (Dominal)	+ 401	+ 401	
Quinine	+ 480	+ 480	
Reserpine	+ 209 — 478		+ 404, 452
Salicylate	— 480	+ 480	
Thenophenopiperidine (Sandostene)	+ 401	+ 401	
Theophylline	± 209		+ 283
Thiopropazate (Dartal)	+ 294	+ 294	+ 286
Trimeprazine	+ 407	+ 407	
Tripelennamine (Pyribenzamine)	+ 209, 294, 401	+ 38, 294, 401	+ 11, 286
Tripolidine	—124	+124	
L-Tryptophan	— 69		
UML	+ 407	— 407	
Urethane	— 287	— 287	+ 285, 287
Yohimbine	— 185 ± 209		

* active on isolated lung[429]	+ protection
** bronchospasm induced by 5-hydroxytryptophan	— no effect
*** inactive on isolated lung[32]	

ANAPHYLACTIC SHOCK AND SEROTONIN IN GUINEA PIGS

It is obvious that several investigators have posed the question whether the symptomatology of anaphylactic shock in the guinea pig might be mediated by a release of serotonin. The interest in this problem is enhanced by the lack of a definitive or generally accepted proof that histamine is the mediator of anaphylactic shock in the guinea pig. Antihistamines protect against anaphylactic shock only when used in relatively high doses[20,140,286,410] at which antiserotonin activity is also present.

An interesting lead was offered by Herxheimer[185,188] who found that development of the resistance to the shock induced by albumin paralleled the resistance to the bronchospasm by serotonin. The same parallelism also occurred in respect to nicotine (ref. 185). However, it has been suggested that nicotine may act as a constrictor of smooth muscle through a release of serotonin[132].

Guinea pigs in anaphylactic shock show an increase of serotonin in lung[219,291,*], but other authors found only minor variations in the serotonin content of lung, intestine and spleen[99]. No changes were found in blood serotonin during anaphylactic shock in guinea pigs[207]. On the other hand it is known that under the same conditions there is an increase of blood histamine[60,139,290] and a decrease of tissue histamine (refs. 61, 94, 290).

Experiments designed to study the physiologically active substances released from lung during anaphylactic shock showed that histamine, but not serotonin or bradykinin, was present in the perfusate[4,140,456]. It is possible that *in vivo* histamine and serotonin, even if released in small quantities, may add their effects[288].

EFFECT OF DRUGS

Table 20 gives a comparative summary of the effect of several drugs as inhibitor of anaphylactic shock, and histamine or serotonin bronchospasm. Consistent with the above-mentioned data is the fact that LSD 25, inhibitor of serotonin but not histamine bronchospasm, does not protect against anaphylactic shock[175,184,187,188]. However, other authors reported that high doses of LSD 25 show a weak anti-anaphylactic activity[342].

On the other hand, in favour of serotonin mediation is the effect of 1,4-dimethyl-7-isopropylazulene, since this compound prevents the effects of serotonin and anaphylactic shock but not that of histamine[138]. However, these results require further confirmation and the parallelism observed may only be coincidental.

The results obtained with antihistamines are difficult to interpret, as has already been mentioned. Mepyramine (pyrilamine) gives protection against the bronchospasm induced by histamine[183,187] and prevents anaphylactic shock[11,175,187] but does not affect serotonin bronchospasm[184,185,187,188].

The depletion of serotonin stores by means of reserpine[459] has been claimed to

---

* The level of serotonin in the lung of guinea pigs is very low in comparison with that in other animal species[465].

prevent anaphylactic shock[404,452] although other authors did not confirm these results[268]. On the other hand, reserpine inhibits the bronchospasm induced by serotonin[209] probably because at high doses reserpine not only depletes tissue serotonin but also blocks the pharmacological properties of the amine[127–129]. It has also been reported that reserpine releases histamine from the lungs of guinea pigs[459].

Splenectomy inhibits anaphylactic shock and partially inhibits serotonin bronchospasm without affecting histamine shock[386].

SEROTONIN AND THE SCHULTZ–DALE REACTION

It has been reported that in the Schultz–Dale reaction (sensitization *in vivo* and anaphylactic reaction *in vitro* in the isolated uterus of the guinea pig) the histamine released accounts only for 10% of the uterine contraction[95]. Mepyramine[383], but not an antiserotonin such as BOL 148 (refs. 107, 383), antagonizes the Schultz–Dale reaction. On the other hand, the isolated uterus of a sensitized guinea pig responds to the antigen even when it is made completely insensitive to histamine[107]. These data favour the hypothesis that other compounds together with histamine are responsible for the symptomatology of the anaphylactic reaction[44a,133].

CUTANEOUS ALLERGIC REACTIONS AND SEROTONIN

The local cutaneous allergic reactions induced by anaphylatoxin[173,370] or by local administration of antigen seem to be related to a release of histamine[163,267,369]. Antihistamines show a protective effect[139,163,175] while chlorpromazine or LSD 25 are inactive[175]. These results are consistent with the fact that the skin of the guinea pig is rich in histamine but not in serotonin[354] and with the observation that serotonin, in contrast to histamine, has only a very weak effect in increasing cutaneous capillary permeability in guinea pigs[175,267,269,*].

In conclusion, the present experimental data may cast some doubt on the classical hypothesis that anaphylactic shock and allergy are mediated by histamine, but they do not indicate serotonin as being primarily responsible for these reactions in guinea pigs.

SEROTONIN AND ALLERGIC REACTIONS IN RABBITS

Humphrey and Jaques, following a previous hypothesis made by others[365], showed that the reaction between antigen and antibody in rabbits results in a release of serotonin[199,201]. This effect occurs *in vitro* and *in vivo*[257,461] and resembles the effect previously reported for histamine[61,62,220,301,314,373,375,388]. The release of serotonin is very rapid; it occurs from platelets[256,457,462] and requires the presence of

---

* During the tuberculin reaction and in some types of delayed allergy, there is an increase of histidine decarboxylase and histamine[109,203–205,392]. The increase of histidine decarboxylase is present also after administration of histamine releasers such as polymyxin B[392,393] and Compound 48/80 (refs. 393, 396). There are no comparable data for 5-hydroxytryptophan decarboxylase. Serotonin[151], in contrast to histamine[105,150], does not increase the phagocytic activity of the monocytes.

calcium ions[198,201,*]. Furthermore the uptake of serotonin from platelets is reduced (ref. 104a). Endotoxin[79,79a] and anaphylatoxin[377a] also release platelet serotonin. It has already been mentioned that rabbit platelets are very rich in serotonin (7.5 μg/10^9 platelets)[200]. As a result of the increased availability of serotonin in the plasma, there is an increase in the excretion of urinary 5-hydroxyindoleacetic acid after anaphylactic shock[259,458].

However, other results demonstrate that serotonin is not the cause of the symptomatology induced by anaphylactic shock in rabbits. The removal of platelets[251], the previous depletion of tissue serotonin by reserpine[110,256,259,459,460,462,**], and the administration of antiserotonin agents such as BOL 148 (ref. 259), UML (ref. 262) or chlorpromazine[252,254,259], do not prevent anaphylaxis. Similar conclusions also apply to the anaphylaxis induced locally in the mesentery of rabbits[261]. On the other hand, rabbits made tolerant to endotoxin show a release of serotonin similar to that present in sensitive animals[77].

In rabbits, irradiation depletes tissue histamine and serotonin and reduces anaphylactic shock[19,258] but there is no causal relationship between the two effects[258]. The problem of anaphylactic shock in rabbits is still unsolved because histamine does not seem to play an important role[245,261,366] either.

Alteration of the capillary permeability in rabbit skin occurs only with high concentrations of serotonin compared to the dose of histamine sufficient to elicit such an effect[415]. Intra-articular injection of serotonin induces connective tissue proliferation in rabbit joints[413a]. Oedema of the nervous tissue may be induced by serotonin[145] and blocked by chlorpromazine or reserpine[49].

Newcastle disease virus induces an oedema of the rabbit cornea similar to that obtained after the local administration of Compound 48/80. A probable participation of serotonin in this reaction is suggested by the fact that LSD 25 protects against both types of oedema, particularly in combination with antihistamines[339,340,***].

Even low concentrations of serotonin stimulate the phagocytic activity of polymorphs and macrophages. It is interesting that this effect is not antagonized by BOL 148 (ref. 337).

HYPERSENSITIVITY IN MICE AND RATS

The administration of *Bordetella pertussis* vaccine in mice or rats induces an increased sensitivity to the toxic effects of various drugs and treatments[236,†] (see Table 21). This increased toxicity is not the result of a non-specific debilitation of the

---

*   In the absence of calcium the effect of serotonin on smooth muscle is considerably decreased[198,475].

**   Reserpine decreases the mortality due to incompatible blood transfusion[406]. However, this effect is not necessarily related to the depletion of serotonin since other conditions such as hypothermia, prevent anaphylaxis[360].

*** Antihistamine agents alone do not prevent the toxic effects of Compound 48/80 (refs. 102, 344–346).

†   The factor responsible is localized in the cellular wall of the microorganism[330,335a]. Vaccines made from other bacteria are not active[231,274,281,350].

TABLE 21

SITUATIONS AND MATERIALS TESTED IN PERTUSSIS-INOCULATED OR ADRENALECTO-
MIZED MICE AND RATS

Treatment	Pertussis vaccine	References	Adrenal-ectomy	References
Anaphylaxis	+	232, 237, 239 278, 279, 281 325, 328, 363 414	+	84, 111, 193 194, 334, 463 464, 476, 477
Cold stress	+	333		
Endotoxins	+	1, 54, 236, 237, 347		
Histamine	+	177, 178, 202 237, 274, 280 328, 331, 345 350, 384	+	73, 148, 169 179, 193, 332
Irradiations	+	378		
Passive anaphylaxis	+	334	+	334
Peptone	+	202		
Pertussis vaccine	+	231, 234, 351	+	231, 234, 351
Pollen of *Lolium perenne*	+	240		
Reduced atmospheric pressure	+	239		
Serotonin	+	235, 237, 328 384	+	53, 123, 193 194, 334
Arthus reaction	—	25		
Carbaminoylcholine	—	231		
Epinephrine	—	270	—	270
Heat stress	—	270	—	270
Metacholine	—	270	—	270
48/80	—	1, 233, 345		
Schultz–Dale reaction	—	108		
Tryptamine			—	125

+ enhanced toxicity.
— toxicity comparable to that of intact animals.

animals because some drugs or treatments are as well tolerated in "pertussis-treated"
mice as in normal animals[264,270].

The conditions required to obtain the hypersensitivity of mice to pertussis vaccine
(age, sex, strain of animals and type of vaccine) are discussed by Kind in a recent
review[236].

It is interesting that among the compounds that increase in toxicity in "pertussis-
treated" mice, are histamine and serotonin, two agents supposed to be involved in
anaphylactic and allergic reactions. Anaphylaxis is also aggravated in "pertussis-

inoculated" mice[282,329] although mice are not considered to be very sensitive to anaphylaxis[474].

Adrenalectomy increases the lethal effects of histamine, serotonin and anaphylaxis (see Table 21) while adrenal corticoids restore the normal resistance of adrenalectomized animals to the three toxic conditions. A simple hypothesis would suggest that agents becoming more toxic in pertussis-treated animals could act through a release of histamine and of serotonin which will become more toxic if there is a concomitant impairment of the adrenocortical functions. This mechanism may operate in the case of rats[274,280], and mice, but not for guinea pigs[274,363,431,*] or rabbits[363].

The increased toxicity of serotonin in mice treated with pertussis vaccine has been reported by several authors[214,235,326,328,362,**]. The LD_{50} for serotonin by the intraperitoneal route decreases from 540 mg/kg in normal mice to 14.5 mg/kg in pertussis-treated mice[214]. Under the same conditions the LD_{50} for histamine falls from 800 mg/kg to 80 mg/kg (ref. 214). Some strains of mice which are not made hypersensitive to histamine by pertussis vaccine[331,349] may still be hypersensitive to serotonin[326,328] and anaphylaxis[167]. The increased sensitivity becomes apparent about one day after treatment with the vaccine, reaches a peak on the fourth day and disappears about the 25th day[236]. Mice dying after histamine or serotonin administration exhibit a symptomatology similar to that described for anaphylactic shock (refs. 326, 463).

A reasonable explanation for this hypersensitivity is still lacking. It was suggested that animals treated with pertussis vaccine or submitted to adrenalectomy do not metabolize histamine or serotonin. Histaminase in rat[336] or mouse lung[241] is blocked after pertussis vaccine treatment. However, it is doubtful whether the inactivation of histaminase may be responsible for the increased toxicity of histamine[6], since pertussis vaccine reduces histaminase in guinea pigs[297] and this species is not sensitive to the effect of pertussis vaccine[274,363,431]. Furthermore, animals treated with pertussis vaccine metabolize exogenous histamine and serotonin to about the same rate as do untreated mice[458].

In adrenalectomized animals that are hypersensitive to serotonin, histamine and anaphylaxis, there are no major changes in the levels of tissue histamine and serotonin. Immediately after adrenalectomy there is a decrease of blood serotonin[306]; there is an increase of tissue histamine[17,34,134,271,295,374,389,391] which is prevented by administration of saline[34,190,374]. The levels of brain, lung, kidney and spleen serotonin of adrenalectomized rats show only minor variations when compared with those from intact animals before or after treatment with serotonin[125]. Monoamine oxidase activity of liver, brain, kidney and lung is also largely comparable in adrenalectomized and normal animals[125].

---

* Also in guinea pigs[228] and rabbits[72] adrenalectomy increases the sensitivity to anaphylactic shock.

** A releaser of histamine and serotonin, Compound 48/80 increases the toxicity of histamine and serotonin in mice[191]. UML counteracts the toxicity of serotonin but not that of Compound 48/80 (ref. 441). On the other hand, heparin protects against Compound 48/80 toxicity but not against serotonin or histamine[192].

TABLE 22

CONDITIONS INCREASING SENSITIVITY TO HISTAMINE AND SEROTONIN

Experimental condition or treatment	Hypersensitivity to	
	histamine (References)	serotonin (References)
Adrenalectomy	169, 179, 332	123, 334
Cold stress		483
Compound 48/80	191, 192	191, 192
Increasing age	349	123
Pertussis vaccine	177, 178, 331, 350	235, 237, 328
Sarcoma 180	348	123
Thyroxine	359, 422	359, 422

These data would suggest that the hypersensitivity is due to an increased pharmacological activity of serotonin rather than to a diminished rate of destruction of the amine. Accordingly Schayer and Ganley[394,395] consider the possibility that the increase of histidine decarboxylase* observed after pertussis vaccine treatment, results in an increased amount of newly formed histamine. This new histamine may produce "irreversible alterations in permeability which can be rendered lethal by an insult such as injection or release of histamine or serotonin". However, Munoz[327] reported that pertussis vaccine increases tissue permeability in some but not in all the strains which become sensitive to histamine.

Cortisone and its congeners prevent the increased toxicity of histamine (refs. 54, 168, 169, 232, 414), serotonin[126,194,232,414] and anaphylactic shock[232,335,414] in adrenalectomized or pertussis vaccine-treated animals. Corticosteroids protect adrenalectomized animals against serotonin with a potency which is proportional to their anti-inflammatory properties[126].

These data suggest an association between the increased toxicity of serotonin or histamine and the increased susceptibility to anaphylaxis in spite of the fact that in one case this was produced by adrenalectomy and in the other by pertussis vaccine treatment**. However recent experiments by Code et al.[63] would suggest that the type of protection exerted by cortisone in pertussis-treated or adrenalectomized mice is related to a different mechanism. On the other hand up to now there is no evidence supporting an impairment of the adrenal function following the pertussis vaccine treatment. The excretion of the 17-ketosteroids and corticosterone is almost normal when the hypersensitivity appears[55,131,238,277,280,334]. This does not exclude the hypothesis that pertussis vaccine is increasing the utilization[334] or the requirement[395] of corticosteroids, inducing in this way a relative peripheral lack of the hormones similar to that present in adrenalectomized animals.

---

* Several treatments, including the administration of serotonin, increase histidine decarboxylase in mouse tissues[390].

** Other conditions are known to alter the sensitivity to serotonin and histamine (see Table 22).

When catecholamines, their precursors, or sympathomimetic agents are given to adrenalectomized rats, they do not prevent the effect[126] of serotonin in these animals. Antiserotonin agents such as the derivatives of lysergic acid[126,326] or cyproheptadine[126] are protectors while chlorpromazine is only weakly effective[126].

ANAPHYLAXIS IN MICE

Anaphylaxis in mice is also prevented by lysergic acid derivatives (LSD 25 and BOL 148) (refs. 116, 438, 439) or chlorpromazine[438,439]. BAS, an analogue of serotonin does not show activity[334]. Contradictory results have been obtained with antihistamines[52,439]. The positive results may be ascribed to their antiserotonin activity[86], since they do not protect intact mice* from lethal doses of histamine[82,298].

Reserpine treatment, which releases in mice only serotonin and not histamine[459], prevents the anaphylactic shock[116,136a,438,439]. The protective effect of reserpine is counteracted by the administration of 5-HTP, but not of DOPA[438,439]. When given alone, 5-HTP increases the sensitivity of mice to anaphylactic shock[439]. The protective effect of reserpine is not seen when mice are pretreated with pertussis vaccine before inducing the anaphylactic shock[438].

These results support the hypothesis that serotonin may be involved in the anaphylactic reactions in mice[136a], but direct experiments are less favourable. For example, it has been reported that after anaphylaxis in mice there is no change in the whole body serotonin content and no variation in the excretion of 5-hydroxyindole-acetic acid[458]. However, in mastocytoma-bearing mice tissue serotonin is decreased after anaphylaxis[91].

An elegant experiment by Fink[106] tends to support the view that the Schultz–Dale reaction** in the isolated uterus of the mouse is not mediated by histamine but by serotonin. This reaction is in fact inhibited by LSD 25 and by reserpine[106]. Furthermore the mouse uterus, particularly in oestrus[51], is very sensitive to serotonin but not to histamine[51,106]. LSD 25 (ref. 106), reserpine[106] and N,N-dimethyl-N'-benzyl-N'-benzoyl-1,3-propanediamine (Ro 2-9102) (ref. 299) are strong inhibitors of the contraction induced by serotonin in mouse uterus.

ENDOTOXIN AND SEROTONIN

Another problem studied in relation to serotonin is the toxicity of endotoxin in mice. BAS[81] and chlorpromazine[249] protect from endotoxin. However, protection has been obtained also with 5-hydroxytryptophan[81] and serotonin[147,249]. Contradictory

---

*   The effect of antihistamines on histamine toxicity is indirectly proportional to the sensitivity of the mice which varies considerably with the strain[5]. The increased toxicity of histamine in adrenalectomized or pertussis vaccine-treated mice is prevented by antihistamine agents[54,178,232].
** This local reaction, in contrast to general anaphylaxis, is not potentiated by pertussis vaccine treatment[108].

results have been obtained with reserpine[56,302] and monoamine oxidase inhibitors (refs. 14, 77a, 81, 249, 302).

In conclusion it is justified to recommend further studies on allergic and hypersensitive reactions in mice based on the hypothesis that serotonin may be a factor of some importance.

ANAPHYLACTOID AND INFLAMMATORY REACTIONS IN RATS

The anaphylactoid reactions in rats cannot be explained on the basis of histamine mediation. It seems therefore worth while to ascertain the validity of the hypothesis that considers serotonin to be chiefly involved in these types of reactions.

Serotonin is very effective in inducing oedema in rats. It has been calculated that 1 μg of serotonin may cause retention of about 2 ml of water in the subcutaneous tissue[22]. The minimal concentration of serotonin able to induce oedema is about 0.1 μg. Thus serotonin appears to be the most effective of the known oedemigenic agents in rats[381,444]. The potency of serotonin is from 10 (ref. 415) to 200 times[380,417] higher than that of histamine, according to the criteria used for the evaluation of the oedema*. The increase of the liquid in the skin after local administration of serotonin is accompanied by an increase of total hexosamines and haemoglobin[247]. Careful studies by means of electron microscopy have shown that serotonin induces leakage of fluid from vessels by disconnecting the endothelial cells along the intercellular junctions[275]. The point of action is on the venous side of the circulation. The maximum effect is on venules of 20–30 μ in diameter and it disappears toward vessels of 75–80 μ or 4–7 μ (ref. 276). It has been suggested that serotonin produces oedema by raising the capillary hydrostatic pressure[161].

The concentrations of serotonin capable of inducing oedema may be considered physiological, since the various cutaneous areas contain sufficient quantity of serotonin to account for the development of a local oedema[357] (see Appendix I, page 241). New-born rats, which have a low level of cutaneous serotonin, are not sensitive to inflammatory agents[85]. The amount of serotonin present in a given area of the skin is in strict relation to the number of mast cells[21,24,80,162,223,354,355,400,409,468,**], although not all the cutaneous serotonin is present in mast cells[413,466].

There is insufficient evidence to assess the possibility that serotonin may be physiologically involved in the regulation of the capillary permeability***.

In Table 23 are summarized some data concerning the release of serotonin and histamine after administration of oedema-producing or inflammatory agents. All the compounds, with the possible exception of formaldehyde, release both serotonin and

---

*   It has been underlined that the results obtained by measuring the intensity of the oedema by paw volume, or increased weight of skin, or spreading of a dye are not always in agreement (refs. 149, 154, 357, 358, 419). Also in mice serotonin induces oedema [455a,462b].

**   The injection of serotonin induces degranulation[13] but not destruction[224] of the mast cells which are also rich in histamine[367].

*** Reserpine does not alter the capillary resistance of the cutaneous vessels of rat[478]

TABLE 23

VARIATIONS OF CUTANEOUS SEROTONIN AND HISTAMINE AFTER ADMINISTRATION OF DIFFERENT AGENTS

Agent	Serotonin	References	Histamine	References
Anaphylaxis	0	385	—	385
Burning	0	15, 16	0	15, 16
Compound 48/80	—	30, 175, 265, 266, 355, 445	—	30, 175, 265, 266, 355, 445
Dextran	—	174, 175, 317, 321, 356, 433	—	174, 175, 356
Egg-white	—	304, 320, 356	—	356
Formaldehyde	0	317, 433		
Histamine	0	320		
Horse serum	—	115, 130	0	115, 130
5-Hydroxytryptamine			—	33, 104, 356
Polymyxin B	—	266, 355, 356, 434	—	266, 355, 356, 435
Tetrabenazine			—	364
Turpentine oil	—	433		
1935 L	—	174, 175	—	173, 174, 175

0   = unchanged
—   = decreased

histamine*. However, the amount of released serotonin may account for the oedema, while the amount of histamine seems to be insufficient to induce an alteration of the permeability. The "histamine liberators" (Compound 48/80, 1935 L, stilbamidine, etc.) also release serotonin from mast cells[8,40]. This release of serotonin must be localized since dextran, polymyxin B and Compound 48/80, in contrast to reserpine, do not increase the urinary elimination of 5-hydroxyindoleacetic acid[210].

Assuming that the cause of the oedema produced by anti-inflammatory agents is the release of serotonin, then a previous depletion of tissue serotonin should prevent the onset of the oedema. Thus animals pretreated with reserpine[357] are still sensitive to the oedema induced by serotonin[18,356], histamine[356], formaldehyde[467], but not to that induced by dextran[70,212], egg-white, Compound 48/80 (ref. 70) and hyaluronidase[296,467] (see Table 24). The route of administration of reserpine[70] or of the inflammatory agent[155] and strain of animals[181] used may be responsible for the failure reported by some authors to confirm the protective effect of reserpine. It is evident that valid conclusions can only be drawn from results obtained by simultaneous measurement of the inflammatory reaction and the presence of amines in the tissues. Reserpine, as already mentioned, releases in rats only serotonin and not histamine[30,31,265]. Polymyxin B, which releases histamine rather than serotonin[435], does not prevent the oedema induced by inflammatory agents[154,353,356,357].

The results obtained with other agents, like Compound 48/80 (refs. 30, 176) are

---

* Ozone exposure develops oedema in rats, increases serotonin[410] and histamine[411] in lung, and decreases serotonin in brain[442]. Also salicylate releases 5-HT in skin[305].

difficult to interpret because both histamine and serotonin are released at the same time*. It is interesting that in rats pertussis vaccine treatment does not increase the local reaction to anaphylactoid agents[164,379,447].

Another method of investigation has been a comparative pharmacological analysis of the inhibition exerted by a number of drugs on the various inflammatory agents. Table 24 shows a summary of the results obtained, although not all the data reported are comparable, because different techniques have been used by the various investigators.

Lysergic acid derivatives are strong inhibitors of the oedema induced by serotonin, but do not inhibit that produced by histamine. Furthermore they prevent dextran and egg-white oedema but not that following formaldehyde[356,455]. Pentamidine inhibits rat-paw oedema induced by serotonin, dextran or egg-white without showing any antihistamine activity[64]. Other antiserotonin agents, such as N,N-dimethyl-N'-benzyl-N'-benzoyl-1,3-propanediamine (Ro 2-9102) and 2-amino-4-(N'-methyl-N-piperazino)-5-(4'-chlorophenylmercapto)pyrimidine (BW 57–301) prevent the oedema produced by dextran or Compound 48/80 at doses without effect on histamine[124]. A quantitative evaluation shows that UML and cyproheptadine are strong inhibitors of the local inflammatory reaction induced by serotonin, but only cyproheptadine is active against histamine. As cyproheptadine is more active than UML in reducing the local reactions induced by dextran and Compound 48/80 in intact and adrenalectomized rats, it is suggested that both histamine and serotonin are involved in the reaction[211]. However, an antihistamine drug such as pyrilamine is only slightly effective in protecting against oedema induced by serotonin, dextran and Compound 48/80 (ref. 380). Antihistamines showing protection against dextran and egg-white oedema have also antiserotonin activity[86]. Furthermore it has been reported that antihistamines prevent oedema by a chemical binding with dextran[217].

Phenothiazines are also strong antagonists of oedema induced by various agents in rats[87,153,215,216].

A survey of Table 24 may lead to the conclusion that compounds effective against serotonin-induced oedema are mostly also antagonistic to other inflammatory agents. The closest parallelism is between the inhibition of serotonin and dextran or egg-white oedema, while a comparison of the activity against serotonin- and formaldehyde-induced oedema clearly shows some disparity. On the whole, experimental evidences would support the hypothesis that serotonin is involved in the oedema and anaphylactoid reactions of rats. More quantitative methods are, however, needed before drawing any definite conclusions. When an anaphylactoid reaction is induced by parenteral administration, it would be useful to examine the alteration of the plasma volume not only in skin but also in other organs. The quantitative method proposed by Beach and Steiner[18] shows that after egg-white administration there is an oedema of the skin, but a decrease of the plasma volume in lung, heart and brain. A note of caution must also be given because the drugs used are not always specific. For instance, we do not

---

* Compound 48/80 releases serotonin in rats but not in rabbits, cats and dogs[3,30].

TABLE 24

EFFECT OF VARIOUS DRUGS ON OEDEMA INDUCED BY 5-HYDROXYTRYPTAMINE AND (
(Numbers indicate references)

Compound	Dextran	Histamine	5-Hydroxytrypta
Acetylsalicylate	+ 89, 243		+ 436
			± 322
ACTH	+ 196	— 448, 471	+ 446
	± 89, 263		— 449
Adrenaline	+ 42, 291	± 291	+ 291
	— 315		
Aminophenazone	— 245		
Aminopyrine	+ 89		+ 96, 322, 436
Antipyrine (Phenazone)	+ 89	+ 451	+ 436, 451
			± 322
Atropine			
1-Benzyl-5-methoxy-2-methyl- tryptamine(BAS)	+ 124, 291	+ 124 — 291	+ 124 ± 291
1-Benzyl-2-methyl-5-methoxy- bufotenine (BAB)	± 124	± 124	± 124
BOL 148	+ 28, 70, 356	— 356	+ 26, 101, 136, 449
BW 57-301	+ 124	— 124	+ 124
Calcium gluconate	+ 89		± 322
Carisoprodol	+ 46		+ 46
Chlorcyclizine			± 230
2-Chloro-10-(3-pyrrolidin- 1′-ylpropyl)phenothiazine (RP 4670)	+ 291, 453	+ 291	+ 291
Chloroquine sulphate	— 450	— 448	+ 446, 449
Chlorpheniramine			+ 428
Chlorpromazine	+ 67, 175, 182, 216, 245, 291, 425	+ 23, 158, 291 ± 448	+ 23, 158, 291, 446, 449
Chlorprothixene	+ 124, 227	+ 124	+ 124
Chlorpyramine (Synopen)			± 322
Chlorthenoxazine (Valmorin)	+ 213		+ 195
Compound 48/80	+ 356 ± 158	— 158, 356	— 158, 356
Cortisone	+ 89, 158, 316 — 67, 263, 315, 432	+ 92 ± 448 — 471	+ 436, 446
Cyproheptadine	+ 211	+ 211	+ 211, 428
Dibenamine	+ 380, 381	— 380, 381	+ 23, 380, 381,

...hyde	Egg-white	Compound 48/80	Hyaluronidase	Others
			— 83, 89	— 462a (Y)
24		+ 124	+ 83, 89	
	+ 59, 291, 479	+ 291	+ 467	+ 420 (B) — 467 (B, T)
	+ 245 + 479		+ 83 + 83, 89 + 89	
				— 368 (B)
	— 291	± 124, 291		— 124 (BR)
		± 124		± 124 (BR)
67	+ 356	+ 70, 356 ± 124	— 356, 467	+ 356 (P) ± 473 (Irr., O) — 124 (BR), 368 (B), 418 (T), 467 (B,T)
		+ 124		— 124 BR)
			+ 83, 89	
	+ 291	+ 291		
	— 450			+ 473 (O) — 462a (Y)
2, 467	+ 68, 245, 291	+ 175, 291	+ 467	+ 227 (T), 368 (B), 425 (T), 473 (O) — 124 (BR)
		+ 124		+ 227 (T) — 124 (BR)
7	+ 195 + 356		+ 356, 467	+ 418 (T), 462a (Y) — 356 (P), 385 (A.S.), 467 (B, T)
4			+ 83, 89	+ 84 (A.S.), 454 (PVP, K) — 35 (B)
	+ 428	+ 211		+ 462a (Y) — 124 (BR, CA, K)
		+ 281, 380	+ 380*	+ 368 (B) — 380 (O)

Continued overleaf

TABLE 24 (*continued*)

Compound	Dextran	Histamine	5-Hydroxytrypta...
6,7-Dichloro-9-γ-dimethylamino-β-oxypropyl-isoalloxazin (MO 242)	+ 47		+ 47
6,7-Dichloro-9-dimethylamino-propyl-isoalloxazin (MO 177)	+ 47		+ 47
Didymium salt of β-acetyl-propionic acid (Helodym 88)			+ 208
2-Dimethylamino-ethanol		+ 74	+ 74
Dimethylaminoethyl-4-chloro-phenoxyacetate			± 437
Dimethylamino-3′-methyl-2′-propyl-10-phenothiazine methoxide (RP 7044)	+ 215, 291	+ 291	+ 291
N,N-Dimethyl-N′-benzyl-N′-benzoyl-1,3-propanediamine (Ro 2-9102)	+ 124	− 124	+ 124
Ergotamine-1-tartrate	+ 455	− 455	+ 455
Formyl-4-antipyrine		− 451	+ 451
Heparin			+ 225
Hydrocortisone	− 7, 263	− 448	− 449
5-Hydroxytryptamine	+ 135 = 341	+ 119a	+ 119, 136
Homochlorocyclizine (SA 97)	+ 124	+ 124	+ 230
Impramine (Tofranil)			+ 322
Indomethacin	+ 124		− 474a
Insulin		− 382	− 382
Iproniazid	+ 136, 402 − 12, 318		− 136, 420
Isocarboxazide (Marplan)	+ 71		+ 71
Isoniazid	+ 88		
LSD 25	+ 175, 424, 455 − 323	± 455	+ 26, 136, 322, 446, 455, 4...
Lysergic acid			− 417
D-Lysergic acid butanolamide			+ 101
Meprobamate	+ 28, 46 ± 45		± 45, 46
Mepyramine	+ 453 ± 175 − 356, 380, 381	+ 380, 381, 415, 448 ± 356	+ 428 ± 136 − 356, 380, 3...
5-Methoxy-2-methyltryptamine	− 291	− 291	± 291
1-Methyl-D-lysergic acid butanolamide (UML)	+ 211 ± 119	± 119 − 211	+ 26, 87, 101, 211
Methylpromazine (RP 4627)	+ 291, 315, 453	+ 291	+ 291
Morphine	+ 89, 136		

dehyde	Egg-white	Compound 48/80	Hyaluronidase	Others
		+ 208		
	+ 74			
	+ 129	+ 291	+ 467	+ 467 (B, O)
		+ 124		— 124 (BR)
	+ 455	+ 455	— 455	+ 455 (TR, K) — 455 (PVP)
			+ 143	— 35 (B), 462a (Y)
		— 124	+ 100	— 420 (B)
	— 474a			+ 124 (A), 474a
		— 382		— 382 (P)
402	— 12	— 420		+ 420 (B) — 12 (O)
				— 368 (B), 385 (A.S.)
455	+ 304, 455	+ 175, 455	+ 455	+ 35 (B), 455 (TR), 462a (Y) — 368 (B), 455 (PVP, K)
356, 467	— 356	+ 175, 415 ± 356 — 152, 380	± 356 — 380*, 467	+ 418 (T) ± 356 (P), 473 (Irr., O) — 385 (A.S.), 462a (Y), 467 (B, T)
	— 291	— 291		
	+ 195	± 211 — 262		+ 462a (Y) — 124 (BR, CA, K)
	+ 291	+ 291		
	+ 304		+ 83, 89	

TABLE 24 (*continued*)

Compound or treatment	Dextran	Histamine	5-Hydroxytryptamine
Nialamide	+ 71		+ 71
Noradrenaline	+ 291	— 291	+ 291
Oestradiol	— 263		— 449
Pentamidine	+ 64	— 64	+ 64
Phenelzine	— 71		— 71
Phenindamine (Thephorin)	+ 243, 291, 315	+ 291	± 291 — 87, 356
Phenobarbital	+ 218a		± 322
Phenylbutazone	+ 45, 47, 48, 89 ± 90, 263 — 67, 118, 424	— 118, 448	+ 45, 47, 48, 43(— 118, 124, 322
γ-Phenylpropylcarbamate	+ 45, 46		+ 45, 46
Polymyxin B	— 356, 466	— 356	— 356
Prednisolone	+ 263	+ 448	+ 446 — 449
Procaine	± 316 — 315		— 380
Promethazine (RP 3277)	+ 171, 175, 215, 216, 263, 356	+ 152, 356, 448	+ 152, 428, 446 ± 136 — 356, 449
Quinine sulphate	+ 89		+ 436
Reserpine	+ 12, 27, 158, 212, 318, 319, 356 ± 136	+ 158 — 356	+ 28, 136 — 158, 356
Salicylate (sodium)	+ 89, 90, 158, 221 ± 263	± 448	+ 88, 221, 446, 4 ± 322
Spinal section	+ 136	— 136	+ 136
STH	— 196	— 448	— 446, 449
Syrosingopine (Su 3118)	+ 28 — 212		+ 28
Thenalidine (1-methyl-4-N- then-2′-ylanilinopiperidine tartrate)	+ 291	+ 291	+ 87, 291, 428
Trimeprazine (RP 6549)	+ 218, 291	+ 291	+ 291, 428
Tripolidine	— 124	+ 124	— 124

+ : inhibition   — : no effect   * : testis extract   B : burning

± : slight inhibition   = : enhancement   A.S. : anaphylactic shock   BR : bradykinin

CA : carrageenin

dehyde	Egg-white	Compound 48/80	Hyaluronidase	Others
	+ 59, 291	+ 291	+ 467	— 420 (B), 467 (B, T)
	+ 64			+ 64 (K)
	+ 291, 356	± 291	+ 356	
7, 89, 90, 24	+ 48, 245, 479		+ 83, 89	+ 35 (B), 124 (CA), 454 (K), 474a — 124 (BR), 454 (PVP), 462a (Y)
46 467	— 356, 466	— 356	— 356, 467	+ 418 (T), 462a (Y), 467 (B, T) — 385 (A.S.)
				— 380 (O)
467	+ 170, 171 ± 356	+ 152 ± 356	+ 356 — 467 + 83, 89	+ 175, 356 (P), 418 (T), 472 (Irr., O) — 368 (B), 467 (B, T) + 473 (Irr., O)
18 467	+ 12, 304, 356	+ 356	+ 296, 467 ± 356	+ 12 (O), 356 (P), 462a (Y) — 110 (A.S.), 385 (A.S.), 418 (T), 467 (B, T)
0	+ 137, 479	+ 221, 309	+ 83, 89	+ 419 (B, T)
	+ 291	+ 291		— 124
	+ 291	+ 291 — 124		— 124 (BR)

Irr. : irradiation   O : ovomucoid   PVP : polyvinylpyrrolidone   TR : tryptamine
K   : kaoline   P : polymyxin B   T : turpentine   Y   : yeast

know enough about the possibility that compounds may interfere with slow-reacting substances or bradykinin suggested to be involved in the anaphylactoid[445] and inflammatory[156,368] reactions.

Also, an interference with adrenal function must be considered, since catecholamines[291] and corticosteroids[436,446] may inhibit, although only in large doses, the oedema produced by serotonin.

ANAPHYLACTIC SHOCK IN RATS

There are few data on anaphylactic shock in rats, because rats are rather resistant to anaphylaxis. It has been reported that reserpine, BOL 148 and mepyramine do not modify anaphylaxis in rats[385], although amine depletion may delay death[156b]. After anaphylaxis, tissue histamine is decreased but serotonin is unchanged[385]. Also a passive anaphylaxis induced in rat skin does not seem to involve serotonin[44] and it is not affected by treatment with BOL 148 (ref. 206). The symptoms present in anaphylactic shock are not similar to those induced by serotonin or histamine[156a].

The importance of serotonin in anaphylactic shock may change in relation with the target organ. For instance when antigen is added to the intestine of pre-sensitized rats, there is a release of histamine but not of serotonin. On the other hand when antigen is added to peritoneal cells, a complete release of serotonin is accompanied by a release of 50% of the histamine. In the uterus, a release of the serotonin is not accompanied by any change of histamine[130,130a]. Rat uterus is sensitive to serotonin but not to histamine[222]. Furthermore the contraction induced by the antigen is inhibited by BOL 148, but not by mepyramine[383]. This picture is exactly the opposite to that reported for the Schultz–Dale reaction on guinea pig uterus[44b].

MISCELLANEOUS

The oedema following burning is inhibited by LSD 25 (ref. 35). However other authors do not agree with these results[368,445]. According to Spector, the inhibition exerted by iproniazid or other MAO inhibitors would be related to an interference with adrenaline rather than with serotonin metabolism[420,421]. Rocha y Silva believes instead that bradykinin may be responsible for the thermic oedema[368].

Data obtained on inflammatory reactions involving proliferation (granuloma) may also suggest a participation of serotonin[120,387,484].

Reserpine increases sensitivity to an infection by *Staphylococcus aureus*[312] and this may be related to the finding that serotonin increases phagocytosis by rat leucocytes[272]. LSD 25 inhibits this effect of serotonin[272]. The toxin of *Staphylococcus aureus* does not release serotonin in the peritoneal fluid[412].

SEROTONIN AND ALLERGIC REACTIONS IN OTHER ANIMAL SPECIES

Unfortunately, the reports in the literature about the metabolism of serotonin during allergic reactions in cats, dogs and humans are scanty.

Anaphylactic shock in dog increases blood serotonin, particularly in the portal vein[403]. This effect is transient (about 30 min), of hepatic origin[385] and it is accompanied by release of histamine[3] and ATP, by a decrease of platelets, hypotension, apnoea[403] and anuria[229]. Serotonin[50,93,303] as does anaphylaxis[58] induces in dogs bronchospasm and pulmonary hypertension[372]. Animals killed after anaphylaxis do not show maior changes in the serotonin content of intestine, liver, lung, skin and spleen[3].

A pretreatment with reserpine prevents the rise of blood serotonin following the anaphylactic shock and prevents the hypotension and the bronchoconstriction. It is interesting that in reserpinized animals there is hyperhistaminaemia[404] despite the reduced symptomatology. LSD 25 has only a weak effect on the anaphylactic reaction in dogs[404].

In cats, anaphylactic shock releases serotonin[97,404], which in its turn may release histamine[103,104]. Cats and dogs show a relatively high level of serotonin in lungs[354] and platelets[200] (see Appendix I).

In horses, 5-HT induces a cutaneous oedema which seems to be indirectly proportional to the number of eosinophils present in a given area of the skin[9]. Eosinophils contain a principle able to decrease the oedema induced by serotonin[10].

The administration of *Escherichia coli* endotoxin decreases blood serotonin (refs. 78, 376, 377)* and increases catecholamines[376,377]. The pressor effects induced by endotoxins are prevented by chlorpromazine[269], dibenamine and LSD 25 (ref. 141).

The toxicity of endotoxin is also reduced by drugs known as antiserotonin agents[338,405]. However, serotonin itself decreases the toxic effects of endotoxin[146].

Normal human subjects do not show bronchoconstriction after intravenous administration or aerosol of serotonin[307,308]. Asthmatic patients (or at least a certain percentage of them) seem to be more sensitive to the bronchial effect of serotonin (refs. 163a, 184, 307, 308, 343). This effect is inhibited by antihistamines[307] which also block in humans the eosinopenia induced by serotonin[371,**]. Serotonin inhibits the capacity of human serum to bind histamine (histaminopexy)[112]. However, there are doubts that the binding of histamine or serotonin to serum proteins is a factor altered in allergic diseases[75].

In man, an increase of blood serotonin and of urinary 5-hydroxyindoleacetic acid has been reported after incompatible transfusion or hypersensitivity to penicillin[403]. An increase of blood histamine has also been reported in similar conditions[253]. However, patients with hay fever do not show alterations in the urinary excretion of 5-HIA (ref. 57).

Patients with rheumatoid arthritis exhibit an exaggerated response (pain, swelling, erythema, cyanosis) to an intradermal or intra-articular injection of serotonin. Histamine, noradrenaline or acetylcholine do not show a similar effect[399,399a]. This reaction is inhibited by lysergic acid derivatives (BOL 148) (refs. 398, 399).

---

* When *E. coli* endotoxin is injected into the superior mesenteric artery of dogs, there is an increase of serotonin in the superior mesenteric vein[76].
** Serotonin induces eosinopenia also in rats[311,423]. An increase in eosinophils is accompanied by an increase of blood serotonin[313].

Alterations in cutaneous permeability in man are not likely to be explained by a local release of serotonin, because both normal[354,443] and pathological[409,466] skins contain only minute amounts of serotonin. The local administration of serotonin does not produce oedema in humans[189]. However, other authors found an increase of capillary permeability inhibited by chlorpromazine[114] and unaffected by iproniazid[113].

The oedema induced in man by a histamine liberator (1935 L) is not accompanied by an increase of urinary 5-HIA excretion[256,260] and is not prevented by reserpine[256]. The increase of the CSF pressure induced by Compound 48/80 in man is antagonized by the administration of an antiserotonin agent, such as butanolamide of 1-methyl-lysergic acid (UML)[408].

It is interesting that a specific serotonin inhibitor, such as UML, shows therapeutic effects in asthmatic and allergic patients[16a,142,163a]. Another powerful antiserotonin agent, cyproheptadine, has also been successfully used in the treatment of allergic disorders[36,44a,361,411a]. However, cyproheptadine shows also a marked antihistamine activity[428].

The therapy of the allergic disorders would probably benefit from the availability of new drugs possessing powerful inhibitory effects against the oedema-inducing and inflammatory properties of serotonin.

REFERENCES

1   ABERNATHY, R. S. AND SPINK, W. W., *J. Immunol.*, 77 (1956) 418.
2   ACHARI, G. AND CHAUDHUDRY, K. D., *Indian Pharmacist*, 7 (1952) 392. (*C.A.*, 47 (1953) 2873 h).
3   ACKASU, A. AND WEST, G. B., *Intern. Arch. Allergy*, 16 (1960) 326.
4   ADO, A. D., *Acta Allergol.*, Suppl. 7 (1960) 73.
5   AMBRUS, J. L., GUTH, P. S., GOLDSTEIN, S., GOLDBERG, M. E. AND HARRISON, J. W. E., *Proc. Soc. Exptl. Biol. Med.*, 88 (1955) 457.
6   ANGELAKOS, E. T. AND LOEW, E. R., *J. Pharmacol. Exptl. Therap.*, 119 (1957) 444.
7   ARCHER, G. T., *Nature*, 184 (1959) (Suppl. 15) 1151.
8   ARCHER, G. T., *Nature*, 190 (1961) 350.
9   ARCHER, R. K., *J. Pathol. Bacteriol.*, 78 (1959) 95.
10 ARCHER, R. K., *Brit. J. Haematol.*, 6 (1960) 229.
11 ARMITAGE, P., HERXHEIMER, H. AND ROSA, L., *Brit. J. Pharmacol.*, 7 (1952) 625.
12 ARRIGONI-MARTELLI, E. AND KRAMER, M., *Boll. soc. ital. biol. sper.*, 34 (1958) 1130.
13 ASBOE-HANSEN, G. AND WEGELIUS, O., *Nature*, 178 (1956) 262.
14 AUSMAN, R. K., DAVIS, R. B., MEEKER, W. R. AND McQUARRIE, D. G., *Federation Proc.*, 19 (1960) 279.
15 BALLANI, G. K., JHA, C. D. AND SANYAL, R. K., *J. Pharm. Pharmacol.*, 11 (1959) 512.
16 BALLANI, G. K., SINHA, Y. K. AND SANYAL, R. K., *J. Pharm. Pharmacol.*, 11 (1959) 192.
16a BALLESTERO, L. H. AND ZMUD, B. S., *Prensa méd. arg.*, 48 (1961) 89.
17 BARTLET, A. L. AND LOCKETT, M. F., *J. Physiol. (London)*, 147 (1959) 51.
18 BEACH, V. L. AND STEINER, B. G., *J. Pharmacol. Exptl. Therap.*, 131 (1961) 400.
19 BEAUMARIAGE, M. L. AND LECOMTE, J., *Compt. rend. soc. biol.*, 151 (1957) 1976.
20 BENDA, P. AND MIRAVET, L. F., *Compt. rend. soc. biol.*, 151 (1957) 2064.
21 BENDITT, E. P., in G. P. LEWIS (Ed.), *5-Hydroxytryptamine*, Pergamon, London, 1958, p. 32.
22 BENDITT, E. P., in G. P. LEWIS (Ed.), *5-Hydroxytryptamine*, Pergamon, London, 1958, p. 127.
23 BENDITT, E. P. AND ROWLEY, D. A., *Science*, 123 (1956) 24.
24 BENDITT, E. P., WONG, R. L., ARASE, M. AND ROEPER, E., *Proc. Soc. Exptl. Biol. Med.*, 90 (1955) 303.
25 BENEDICT, A. A. AND TIPS, R. L., *Proc. Soc. Exptl. Biol. Med.*, 87 (1954) 618.
26 BERDE, B., DOEPFNER, W. AND CERLETTI, A., *Helv. Physiol. Acta*, 18 (1960) 537.
27 BERTÉ, F., *Boll. soc. ital. biol. sper.*, 37 (1961) 106.
28 BERTÉ, F. AND CREMA, A., to be published.
29 BHATTACHARYA, B. K., *Arch. intern. pharmacodyn.*, 103 (1955) 357.
30 BHATTACHARYA, B. K. AND LEWIS, G. P., *Brit. J. Pharmacol.*, 11 (1956) 202.
31 BHATTACHARYA, B. K. AND LEWIS, G. P., *Brit. J. Pharmacol.*, 11 (1956) 411.
32 BHATTACHARYA, B. K. AND TANACKOVIC, D. A., *Arch. intern. pharmacodyn.*, 104 (1956) 275.
33 BINET, L. AND QUIVY, D., *Compt. rend.*, 247 (1958) 1153.
34 BJURÖ, T. AND WESTLING, H., *Nature*, 184 (1959) 1241.
35 BOBALIK, G. R. AND BASTIAN, J. W., *Federation Proc.*, 20 (1961) 134.
86 BODI, T., SIEGLER, P. E., BROWN, E. B., GERSHENFELD, M. A. AND NODINE, J. H., *Ann. Allergy*, 19 (1961) 386.
37 BORÉUS, L. O., *Acta Physiol. Scand.*, 52 (1961) 320.
38 BOVET, D. AND BOVET-NITTI, F., *Structure et Activité Pharmacodynamique des Médicaments du Système Nerveux Végétatif*, S. Karger, Bâle, 1948.
39 BOVET, D. AND WALTHERT, *Ann. pharm. franç.*, (1944) Suppl. (quoted by 453).
40 BRAY, R. E. AND VAN ARSDEL JR., P. P., *Proc. Soc. Exptl. Biol. Med.*, 106 (1961) 255.
41 BRECHT, K. AND JESCHKE, D., *Naturwissenschaften*, 47 (1960) 20.
41a BRECHT, K. AND JESCHKE, D., *Arch. ges. Physiol., Pflüger's*, 274 (1962) 385.
42 BRIOT, M. AND HALPERN, B. N., *Compt. rend. soc. biol.*, 146 (1952) 21.
43 BROCKLEHURST, W. E., in G. P. LEWIS (Ed.), *5-Hydroxytryptamine*, Pergamon, London, 1958, p. 172.
44 BROCKLEHURST, W. E., HUMPHREY, J. H. AND PERRY, W. L. M., *J. Physiol. (London)*, 150 (1960) 489.
44a BRUUN, E. AND LETMAN, H., *Acta Allergol.*, 17 (1962) 343.
45 BÜCH, O., *Arch. intern. pharmacodyn.*, 123 (1959) 140.
46 BÜCH, O., *Arch. exptl. Pathol. Pharmakol., Naunyn-Schmiedeberg's*, 238 (1960) 92.

47 BÜCH, O., MOLNAR, I. AND WAGNER-JAUREGG, TH., *Helv. Physiol. Pharmacol. Acta*, 18 (1960) C9.
48 BÜCH, O. AND WAGNER-JAUREGG, TH., *Arzneimittel-Forsch.*, 10 (1960) 834.
49 BULLE, P. H., *Proc. Soc. Exptl. Biol. Med.*, 94 (1957) 553.
50 BULLE, P. H., *Federation Proc.*, 18 (1959) 1472.
51 BURDON, K. L., OZKARAGOZ, K., KAUFMAN, H. S. AND MCGOVERN, J. P., *Ann. Allergy*, 18 (1960) 972.
52 CAMERON, J., *Brit. J. Exptl. Pathol.*, 37 (1956) 470.
53 CANAL, N. AND MAFFEI FACCIOLI, A., *Boll. soc. ital. biol. sper.*, 34 (1958) 787.
54 CHEDID, M. L., *Compt. rend. soc. biol.*, 148 (1954) 1039.
55 CHEDID, M. L., *Ann. endocrinol. (Paris)*, 15 (1954) 746.
56 CHEDID, M. L. AND PARAUT, M., *Ann. endocrinol. (Paris)*, 22 (1961) 117.
57 CHRISTENSSON, B. AND EKWALL, B., *Acta Allergol.*, 15 (1960) 425.
58 CIRSTEA, M., *J. physiol. (Paris)*, 52 (1960) 847.
59 CLARK, W. G. AND MACKAY, E. M., *Proc. Soc. Exptl. Biol. Med.*, 71 (1949) 86.
60 CODE, C. F., *Am. J. Physiol.*, 127 (1939) 78.
61 CODE, C. F., *Ann. Allergy*, 2 (1944) 457.
62 CODE, C. F., *Physiol. Revs.*, 32 (1952) 47.
63 CODE, C. F., CODY, D. T. AND KENNEDY, J. C., *J. Physiol. (London)*, 159 (1961) 61 P.
64 COHEN, Y. AND BOISMARE, F., *Compt. rend.*, 253 (1961) 2161.
65 COLLIER, H. O. J., HOLGATE, A., SCHACHTER, M. AND SHORLEY, P. G., *J. Physiol. (London)*, 149 (1959) 54 P.
66 COLLIER, H. O. J., HOLGATE, J. A., SCHACHTER, M. AND SHORLEY, P. G., *Brit. J. Pharmacol.*, 15 (1960) 290.
67 COURVOISIER, S. AND DUCROT, R., *Arch. intern. pharmacodyn.*, 102 (1955) 33.
68 COURVOISIER, S., FOURNEL, J., DUCROT, R., KOLSKY, M. AND KOLTSCHET, P., *Arch. intern. pharmacodyn.*, 92 (1953) 305.
69 COURVOISIER, S. AND LEAN, O., in P. B. BRADLEY, P. DENIKER AND C. RADOUCO-THOMAS (Eds.), *Neuropsychopharmacology*, Elsevier, Amsterdam, 1959, p. 303.
69a CRAPS, L., *Intern. Arch. Allergy*, 20, Suppl. 2 (1962) 1.
70 CRAPS, L. AND INDERBITZIN, TH., *Intern. Arch. Allergy*, 18 (1961) 268.
71 CREMA, A. AND BERTÈ, F., *Boll. soc. ital. biol. sper.*, 36 (1960) 1069.
72 CRIEP, L. H., MAYER, L. D. AND MENCHACA, O. E. L., *J. Allergy*, 22 (1951) 314.
73 CRIVELLARI, C. A., *Am. J. Physiol.*, 81 (1927) 414.
74 CRONHEIM, G. E. AND TOEKES, I. M., *J. Pharmacol. Exptl. Therap.*, 127 (1959) 167.
75 CRUCHAUD, A., GIRARD, J. P., DOMINÉ, E. AND MICHELI, H., *Intern. Arch. Allergy*, 19 (1961) 65.
76 DAVIS, R. B., AUSMAN, R. K., MEEKER, W. R., GEMMILL, S. J. AND AUST, J. B., *Federation Proc.*, 19 (1960) 158.
77 DAVIS, R. B., BAILEY, W. L. AND HANSON, N. P., *J. Lab. Clin. Med.*, 58 (1961) 811.
77a DAVIS, R. B., BROWN, B. W., MEEKER JR., W. R., AUSMAN, R. K. AND BAILEY, W. L., *Proc. Soc. Exptl. Biol. Med.*, 109 (1952) 412.
78 DAVIS, R. B., MCQUARRIE, D. G. AND MEEKER, W. R., *Federation Proc.*, 18 (1959) 211.
79 DAVIS, R. B., MEEKER JR., W. R. AND BAILEY, W. L., *Federation Proc.*, 20 (1961) 261.
79a DAVIS, R. B., MEEKER JR., W. R. AND BAILEY, W. L., *Proc. Soc. Exptl. Biol. Med.*, 108 (1961) 774.
80 DAY, S. M. AND GREEN, J. P., *Federation Proc.*, 18 (1959) 1505.
81 DES PREZ, R. M., FALLON, N. AND HOOK, E. W., *Proc. Soc. Exptl. Biol. Med.*, 107 (1961) 529.
82 DEUTSCH, S. AND LOEW, E. R., *Federation Proc.*, 13 (1954) 348.
83 DEWES, R., *Arch. intern. pharmacodyn.*, 104 (1955) 19.
84 DEWS, P. B. AND CODE, C. F., *J. Immunol.*, 70 (1953) 199.
85 DIXON, J. B., *J. Physiol. (London)*, 147 (1959) 144.
86 DOEPFNER, W. AND CERLETTI, A., *Intern. Arch. Allergy*, 10 (1957) 348.
87 DOEPFNER, W. AND CERLETTI, A., *Intern. Arch. Allergy*, 12 (1958) 89.
88 DOMENJOZ, R., *Schweiz. med. Wochschr.*, 82 (1952) 1023.
89 DOMENJOZ, R., *Arch. exptl. Pathol. Pharmakol., Naunyn-Schmiedeberg's*, 225 (1955) 14.
90 DOMENJOZ, R., THEOBALD, W. AND MÖRSDORF, K., *Arzneimittel-Forsch.*, 5 (1955) 488.
91 DONALDSON JR., R. M., MALKIEL, S. AND GRAY, S. J., *Proc. Soc. Exptl. Biol. Med.*, 103 (1960) 261.
92 DOUGHERTY, T. F. AND SCHNEEBELI, G. L., *Ann. N.Y. Acad. Sci.*, 61 (1955) 328.
93 DOUGLAS, W. W. AND TOH, C. C., *J. Physiol. (London)*, 120 (1953) 311.

94   DRAGSTED, C. A., *J. Allergy*, 16 (1945) 69.
95   DWORETZKY, M., *Am. J. Physiol.*, 197 (1959) 31.
96   ECKHARDT, E. T., THOMAS, G. B. AND GOVIER, W. M., *Proc. Soc. Exptl. Biol. Med.*, 98 (1958) 211.
97   ENGELHARDT, G. AND ROSER, F., *Arch. exptl. Pathol. Pharmakol., Naunyn-Schmiedeberg's*, 230 (1957) 90.
98   ENGELHARDT, G. AND SCHWABE, U., *Klin. Wochschr.*, 38 (1960) 145.
99   ENGELHARDT, G. AND SCHWABE, U., *Arch. exptl. Pathol. Pharmakol., Naunyn-Schmiedeberg's*, 239 (1960) 170.
100   EZER, E. AND SZPORNY, L., *Acta Physiol. Acad. Sci. Hung.*, 20 (1961) 171.
101   FANCHAMPS, A., DOEPFNER, W., WEIDMANN, H. AND CERLETTI, A., *Schweiz. med. Wochschr.*, 90 (1960) 1040.
102   FEINBERG, S. M. AND STERNBERGER, L. A., *J. Allergy*, 26 (1955) 170.
103   FELDBERG, W. AND SMITH, A. N., *J. Physiol. (London)*, 122 (1953) 62 P.
104   FELDBERG, W. AND SMITH, A. N., *Brit. J. Pharmacol.*, 8 (1953) 406.
104a FICHERA, C., *Boll. soc. ital. biol. sper.*, 38 (1962) 1286.
105   FINDLAY, G. M., *J. Pathol. Bacteriol.*, 31 (1928) 633.
106   FINK, M. A., *Proc. Soc. Exptl. Biol. Med.*, 92 (1956) 673.
107   FINK, M. A. AND GARDNER, C. E., *Proc. Soc. Exptl. Biol. Med.*, 97 (1958) 554.
108   FINK, M. A. AND ROTHLAUF, M. V., *Proc. Soc. Exptl. Biol. Med.*, 90 (1955) 447.
109   FISHER, J. P. AND COOKE, R. A., *J. Allergy*, 29 (1958) 396.
110   FISHER, P. AND LECOMTE, J., *Compt. rend. soc. biol.*, 150 (1956) 1026.
111   FLASHMAN, D., *J. Infectious Diseases*, 38 (1926) 461.
112   FLAVIAN, N. AND PARROT, J. L., *Compt. rend.*, 250 (1960) 1753.
113   FONTANINI, F., PRATI, P. L. AND BARBIERI, U., *Boll. soc. med. chir. Modena*, 58 (1958) 240.
114   FONTANINI, F., PRATI, P. L. AND BARBIERI, U., *Arch. ital. sci. farmacol.*, 8 (1958) 302.
115   FORMANEK, K. AND HÖLLER, H., *Arch. exptl. Pathol. Pharmakol., Naunyn-Schmiedeberg's*, 237 (1959) 430.
116   FOX JR., C. L., EINBINCER, J. M. AND NELSON, C. T., *Am. J. Physiol.*, 192 (1958) 241.
117   FRAHM, M., *Arch. exptl. Pathol. Pharmakol., Naunyn-Schmiedeberg's*, 232 (1957) 286.
118   FRANCHIMONT, P., LECOMTE, J. AND VAN CAUWENBERGE, H., *Compt. rend. soc. biol.*, 154 (1960) 2383.
119   FRANCHIMONT, P., LECOMTE, J. AND VAN CAUWENBERGE, H., *Compt. rend. soc. biol.*, 155 (1961) 180.
119a FRANCHIMONT, P., LECOMTE, J. AND VAN CAUWENBERGE, H., *Compt. rend. soc. biol.*, 156 (1962) 400.
120   FRANCHIMONT, P., VAN CAUWENBERGE, H. AND LECOMTE, J., *Compt. rend. soc. biol.*, 155 (1961) 432.
121   FREYBURGER, W. A., GRAHAM, B. E., RAPPORT, M. M., SEAY, P. H., GROVIER, W. H., SWOAP, O. F. AND VAN DER BROOK, M. J., *J. Pharmacol. Exptl. Therap.*, 105 (1952) 80.
122   GADDUM, J. H. AND STEPHENSON, R. P., *Brit. J. Pharmacol.*, 13 (1958) 493.
123   GARATTINI, S., GAIARDONI, P., MORTARI, A. AND PALMA, V., *Nature*, 190 (1961) 540.
124   GARATTINI, S., JORI, A. AND BONACCORSI, A., unpublished results.
125   GARATTINI, S., LAMESTA, L., MORTARI, A., PALMA, V. AND VALZELLI, L., *J. Pharm. Pharmacol.*, 13 (1961) 385.
126   GARATTINI, S., NANNI, E. AND PALMA, V., to be published.
127   GARATTINI, S. AND VALZELLI, L., *Boll. soc. ital. biol. sper.*, 32 (1956) 288.
128   GARATTINI, S. AND VALZELLI, L., *Boll. soc. ital. biol. sper.*, 32 (1956) 292.
129   GARATTINI, S. AND VALZELLI, L., *Boll. soc. ital. biol. sper.*, 32 (1956) 295.
130   GARCIA AROCHA, H., *Federation Proc.*, 18 (1959) 50.
130a GARCIA-AROCHA, H., *Can. J. Biochem. Physiol.*, 39 (1961) 403.
131   GAUTHIER, G. F., LOEW, E. R. AND JENKINS, H. J., *Proc. Soc. Exptl. Biol. Med.*, 90 (1955) 726.
132   GEIGER, W. B. AND ALPERS, H. S., *Science*, 125 (1957) 1141.
133   GEIGER, W. B. AND ALPERS, H. S., *J. Allergy*, 30 (1959) 316.
134   GEIRINGER, E. AND HARDWICK, D. C., *J. Physiol. (London)*, 119 (1953) 410.
135   GEORGES, G. AND HEROLD, M., *Compt. rend. soc. biol.*, 151 (1957) 695.
136   GEORGES, G., HEROLD, M. AND CAHN, J., *Compt. rend. soc. biol.*, 152 (1958) 1666.
136a GERSHON, M. D. AND ROSS, L. L., *J. Exptl. Med.*, 115 (1962) 367.
137   GIBERTI, A., PONZONI, R. AND PIRONDINI, G. P., *Arch. sci. biol. (Bologna)*, 36 (1952) 219.

138 GIERTZ, H. AND HAHN, F., *Arzneimittel-Forsch.*, 9 (1959) 553.
139 GIERTZ, H., HAHN, F., JURNA, I. AND SCHMUTZLER, W., *Arch. exptl.Pathol. Pharmakol., Naunyn-Schmiedeberg's*, 242 (1961) 65.
140 GIERTZ, H., HAHN, F., OPFERKUCH, W. AND SCHMUTZLER, W., *Arch. exptl. Pathol. Pharmakol., Naunyn-Schmiedeberg's*, 242 (1961) 42.
141 GILBERT, R. P., *Proc. Soc. Exptl. Biol. Med.*, 100 (1959) 346.
142 GIRARD, J. P., *Helv. Med. Acta*, 28 (1961) 476.
143 GLICK, D. AND SYLVEN, B., *Science*, 113 (1951) 388.
144 GÖING, H. AND MICKE, H., *Arch. exptl. Pathol. Pharmakol., Naunyn-Schmiedeberg's*, 241 (1961) 309.
145 GOMIRATO, G. AND ZANALDA, A., *Arch. sci. med.*, 106 (1958) 323.
146 GORDON, P. AND LIPTON, M. A., *Federation Proc.*, 16 (1957) 301.
147 GORDON, P. AND LIPTON, M. A., *Proc. Soc. Exptl. Biol. Med.*, 105 (1960) 162.
148 GOTTESMAN, J. M. AND GOTTESMAN, J., *J. Exptl. Med.*, 47 (1928) 503.
149 GÖZSY, B. AND KÁTÓ, L., *Nature*, 178 (1956) 1352.
150 GÖZSY, B. AND KÁTÓ, L., *Can. J. Biochem. Physiol.*, 34 (1956) 571.
151 GÖZSY, B. AND KÁTÓ, L., *Science*, 125 (1957) 934.
152 GÖZSY, B. AND KÁTÓ, L., *J. Physiol. (London)*, 139 (1957) 1.
153 GÖZSY, B. AND KÁTÓ, L., *Arch. intern. pharmacodyn.*, 128 (1960) 75.
154 GÖZSY, B. AND KÁTÓ, L., *Ind. J. Med. Research*, 48 (1960) 115.
155 GÖZSY, B. AND KÁTÓ, L., *Ann. N.Y. Acad. Sci.*, 88 (1960) 43.
156 GÖZSY, B. AND KÁTÓ, L., *Intern. Arch. Allergy*, 19 (1961) 168.
156a GÖZSY, B. AND KÁTÓ, L., *Intern. Arch. Allergy*, 21 (1962) 138.
156b GÖZSY, B. AND KÁTÓ, L., *Intern. Arch. Allergy*, 21 (1962) 151.
157 GRAHAM, X., *J. Fac. Med. Baghdad, Iraq*, 18 (1954) 1.
158 GREEF, K. AND COUTZEN, C., *Arch. exptl. Pathol. Pharmakol., Naunyn-Schmiedeberg's*, 239 (1960) 35.
159 GREEN, A. F., *Brit. J. Pharmacol.*, 8 (1953) 171.
160 GYERMEK, L., *Pharmacol. Revs.*, 13 (1961) 399.
161 HADDY, F. J., *Federation Proc.*, 19 (1960) 56.
162 HAGEN, P., *Can. J. Biochem. Physiol.*, 39 (1961) 639.
163 HAHN, F. AND OBERDORF, A., *Z. Immunitätsforsch*, 107 (1950) 528.
163a HAJÓS, M. K., *Acta Allergol.*, 17 (1962) 358.
164 HALKIN, F., *Compt. rend. soc. biol.*, 154 (1960) 1668.
165 HALPERN, B. N., *Arch. intern. pharmacodyn.*, 68 (1942) 339.
166 HALPERN, B. N., *Bull. soc. chim. biol.*, 29 (1947) 309.
167 HALPERN, B. N., *Compt. rend. soc. biol.*, 146 (1952) 1996.
168 HALPERN, B. N., BENACERRAF, B. AND BRIOT, M., *Brit. J. Pharmacol.*, 7 (1952) 287.
169 HALPERN, B. N., BENACERRAF, B. AND BRIOT, M., *Proc. Soc. Exptl. Biol. Med.*, 79 (1952) 37.
170 HALPERN, B. N. AND BRIOT, M., *Arch. intern. pharmacodyn.*, 82 (1950) 247.
171 HALPERN, B. N. AND BRIOT, M., *Arch. intern. pharmacodyn.*, 91 (1952) 291.
172 HALPERN, B. N. AND BRIOT, M., *Compt. rend. soc. biol.*, 148 (1954) 959.
173 HALPERN, B. N., BRIOT, M. AND NEVEN, TH., *Compt. rend. soc. biol.*, 148 (1954) 308.
174 HALPERN, B. N., LIACOPOULOS, P. AND LIACOPOULOS-BRIOT, M., *Compt. rend. soc. biol.*, 151 (1957) 1692.
175 HALPERN, B. N., LIACOPOULOS, P. AND LIACOPOULOS-BRIOT, M., *Arch. intern. pharmacodyn.*, 119 (1959) 56.
176 HALPERN, B. N., NEVEN, T. AND BRANNELLEC, A., *Compt. rend. soc. biol.*, 153 (1959) 14.
177 HALPERN, B. N. AND ROUX, J., *Compt. rend. soc. biol.*, 143 (1949) 923.
178 HALPERN, B. N. AND WOOD, D. R., *Brit. J. Pharmacol.*, 5 (1950) 510.
179 HALPERN, B. N. AND WOOD, D. R., *Compt. rend.*, 230 (1950) 138.
180 HANSEN, K. AND ZIPF, H. F., *Arch. exptl. Pathol. Pharmakol., Naunyn-Schmiedeberg's*, 240 (1960) 253.
181 HARRIS, J. M. AND WEST, G. B., *Nature*, 191 (1961) 399.
182 HERTTING, G. AND STOKLASKA, E., *Arch. exptl. Pathol. Pharmakol., Naunyn-Schmiedeberg's*, 237 (1959) 423.
183 HERXHEIMER, H., *J. Physiol. (London)*, 120 (1953) 65P.
184 HERXHEIMER, H., *J. Physiol. (London)*, 122 (1953) 49P.

185 HERXHEIMER, H., *J. Physiol. (London)*, 128 (1955) 435.
186 HERXHEIMER, H., *Brit. J. Pharmacol.*, 10 (1955) 160.
187 HERXHEIMER, H., *Arch. intern. pharmacodyn.*, 106 (1956) 371.
188 HERXHEIMER, H., in G. P. LEWIS (Ed.), *5-Hydroxytryptamine*, Pergamon, London, 1958, p. 163.
189 HERXHEIMER, H. AND SCHACHTER, M., *Nature*, 183 (1959) 1510.
190 HICKS, R. AND WEST, G. B., *Nature*, 182 (1958) 401.
191 HIGGINBOTHAM, R. D., *Federation Proc.*, 18 (1959) 572.
192 HIGGINBOTHAM, R. D., *Proc. Soc. Exptl. Biol. Med.*, 102 (1959) 4.
193 HIGGINBOTHAM, R. D., *Texas Repts. Biol. and Med.*, 18 (1960) 358.
194 HIGGINBOTHAM, R. D., *Federation Proc.*, 19 (1960) 141.
195 HILLEBRECHT, J., *Arzneimittel-Forsch.*, 9 (1959) 625.
196 HOENE, R., *Ann. Acfas*, 18 (1952) 70 (*C.A.*, 48 (1954) 13103 b).
197 HOLGATE, J. A. AND WARNER, B. T., *Brit. J. Pharmacol.*, 15 (1960) 561.
198 HUIDOBRO, H. AND VALETTE, G., *Compt. rend.*, 250 (1960) 1375.
199 HUMPHREY, J. H. AND JAQUES, R., *J. Physiol. (London)*, 119 (1953) 43P.
200 HUMPHREY, J. H. AND JAQUES, R., *J. Physiol. (London)*, 124 (1954) 305.
201 HUMPHREY, J. H. AND JAQUES, R., *J. Physiol. (London)*, 128 (1955) 9.
202 HUNDER, G. AND SPINK, W. W., *Proc. Soc. Exptl. Biol. Med.*, 95 (1957) 55.
203 INDERBITZIN, T., *Intern. Arch. Allergy*, 7 (1955) 140.
204 INDERBITZIN, T., *Intern. Arch. Allergy*, 8 (1956) 150.
205 INDERBITZIN, T., *Intern. Arch. Allergy*, 9 (1956) 146.
206 INDERBITZIN, T. AND CRAPS, L., *Dermatologica*, 114 (1957) 208.
207 INOUE, T. AND KURIAKI, K., *Compt. rend. soc. biol.*, 151 (1957) 1470.
208 JANCSO, N., *J. Pharm. Pharmacol.*, 13 (1961) 577.
209 JAQUES, R., BEIN, H. J. AND MEIER, R., *Helv. Physiol. Acta*, 14 (1956) 269.
210 JASMIN, G., *Proc. Soc. Exptl. Biol. Med.*, 105 (1960) 581.
211 JORI, A., BENTIVOGLIO, A. P. AND GARATTINI, S., *J. Pharm. Pharmacol.*, 13 (1961) 617.
212 JORI, A. AND LEONARDI, A., *Atti soc. lombarda sci. med. e biol.*, 14 (1959) 274.
213 KADATZ, R., *Arzneimittel-Forsch.*, 7 (1957) 651.
214 KALLOS, P. AND KALLOS-DEFFNER, L., *Intern. Arch. Allergy*, 11 (1957) 237.
215 KÁTÓ, L. AND GÖZSY, B., *J. Pharmacol. Exptl. Therap.*, 129 (1960) 231.
216 KÁTÓ, L. AND GÖZSY, B., *Toxicol. Appl. Pharmacol.*, 2 (1960) 144.
217 KÁTÓ, L. AND GÖZSY, B., *Federation Proc.*, 20 (1961) 256.
218 KÁTÓ, L. AND GÖZSY, B., *Toxicol. Appl. Pharmacol.*, 3 (1961) 145.
218a KÁTÓ, L. AND GÖZSY, B., in J. M. BORDELEAU (Ed.), *Extrapyramidal System and Neuroleptics*, Éditions Psychiatriques, Montreal, 1961, p. 165.
219 KATO, R., MARIANI, L. AND VALZELLI, L., *Atti soc. lombarda sci. med. e biol.*, 13 (1958) 297.
220 KATZ, G., *Science*, 91 (1940) 221.
221 KELEMEN, E., *Brit. J. Pharmacol.*, 12 (1957) 28.
222 KELLAWAY, C. H., *Brit. J. Exptl. Pathol.*, 11 (1930) 72.
223 KELLER, R., *Helv. Physiol. Acta*, 15 (1957) 371.
224 KELLER, R., *Intern. Arch. Allergy*, 11 (1957) 328.
225 KELLER, R., *Arzneimittel-Forsch.*, 8 (1958) 390.
226 KELLER, R., *Schweiz. med. Wochschr.*, 90 (1960) 503.
227 KEMPER, F., *Arzneimittel-Forsch.*, 10 (1960) 777.
228 KEPINOW, L., *Compt. rend. soc. biol.*, 87 (1922) 327.
229 KIERSZ, J., *Int. Arch. Allergy*, 10 (1957) 376.
230 KIMURA, E. T., YOUNG, P. R. AND RICHARDS, R. K., *J. Allergy*, 31 (1960) 237.
231 KIND, L. S., *J. Immunol.*, 70 (1953) 411.
232 KIND, L. S., *J. Allergy*, 24 (1953) 52.
233 KIND, L. S., *J. Allergy*, 26 (1955) 507.
234 KIND, L. S., *J. Immunol.*, 77 (1956) 115.
235 KIND, L. S., *Proc. Soc. Exptl. Biol. Med.*, 95 (1957) 200.
236 KIND, L. S., *Bacteriol. Rev.*, 22 (1958) 173.
237 KIND, L. S., *J. Immunol.*, 82 (1959) 32.
238 KIND, L. S. AND GADSDEN, R. H., *Proc. Soc. Exptl. Biol. Med.*, 84 (1953) 373.
239 KIND, L. S. AND GALLEMORE, J. I., *Proc. Soc. Exptl. Biol. Med.*, 92 (1956) 345.
240 KIND, L. S. AND ROESNER, L., *Proc. Soc. Exptl. Biol. Med.*, 100 (1959) 808.

241 KIND, L. S. AND WOODS, E. F., *Proc. Soc. Exptl. Biol. Med.*, 84 (1954) 601.
242 KING, T. O., *Arch. intern. pharmacodyn.*, 110 (1957) 71.
243 KITCHIN, H. W., SMITH, W. Y., WALLACE, G. L., PERKINS, M. AND MORRISON, J. L., *Arch. intern. pharmacodyn.*, 99 (1954) 17.
244 KONZETT, H., *Brit. J. Pharmacol.*, 11 (1956) 289.
245 KRAMER, M., *Arch. exptl. Pathol. Pharmakol., Naunyn-Schmiedeberg's*, 228 (1956) 302.
246 KRUEGER, A. P. AND SMITH, R. F., *J. Gen. Physiol.*, 43 (1960) 533.
247 LAPIÈRE, C. M., *Compt. rend. soc. biol.*, 155 (1961) 1145.
248 LAPIÈRE, C. M., LECOMTE, J. AND VAN CAUWENBERGE, H., *Compt. rend. soc. biol.*, 153 (1959) 1088.
249 LASKER, S. E. AND FOX JR., C. L., *Federation Proc.*, 19 (1960) 60.
250 LECOMTE, J., *Arch. intern. physiol. et biochim.*, 64 (1956) 223.
251 LECOMTE, J., *Arch. intern. physiol. et biochim.*, 64 (1956) 302.
252 LECOMTE, J., *Arch. intern. physiol. et biochim.*, 64 (1956) 554.
253 LECOMTE, J., *Intern. Arch. Allergy*, 9 (1956) 250.
254 LECOMTE, J., *Arch. intern. physiol. et biochim.*, 65 (1957) 451.
255 LECOMTE, J., *Acta Allergol.*, 15 (1960) 61.
256 LECOMTE, J., *Acta Allergol.*, Suppl. 7 (1960) 81.
257 LECOMTE, J. AND BEAUMARIAGE, M. L., *Compt. rend. soc. biol.*, 151 (1957) 2234.
258 LECOMTE, J., BEAUMARIAGE, M. L., SALMON, J. AND DUPONT, M., *Arch. intern. physiol. et biochim.*, 66 (1958) 404.
259 LECOMTE, J. AND FISCHER, P., *Arch. intern. physiol. et biochim.*, 66 (1958) 50.
260 LECOMTE, J. AND FISCHER, P., *Acta Allergol.*, 12 (1958) 240.
261 LECOMTE, J. AND HUGUES, J., *Intern. Arch. Allergy*, 8 (1955) 72.
262 LECOMTE, J. AND LAPIÈRE, C. M., *Compt. rend. soc. biol.*, 154 (1960) 2386.
263 LECOMTE, J., VAN CAUWENBERGE, H. AND VLIERS, M., *Arch. intern. pharmacodyn.*, 121 (1959) 65.
264 LEWIS, J. T., *Rev. asoc. méd. arg.*, 33 (1920) 629.
265 LEWIS, G. P. (Ed.), *5-Hydroxytryptamine*, Pergamon, London, 1958, p. 26.
266 LEWIS, G. P. (Ed.), *5-Hydroxytryptamine*, Pergamon, London, 1958, p. 35.
267 LIACOPOULOS, P., HALPERN, B. N. AND LIACOPOULOS-BRIOT, M., *Compt. rend. soc. biol.*, 151 (1957) 1661.
268 LIBRO, V. AND MARIANI, L., *Atti accad. med. lombardo*, 16 (1961) 244.
269 LILLEHEI, R. C. AND MACLEAN, L. D., *Ann. Surg.*, 148 (1958) 513.
270 LOEW, E. R. AND WOODMAN, E., *Am. J. Physiol.*, 187 (1956) 615.
271 LUCAS, G. H. W., *Am. J. Physiol.*, 77 (1926) 114.
272 LUDANY, G., VAJDA, J., RIGÓ, J. AND HAN TU VU, *Acta Physiol. Acad. Sci. Hung.*, 14 (1958) 371.
273 MACHAFFIE, R. A., BARAK, A. J. AND O'BRIEN, R. L., *J. Allergy*, 29 (1958) 545.
274 MAITLAND, H. G., KOHN, R. AND MCDONALD, A. D., *J. Hyg.*, 53 (1955) 196.
275 MAJNO, G. AND PALADE, G. E., *J. Biophys. Biochem. Cytol.*, 11 (1961) 571.
276 MAJNO, G., PALADE, G. E. AND SCHOEFL, G. I., *J. Biophys. Biochem. Cytol.*, 11 (1961) 607.
277 MALKIEL, S., *J. Allergy*, 27 (1956) 445.
278 MALKIEL, S. AND HARGIS, B. J., *Proc. Soc. Exptl. Biol. Med.*, 80 (1952) 122.
279 MALKIEL, S. AND HARGIS, B. J., *Proc. Soc. Exptl. Biol. Med.*, 81 (1952) 109.
280 MALKIEL, S. AND HARGIS, B. J., *Proc. Soc. Exptl. Biol. Med.*, 81 (1952) 689.
281 MALKIEL, S. AND HARGIS, B. J., *J. Allergy*, 23 (1952) 352.
282 MALKIEL, S., HARGIS, B. J. AND FEINBERG, S. M., *J. Immunol.*, 71 (1953) 311.
283 MANCINI, L. AND FRANCESCHINI, G., *Minerva med.*, 2 (1955) 1436.
284 MANNAIONI, P. F., LEVI, R. AND GIOTTI, A., *Boll. soc. ital. biol. sper.*, 37 (1961) 1140.
285 MARIANI, L., *Congresso Soc. Farmaceutica Med. Latino*, 1959.
286 MARIANI, L., *Folia Allergol. (Rome)*, 6 (1959) 523.
287 MARIANI, L., *Atti accad. med. lombarda*, 15 (1960) 159.
288 MARIANI, L., *Atti accad. med. lombarda*, 15 (1960) 165.
289 MARIANI, L., *Atti accad. med. lombarda*, 16 (1961) 237.
290 MARIANI, L., LIBRO, V. AND FORNASARI, F., *Atti accad. med. lombarda*, 16 (1961) 59.
291 MARIANI, L. AND VALZELLI, L., *Boll. soc. ital. biol. sper.*, 34 (1958) 1169.
292 MARIANI, L. AND VERTUA, R., *Ricerca sci.*, 30 (1960) 1552.
293 MARIANI, L. AND VERTUA, R., *Minerva nucl.*, 5 (1961) 90.
294 MARIANI, L. AND VILLANI, R., *Boll. soc. ital. biol. sper.*, 35 (1959) 1598.

295 MARSHALL, P. B., *J. Physiol. (London)*, 102 (1943) 180.
296 MATHIES, H., *Med. Exptl.*, 4 (1961) 12.
297 MATSUI, T. AND KUWAJIMA, Y., *Nature*, 184 (1959) Suppl. 4, 199.
298 MAYER, R. L. AND BROUSSEAU, D., *Proc. Soc. Exptl. Biol. Med.*, 63 (1946) 187.
299 MCGOVERN, J. P., OZKARAGOZ, K., HENSEL, A. E. AND BURDON, K. L., *Ann. Allergy*, 18 (1960) 1342.
300 MCGOVERN, J. P., OZKARAGOZ, K., HENSEL, A. E. AND BURDON, K. L., *J. Allergy*, 32 (1961) 321.
301 MCINTIRE, F. C., ROTH, L. W. AND RICHARDS, R. K., *Am. J. Physiol.*, 159 (1949) 332.
302 MCLEAN, R. A. AND BERRY, L. J., *Proc. Soc. Exptl. Biol. Med.*, 105 (1960) 91.
303 MEDAKOVIC, M., *Acta Med. Iugoslav.*, 12 (1958) 293.
304 MEDAKOVIC, M., *Arch. intern. physiol. et biochim.*, 67 (1959) 294.
305 MEDAKOVIC, M. AND RADMANOVIC, B., *J. Pharm. Pharmacol.*, 12 (1960) 695.
306 MEDAKOVIC, M. AND SPUZIC, I., *Nature*, 183 (1959) 1685.
307 MICHELSON, A. L. AND HOLLANDER, W., *J. Clin. Invest.*, 35 (1956) 724.
308 MICHELSON, A. L., HOLLANDER, W. AND LOWELL, F. C., *J. Lab. Clin. Med.*, 51 (1958) 57.
309 MILES, A. A. AND WILHELM, D. L., *Brit. J. Exptl. Pathol.*, 36 (1955) 71.
310 MILES, A. A., *Federation Proc.*, 20 (1961) 141.
311 MILKOVIC, S. AND SUPEK, Z., *Arch. exptl. Pathol. Pharmakol., Naunyn-Schmiedeberg's*, 228 (1956) 146.
312 MISHRA, B. P. AND SANYAL, R. K., *J. Pharm. Pharmacol.*, 11 (1959) 127.
313 MITCHELL, R. G. AND CASS, R., *J. Clin. Invest.*, 38 (1959) 595.
314 MONGAR, J. L. AND SCHILD, H. O., *J. Physiol. (London)*, 135 (1957) 320.
314a MONGAR, J. L. AND SCHILD, H. O., *Physiol. Revs.*, 42 (1962) 226.
315 MORRISON, J. L., BLOOM, W. L. AND RICHARDSON, A. P., *J. Pharmacol. Exptl. Therap.*, 101 (1951) 27.
316 MORRISON, J. L., RICHARDSON, A. P. AND BLOOM, W. L., *Arch. intern. pharmacodyn.*, 88 (1951) 98.
317 MÖRSDORF, K., *Med. Exptl.*, 1 (1959) 87.
318 MÖRSDORF, K., *Arch. exptl. Pathol. Pharmakol., Naunyn-Schmiedeberg's*, 236 (1959) 46.
319 MÖRSDORF, K., *Arch. exptl. Pathol. Pharmakol., Naunyn-Schmiedeberg's*, 240 (1960) 29.
320 MÖRSDORF, K., *Med. Exptl.*, 4 (1961) 293.
321 MÖRSDORF, K., *Med. Exptl.*, 4 (1961) 345.
322 MÖRSDORF, K. AND BODE, J. J., *Arch. intern. pharmacodyn.*, 118 (1959) 292.
323 MÖRSDORF, K. AND FEHRES, L., *Med. Exptl.*, 1 (1959) 58.
324 MÖRSDORF, K., STENGER, E. G., THEOBALD, W. AND DOMENJOZ, R., *Arzneimittel-Forsch,.* 5 (1955) 314.
325 MOTA, I., *Nature*, 182 (1958) 1021.
326 MUNOZ, J., *Proc. Soc. Exptl. Biol. Med.*, 95 (1957) 328.
327 MUNOZ, J., *Federation Proc.*, 19 (1960) 158.
328 MUNOZ, J. AND GREENWALD, M. A., *Federation Proc.*, 16 (1957) 427.
329 MUNOZ, J. AND MAUNG, M., *Proc. Soc. Exptl. Biol. Med.*, 106 (1961) 70.
330 MUNOZ, J., RIB, E. AND LARSON, C. L., *J. Immunol.*, 83 (1959) 496.
331 MUNOZ, J. AND SCHUCHARDT, L. F., *J. Allergy*, 24 (1953) 330.
332 MUNOZ, J. AND SCHUCHARDT, L. F., *J. Allergy*, 25 (1954) 125.
333 MUNOZ, J. AND SCHUCHARDT, L. F., *Proc. Soc. Exptl. Biol. Med.*, 94 (1957) 186.
334 MUNOZ, J., SCHUCHARDT, L. F. AND VERWEY, W. F., *J. Immunol.*, 80 (1958) 77.
335 NELSON, C. T., FOX, C. L. AND FREEMAN, E. B., *Proc. Soc. Exptl. Biol. Med.*, 75 (1950) 181.
335a NIWA, M., *J. Biochem. (Tokyo)*, 51 (1962) 222.
336 NIWA, M., YAMADEYA, Y., MATSUI, T. AND KUWAJIMA, Y., *Nature*, 183 (1959) 755.
337 NORTHOVER, B. J., *J. Pathol. Bacteriol.*, 82 (1961) 355.
338 NOYES, H. E., SANFORD, J. P. AND NELSON, R. M., *Proc. Soc. Exptl. Biol. Med.*, 92 (1956) 617.
339 OH, J. O. AND EVANS, C. A., *Bacteriol. Proc. (Soc. Am. Bacteriologists)*, 59th General Meeting, (1959) 145.
340 OH, J. O. AND EVANS, C. A., *Virology*, 10 (1960) 127.
341 PALAZZOADRIANO, M., *Arch. ital. sci. farmacol.*, 10 (1960) 3.
342 PALLOTTA, A. J. AND WARD, J. W., *J. Pharmacol.*, 119 (1957) 174.
343 PANZANI, R. AND PASCAL, N., *Intern. Arch. Allergy*, 18 (1961) 199.
344 PAPACOSTAS, C. A. AND LOEW, E. R., *Proc. Soc. Exptl. Biol. Med.*, 100 (1959) 604.

345 PAPACOSTAS, C. A., LOEW, E. R. AND WEST, G. B., *Arch. intern. pharmacodyn.*, 120 (1959) 353.
346 PAPACOSTAS, C. A., MOZDEN, P. AND LOEW, E. R., *Proc. Soc. Exptl. Biol. Med.*, 97 (1958) 297.
347 PARFENTJEV, I. A., *Yale J. Biol. Med.*, 27 (1954) 46.
348 PARFENTJEV, I. A., *Proc. Am. Assoc. Cancer Research*, 2 (1955) 38.
349 PARFENTJEV, I. A., *Proc. Soc. Exptl. Biol. Med.*, 89 (1955) 297.
350 PARFENTJEV, I. A. AND GODDLINE, M. A., *J. Pharmacol. Exptl. Therap.*, 92 (1948) 411.
351 PARFENTJEV, I. A., GODDLINE, M. A. AND VIRION, M. E., *J. Bacteriol.*, 53 (1947) 603.
352 PARKER, J. M., MCCOLL, J. D. AND FERGUSON, J. W. K., *J. Pharmacol. Exptl. Therap.*, 118 (1956) 359.
353 PARRATT, J. R. AND WEST, G. B., *J. Physiol. (London)*, 135 (1956) 10P.
354 PARRATT, J. R. AND WEST, G. B., *J. Physiol. (London)*, 137 (1957) 169.
355 PARRATT, J. R. AND WEST, G. B., *J. Physiol. (London)*, 137 (1957) 179.
356 PARRATT, J. R. AND WEST, G. B., *J. Physiol. (London)*, 139 (1957) 27.
357 PARRATT, J. R. AND WEST, G. B., *J. Physiol. (London)*, 140 (1958) 105.
358 PARRATT, J. R. AND WEST, G. B., *Brit. J. Pharmacol.*, 13 (1958) 65.
359 PARRATT, J. R. AND WEST, G. B., *Intern. Arch. Allergy*, 16 (1960) 288.
360 PASTEUR VALERY-RADOT, L., MAURIC, G. AND HUGO, A., *Compt. rend. soc. biol.*, 108 (1931) 649.
361 PIERI, A., *Gazz. med. ital.*, 120 (1961) 129.
362 PITTMAN, M., *Federation Proc.*, 16 (1957) 867.
363 PITTMAN, M. AND GERMUTH, F. G., *Proc. Soc. Exptl. Biol. Med.*, 87 (1954) 425.
364 QUIVY, D., *Compt. rend. soc. biol.*, 154 (1960) 294.
365 RAND, M. AND REID, G., *Australian J. Exptl. Biol. Med. Sci.*, 30 (1952) 153.
366 REUSE, J. J., *Arch. intern. pharmacodyn.*, 78 (1949) 363.
367 RILEY, J. F. AND WEST, G. B., *J. Physiol. (London)*, 120 (1953) 528.
368 ROCHA Y SILVA, M. AND ANTONIO, A., *Med. Exptl.*, 3 (1960) 371.
369 ROCHA Y SILVA, M. AND ARONSON, M., *Brit. J. Exptl. Pathol.*, 33 (1952) 577.
370 ROCHA Y SILVA, M., ARONSON M. AND BIER, O. G., *Nature*, 157 (1951) 801.
371 ROSA, L., CENACCHI, G. C. AND BERAMI, G., *Acta Allergol.*, 12 (1958) 427.
372 ROSE, J. C., *J. Clin. Invest.*, 36 (1957) 924.
373 ROSE, B. AND BROWNE, J. S. L., *J. Immunol.*, 41 (1941) 403.
374 ROSE, B. AND BROWNE, J. S. L., *Am. J. Physiol.*, 131 (1941) 589.
375 ROSE, B. AND WEIL, P., *Proc. Soc. Exptl. Biol. Med.*, 42 (1939) 494.
376 ROSENBERG, J. C., LILLEHEI, R. C., LONGERBEAM, J. AND ZIMMERMANN, B., *Ann. Surg.*, 154 (1961) 611.
377 ROSENBERG, J. C., LILLEHEI, R. C., MORAN, W. H. AND ZIMMERMANN, B., *Proc. Soc. Exptl. Biol. Med.*, 102 (1959) 335.
377a ROTHSCHILD, A. M. AND GARCIA LIMA, E., *Experientia*, 19 (1963) 19.
378 ROWEN, M., MOOS, W. W. AND SAMTER, M., *Proc. Soc. Exptl. Biol. Med.*, 88 (1955) 548.
379 ROWLEY, D. A., *Lancet*, 268 (1955) 232.
380 ROWLEY, D. A., *Federation Proc.*, 18 (1959) 503.
381 ROWLEY, D. A. AND BENDITT, E. P., *J. Exptl. Med.*, 103 (1956) 399.
382 SANYAL, R. K., SPENCER, P. S. J. AND WEST, G. B., *Nature*, 184 (1959) 2020.
383 SANYAL, R. K. AND WEST, G. B., *Nature*, 180 (1957) 1417.
384 SANYAL, R. K. AND WEST, G. B., *J. Physiol. (London)*, 142 (1958) 571.
385 SANYAL, R. K. AND WEST, G. B., *J. Physiol. (London)*, 144 (1958) 525.
386 SANYAL, R. K. AND WEST, G. B., *J. Pharm. Pharmacol.*, 11 (1959) 17.
387 SAXENA, P. N., *Arch. intern. pharmacodyn.*, 126 (1960) 228.
388 SCHACHTER, M., *Brit. J. Pharmacol.*, 8 (1953) 412.
389 SCHAYER, R. W., *Am. J. Physiol.*, 187 (1956) 63.
390 SCHAYER, R. W., *Am. J. Physiol.*, 198 (1960) 1187.
391 SCHAYER, R. W., DAVIS, K. J. AND SMILEY, R. L., *Am. J. Physiol.*, 182 (1955) 54.
392 SCHAYER, R. W. AND GANLEY, O. H., *Am. J. Physiol.*, 197 (1959) 721.
393 SCHAYER, R. W. AND GANLEY, O. H., *Federation Proc.*, 18 (1959) 137.
394 SCHAYER, R. W. AND GANLEY, O. H., *Federation Proc.*, 19 (1960) 140.
395 SCHAYER, R. W. AND GANLEY, O. H., *J. Allergy*, 32 (1961) 204.
396 SCHAYER, R. W., ROTHSCHILD, Z. AND BIZONY, P., *Am. J. Physiol.*, 196 (1959) 295.
397 SCHEIFFARTH, F., ZICHA, L., SCHOTT, G., SCHMID, E. AND ALMS, U., *Arzneimittel-Forsch.*, 11 (1961) 595.

398 SCHERBEL, A. L. AND HARRISON, J. W., *Clin. Research Proc.*, 6 (1958) 402.
399 SCHERBEL, A. L. AND HARRISON, J. W., *Angiology*, 10 (1959) 29.
399a SCHERBEL, A. L. AND HARRISON, J. W., *Circulation*, 19 (1959) 777.
400 SCHINDLER, R., *Biochem. Pharmacol.*, 1 (1958) 323.
401 SCHMID, E., ZICHA, L., SCHEIFFARTH, F. AND BÜTTNER, O., *Arzneimittel-Forsch.*, 9 (1959) 474.
402 SETNIKAR, I., SALVATERRA, M. AND TEMELCOU, O., *Brit. J. Pharmacol.*, 14 (1959) 484.
403 SHIBUSAWA, K., TOKUZAWA, K., KISHI, S., KAJIYA, Y., KASUGA, H. AND FUJIWARA, S., *Gumna J. Med. Sci. (Japan)*, 7 (1958) 91.
404 SHIBUSAWA, K., TOKUZAWA, K., KISHI, S., KAJIYA, Y., KASUGA, H. AND FUJIWARA, S., *Gumna J. Med. Sci. (Japan)*, 7 (1958) 101.
405 SHIMAMOTO, T., INOUE, M., KONISHI, T. AND IWAHARA, S., *Arch. intern. pharmacodyn.*, 121 (1959) 342.
406 SHIMAMOTO, T., INOUE, M. AND YOKOYAMA, M., *Proc. Japan Acad.*, 34 (1958) 548.
407 SICUTERI, F., FRANCHI, G. AND MICHELACCI, S., *Med. Exptl.*, 3 (1960) 89.
408 SICUTERI, F., MICHELACCI, S. AND FRANCHI, G., *Intern. Arch. Allergy*, 15 (1959) 291.
409 SJOERDSMA, A., WAALKES, T. P. AND WEISSBACH, H., *Science*, 125 (1957) 1202.
410 SKILLEN, R. G., THIENES, C. H., CANGELOSI, J. AND STRAIN, L., *Proc. Soc. Exptl. Biol. Med.*, 107 (1961) 178.
411 SKILLEN, R. G., THIENES, C. H., CANGELOSI, J. AND STRAIN, L., *Proc. Soc. Exptl. Biol. Med.*, 108 (1961) 121.
411a SMELLIE, H. AND FRY, L., *Acta Allergol.*, 17 (1962) 352.
412 SMITH, D. D. AND MILES, A. A., *Brit. J. Exptl. Pathol.*, 41 (1960) 305.
413 SMITH, D. E. AND LEWIS, Y. S., *J. Exptl. Med.*, 113 (1961) 683.
413a SMYTH, C. J., GUM, O. B. AND HAMILTON, P. K., *J. Lab. Clin. Med.*, 60 (1962) 492.
414 SOLOTOROVSKY, M. AND WINSTEN, S., *J. Immunol.*, 72 (1954) 177.
415 SPARROW, E. M. AND WILHELM, D. L., *J. Physiol. (London)*, 137 (1957) 51.
416 SPECTOR, W. G., *Pharmacol. Revs.*, 10 (1958) 475.
417 SPECTOR, W. G. AND WILLOUGHBY, D. A., *J. Pathol. Bacteriol.*, 74 (1957) 57.
418 SPECTOR, W. G. AND WILLOUGHBY, D. A., *Nature*, 181 (1958) 708.
419 SPECTOR, W. G. AND WILLOUGHBY, D. A., *Nature*, 182 (1958) 949.
420 SPECTOR, W. G. AND WILLOUGHBY, D. A., *Nature*, 186 (1960) 162.
421 SPECTOR, W. G. AND WILLOUGHBY, D. A., *Nature*, 192 (1961) 489.
422 SPENCER, P. S. J. AND WEST, G. B., *Brit. J. Pharmacol.*, 17 (1961) 137.
423 STEINER, F. A. AND HEDINGER, C., *Experientia*, 12 (1956) 109.
424 STENGER, E. G., *Arzneimittel-Forsch.*, 8 (1958) 693.
425 STENGER, E. G., *Arzneimittel-Forsch.*, 9 (1959) 190.
426 STERN, P., *Bulletin Sci.*, 3 (1957) 74.
427 STERN, P. AND MILINE, R., *Arzneimittel-Forsch.*, 6 (1956) 445.
428 STONE, C. A., WENGER, H. C., LUDDEN, C. T., STAVORSKI, J. M. AND ROSS, C. A., *J. Pharmacol. Exptl. Therap.*, 131 (1961) 73.
429 STORMORKEN, H., *Arch. intern. pharmacodyn.*, 119 (1959) 232.
430 STORMORKEN, H., *Arch. intern. pharmacodyn.*, 119 (1959) 238.
431 STRONK, M. G. AND PITTMAN, M., *J. Infectious Diseases*, 96 (1955) 152.
432 STUCKI, J. AND THOMPSON, C., *Am. J. Physiol.*, 193 (1958) 275.
433 SUPEK, Z., JOVIĆ, M. AND KEČKEŠ, S., *Arzneimittel-Forsch.*, 11 (1961) 132.
434 TELFORD, J. M. AND WEST, G. B., *Brit. J. Pharmacol.*, 15 (1960) 532.
435 TELFORD, J. M. AND WEST, G. B., *J. Pharm. Pharmacol.*, 12 (1960) 254.
436 THEOBALD, W. AND DOMENJOZ, R., *Arzneimittel-Forsch.*, 8 (1958) 18.
437 THUILLIER, J. AND NAKAJIMA, H., in E. ROTHLIN (Ed.), *Neuropsychopharmacology*, Elsevier, Amsterdam, 1961, Vol. 2, p. 62.
438 TOKUDA, S. AND WEISER, R. S., *Federation Proc.*, 19 (1960) 141.
439 TOKUDA, S. AND WEISER, R. S., *J. Immunol.*, 86 (1961) 292.
440 TORTORICI, G., *Patol. sper.*, 43 (1955) 137.
441 TRAPOLD, J. H., MULFORD, R. AND GROHMAN, D., *Federation Proc.*, 19 (1960) 53.
442 TROQUET, J. AND LECOMTE, J., *Compt. rend. soc. biol.*, 153 (1959) 1886.
443 UDENFRIEND, S., in G. P. LEWIS (Ed.), *5-Hydroxytryptamine*, Pergamon, London, 1958, p. 36.
444 UNGAR, G., KOBRIN, S. AND SEZESNY, B. R., *Arch. intern. pharmacodyn.*, 123 (1959) 71.
445 UVNÄS, B., *Chemotherapia*, 3 (1961) 137.

446 VAN CAUWENBERGE, H., LAPIÈRE, C. M., HALKIN, F. AND LECOMTE, J., *Compt. rend. soc. biol.*, 153 (1959) 189.

447 VAN CAUWENBERGE, H., LAPIÈRE, C. M. AND LECOMTE, J., *Compt. rend. soc. biol.*, 154 (1960) 440.,

448 VAN CAUWENBERGE, H., LAPIÈRE, C. M., LECOMTE, J. AND HALKIN, F., *Compt. rend. soc. biol.*, 153 (1959) 514.

449 VAN CAUWENBERGE, H., LAPIÈRE, C. M., LECOMTE, J. AND RENSON, J., *Compt. rend. soc. biol.*, 152 (1958) 1848.

450 VAN CAUWENBERGE, H., LECOMTE, J. AND LAPIÈRE, C. M., *Compt. rend. soc. biol.*, 152 (1958) 1405.

451 VAN CAUWENBERGE, H., LECOMTE, J. AND LAPIÈRE, C. M., *Compt. rend. soc. biol.*, 153 (1959) 1292.

452 VERACE, V. AND DI TRAPANI, G., *Arch. sci. med.*, 106 (1958) 829.

453 VIAUD, P., *J. Pharm. Pharmacol.*, 6 (1954) 361.

454 VOGEL, G. AND MAREK, M. L., *Arzneimittel-Forsch.*, 11 (1961) 356.

455 VOGEL, G. AND MAREK, M. L., *Arzneimittel-Forsch.*, 11 (1961) 1051.

455a VOGIN, E. E., ROSSI, G. V., CHASE, G. D. AND OSOL, A., *Experientia*, 18 (1962) 191.

456 VON HAHN, F., GIERTZ, H. AND SCHMUTZLER, W., *Intern. Arch. Allergy*, 18 (1961) 62.

456a VOTAVA, Z. AND LAMPLOVA, I., in E. ROTHLIN (Ed.), *Neuropsychopharmacology*, Elsevier, Amsterdam, 1961, Vol. 2, p. 68.

457 WAALKES, T. P. AND COBURN, H., *J. Allergy*, 30 (1959) 394.

458 WAALKES, T. P. AND COBURN, H., *J. Allergy*, 31 (1960) 151.

459 WAALKES, T. P., COBURN, H. AND TERRY, L. L., *J. Allergy*, 30 (1959) 408.

460 WAALKES, T. P. AND WEISSBACH, H., *Proc. Soc. Exptl. Biol. Med.*, 93 (1956) 394.

461 WAALKES, T. P., WEISSBACH, H., BOZICEVICH, J. AND UDENFRIEND, S., *J. Clin. Invest.*, 36 (1957) 1115.

462 WAALKES, T. P., WEISSBACH, H., BOZICEVICH, J. AND UDENFRIEND, S., *Proc. Soc. Exptl. Biol. Med.*, 95 (1957) 479.

462a WALZ. D. T., DiMARTINO, M. AND TEDESCHI, R. E., *Federation Proc.*, 22 (1963) 2287.

462b WEIS, J., *Med. exptl.*, 8 (1963) 1.

463 WEISER, R. S., GOLUB, O. J. AND HAMRE, D. M., *J. Infectious Diseases*, 68 (1941) 97.

464 WEISER, R. S., GOLUB, O. J. AND HAMRE, D. M., *J. Infectious Diseases*, 68 (1941) 187.

465 WEISSBACH, H., WAALKES, T. P. AND UDENFRIEND, S., *Science*, 125 (1957) 235.

466 WEST, G. B., *Intern. Arch. Allergy*, 10 (1957) 257.

467 WEST, G. B., *Intern. Arch. Allergy*, 11 (1957) 312.

468 WEST, G. B., *J. Pharm. Pharmacol.*, 11 (1959) 513.

469 WEST, G. B., *Intern. Arch. Allergy*, 18 (1961) 56.

470 WESTERMANN, E., BALZER, H. AND KNELL, J., *Arch. exptl. Pathol. Pharmakol., Naunyn-Schmiedeberg's*, 234 (1958) 194.

470a WILHELM, D. L., *Pharmacol. Revs.*, 14 (1962) 251.

471 WILHELMI, G., *Medizinische*, 21 (1952) 1591.

472 WILLOUGHBY, D. A., *J. Physiol. (London)*, 148 (1959) 42P.

473 WILLOUGHBY, D. A., *Nature*, 184 (1959) 1156.

474 WILSON, G. S. AND MILES, A. A., *Topley and Wilson's Principles of Bacteriology and Immunology*, 4th ed., Williams and Wilkins, Baltimore, 1955.

474a WINTER, C. A., RISLEY, E. A. AND NUSS, G. W., *Federation Proc.*, 22 (1963) 543.

475 WOOLLEY, D. W., *Proc. Natl. Acad. Sci. U.S.*, 44 (1958) 197.

476 WYMAN, L. C., *Am. J. Physiol.*, 89 (1929) 356.

477 WYMAN, L. C. AND SUDEN, C., *Am. J. Physiol.*, 94 (1930) 579.

478 ZAMBONI, P., *Arch. intern. pharmacodyn.*, 113 (1957) 203.

479 ZANUSSI, C., MAZZEI, D. AND TASCHINI, P., *Boll. ist. sieroterap. milan.*, 32 (1953) 414.

480 ZICHA, L., SCHEIFFARTH, F., SCHMID, E., ALMS, U. AND SCHOTT, G., *Arzneimittel-Forsch.*, 11 (1961) 598.

481 ZICHA, L., SCHEIFFARTH, F., SCHMID, E., KOSCHERA, H. AND GRAF, N., *Arzneimittel-Forsch.*, 10 (1960) 728.

482 ZICHA, L., SCHMID, E., SCHEIFFARTH, F., GRAF, N. AND KOSCHERA, H., *Arzneimittel-Forsch.*, 10 (1960) 831.

483 ZILBERSTEIN, R. M., *Nature*, 185 (1960) 249.

484 ZILELI, T., CHAPMAN, L. AND WOLFF, H. G., *Arch. intern. pharmacodyn.*, 136 (1962) 463.

Chapter 7

SEROTONIN AND RENAL FUNCTION

As has already been mentioned 5-hydroxytryptamine is excreted in the urine either unchanged or as one of its metabolites. The serotonin coming from the platelets, as well as the locally formed serotonin, is excreted through the kidneys. It has been observed that the kidneys have some tryptophan 5-hydroxylase activity[38] and that the 5-hydroxytryptophan formed is thereafter decarboxylated[5,29,30,147,148]. Monoamine oxidase activity is also present in the kidneys[82,148], the cortical part being more active than the medulla[25]. The richest site of monoamine oxidase activity is the proximal part of the convoluted tubules[25], while the glomeruli are without such activity[94].

ANTIDIURETIC ACTIVITY OF SEROTONIN

In various animal species, including dog, rat, rabbit and guinea pig, serotonin decreases the rate of urinary excretion[52,61]. This observation led Erspamer and others to suggest that 5-hydroxytryptamine should be considered as a renal hormone, important for the regulation of urine flow in normal and pathological conditions (refs. 60–62, 66, 139). Furthermore, serotonin has been identified[74] with the "stable antidiuretic substance" of the serum[89].

In rats serotonin reduces the volume of urine at a dose as small as 4 μg/kg (ref. 61, p. 460). According to Erspamer, this dose must be considered "physiological" because it is approximately the amount of serotonin destroyed each hour per kg body weight by monoamine oxidase (see Chapter 4). This effect of serotonin is particularly evident in hydrated rats[52,58,*], and when the amine is given subcutaneously rather than intravenously[7,45] or intraperitoneally[73]. Serotonin is not antidiuretic when given by the oral route. Van Arman[137] obtained reduced diuresis by serotonin with 10 μg/kg while Corcoran *et al.*[40] had to use large doses (130 μg/kg) before observing the same effect. It is, however, generally accepted that depending from the experimental conditions, it is possible to obtain the antidiuretic effect with very low doses of serotonin. In order to study antiserotonin agents, a suitable dose is 100 μg/kg of serotonin which will reduce the urinary excretion by about 50% for 2 hours[73].

The reduction of urine excretion by serotonin is accompanied by a reduction of the chloride excretion while the antidiuresis by extracts of the posterior pituitary increases the chloride elimination[67].

Urine osmotic activity[46] and creatinine excretion[73] are also reduced by serotonin.

---

* Administration of water or saline results in an increased excretion of 5-hydroxyindoleacetic acid[20a].

Mannitol[46] and *p*-aminohippuric acid[52,73] clearances are similarly reduced by serotonin. Thus the amine decreases the glomerular filtration rate (GFR).

Serotonin is active not only in water-loaded rats but also in counteracting the diuresis induced by xanthines, mercurials, salts and urea[56,70]. Repeated administration of serotonin induces tolerance to its antidiuretic effect[68].

In dogs, serotonin is less antidiuretic than in rats. However, 10–20 μg/kg given intravenously will elicit a definite antidiuresis[2]. In contrast to rats, a dose of 74 μg/kg of serotonin given subcutaneously will not show any effect on the urine excretion[2]. Other authors studied antidiuresis in dogs using considerably higher doses of serotonin[14,15,123] than those previously mentioned. Before observing a reduction of urine excretion, there is a decrease of the elimination of sodium[24,100,101], chloride[15,100,101], and potassium[100,101] although Corcoran *et al.* did not find a clear effect on the electrolyte excretion[40]. As the dose of serotonin is increased, both the antidiuretic effect and the renal plasma flow (RPF) increase[123,134] while the glomerular filtration rate (GFR) decreases[71,123,131,134].

In other animal species there is little information concerning the antidiuretic effect of serotonin. Cats[35,76], rabbits[52], guinea pig[52,99] and mice[143] show antidiuresis with serotonin. Cats[76] require doses of 5–50 μg/kg i.v., while the minimum effective dose of serotonin in mice[143] is about 0.5 mg/kg. In mice, there is also a reduction of chloride excretion[143]. In guinea pigs, histamine is more antidiuretic than serotonin[99].

Serotonin does not influence the passage of water through the skin of frogs[59], although this tissue is rich in serotonin[75].

In man there are contradictory results. On the one hand 10–20 μg/kg/min given intravenously decrease the *p*-aminohippuric acid[115,129] and the mannitol[128] clearances. A single injection of 1–2 mg i.v. of serotonin is also effective in reducing sodium and chloride excretion[91,92]. Decrease of glomerular filtration rate requires a dose of 10 mg of serotonin[9,34]. The same dose induces antidiuresis[6]. Similar results have been obtained in subjects submitted to a water load before serotonin administration[8,10].

On the other hand, other authors did not find any effect on diuresis when serotonin was given in doses of 5–10 mg intravenously or subcutaneously (refs. 31, 33, 34, 39, 41, 111, 140). No antidiuretic activity was present in nephrosclerotic patients receiving serotonin[19], while a marked effect was observed in patients with diabetes insipidus[9,10].

Carcinoid patients have a high level of blood serotonin but do not show consistent changes in the renal functions[98,132,133]. Exogenous serotonin was found to be antidiuretic[132] and it also decreased the clearances[128] in these patients.

MECHANISM OF ACTION OF SEROTONIN ANTIDIURESIS

It is possible that the antidiuresis induced by serotonin in different animal species may be produced by different mechanisms. Only on rats and dogs are sufficient data available to justify discussion of this problem.

Administration of *large* doses of serotonin stimulates the posterior hypophysis to release antidiuretic hormone (ADH)[93], but this is not a feasible explanation for the antidiuretic effect of serotonin because: (*a*) no release of ADH has been observed after administration of antidiuretic doses of serotonin[137]; (*b*) serotonin is still antidiuretic if injected in hypophysectomized rats[83] and dogs[15]; (*c*) serotonin decreases the excretion of chloride while posthypophysis extracts increase the elimination of this ion[67,143].

It has been suggested that the contraction of the ureters occurring after administration of serotonin[3,4,119] may lead to the formation of reflexes which tend to inhibit renal activity. However, serotonin exerts its antidiuretic effect on the denervated kidney[2,49,119].

Other authors believe that the pain induced by subcutaneous injections of serotonin[11] may be responsible for the antidiuretic effect through a release of ADH[1,7]. This is questionable because of the reasons already mentioned. Moreover, Erspamer and Correale[67] did not find painful reactions with the concentrations of serotonin active on diuresis.

Similarly, the hypothesis has to be rejected that the effect of serotonin on diuresis is the result of an effect on the general blood pressure, because it has been shown that the antidiuretic effect of serotonin occurs at doses not affecting the general circulation[46,49,64,116,134,137].

A more likely suggestion was made by Benditt, who found that 0.3 μg of serotonin injected subcutaneously in rats induces oedema and sequestrates about 0.5 ml of water in the skin[20]. This may explain the fact that serotonin is more active when injected subcutaneously than intravenously or intraperitoneally.

In their early work, Erspamer and Ottolenghi[73] explained the antidiuretic effect of serotonin in rats by stating that "5-HT acts through a constriction of the afferent vascular bed of the glomeruli, fall of the intraglomerular hydrostatic pressure and decrease of the blood flow through the peritubular capillary network". The hypothesis that serotonin is antidiuretic because of a direct effect on the kidney is still valid although some objections to this concept have been raised.

The administration of serotonin induces vasoconstriction of the renal circulation[65,72,114,117,*] at doses not affecting the general circulation[116,**]. The vasoconstriction is also present in isolated perfused kidney[117], in denervated kidney[35] and it is even increased after section of the spinal cord[115]. The effect of serotonin on the renal circulation is less marked than that of adrenaline[35,42].

It is difficult to assess whether the whole antidiuretic activity of serotonin may be explained on the basis of the vascular effect on kidney. In rats, there is a constriction of the afferent vascular bed of the glomerulus[65,72], but this effect has not been detected in dogs[110]. According to other authors, doses of serotonin which are antidiuretic in dogs[14] or in cats [35] do not affect the renal circulation.

---

* The reduced toxicity of mercury salts after treatment with serotonin is explained on the basis that they are partially excluded from the renal circulation[57].
** In rabbits and guinea pigs, serotonin constricts pulmonary vessels[141] without any effect on renal circulation[142].

TABLE 25

EFFECT OF VARIOUS AGENTS ON THE RENAL ACTIVITIES OF SEROTONIN
(Numbers indicate references)

| | | *Effect of serotonin* | |
Compound	*Antidiuresis*	*Renal blanching or necrosis*	*Contraction of ureter or bladder*
Acetylcholine			= 135
Adenosine	− 56		
Adenosine triphosphate			= 135
Adrenaline		− 27, = 97	
5-Amino-2-methylgramine	− 63		
Antazoline	− 54		
Atropine	− 56	− 27	− 32, 86
Azacyclonol		− 27	
BAS (Benanserin)	− 91, 108	= 108	− 37
BOL 148	+ 46, − 44a, 128	+ 97	+ 85, 130
5-Chloro-2-methylgramine	± 63		
Chlorpheniramine		− 27	+ 37
Chlorpromazine	+ 43, 44, 81, 105 ± 54	− 27	
Chlorthiazide		− 97	
Cyproheptadine	± 81a		
Dibenamine	+ 53, 58, 137, 138	+ 97	+ 58
Dicyclomine		+ 27	
Dihydroergotamine	+ 58		
Diphenhydramine	+ 54		
Gramine	+ 55		
Harmaline	− 55		
Harmine	− 55		
Hexamethonium	− 56		− 32
Histamine	− 56, = 99	− 97	= 99
Hydralazine	− 51, 56	+ 27, 97, 112, 113	
Iproniazid			− 37
Isopropylnoradrenaline	+ 81	+ 27	+ 37
LSD 25 (Lysergic acid diethyl-amide)	+ 44a, 46	+ 27, 97	+ 37
Mephentermine		− 27	
Meprobamate		− 97	
Methacholine		− 27	
Methapyrilene		− 27	
Methoxamine		− 27	
2-Methyl-3-ethyl-5-dimethyl-aminoindole	±63		
Methylphenetidate		− 27	
Morphine			+ 85
5-Nitro-2-methylgramine	− 63		
Noradrenaline	+ 58	= 27, − 97	
DL-Norsynephrine	+ 58		
Oestradiol		= 97	
Papaverine	− 56		
Pentamethonium	− 56		
Pentobarbital		− 97	
Phenergan	+ 54		− 32

TABLE 25 (*continued*)

Compound	Effect of serotonin		
	Antidiuresis	Renal blanching or necrosis	Contraction of ureter or bladder
Pheniramine	+ 54		
Phenoxybenzamine		— 27	
Phentolamine	— 81a		
Piperoxan	— 58		
Posthypophysary extract	= 56		
Pyribenzamine	+ 54, —84		
Pyrilamine	± 54		
Reserpine	± 81	— 97	± 37
Salicylate		= 97	
Sodium glutamate		+ 27	
Sodium nitrate	— 56	+ 27	
Testosterone		+ 97	
Tetramethylammonium iodide	— 56		
Theophylline	+ 56		
Thyroxine		— 97	
Tolazoline	— 58		
Tyramine	+ 58		
UML (1-Methyllysergic acid butanolamide)	+ 44a, 81a		
Urea	+ 56		
Yohimbine	— 58		

+ antagonism
— ineffective
± doubtful results
= increase

It is possible that the effect of serotonin on diuresis may be the result of a decreased filtration combined with a facilitation of the water absorption in renal tubules[134,137,138].

Very little is known on the biochemical effects of serotonin on kidney metabolism. It has been reported that renal glutaminase is not affected, carbonic anhydrase is depressed and ATPase increased[96].

EFFECT OF SEROTONIN ANALOGUES OR ANTAGONISTS ON DIURESIS

5-Hydroxytryptophan[5,69], 5-acetoxy-N-acetyltryptophan and 5-hydroxy-N-acetyltryptophan[69a] induce antidiuresis only at high concentrations and 5-hydroxyindoleacetic acid is not active in this respect[18]. L-Tryptophan, but not D-tryptophan, is antidiuretic probably through the formation of serotonin[21]. 5-Methoxytryptamine retains about 25–40% of the antidiuretic activity of serotonin[50,73]. Tryptamine and indole derivatives are usually much less effective than serotonin[21a,50,55,63,73]. 6 mg/kg of 4-hydroxytryptamine is less effective as an antidiuretic than 40 μg/kg of serotonin[69]. BAS is antidiuretic and does not prevent the antidiuretic effect of serotonin[90,108].

Several gramine derivatives display only a negligible influence on serotonin anti-diuresis[63].

Lysergic acid derivatives are antagonists of the antidiuretic action of serotonin (refs. 36, 44a, 46) and of the constrictor effect of serotonin on renal circulation[117]. Chlorpromazine also prevents the effect of serotonin[43,44,81,83,105] but not the antidiuresis by ADH[43,44].

The antagonism exerted by dibenamine lasts about seven days[53] and is more evident when dibenamine is given intraperitoneally rather than subcutaneously[137].

Antihistamines[54] are antagonists but atropine, ganglionic blockade or vaso-dilators[56] do not prevent the effect of serotonin.

Serotonin antagonists have been reviewed by Gyermek[87] and a list of compounds which have been tested on serotonin antidiuresis is given in Table 25. Furthermore, serotonin antagonists may exert a diuretic activity[135a].

CHANGES IN DIURETIC ACTIVITY WHICH MAY BE EXPLAINED BY AN EFFECT OF SEROTONIN

During extracorporeal circulation, the onset of oliguria[144] and pathological changes in kidney[127] have been observed. The presence of a vasoconstrictor agent has been detected in extravasated blood[28] and identified as serotonin[79,126] coming from the destruction of the platelets[79,125].

Post-operative oliguria may also be related, not only to the release of ADH[48], but also to an increase of serotonin in serum[80].

After *burning* of the skin in the dog, there is a marked oliguria[13] and the appearance in serum of a vasoconstrictor agent[121,122]. This vasoconstrictor agent is present also when blood is heated[17]. After burning the skin, the serotonin content of kidney is reduced[120]. Lysergic acid diethylamide[16] and BAS[120] do not prevent the antidiuresis induced by burning. However, these results do not exclude serotonin as an explanation of the oliguria after burning, since LSD is antidiuretic in dogs[16] and BAS does not prevent the antidiuresis by serotonin[91,108].

On the other hand, when hot water is injected intraperitoneally, there is an increase of kidney serotonin and a block of the diuresis[105]. This effect is present in hypophysectomized rats[5] and is inhibited by chlorpromazine[83,105]. Treatment for several days with reserpine, a serotonin releaser, prevents the anuria induced by hot water administration[22,104].

Reserpine itself shows antidiuretic properties which are at least partly due to a release of ADH[22,23]. However, chlorpromazine prevents the antidiuretic effect of reserpine[104] without showing any effect on ADH-induced antidiuresis[36a,43,44]. Like serotonin, reserpine induces sodium and chloride retention[92a,95,118] and counteracts the natriuretic effect of hydrochlorothiazide and theophylline[109]. Reserpine prevents also the antidiuresis following the administration of dextran[84]. The effect of dextran is probably due to the release of serotonin[84].

Vincamine releases serotonin and decreases diuresis[99a].

Finally, *electroshock* decreases urine excretion and lowers the stores of intestinal serotonin[136].

RENAL TOXICITY BY SEROTONIN

The administration of serotonin in large single doses[97] (50 mg/kg) or small repeated doses[77] (1 mg/kg) induces pathological lesions of the kidney in rats. The sequence of events is arteriolar spasm, ischaemia and necrosis[12,102]. The necrosis is mainly localized to the cortex[88,106,107]. The glomeruli are usually not involved[27] while the tubules show dilatation, degeneration and necrosis[47,102,146] with the presence of eosinophilic material[27]. These lesions are also induced by serotonin in mononephrectomized rats[124] and increase in severity when serotonin is administered to traumatized animals[145] or to rats fed a diet deficient in choline[103]. Pregnancy does not aggravate the renal lesions induced by serotonin[146].

The effect of serotonin in these doses is rather specific because other organs do not show necrosis[77]. But with increasing doses of serotonin, necrosis also appears in the submaxillary gland, stomach[97], tail and tips of digits[103].

Page suggested that the renal cortical necrosis induced by serotonin may be similar to that found in pregnant women who died after the occurrence of severe abruptio placentae[112].

Adrenaline[102], 5-hydroxytryptophan[20,77] and reserpine[20] induce a tubular necrosis similar to that produced by serotonin.

Lysergic acid derivatives[27,97], chlorpromazine[97], dibenamine[97] and hydralazine[27,97] prevent the renal lesions produced by serotonin. A number of other compounds tested for ability to prevent this effect of serotonin is listed in Table 25.

CONTRACTION OF URETERS AND OF URINARY BLADDER

Isolated pig *ureter* is very sensitive to serotonin. There is an increase of tonus and motility at a concentration of serotonin[32] less than 10^{-7} g. Repeated exposures to serotonin induce tachyphylaxis which is counteracted by intercalating administrations of noradrenaline or cocaine, but not of acetylcholine or histamine[37]. Isolated guinea pig ureter is more sensitive to histamine than serotonin[99]. The tachyphylaxis to serotonin is prevented by histamine or adrenaline[99].

Isolated dog ureter is sensitive to histamine but not to serotonin up to a concentration of 10^{-3} g (refs. 26, 135). However, serotonin is effective when the dog ureter is left *in situ*[3,4].

Phenergan, hexamethonium or atropine do not abolish the effect of serotonin on pig ureter[32] while lysergic acid diethylamide, chlorpromazine and isopropylnoradrenaline are powerful inhibitors[37].

Also the *urinary bladder* responds with a contraction in the presence of serotonin[66,78]. Dog urinary bladder is very sensitive *in vitro*[52] (less than 10^{-7} g) and *in situ*[51,52] (10 μg/kg) while rabbit, guinea pig and rat bladder are less sensitive[52].

According to Gyermek, serotonin is about twice as effective as histamine and about 20 times more effective than acetylcholine on dog bladder [86]. Atropine and ganglionic blockade will not prevent serotonin activity on bladder [86], while BOL, dibenamine and morphine are effective antagonists[58,85]. BOL counteracts also the stimulation of the bladder induced by serotonin in carcinoid patients[130].

A list of compounds tested for antagonism to serotonin action on ureter or urinary bladder is reported in Table 25.

REFERENCES

1 ABRAHAMS, V. C. AND PICKFORD, M., *J. Physiol. (London)*, 126 (1954) 329.
2 ABRAHAMS, V. C. AND PICKFORD, M., *Brit. J. Pharmacol.*, 11 (1956) 35.
3 ABRAHAMS, V. C. AND PICKFORD, M., *Brit. J. Pharmacol.*, 11 (1956) 44.
4 ABRAHAMS, V. C. AND PICKFORD, M., *Brit. J. Pharmacol.*, 11 (1956) 50.
5 AIRAKSINEN, M. M. AND UUSPÄÄ, V. J., *Biochem. Pharmacol.*, 8 (1961) 48.
6 AMBANELLI, U. AND SALVI, G., *Giorn. clin. med. (Parma)*, 40 (1959) 1499.
7 AMES, R. G. AND VAN DYKE, H. B., *Endocrinology*, 50 (1952) 350.
8 ANNONI, G., LEONE, M., LONGARETTI, A., LONGO, G. AND LUCCHELLI, P. D., *Clin. terap.*, 8 (1955) 125.
9 ANNONI, G. AND LUCCHELLI, P. D., *Farmaco (Pavia)*, Ed. sci., 12 (1957) 362.
10 ANNONI, G., LUCCHELLI, P. D. AND BARBIERA, G., *Farmaco (Pavia)*, Ed. sci., 10 (1955) 306.
11 ARMSTRONG, D., DRY, R. M. L., KEELE, C. A. AND MARKHAM, J. W., *J. Physiol. (London)*, 120 (1953) 326.
12 BALLERINI, G. AND CANTELLI, T., *Riv. anat. patol. e oncol.*, 11 (1956) 14.
13 BARAC, G., *Compt. rend. soc. biol.*, 140 (1946) 580.
14 BARAC, G., *Arch. intern. physiol.*, 61 (1953) 403.
15 BARAC, G., *Compt. rend. soc. biol.*, 149 (1955) 1523.
16 BARAC, G., *Compt. rend. soc. biol.*, 151 (1957) 1027.
17 BARAC, G., *Compt. rend. soc. biol.*, 152 (1958) 883.
18 BARAC, G., *Compt. rend. soc. biol.*, 155 (1961) 1732.
19 BASSANI, A. AND FANTINI, S., *Giorn. gerontol.*, 4 (1956) 131.
20 BENDITT, E. P., in G. P. LEWIS (Ed.), *5-Hydroxytryptamine*, Pergamon, London, 1958, p. 127.
20a BERTACCINI, G. AND ERSPAMER, V., *J. Pharm. Pharmacol.*, 14 (1962) 687.
21 BERTACCINI, G. AND NOBILI, M. B., *Brit. J. Pharmacol.*, 17 (1961) 519.
21a BERTACCINI, G. AND ZAMBONI, P., *Arch. intern. pharmacodyn.*, 133 (1961) 138.
22 BERTELLI, A., *Boll. soc. ital. biol. sper.*, 32 (1956) 826.
23 BERTELLI, A., *Boll. soc. ital. biol. sper.*, 32 (1956) 829.
24 BLACKMORE, W. P., *Federation Proc.*, 16 (1957) 282.
25 BLASCHKO, H. AND HELLMANN, K., *J. Physiol. (London)*, 122 (1953) 419.
26 BORGSTEDT, H. H., *Federation Proc.*, 18 (1959) 370.
27 BROSMAN, S. A., BRADFORD, P. F. AND HUGHES, F. W., *Am. J. Clin. Pathol.*, 32 (1959) 457.
28 BRULL, L. AND LOUIS BAR, D., *Arch. intern. physiol. et biochim.*, 65 (1957) 470.
29 BUZARD, J. A. AND NYTCH, P. D., *J. Biol. Chem.*, 227 (1957) 225.
30 BUZARD, J. A. AND NYTCH, P. D., *J. Biol. Chem.*, 229 (1957) 409.
31 CALÌ, G., LOMEO, G. AND BOMPIANI, G. D., *Rass. fisiopatol. clin. e terap.*, 27 (1955) 1077.
32 CARRETTI, D., MALAGUTI, G. AND VECCHIATI, R., *Boll. soc. med. chir. Modena*, 55 (1955) 203.
33 CARRETTI, D. AND RONCHETTI, G., *Boll. soc. med. chir. Modena*, 55 (1955) 581.
34 CENACCHI, G. C. AND CARIANI, A., *Boll. soc. ital. biol. sper.*, 34 (1958) 569.
35 CERLETTI, A., CARPI, A. AND ROTHLIN, E., *Helv. Physiol. Acta*, 13 (1955) C8.
36 CERLETTI, A. AND KONZETT, H., *Arch. exptl. Pathol. Pharmakol., Naunyn-Schmiedeberg's*, 228 (1956) 146.
36a CHAUDHURY, R. R., CHAUDHURI, M. R. AND LU, F. C., *Can. J. Biochem. Physiol.*, 40 (1962) 1465.
37 CIMA, G. AND FRESCHI, C., *Boll. soc. ital. biol. sper.*, 33 (1957) 867.
38 COOPER, J. C. AND MELCER, I., *J. Pharmacol. Exptl. Therap.*, 132 (1961) 265.
39 CORÀ, D., ABRIGNANI, F. AND DE BIASI, S., *Minerva nefrol.*, 4 (1957) 21.
40 CORCORAN, A. C., MASSON, G. C., DEL GRECO, F. AND PAGE, J. H., *Arch. intern. pharmacodyn.*, 97 (1954) 483.
41 CURTI, P. C. AND FORNAROLI, P. *Farmaco (Pavia)*, Ed. sci., 10 (1955) 965.
42 CUYPERS, Y., NIZET, A. AND BARAC, G., *Compt. rend. soc. biol.*, 151 (1957) 1768.
43 DASGUPTA, S. R., *Federation Proc.*, 16 (1957) 290.
44 DASGUPTA, S. R., *Arch. intern. pharmacodyn.*, 112 (1957) 264.
44a DE CARO, G., *Arch. intern. pharmacodyn.*, 141 (1963) 54.
45 DEL GRECO, F., MASSON, G. M. C. AND CORCORAN, A. C., *Federation Proc.*, 15 (1956) 415.
46 DEL GRECO, F., MASSON, G. M. C. AND CORCORAN, A. C., *Am. J. Physiol.*, 187 (1956) 509.
47 DOLCINI, H. A., ZAIDMAN, I., LICHTENBERG, F. AND GRAY, S. J., *Am. J. Physiol.*, 199 (1960) 1153.

48 EISEN, V. D. AND LEWIS, A. A. G., *Lancet*, 267 (1954) 361.
49 EMANUEL, D. A., SCOTT, J., COLLINS, R. AND HADDY, F. J., *Federation Proc.*, 17 (1958) 42.
50 ERSPAMER, V., *Nature*, 170 (1952) 281.
51 ERSPAMER, V., *A.M.A. Arch. Internal Med.*, 90 (1952) 505.
52 ERSPAMER, V., *Ricerca sci.*, 22 (1952) 694.
53 ERSPAMER, V., *Ricerca sci.*, 22 (1952) 1568.
54 ERSPAMER, V., *Ricerca sci.*, 22 (1952) 2148.
55 ERSPAMER, V., *Ricerca sci.*, 23 (1953) 1203.
56 ERSPAMER, V., *Ricerca sci.*, 23 (1953) 2250.
57 ERSPAMER, V., *Experientia*, 9 (1953) 186.
58 ERSPAMER, V., *Arch. intern. pharmacodyn.*, 93 (1953) 293.
59 ERSPAMER, V., *Acta Pharmacol. Toxicol.*, 10 (1954) 1.
60 ERSPAMER, V., *Experientia*, 10 (1954) 471.
61 ERSPAMER, V., *Pharmacol. Revs.*, 6 (1954) 425.
62 ERSPAMER, V., *Rend. sci. Farmitalia*, 1 (1954) 79.
63 ERSPAMER, V., *Science*, 121 (1955) 369.
64 ERSPAMER, V. AND ASERO, B., *Ricerca sci.*, 21 (1951) 2132.
65 ERSPAMER, V. AND ASERO, B., *Nature*, 169 (1952) 800.
66 ERSPAMER, V. AND ASERO, B., *J. Biol. Chem.*, 200 (1953) 311.
67 ERSPAMER, V. AND CORREALE, P., *Arch. intern. pharmacodyn.*, 101 (1955) 99.
68 ERSPAMER, V., CORREALE, P. AND FIORE-DONATI, L., *Arch. intern. pharmacodyn.*, 106 (1956) 122.
69 ERSPAMER, V., GLÄSSER, A. AND MANTEGAZZINI, P., *Experientia*, 16 (1960) 505.
69a ERSPAMER, V. AND NOBILI, M. B., *Arch. intern. pharmacodyn.*, 139 (1962) 433.
70 ERSPAMER, V. AND OTTOLENGHI, A., *Experientia*, 6 (1950) 428.
71 ERSPAMER, V. AND OTTOLENGHI, A., *Experientia*, 7 (1951) 191.
72 ERSPAMER, V. AND OTTOLENGHI, A., *Experientia*, 8 (1952) 31.
73 ERSPAMER, V. AND OTTOLENGHI, A., *Arch. intern. pharmacodyn.*, 93 (1953) 177.
74 ERSPAMER, V. AND SALA, G., *Brit. J. Pharmacol.*, 9 (1954) 31.
75 ERSPAMER, V. AND VIALLI, M., *Nature*, 167 (1951) 1033.
76 FASTIER, F. N. AND WAAL, H., *Brit. J. Pharmacol.*, 12 (1957) 484.
77 FIORE-DONATI, L. AND ERSPAMER, V., *Am. J. Pathol.*, 33 (1957) 895.
78 FREYBURGER, W. A., GRAHAM, B. E., RAPPORT, M. M., SEAY, P. H., GOVIER, W. M., SWOAP,
    O. F. AND VAN DER BROOK, M. J., *J. Pharmacol. Exptl. Therap.*, 105 (1952) 80.
79 FRICK, M. H., *Nature*, 187 (1960) 609.
80 FRICK, M. H., VIRKKULA, L. AND PAASONEN, M. K., *Surgery*, 50 (1961) 447.
81 GARATTINI, S. AND VALZELLI, L., *Boll. soc. ital. biol. sper.*, 32 (1956) 295.
81a GARATTINI, S., unpublished results.
82 GIORDANO, C., BLOOM, J. AND MERRILL, J. P., *Experientia*, 16 (1960) 346.
83 GIULIANI, G., MARIANI, L. AND VILLANI, R. *Boll. soc. ital. biol. sper.*, 36 (1960) 1412.
84 GREEFF, K. AND CONTZEN, C., *Arch. exptl. Pathol. Pharmakol.*, *Naunyn-Schmiedeberg's*, 239
    (1960) 35.
85 GYERMEK, L., *Pharmacologist*, 2 (1960) 89.
86 GYERMEK, L., *Am. J. Physiol.*, 201 (1961) 325.
87 GYERMEK, L., *Pharmacol. Revs.*, 13 (1961) 399.
88 HEDINGER, C. AND LANGEMANN, H., *Schweiz. med. Wochschr.*, 85 (1955) 541.
89 HELLER, H., *Ciba Foundation Coll. Endocrinol.*, 4 (1952) 463.
90 HOLLANDER, W. AND MICHELSON, A. L., *J. Clin. Invest.*, 35 (1956) 712.
91 HOLLANDER, W., MICHELSON, A. L. AND WILKINS, R. W., *Circulation*, 16 (1957) 246.
92 HULET, W. H. AND PERERA, G. A., *Proc. Soc. Exptl. Biol. Med.*, 91 (1956) 512.
92a KHAZAN, N., ADIR, J., PFEIFER, Y. AND SULMAN, F. G., *Proc. Soc. Exptl. Biol. Med.*, 109
    (1962) 32.
93 KIVALO, E., RINNE, U. K. AND MARJANEN, P., *Ann. Med. Internae Fenniae*, 46 (1957) 191.
94 KOELLE, G. B. AND VALK JR., A. DE T., *J. Physiol. (London)*, 126 (1954) 434.
95 KROGSGARD, A. R., *Acta Med. Scand.*, 154 (1956) 41.
96 KURIAKI, K. AND BABA, N., *Compt. rend. soc. biol.*, 153 (1959) 1638.
97 JASMIN, G. AND BOIS, P., *Lab. Invest.*, 9 (1960) 503.
98 LEMBECK, F., in G. P. LEWIS (Ed.), *5-Hydroxytryptamine*, Pergamon, London, 1958, p. 147.
99 LIBRO, D., LIBRO, E. AND COPPI, F., *Atti accad. med. Lombarda*, 15 (1960) 184.

99a LINÉT, O., KREJČÍ, I. AND HÁVA, M., *Acta Biol. Med. Germ.*, 9 (1962) 158.

100 LITTLE, J. M., ANGELL, E. A., HUFFMAN, W. AND BROOKS W., *Federation Proc.*, 19 (1960) 367.

101 LITTLE, J. M., ANGELL, E. A., HUFFMAN, W. AND BROOKS, W., *J. Pharmacol. Exptl. Therap.*, 131 (1961) 44.

102 MACDONALD, R. A., *Am. J. Pathol.*, 35 (1959) 297.

103 MACDONALD, R. A., ROBBINS, S. L. AND MALLORY, G. K., *A.M.A.Arch. Pathol.*, 65 (1958) 369.

104 MARIANI, L., BREDA, G. AND VILLANI, R., *Boll. soc. ital. biol. sper.*, 37 (1961) 633.

105 MARIANI, L. AND VALZELLI, L., *Boll. soc. ital. biol. sper.*, 34 (1958) 1167.

106 MORLUNCHI, C., PICOTTI, T. AND CHIODI, G. *Boll. Soc. med. chir. Pisa*, 24 (1956) 520.

107 MORLUNCHI, C., ROVATI, V. AND PICOTTI, T., *Boll. Soc. med. chir. Pisa*, 24 (1956) 279.

108 MURELLI, B., VALSECCHI, A. AND VALZELLI, L., *Boll. soc. ital. biol. sper.*, 33 (1957) 859.

109 NECHAY, B. R. AND SANNER, E., *Acta Pharmacol. Toxicol.*, 18 (1961) 339.

110 NIZET, E. L., WILSEN, G. AND BARAC, G., *Arch. intern. pharmacodyn.*, 96 (1953) 76.

111 NOTTER, B., *Schweiz. med. Wochschr.*, 86 (1956) 481.

112 PAGE, E. W. AND GLENDENING, M. B., *Obstet. Gynecol.*, 5 (1955) 781.

113 PAGE, E. W. AND GLENDENING, M. B., *Am. J. Med.*, 19 (1955) 285.

114 PAGE, I. H., *J. Pharmacol. Exptl. Therap.*, 105 (1952) 58.

115 PAGE, I. H., in G. P. LEWIS (Ed.), *5-Hydroxytryptamine*, Pergamon, London, 1958, p. 93.

116 PAGE, I. H. AND MCCUBBIN, J. W., *Am. J. Physiol.*, 173 (1953) 411.

117 PASSOW, H., SCHNIEWIND, H. AND WEISS, C., *Arch. exptl. Pathol. Pharmakol., Naunyn-Schmiedeberg's*, 240 (1960) 179.

118 PERERA, G. A., *J. Am. Med. Assoc.*, 159 (1955) 439.

119 PICKFORD, M., in G. P. LEWIS (Ed.), *5-Hydroxytryptamine*, Pergamon, London, 1958, p. 109.

120 RENSON, J., BOUNAMEAUX, Y. AND BARAC, G., *Arch. intern. pharmacodyn.*, 119 (1959) 492.

121 ROVATI, V., MORLUNCHI, C. AND PICOTTI, T., *Boll. Soc. med. chir. Pisa*, 24 (1956) 295.

122 ROVATI, V., MORLUNCHI, C. AND PICOTTI, T., *Boll. Soc. med. chir. Pisa*, 24 (1956) 479.

123 SALA, G. AND CASTEGNARO, E., *Proc. Soc. Exptl. Biol. Med.*, 82 (1953) 621.

124 SALGADO, E. AND GREEN, D. M., *Am. J. Physiol.*, 183 (1955) 657.

125 SARAJAS, H. S. S., KRISTOFFERSSON, R. AND FRICK, M. H., *Nature*, 184 (1959) 1127.

126 SARAJAS, H. S. S., KRISTOFFERSSON, R. AND FRICK, M. H., *Am. J. Physiol.*, 197 (1959) 1195.

127 SARAJAS, H. S. S. AND SAURE, L., *Nature*, 185 (1960) 768.

128 SCHNECKLOTH, R. E., PAGE, I. H. AND CORCORAN, A. C., *Clin. Research*, 5 (1957) 308.

129 SCHNECKLOTH, R. E., PAGE, I. H. AND CORCORAN, A. C., *J. Lab. Clin. Med.*, 50 (1957) 950.

130 SCHNECKLOTH, R. E., PAGE, I. H., DEL GRECO, F. AND CORCORAN, A. C., *Circulation*, 16 (1957) 523.

131 SELKURT, E. E., BRANDFONBRENER, M. AND GELLER, H. M., *Am. J. Physiol.*, 170 (1953) 61.

132 SINCLAIR, I. S. R., *Proc. Roy. Soc. Med.*, 50 (1947) 447.

133 SMITH, A. N., NYLUS, L. M., DALGLIESH, C. E., DUTTON, R. W., LENOX, B. AND MACFARLANE, P. S., *Scot. Med. J.*, 2 (1957) 24.

134 SPINAZZOLA, A. J. AND SHERROD, T. R., *J. Pharmacol. Exptl. Therap.*, 119 (1957) 114.

135 SPIRITO, A. AND TEODORO, U., *Inform. med. (Genoa), Sez. clin. sci.*, 14 (1959).

135a TRAPOLD, J. H. AND BRINER, U., *Federation Proc.*, 21 (1962) 430.

136 VALSECCHI, A. AND VALZELLI, L., *Boll. soc. ital. biol. sper.*, 33 (1957) 855.

137 VAN ARMAN, C. G., *Arch. intern. pharmacodyn.*, 108 (1956) 356.

138 VAN ARMAN, C. G. AND JENKINS, B. A., *J. Pharmacol. Exptl. Therap.*, 116 (1956) 59.

139 VERCELLONE, A. AND ASERO, B., *Farmaco (Pavia), Ed. sci.*, 9 (1954) 285.

140 VERCILLO, L. AND MAZZETTI, G. M., *Minerva med.*, 48 (1957) 1217.

141 VIRTAMA, P. AND JÄNKÄLA, E., *Angiology*, 11 (1961) 77.

142 VIRTAMA, P. AND JÄNKÄLA, E., *Angiology*, 12 (1961) 372.

143 WAAL, H. J. AND VEALE, A. M. O., *Proc. Univ. Otago Med. School*, 34 (1956) 12.

144 WATKINS, E., in J. G. ALLEN (Ed.), *Extracorporeal Circulation*, Charles C. Thomas, Springfield, Ill., 1958.

145 WAUGH, D. AND BESCHEL, H., *Am. J. Pathol.*, 39 (1961) 547.

146 WAUGH, D. AND PEARL, M. J., *Am. J. Pathol.*, 36 (1960) 431.

147 WEISSBACH, H., BOGDANSKI, D. F., REDFIELD, B. G. AND UDENFRIEND, S., *J. Biol. Chem.*, 227 (1957) 617.

148 WENZEL, D. G. AND BECKLOFF, G. N., *J. Am. Pharm. Assoc.*, 47 (1958) 844.

Chapter 8

INFLUENCE OF SEROTONIN ON ENDOCRINE AND METABOLIC ACTIVITIES

Serotonin induces a number of endocrine and metabolic effects which may be relevant for the interpretation of some of its pharmacological actions. The data available in this field are frequently isolated observations without any lead to the understanding of the mechanism of action. Here is a field deserving further investigation particularly having the aim to distinguish the specific effects of serotonin from the unspecific reactions due to the administration of high doses of the amine, which may disturb many homeostatic mechanisms of the body. In this respect the reader should be aware of the fact that most of the data reported hereafter are obtained with doses of serotonin which are likely to interfere with the cardio-circulatory system and with the thermoregulatory processes.

EFFECTS ON PITUITARY ADRENOCORTICOTROPHIC HORMONE SECRETION AND ON ADRENALS

The stimulation exerted by serotonin on *ACTH secretion* from the pituitary gland has been studied by several authors. The proof for the increased secretion of ACTH is indirect but consistent, and it is based mostly on the evaluation of adrenocortical functions. Adrenal ascorbic acid is decreased in rats after serotonin treatment[26,62,91,92,113,237,282,*] without any change in liver ascorbic acid[282]. This effect, as well as the decrease of adrenal cholesterol[92], is present also in animals pretreated with nembutal or morphine[92] suggesting that serotonin does not act merely as a stressor agent. Hypophysectomized rats do not respond to serotonin administration with a depletion of adrenal ascorbic acid[228,236,237]. Since the effect of exogenous ACTH in hypophysectomized animal is not potentiated by the administration of serotonin[114], it seems logical to extrapolate that the effect of serotonin on the adrenal is mediated through a release of ACTH. However, the effect of serotonin is not a direct one on pituitary gland. In fact the depletion of adrenal ascorbic acid induced by serotonin is prevented by prednisone treatment, hypothalamic lesions[304] or midbrain section[121].

Other signs of adrenal stimulation have been found in various animal species: rat[91,313,314], mouse[133,134], guinea pig[4,277], cockerel[239], and include increase of adrenal weight[21,46,47], hypertrophy of preputial glands in adrenalectomized animals[91] and

---

* 0.25 mg/kg i.p. or 0.01 mg/kg i.v. [236].

eosinophil fall[26,133,223,313,314]. In humans 10 mg i.v. of serotonin do not affect the number of circulating eosinophils[284].

In particular conditions, serotonin prevents the adrenal effects induced by stressor agents[113,114]. The depletion of adrenal ascorbic acid by serotonin is potentiated in the presence of subactive doses of exogenous adrenaline[62], while the eosinophil effect is increased by the administration of cortisone[133,134] and it is inhibited[223,*] by LSD-25. In dogs the decrease of eosinophils is particularly marked when serotonin is injected intracisternally[26].

The precursor of serotonin, 5-hydroxytryptophan, has been reported as ineffective in stimulating adrenocortical functions[285,289].

Monoamine oxidase inhibitors also stimulate adrenals[66,70,114] but it is unlikely that they act through an increase of serotonin in hypothalamic structures. Reserpine, a releaser of serotonin, also increases the secretion of pituitary ACTH[36,281,340], induces adrenal hypertrophy[20,79,84,112,119,162,163,322,336,**], depletion of adrenal ascorbic acid[36,162] and corticosterone[77], increase of urinary corticosteroid excretion (refs. 63, 64, 79, 112, 162), increase of plasma corticosterone[36,78a,229] and fall of eosinophils[120]. Some of these effects are not present in hypophysectomized animals[36] and disappear during a repeated treatment with reserpine[162,281]. Furthermore, reserpine does not abolish a subsequent effect of ACTH[120] or stressor agents[229]. The effect of reserpine is prevented by monoamine oxidase inhibitors[77,208]. Tetrabenazine, a short-acting reserpine-like agent, also stimulates adrenocortical function in rats[78]. The question whether the stimulation of ACTH secretion induced by reserpine may be due to a release of serotonin in brain is still unsolved[36,238].

Besides its effect of ACTH release, serotonin may exert a number of *direct effects on adrenals*. This may be connected with the fact that serotonin and 5-hydroxytryptophan are taken up by adrenal medullary granules[27]. Rabbit adrenal shows a smaller serotonin uptake than dog, cow and rat adrenal[59a].

Serotonin added *in vitro* to the adrenals of rabbits, guinea pigs, rats or cows[277] increases the secretion of corticosteroids as revealed by the tetrazolium-blue reaction[59a,277,278]. A similar effect is shown by ACTH, while reserpine and noradrenaline are much less effective[277].

The arterial perfusion of the adrenal gland of hypophysectomized dogs with serotonin results in a stimulation of the release of hydrocortisone and corticosterone. A similar effect was seen with ACTH but not with 5-hydroxytryptophan or reserpine[330]. It was suggested that this effect of serotonin may be mediated through an increased production of 3′,5′-AMP[330]. Serotonin is indeed able to stimulate the synthesis of 3′,5′-AMP in the liver fluke *Fasciola hepatica*[204], and 3′,5′-AMP has been shown to stimulate adrenal cortical secretion in the absence of ACTH[137,141].

Aldosterone and another unknown fraction (X_2) are increased when serotonin is added to rat adrenal glands *in vitro*[60], conditions under which corticosterone

---

* LSD alone stimulated the pituitary–adrenal axis[280].

** In mouse reserpine causes a reduction of the cortex volume[53].

TABLE 26

TOXICITY OF SEROTONIN IN ADRENALECTOMIZED ANIMALS

Species	Number of animals	Av. body weight (g)	Experimental condition	Route of administration	LD_{50} (mg/kg)	95% confidence limits
Rat	34	204	Intact	s.c.	>300	—
	210	187	Adrenalectomized 72 h earlier*	s.c.	14.5	16.4–12.8
	41	151	Intact (young)	i.v.	170	204.0–141.6
	41	427	Intact (old)	i.v.	46	61.2–34.5
	35	177	Adrenalectomized 72 h earlier (young)	i.v.	5.8	6.7–5.0
	27	397	Adrenalectomized 72 h earlier (old)	i.v.	1.2	1.7–0.9
	28	157	Partial hepatectomy 48 h earlier	i.v.	155	217.0–111.0
Mouse	60	10	Intact (young)	i.v.	245	303.8–197.5
	38	34	Intact (old)	i.v.	148	198.3–110.4
	56	19	Intact (adult)	i.v.	223	285.4–174.0
	20	19	Intact (adult)	i.p.	>600	—
	20	19	Intact (adult)	s.c.	>600	—
	67	18	Adrenalectomized 72 h earlier (adult)	i.v.	0.9	1.4–0.5
	87	19	Adrenalectomized 72 h earlier (adult)	i.p.	1.9	4.8–0.7
	45	18	Adrenalectomized 72 h earlier (adult)	s.c.	19	23.4–15.4

* Adrenalectomized animals were kept drinking saline.

secretion would not be stimulated[60,167]. Moreover, serotonin stimulates aldosterone secretion in hypophysectomized dogs[87,330]. In this respect it is interesting that a derivative of serotonin (adrenoglomerulotropin) (see formula) found in the pineal gland has been demonstrated to be a 100 times more active than serotonin in increasing aldosterone secretion[87].

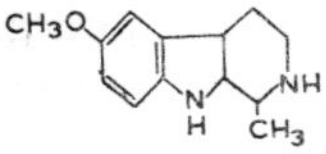

Fig. 22. Adrenoglomerulotropin.

EFFECT OF ADRENAL HORMONES AND OF ADRENALECTOMY ON SEROTONIN

Not only does serotonin influence the secretion of corticosteroids, but the latter may also affect the stores of serotonin. For example, cortisone is able to reduce the concentration of serotonin in rat skin[140]. Moreover when tissues (skin, intestine, pylorus) are depleted of serotonin by means of polymyxin B or Compound 48/80, the recovery times required to reach the normal levels of serotonin are longer when cortisone is given[48,139]. This effect is elicited by ACTH, but not by deoxycorticosterone[48]. However, ACTH does not deplete platelet serotonin in humans[287] or intestine serotonin in rats[273].

Cortisone does not affect the excretion of 5-hydroxyindoleacetic acid in rat urine[48] although this metabolite is increased in the urine of adrenalectomized rats[83,267a].

Adrenalectomized animals drinking *saline* show a normal concentration of tissue serotonin (brain, spleen, lung and kidney)[110,223,343], and a loading with serotonin results in a concentration of tissue serotonin comparable to that of intact animals[110]. On the other hand adrenalectomized rats drinking *water* show high levels of tissue serotonin[140].

Monoamine oxidase activity in rat liver, lung, brain and kidney[110] and the depletion of serotonin by reserpine[252] is also similar in intact or adrenalectomized animals. Blood serotonin is decreased immediately after adrenalectomy[214] and is increased after stress by cold[124]. Hypophysectomy does not affect gastrointestinal serotonin[273].

As already discussed in Chapter 6, adrenalectomized rats become more sensitive to serotonin[14,106] (see Table 26) and to reserpine[6,108]. Some pharmacological properties of serotonin are increased in adrenalectomized rats including hypothermia (refs. 110, 143) and hypotension[110], but not smooth muscle contraction[254]. On the other hand the hyperglycaemia induced by serotonin[57,144] or by reserpine[316] in rabbits and dogs is abolished after adrenalectomy. This is correlated with the fact that serotonin releases adrenaline from the adrenal glands[144,270,315].

There are differences in the toxicity of serotonin in various strains of rats[108] and mice[214a] which seem to be related to the plasma corticosterone level rather than to the tissue monoamine oxidase activity[108]. Glucocorticoids in relatively large doses protect adrenalectomized rats from the increased toxicity of serotonin[107,143,235] (for details see Chapter 6). Under these experimental conditions[107] lysergic acid derivatives and cyproheptadine also prevent the toxic effects of serotonin but catecholamines do not.

EFFECTS ON THE THYROID

Serotonin induces in rats a decrease of the thyroid activity as shown by means of radioactive iodine utilization[286,306,345,*]. Prolonged administration of 5-hydroxytryptamine or 5-hydroxytryptophan in rabbits results in a hypertrophy of the thyroid gland accompanied by signs of increased function[44]. These effects are difficult to reconcile with other observations such as the inhibition by serotonin, 5-hydroxytryptophan and 5-hydroxyindoleacetic acid, of the increase of oxygen consumption due to thyroxine[189]. The degradation of [^{131}I]thyroxine is inhibited by the three 5-hydroxyindoles *in vitro* but not *in vivo*[105]. Melatonin (*N*-acetyl-5-methoxytryptamine) reduces the weight of the thyroid and prevents hypertrophy induced by methylthiouracil[17a].

In addition there is evidence that reserpine produces effects on thyroid function, ranging from inhibition of secretion of thyrotrophin[29,161,262,342] and decrease of ^{131}I in the thyroid gland[96,209] to inhibition of the peripheral effects of thyroxine[68,169]. There are no data to decide whether these effects of reserpine are related to the release of serotonin. Furthermore, some authors question the specificity of the reserpine activity on thyroid[69] or do not find any clear effect[20,101,123]. However, reserpine has been reported as a useful adjunct in the management of severe hyperthyroidism[42].

It is interesting to recall that the rat thyroid gland contains a high concentration (refs. 249, 250) of serotonin (about 8 μg/g). Iproniazid or 5-hydroxytryptophan alone do not change the serotonin level in thyroid, but the combination of the two drugs results in a marked increase of thyroidal serotonin[250]. Guanethidine, α-methyl-DOPA and reserpine release thyroidal serotonin[194c]. Antithyroid drugs, as metimazole and potassium thiocyanate, do not influence the concentration of thyroid serotonin in rats[251]. Thyroidectomy is also ineffective in changing serotonin level in stomach and intestine[273]. Treatment with thyroxine or thyroid extracts increases the level of tissue serotonin[312a] and enhances in mice or rats the toxicity of histamine, serotonin (ref. 312), reserpine[85] and monoamine oxidase inhibitors[45].

The toxicity of serotonin is not altered by a pretreatment with antithyroid drugs, while the enhanced toxicity of serotonin in hyperthyroid rats is counteracted by cortisone[312].

The relationships between thyroid function and serotonin metabolism are not well defined[75,145,305]. Only small variations of 5-hydroxytryptophan decarboxylase, serotonin and monoamine oxidase in rat heart have been observed after thyroid feeding or propylthiouracil treatment[303]. These differences do not have the same direction and intensity in the two sexes[303].

---

* Other authors reported an increased uptake of ^{131}I by the thyroid and by the serum protein-bound iodine after serotonin treatment[331].

It has also been reported that some hyperthyroid patients show an increase in the excretion of urinary 5-hydroxyindoleacetic acid[302] while some hypothyroid patients show a low excretion of this substance[136].

Methysergide (UML) has been reported to be active on the ocular symptoms of hyperthyroidism (Graves' disease)[191a].

EFFECTS ON GROWTH PROCESSES

A participation of serotonin in the growth processes has been suggested by analogy with a similar role proposed for histamine[151,154] and because of its chemical similarities with the auxine, indolylacetic acid[341].

Only preliminary and contradictory results are available in vertebrates[240,339]. A direct effect on cellular reproduction has been observed by measuring the number of liver mitoses in young rats treated with serotonin. This effect does not seem to depend on the presence of liver necrosis[192].

The effect of serotonin on tumour growth has also been investigated. The growth of a sarcoma transplanted into inbred Slonaker rats[291] is stimulated by serotonin[286,292] and is inhibited by serotonin antagonists like lysergic diethylamide derivatives[290a,293]. Serotonin also increases the percentage of takes and the size of Walker carcinosarcoma growing in difficult conditions[59]. Serotonin decreases oxygen tension in this tumour[51]. However, other tumours such as the sarcoma 180 in mice, are inhibited by serotonin[267] while a leukaemia, L 1210, sensitive to reserpine*, is not affected by 5-hydroxytryptamine[122]. It should be recalled that serotonin in small concentrations reduces the food intake[308] and this effect may reduce tumour growth. 5-Hydroxytryptophan decarboxylase and monoamine oxidase are present in Morris hepatoma 5123 but are absent in other hepatomas. No correlation has been observed between cancerogenesis by 3'-methyldiaminoazobenzene and changes in these enzymes[164,165].

The observation that in rabbits the administration of STH increases the urinary excretion of 5-hydroxyindoleacetic acid[173,174], suggests that growth hormone interferes with serotonin metabolism.

The metamorphosis of *Bufo bufo* and *Discoglossus pictus* is reduced by serotonin and this effect is counteracted by RNA[245]. However, the metamorphosis of the frog, *Rana temporaria*, is not affected by serotonin[157] and is accelerated by reserpine[158].

In the plant kingdom serotonin is stimulant on the "avena test"[37,174,175,242,244], increases the growth of *maize*[242,243], *Pisum sativum* roots[37], and the weight of potato discs[259].

The effect on potato is related to changes in ion content[260] and is further increased by the presence of potassium chloride[261]. The inhibitory effect of large doses of indolylacetic acid on maize roots is counteracted by serotonin[243]. Serotonin does not affect the growth of *mesocotyles*[248] and of *Lens culinaria*[259].

---

* Reserpine also decreases the growth of sarcoma 37 in mice[19] but enhances cancerogenesis in rats[147].

EFFECTS ON PREGNANCY

Serotonin, when administered in large doses to pregnant animals, induces death of foetuses. This has been observed in rats[32a,49,206,334,339a], mice[95,264,265] and rabbits[234,256], and is more evident when serotonin is given in the early or late phase of pregnancy (refs. 65, 264, 265). Probably the death of foetuses is in relation to circulatory effects including decreased passage of substances into the placenta and foetus[275a], fall in blood pressure, necrosis of yolk sac, decreased circulation of the placenta[65,275a], uterus contraction[275a], placental detachment[234] and vasoconstriction of the umbilical vessels[8,81,256,275a]. The death of the foetuses seems to occur when the level of serotonin in placenta* shows at least a threefold increase[266].

The effect of serotonin on the survival of foetuses is prevented by progesterone (ref. 32a), LSD and derivatives[95,206], chlorpromazine, levomepromazine and meprobamate[206].

5-Hydroxytryptophan does not induce intrauterine death but decreases the growth of foetuses[337]. Monoamine oxidase inhibitors suppress fertility in rats (refs. 94, 310, 311, 324, 336a, 337) and mice[264,265,310], although they differ from serotonin in being active only in the first phase of pregnancy[94,264,265]. Monoamine oxidase inhibitors[30,190] and serotonin[66a,190] prevent pseudopregnancy and deciduomata. Serotonin however has no effect on deciduomata formation in rabbits and increases it in hamsters[66a].

In women[168], but not in rats[5], serotonin excretion in urine is increased 3–4 times at the beginning and at the end of pregnancy. The administration of serotonin in pregnant women has been said to induce labour[327], but this effect has not been observed by another author[111].

EFFECTS ON MAMMARY GLAND

Serotonin induces hypertrophy of the mammary gland with hyperplasia of the ducts and lobular differentiation in intact, ovariectomized[263] but not hypophysectomized rats[217]. After a pretreatment with oestrogens, serotonin will not only induce the growth of mammary gland but also the onset of lactation in rats[182,218] and rabbits (refs. 156, 220). In lactating animals, serotonin prolongs the mammary secretion after litter removal[218].

These effects are probably caused by a stimulus to the pituitary to release more prolactin[220] and are present also when serotonin is injected into the hypothalamus[156a]. The specificity of the lactation induced by serotonin is questionable because adrenaline, acetylcholine[216,220], chlorpromazine[317], electrical stimulation of uterine cervix[205,219] and nonspecific stresses[247] increase the secretion of prolactin "in amounts sufficient to induce lactation in estrogen-primed rats" (ref. 246). Effects similar to those described for serotonin have also been observed after reserpine treatment (refs. 3, 20, 72, 138, 210, 215, 231, 283, 321), although differences according to the strain of animal used have been reported[319,320]. Other authors have also postulated an effect

---

* Placenta contains measurable amounts of serotonin[43].

of reserpine on prolactin secretion[16,159,183]. Breast growth and lactation have been observed in patients treated with chlorpromazine[207] or reserpine[142,307].

EFFECTS ON GONADS

Serotonin induces atrophy of the ovaries, delay in the opening of the vagina and in the onset of vaginal oestrus in rats and mice[32,274]. The oxygen tension in testes is decreased after serotonin administration in anaesthetized rats[50]. Gonadectomy in male rats or administration of testosterone does not change the gastrointestinal serotonin[273].

The effect of oestradiol on the uterine weight is enhanced by the administration of serotonin[344]. Monoamine oxidase inhibitors (iproniazid and nialamide) also induce atrophy of the ovaries[274,323], but not of the testicles[299], and delay in sexual development in female rats[299,325].

Reserpine shows a variety of effects on the pituitary-genital functions[18,161,309,321] including inhibition of the oxytocin release[231] and of ovulation[17] in rats, retardation of growth of testicles in birds[7,130–132].

EFFECTS ON MELANOPHORES

Serotonin seems to be involved in the control of pigmentation. Melanophores of *Octopus vulgaris*[152] and of the fish *Serranus scriba*[255] are contracted by serotonin. On the other hand melanophores of intact[153] and hypophysectomized[67] frogs are dispersed by serotonin. A similar effect is induced by reserpine[160] and an opposite one by monoamine oxidase inhibitors[290] in intact frogs.

A number of melanophore-expanding agents are inhibited by serotonin. Examples of this effect are LSD and its congeners and reserpine in *Lebister reticulatus*[22,23,52], LSD in *Phoxinus phoxinus* L.[318]. Darkening induced in *Phoxinus* by chlorpromazine and imipramine is however not inhibited by serotonin[318]. A melanocyte-stimulating principle present in the dorsal skin secretion of *Xenophus laevis* is believed to be serotonin[329].

In the pineal gland of vertebrates a very powerful inhibitor of melanophore dispersion has been discovered. This agent called melatonin (*N*-acetyl-5-methoxytryptamine)[170,185–188] is metabolically related to serotonin[10,166a]. Serotonin is also present in the pineal glands of oxen[117], monkeys and humans[118,*].

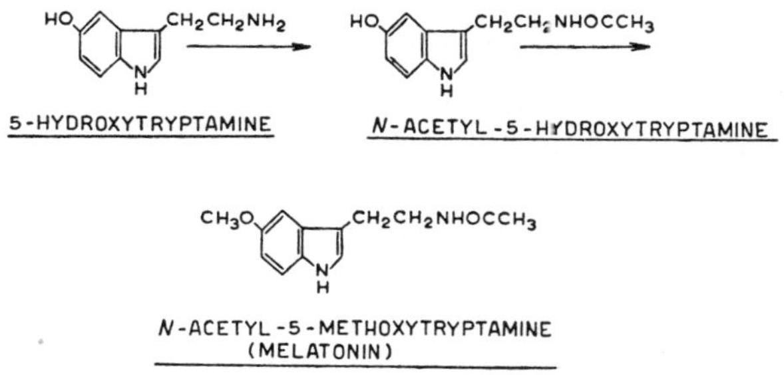

Fig. 23.
Derivation of melatonin.

* A review on the localization of monoamines in the mammalian pineal gland was recently written by Owman[248a].

EFFECTS ON OXYGEN CONSUMPTION AND TISSUE OXYGEN TENSION

Oxygen consumption is lowered after administration of serotonin in rats[97–100,269], but not in guinea pigs, rabbits and dogs[269]. This effect runs parallel to the onset of hypothermia[97] and is counteracted by increasing the room temperature[98]. Young rats are more sensitive than old rats to the lowering of oxygen consumption induced by serotonin[97]. Adrenalectomy[100] and chlorpromazine[98] do not influence this effect, on the contrary it is prolonged by iproniazid[100]. Reserpine acts like serotonin and decreases oxygen uptake in mice[288]. When serotonin is injected intravenously in rats, the oxygen uptake *in vitro* of the excised tissues is, however, increased[222]. The increase of oxygen consumption observed in rats submitted to the excision of the large intestine is not due to changes of tissue serotonin[25].

Direct measurement *in vivo* shows that the tissue oxygen tension is lowered after serotonin administration. This effect has been observed in subcutaneous tissue[50], spleen[51,74], bone marrow[51,74], muscle[50,51,74], testicle[50,51], brain[73] and tumour[51]. The decrease of tissue oxygen tension induced by serotonin is counteracted by LSD[74], UML[51], and cyproheptadine[51]. 5-Hydroxytryptophan[51] and reserpine[50] show the same effect as serotonin. Probably in relation with this property, serotonin prevents the convulsions and the toxicity by oxygen hyperpressure[171,172,174,175]. Cater *et al.*[51a] find that serotonin will prevent the rise of oxygen tension in rat and mouse tumours even when the animals breathe oxygen under a pression of 5 atm abs., and this effect is abolished by cyproheptadine.

SEROTONIN AS A RADIOPROTECTOR

Serotonin is a protector against the lethal effects of total-body irradiation (refs. 11, 12, 55, 82, 129, 176, 177a, 180, 221, 268, 271). In mice the radioprotective doses of serotonin range between 10 (ref. 178) and 90 mg/kg (ref. 74) according to the dose of X-irradiation given. The protection occurs only when serotonin is administered 30 min before or together with irradiation[178]. The radioprotective effect of serotonin has been observed also in rats[34,35] and guinea pigs[193–194b] but the mechanism of action has still to be elucidated*. Serotonin is active *in vitro* in protecting the polymer irradiated in oxygenated solution[2] and this suggests a competitive effect on the production of free radicals. The lowering of oxygen tension may be an important factor (refs. 33a, 50, 74, 278a) in explaining the radioprotective effect of serotonin. It is interesting in this respect that this radioprotection is only prevented by increasing the oxygen pressure to 4–5 atm[33b,34,74]. The hypothermia[13,33a] and the vasoconstriction[76] induced by serotonin are probably not responsible for the radioprotective effect and the external warming does not prevent serotonin radioprotection[74].

Some indirect evidences point out that irradiation may affect the metabolism of

---

* A review article on the possible mechanisms of the radioprotective effect of serotonin has been recently published[76].

serotonin. Serotonin in the hypothalamus[272], blood and spleen[86] is decreased after X-irradiation. Also the number of enterochromaffin cells of guinea pig and rat duodenum was reduced[238a]. This may be partly explained by an impairment of the 5-hydroxytryptophan decarboxylase activity[177] and by an increased excretion of 5-hydroxyindole derivatives in urine[35]. However, the lowering of tissue serotonin by α-methyl-DOPA does not influence the survival of animals after irradiation[150a]. Furthermore, no change of urinary 5-hydroxyindoleacetic acid has been observed in female patients under X-ray treatment[28].

The radioprotective effect of serotonin is shared by 5-hydroxytryptophan (refs. 33a, 54, 179, 271), other indole derivatives[76,194b] and reserpine[33a,135,178,181], but not by some analogues of serotonin as 6-hydroxy- or 7-hydroxytryptamine[75a], 5-methoxytryptamine, bufotenine, bufotenidine[271], α-methyltryptamine and α-ethyltryptamine[54].

Serotonin radioprotection is potentiated by aminoethylthiouronium* and by cysteamine[333] and is prevented by LSD[33,74,179], BOL[33,179,268], UML[75a,76] and BAS-phenol[33,33a], but not by atropine, dibenzyline and antihistamines[33]. Monoamine oxidase inhibitors prolong the radioprotective effect of serotonin[76].

At variance with the above-mentioned data, other authors found that bufotenine is a radioprotector in guinea pigs and that UML did not prevent the protection exerted by serotonin[194b].

EFFECTS ON CARBOHYDRATE METABOLISM

Serotonin influences the carbohydrate metabolism in various ways according to the animal species considered. In mice, serotonin induces hyperglycaemia[93], while in rats after oral[227], subcutaneous and intraperitoneal[150] administration of serotonin hypoglycaemia has been observed with an increase of blood lactic and pyruvic acid (refs. 115, 116). However, intravenous injection of serotonin in both normal or adrenal-demedullated rats elicits hyperglycaemia[61].

In rabbits, no effects on blood sugar have been seen with serotonin with doses up to 5 mg/kg[241,335] but with 30 mg/kg or more the usual response is hyperglycaemia[61,211]. However, various types of hyperglycaemia as after alloxan, chlorpromazine and glucose loading in rabbits are transformed to hypoglycaemia by serotonin[241]. In dogs, low doses up to 10 µg/kg do not change the level of blood glucose[301,338] but with higher doses (0.2–0.5 mg/kg) there is hypoglycaemia[300], while an infusion of serotonin results in an increase of blood sugar[57]. The administration of serotonin in the pancreatic vein induces systemic hyperglycaemia[104]. This is in apparent contrast with the observation that doses of serotonin without effect in intact dogs produce hyperglycaemia in pancreatectomized animals[301,338,**]. On the other hand doses of serotonin inducing

---

* These data were not confirmed by other authors[55].

** The pancreas in dogs and other species has a low level of serotonin but a high 5-hydroxytryptophan decarboxylase and monoamine oxidase activity[338]. Serotonin is present also in the pancreas of other animal species[86a].

hyperglycaemia are inactive in adrenalectomized dogs[56,57]. This shows that the effect of serotonin in dogs is probably the resultant of an interference with insulin and the catecholamines at least.

Further evidence that the increase of blood sugar induced by serotonin is mediated through a release of adrenaline* from the adrenals is given by the observation that dihydroergotamine[315] is an antagonist of the serotonin-hyperglycaemia in dogs[104] and rabbits[212].

5-Hydroxytryptophan is hypoglycaemic in rats[150] and hyperglycaemic in rabbits (ref. 166), while 5-hydroxyindoleacetic acid is without effect in rats[226].

Monoamine oxidase inhibitors show a variety of effects[71,150,213] while reserpine and deserpidine are always hyperglycaemic[56,58,89,104,175a,211,279]. However dihydroergotamine and homochlorcyclizine prevent serotonin-hyperglycaemia but not reserpine-hyperglycaemia[58,104].

The hyperglycaemia induced by serotonin in dogs is accompanied by a decrease of liver glycogen and by an increase of phosphorylase activity[56,57]. These effects are also abolished by adrenalectomy suggesting that catecholamines are responsible[56,57]. However, phosphorylase activity is increased in rat adipose tissue when serotonin is added *in vitro*[328] suggesting that serotonin may have also a direct effect on the enzyme. In heart and brain serotonin does not affect phosphorylase activity[18a,183a,208a].

In rats, serotonin prevents the delayed rise in liver glycogen following adrenaline administration[125] and lowers glycogen in diaphragm without changing phosphorylase activity[184]. The phosphate (^{32}P) uptake is increased in liver and diaphragm and decreased in spleen and testicle[191,**].

Ox anterior pituitary slices oxidize [1-^{14}C] glucose (but not [6-^{14}C] glucose) and this effect is stimulated by serotonin[15]. Since this activity of serotonin is blocked by monoamine oxidase inhibitors, it is suggested that serotonin acts in this case through the formation of 5-hydroxyindoleacetic acid[14]. Serotonin shows the enhancement of the glucose oxidation only in the hypophysis but not in liver and testicle[15]. However, increased formation of CO_2 from labelled glucose has been observed in liver of mice[332].

In thyroid, serotonin increases the oxidation of [1-^{14}C]- and [6-^{14}C] glucose[253]. This stimulation is probably related to an increased oxidation of TPNH[253] since the formation of TPN is a limiting factor in the oxidation of glucose through the hexose monophosphate pathway in thyroid[88].

Other types of effects induced by serotonin on carbohydrate metabolism are described in invertebrates. Excised gills of *Mytilus edulis* and *Modiolus demissus* are stimulated in their ciliary activity[108] and respiration[233] by serotonin. In anaerobiosis serotonin increases glycogen breakdown and acid production[224,232a,233], while in aerobiosis there is an increase of CO_2 production with only a small effect on glycogenolysis

---

* The disappearance of [^{3}H]adrenaline in mice[9] and the uptake of [^{3}H]noradrenaline in cats[70a] are not affected by serotonin.

** Urinary excretion of 5-hydroxyindoleacetic acid in diabetic patients is normal[340a].

and on acid production[232]. Phosphorylase activity is not influenced by serotonin[232]. LSD mimics the effects of serotonin, while BOL is an antagonist[232,232a,233]. The effect on carbohydrate metabolism by serotonin is related with the acceleration of the lateral cilia of *Mytilus*[1,126–128] and this is thought to be physiological since *Mytilus edulis* contains serotonin[1a] and the enzymes for synthesis[127a,225] and degradation[31] of the amine.

The rhythmical activity of liver fluke, *Fasciola hepatica*, is stimulated by serotonin[55,200], and this effect is accompanied by an increase of glucose uptake, glycogen breakdown[198] and lactic acid production[197]. The increased glycolysis, according to the research of Mansour *et al.*, is due to increased synthesis of cyclic 3′,5′-AMP[202,204] and the consequent activation of phosphofructokinase[199,201–203]. BOL inhibits this effect of serotonin[197,200].

EFFECTS ON ELECTROLYTES

Sodium is unchanged in rat plasma after treatment with serotonin[46] but the transfer across the placenta in mice is inhibited[275,275a]. The sodium uptake from slices of red beetroot is also inhibited[258]. Sodium retention has been observed in humans treated with serotonin[144]. This problem has been discussed already in the previous chapter as far as renal excretion is concerned.

Potassium output from the nonhepatic splanchnic area is increased by serotonin in dogs[300].

Selye has recently described an interesting effect called "calciphylaxis"[296,298]. In rats pretreated with dihydrotachysterol the administration of serotonin induces calciphylactic muscular dystrophy[297] and calcification of the submaxillary salivary glands[295,296,*]. It may be interesting in this respect to observe that serotonin facilitates the intestinal absorption of calcium (^{45}Ca)[109].

MISCELLANEOUS

Serotonin prevents hepatic damage induced by carbon tetrachloride[90] or by allyl alcohol[80] and protects rat liver mitochondria from swelling[48a]. In humans the liver uptake of bromsulphalein is increased after serotonin treatment[326].

Liver tryptophan pyrrolase (TPO) is inhibited *in vitro* by serotonin[103] but is increased *in vivo* although to a lesser extent than by tryptophan[103,276] or reserpine[41a]. This increase does not occur in adrenalectomized rats[103].

5-Hydroxytryptophan shows a similar effect in rats[276]. These effects are probably due to a stimulation by serotonin of corticoid release from the adrenals, because TPO seems to be under the control of cortisone[36,195]. α-Ketoglutarate transaminase is also increased by serotonin[276] while tyrosine transaminase[149] and tyrosinase[294] are decreased.

---

* It may be recalled that submaxillary salivary gland contains serotonin[148] which affects the electrically induced salivation[38–40,257].

References p. 161

Serotonin is a powerful inhibitor of lipid peroxide formation in rat tissues[24] and is a good electron donor[155].

In contrast to other agents that increase phosphorylase activity in rat adipose tissue, serotonin does not release free fatty acids in rats[328].

Rats after parathyroidectomy maintain a normal level of serotonin in the gastrointestinal tract[273].

Serotonin aggravates the osteolathyrism induced in rats by aminoacetonitrile. A serotonin antagonist, UML, decreases the severity of this osteolathyrism[102].

REFERENCES

1   AIELLO, E. L., *Biol. Bull.*, 113 (1957) 325.
1a AIELLO, E. L., *J. Cellular Comp. Physiol.*, 60 (1962) 17.
2   ALEXANDER, P., BACQ, Z. M., COUSENS, S. F., FOX, M., HERVE, A. AND LAZAR, J., *Radiation Research*, 2 (1955) 392.
3   ALLOITEAU, J. J., *Compt. rend.*, 250 (1960) 4459.
4   AMANTE, S., *Gazz. intern. med. e chir.*, 53 (1958) 2637.
5   ANGERVALL, L., ENERBÄCK, L. AND WESTLING, H., *Experientia*, 16 (1960) 209.
6   ASHWIN, J. G., *Proc. Soc. Exptl. Biol. Med.*, 104 (1960) 188.
7   ASSENMACHER, I., TIXIER-VIDAL, A. AND BAYLÉ, J. D., *Compt. rend. soc. biol.*, 155 (1961) 2235.
8   ASTRÖM, A. AND SAMELIUS, U., *Brit. J. Pharmacol.*, 12 (1957) 410.
9   AXELROD, J. AND TOMCHICK, R., *Nature*, 184 (1959) 2027.
10 AXELROD, J. AND WEISSBACH, H., *Science*, 131 (1960) 1312.
11 BACQ, Z. M., *Acta Radiol.*, 41 (1954) 47.
12 BACQ, Z. M. AND HERVE, A., *Schweiz. med. Wochschr.*, 82 (1952) 1018.
13 BACQ, Z. M. AND LIEBERG-HUTTER, S., *J. Physiol. (London)*, 145 (1958) 52P.
14 BARONDES, S. H., *J. Biol. Chem.*, 237 (1962) 204.
15 BARONDES, S. H., JOHNSON, P. AND FIELD, J. B., *Endocrinology*, 69 (1961) 809.
16 BARRACLOUGH, C. A., *Anat. Record*, 127 (1957) 262.
17 BARRACLOUGH, C. A. AND SAWYER, C. H., *Endocrinology*, 61 (1957) 341.
17a BASCHIERI, L., DE LUCA, F., CRAMAROSSA, L., DE MARTINO, C., OLIVERIO, A. AND NEGRI, M., *Experientia*, 19 (1963) 15.
18 BEIN, H. J., *Pharmacol. Revs.*, 8 (1956) 435.
18a BELFORD, J. AND FEINLEIB, M. R., *Biochem. Pharmacol.*, 6 (1961) 189.
19 BELKIN, M. AND HARDY, W. G., *Science*, 125 (1957) 233.
20 BENSON, G. K., *Proc. Soc. Exptl. Biol. Med.*, 99 (1958) 550.
21 BENSON, G. K., *Proc. Soc. Exptl. Biol. Med.*, 103 (1960) 132.
22 BERDE, B. AND CERLETTI, A., *Helv. Physiol. Acta*, 14 (1956) 325.
23 BERDE, B. AND CERLETTI, A., *Z. ges. exptl. Med.*, 129 (1957) 149.
24 BERNHEIM, M. L. C., OTTOLENGHI, A. AND BERNHEIM, F., *Biochim. Biophys. Acta*, 23 (1957) 431.
25 BERTACCINI, G., NOBILI, M. B. AND ZAMBONI, P., *Arch. ital. sci. farmacol.*, 9 (1959) 394.
26 BERTELLI, A., CANTONE, G. AND MARTINI, L., *Atti soc. lombarda sci. med. e biol.*, 9 (1954) 10.
27 BERTLER, Å., ROSENGREN, A. M. AND ROSENGREN, E., *Experientia*, 16 (1960) 418.
28 BETTENDORF, G., MOJE, A. P. AND SCHMERMUND, H. J., *Strahlentherapie*, 117 (1962) 370.
29 BIERWAGEN, H. E. AND SMITH, D. L., *Proc. Soc. Exptl. Biol. Med.*, 100 (1959) 108.
30 BIGNAMI, G., *Compt. rend.*, 250 (1960) 3731.
31 BLASCHKO, H. AND MILTON, A., *Brit. J. Pharmacol.*, 15 (1960) 42.
32 BOTROS, M. AND ROBSON, J. M., *J. Endocrinol.*, 20 (1960) 10.
32a BOTROS, M., LINDSAY, D., POULSON, E. AND ROBSON, J. M., *Biochem. Pharmacol.*, 8 (1961) 102.
33 BRENK, H. A. S. VAN DEN, AND ELLIOT, K., *Nature*, 182 (1958) 1506.
33a BRENK, H. A. S. VAN DEN, AND HAAS, M., *Intern. J. Radiation Biol.*, 3 (1961) 73.
33b BRENK, H. A. S. VAN DEN, AND JAMIESON, D., *Intern. J. Radiation Biol.*, 4 (1962) 379.
34 BRENK, H. A. S. VAN DEN, AND MOORE, R., *Nature*, 183 (1959) 1530.
35 BRINKMANN, R. AND LAMBERTS, H. B., Immediate and Low Level Effects of Ionizing Radiations. (Conference held in Venice, June 1959). *Intern. J. Radiation Biol.*, (1960) Suppl.
36 BRODIE, B. B., MAICKEL, R. P. AND WESTERMANN, E. O., in S. S. KETY AND J. ELKES (Eds.), *Regional Neurochemistry*, Pergamon, London, 1961, p. 351.
37 BULARD, C. AND LEOPOLD, A. C., *Compt. rend. soc. biol.*, 154 (1960) 1432.
38 CALDARERA, G., RIVA SANSEVERINO, E. AND URBANO, A., *Boll. soc. ital. biol. sper.*, 37 (1961) 1080.
39 CALDARERA, G., RIVA SANSEVERINO, E. AND URBANO, A., *Boll. soc. ital. biol. sper.*, 37 (1961) 1083.
40 CALDARERA, G., RIVA SANSEVERINO, E. AND URBANO, A., *Boll. soc. ital. biol. sper.*, 37 (1961) 1085.
41 CANAL, N. AND MAFFEI-FACCIOLI, A., *Boll. soc. ital. biol. sper.*, 34 (1958) 787.
41a CANAL, N. AND MAFFEI-FACCIOLI, A., *Naturwissenschaften*, 46 (1959) 494.

42 CANARY, J. J., SCHAAF, M., DUFFY, B. J. AND KYLE, L. H., *New Engl. J. Med.*, 257 (1957) 435.
43 CANTONE, G. AND MARTINI, L., *Endocrinol. e sci. costituz.*, 23 (1955) 1.
44 CARLIER, J., LEIEUNE-LEDANT, G. AND KEIL, CH., *J. physiol. (Paris)*, 52 (1960) 42.
45 CARRIER, R. N. AND BUDAY, P. V., *Nature*, 191 (1961) 1107.
46 CASELLA, C. AND RAPUZZI, G., *Boll. soc. ital. biol. sper.*, 33 (1957) 722.
47 CASELLA, C. AND RAPUZZI, G., *Boll. soc. ital. biol. sper.*, 33 (1957) 1569.
48 CASS, R. AND MARSHALL, P. B., *Arch. intern. pharmacodyn.*, 136 (1962) 311.
48a CASU, A., *Experientia*, 16 (1960) 489.
49 CATALDI, G., *Riv. ital. ginecol.*, 39 (1956) 3.
50 CATER, D. B., GARATTINI, S., MARINA, F. AND SILVER, I. A., *Proc. Roy. Soc. London, B*, 155 (1961) 136.
51 CATER, D. B., GRIGSON, C. M. B. AND WATKINSON, D. A., *Acta Radiol.*, 58 (1962) 401.
51a CATER, D. B., SCHOENIGER, E. L. AND WATKINSON, D. A., *Lancet*, 283 (1962) 381.
52 CERLETTI, A. AND BERDE, B., *Experientia*, 11 (1955) 312.
53 CHIRVAN-NIA, P. AND DELOST, P., *Compt. rend. soc. biol.*, 155 (1961) 1516.
54 COEN, E. D. AND WHITEHEAD, R. W., *Federation Proc.*, 21 (1962) 423a.
54a COESSENS, R., *Compt. rend. soc. biol.*, 156 (1962) 974.
55 COHEN, A. AND COHEN, L., *Brit. J. Radiol.*, 35 (1962) 200.
56 COLOMBO, J. P., WEBER, W., GUIDOTTI, G., KANAMEISHI, D. AND FOA', P., *Atti I Symposium sull'Estere di Cori e Glucidi Fosforilati, Milano*, 16–18 settembre, 1960.
57 COLOMBO, J. P., WEBER, J. W., GUIDOTTI, G., KANAMEISHI, D. AND FOA', P., *Endocrinology*, 67 (1960) 693.
58 COLOMBO, J. P., WEBER, J. W., KANAMEISHI, D. AND FOA', P., *Endocrinology*, 67 (1960) 248.
59 COMVALINS, N., *Surg. Forum*, 11 (1960) 68.
59a CONNORS, M. AND ROSENKRANTZ, H., *Endocrinology*, 71 (1962) 407.
60 CORMIER, M. AND JOUAN, P., *Compt. rend.*, 254 (1962) 3444.
61 CORRELL, J. T., LYTH, L. F., LONG, S. AND VANDERPOEL, J. C., *Am. J. Physiol.*, 169 (1952) 537.
62 COSTA, E. AND ZETLER, G., *Proc. Soc. Exptl. Biol. Med.*, 98 (1958) 249.
63 CO TUI, F., BRINITZER, W., ORR, A. AND ORR, E., *Psychiat. Quart.*, 34 (1960) 47.
64 CO TUI, F., RILEY, E. AND ORR, A., *J. Clin. Exptl. Psychopathol.*, 17 (1956) 142.
65 CRAIG, J. H., *Federation Proc.*, 21 (1962) 276b.
66 CRANE, G. E. AND WOLFMAN, M., *J. Nervous Mental Disease*, 130 (1960) 134.
66a CZYBA, J. C. AND CHIRIS, M., *Compt. rend. soc. biol.*, 156 (1962) 856.
67 DAVEY, K. G., *Nature*, 183 (1959) 1271.
68 DE FELICE, E. A., SMITH, T. C. AND DEARBORN, E. H., *Proc. Soc. Exptl. Biol. Med.*, 94 (1957) 171.
69 DELOST, P. AND PRULHIÈRE, N., *Compt. rend. soc. biol.*, 155 (1961) 2356.
70 DE MAIO, D., *Acta Neurol. (Naples)*, 14 (1959) 761.
70a DENGLER, H. J., SPIEGEL, H. E. AND TITUS, E. O., *Nature*, 191 (1961) 816.
71 DE SCHAEPDRYVER, A. F. AND PREZIOSI, P., *Arch. intern. pharmacodyn.*, 119 (1959) 506.
72 DESCLIN, L., *Compt. rend. soc. biol.*, 151 (1957) 1774.
73 DI STEFANO, V., LEARY, D. E. AND FELDMAN, I., *Federation Proc.*, 15 (1956) 417.
74 DOULL, J. AND TRICOW, B. J., *Federation Proc.*, 20 (1961) 400c.
75 DUBNICK, B., LEASON, G. A. AND LEVERETT, R., *Pharmacologist*, 2 (1960) 67.
75a DUKOR, P., *Experientia*, 18 (1962) 513.
76 DUKOR, P., *Strahlentherapie*, 117 (1962) 330.
77 EECHAUTE, W., LACROIX, E. AND GOESSENS, R., *Arch. intern. physiol. et biochim.*, 70 (1962) 117.
78 EECHAUTE, W., LACROIX, E. AND LEUSEN, I., *Experientia*, 18 (1962) 233.
78a EECHAUTE, W., LACROIX, E., LEUSEN, I. AND BOUCKAERT, J. J., *Arch. intern. pharmacodyn.*, 139 (1962) 403.
79 EGDAHL, R. H., RICHARDS, J. B. AND HUME, D. M., *Science*, 123 (1953) 418.
80 EGER, W., *Naturwissenschaften*, 45 (1958) 442.
81 ELIASSON, R. AND ASTRÖM, A., *Acta Pharmacol. Toxicol.*, 11 (1955) 254.
82 ELTGEN, D., KOCH, R. AND LANGENDORFF, H., *Strahlentherapie*, 114 (1961) 118.
83 ENERBÄCK, L., *Endocrinology*, 67 (1960) 717.
84 EPSTEIN, R., ERIKSON, L. B. AND REYNOLDS, S. R. M., *Endocrinology*, 66 (1960) 167.
85 ERSHOFF, B. H., *Proc. Soc. Exptl. Biol. Med.*, 99 (1958) 189.

86   ERSHOFF, B. H. AND GAL, E. M., *Proc. Soc. Exptl. Biol. Med.*, 108 (1961) 160.
86a FALCK, B. AND HELLMAN, B., *Experientia*, 19 (1963) 139.
87   FARRELL, G. AND McISAAC, W. M., *Arch. Biochem. Biophys.*, 94 (1961) 543.
88   FIELD, J. B., PASTAN, I. AND HERRING, B., *Biochim. Biophys. Acta*, 50 (1961) 513.
89   FINGER, K. F., PEREIRA, J. N. AND SCHNEIDER, J. A., *Federation Proc.*, 21 (1962) 175f.
90   FIORE-DONATI, L. AND CHIECO-BIANCHI, L., *Boll. soc. ital. biol. sper.*, 34 (1958) 615.
91   FIORE-DONATI, L., POLLICE, L. AND CHIECO-BIANCHI, L., *Experientia*, 15 (1959) 193.
92   FISCHER, P., RENSON, J. AND CICCARONE, P., *Arch. intern. physiol. et biochim.*, 67 (1959) 147.
93   FISHEL, C. W. AND SZENTIVANYI, A., *Federation Proc.*, 21 (1962) 271c.
94   FLÜCKIGER, E. AND SALZMANN, R., *Experientia*, 17 (1961) 130.
95   FLÜCKIGER, E. AND SALZMANN, R., *Experientia*, 17 (1961) 131.
96   FLÜHMAN, C. F., *Proc. Soc. Exptl. Biol. Med.*, 93 (1956) 615.
97   FÖLDES, I., HARANGHY, L., CSÖTÖRTÖK, L. AND BEREGI, E., *Gerontologia*, 5 (1961) 19.
98   FÖLDES, I. AND KOMLÓS, E., *Arch. intern. pharmacodyn.*, 120 (1959) 121.
99   FÖLDES, I., KOMLÓS, E. AND PÁLINKÁS, J., *Acta Physiol. Acad. Sci. Hung.*, 18 (1961) 105.
100 FÖLDES, I. AND PÁLINKÁS, J., *Arch. intern. pharmacodyn.*, 127 (1960) 331.
101 FORD, R. V., LIVERSAY, S. I., MILLER, W. R. AND MOYER, J. H., *Med. Record*, 47 (1953) 608.
102 FRANCHIMONT, P., LEFEBVRE, P. AND VAN CAUWENBERGE, H., *Compt. rend. soc. biol.*, 155 (1961) 427.
103 FRIEDEN, E., WESTMARK, G. W. AND SCHOR, J. M., *Arch. Biochem. Biophys.*, 92 (1961) 176.
104 GALANSINO, G., D'AMICO, G., KANAMEISHI, D., BERLINGER, F. G. AND FOA', P. P., *Am. J. Physiol.*, 198 (1960) 1059.
105 GALTON, V. A. AND INGBAR, S. H., *Endocrinology*, 68 (1961) 435.
106 GARATTINI, S., GAIARDONI, P., MORTARI, A. AND PALMA, V., *Nature*, 190 (1961) 540.
107 GARATTINI, S., GIACHETTI, A., NANNI, E. AND PALMA, V., unpublished data.
108 GARATTINI, S., GIACHETTI, A., NANNI, E., PALMA, V. AND RE, R., unpublished data.
109 GARATTINI, S., GROSSI, E., PAOLETTI, P., PAOLETTI, R. AND POGGI, M., *Nature*, 191 (1961) 185.
110 GARATTINI, S., LAMESTA, L., MORTARI, A., PALMA, V. AND VALZELLI, L., *J. Pharm. Pharmacol.*, 13 (1961) 385.
111 GARRETT, W. J., *Arch. intern. pharmacodyn.*, 117 (1958) 435.
112 GAUNT, R., RENZI, A. A., ANTONCHAK, N., MILLER, G. J. AND GILMAN, M., *Ann. N.Y.Acad. Sci.*, 59 (1954) 22.
113 GEORGES, G., *Compt. rend. soc. biol.*, 151 (1957) 692.
114 GEORGES, G. AND HEROLD, M., *Compt. rend. soc. biol.*, 152 (1958) 436.
115 GEY, K. F. AND PLETSCHER, A., *Helv. Physiol. Acta*, 18 (1960) C70.
116 GEY, K. F. AND PLETSCHER, A., *Experientia*, 17 (1961) 25.
117 GIARMAN, N. J. AND DAY, M., *Biochem. Pharmacol.*, 1 (1958) 235.
118 GIARMAN, N. J. AND FREEDMAN, D. X., *Nature*, 186 (1960) 480.
119 GIROD, C., *Compt. rend. soc. biol.*, 155 (1961) 1628.
120 GIROD, C. AND SLIMANE TALEB, S., *Compt. rend. soc. biol.*, 151 (1957) 1158.
121 GIULIANI, G., MARTINI, L. AND PECILE, A., *Acta Endocrinol.*, 35 (1960) 37.
122 GOLDIN, A., BURTON, R. M., HUMPHREYS, S. R. AND VENDITTI, J. M., *Science*, 125 (1957) 156.
123 GOODMAN, J. R., FLORSHEIM, W. H. AND TEMPEREAN, C. E., *Proc. Soc. Exptl. Biol. Med.*, 90 (1955) 196.
124 GORDON, P., *Nature*, 191 (1961) 183.
125 GORDON, P., HADDY, F. AND LIPTON, M., *Federation Proc.*, 18 (1959) 397.
126 GOSSELIN, R. E., *J. Cellular Comp. Physiol.*, 58 (1961) 17.
127 GOSSELIN, R. E. AND ERNST, H. M., *Abstracts, Fall Meeting Am. Soc. Pharmacol. Exptl. Therap., Ann Arbor, Mich.*, 1958, p. 15.
127a GOSSELIN, R. E., MOORE, K. E. AND MILTON, A. S., *J. Gen. Physiol.*, 46 (1962) 277.
128 GOSSELIN, R. E. AND O'HARA, G., *J. Cellular Comp. Physiol.*, 58 (1961) 1.
129 GRAY, J. L., TEW, J. T. AND JENSEN, H., *Proc. Soc. Exptl. Biol. Med.*, 80 (1952) 604.
130 HAGEN, P., *J. Physiol. (London)*, 152 (1960) 15P.
131 HAGEN, P. AND WALLACE, A. C., *Federation Proc.*, 19 (1960) 168.
132 HAGEN, P. AND WALLACE, A. C., *Brit. J. Pharmacol.*, 17 (1961) 267.
133 HALBERG, F., *Am. J. Physiol.*, 179 (1954) 309.
134 HALBERG, F., VISSCHER, M. B. AND BITTNER, J. J., *Am. J. Physiol.*, 179 (1954) 642.

135 HALEY, T. J., FLESHER, A. M. AND MAVIS, L., *Nature*, 192 (1961) 1309.
136 HAVERBACK, B. J., SJOERDSMA, A. AND TERRY, L. L., *New Engl. J. Med.*, 255 (1956) 270.
137 HAYNES JR., R. C., *J. Biol. Chem.*, 233 (1958) 1220.
138 HERLANT, M. AND PASTEELS, J. L., *Compt. rend.*, 249 (1959) 2625.
139 HICKS, R. AND WEST, G. B., *Nature*, 181 (1958) 1342.
140 HICKS, R. AND WEST, G. B., *Nature*, 182 (1958) 401.
141 HILTON, J. G., NEDELJKOVIC, R. I. AND DERMKSIAN, G., *Acta Endocrinol.*, 51 (1960) Suppl. 705.
142 HIMWICH, H. E., *Science*, 127 (1958) 59.
143 HOFFMAN, R. A., *Am. J. Physiol.*, 196 (1959) 876.
144 HOLTZ, P., BALZER, H. AND WESTERMANN, E., *Arch. exptl. Pathol. Pharmakol., Naunyn-Schmiedeberg's*, 231 (1957) 361.
145 HOPSU, V. K. AND KARINKANTA, H., *Ann. Med. Exptl. et Biol. Fenniae (Helsinki)*, 40 (1962) 1.
146 HULET, W. H. AND PERERA, G. A., *Proc. Soc. Exptl. Biol. Med.*, 91 (1956) 512.
147 HURST, L., LACASSAGNE, A. AND ROSENBERG, A. J., *Compt. rend. soc. biol.*, 152 (1958) 441.
148 IACHELLO, R., RIVA SANSEVERINO, E. AND URBANO, A., *Boll. soc. ital. biol. sper.*, 38 (1962) 516.
149 JACOBY, G. A. AND LA DU, B. N., *Federation Proc.*, 21 (1962) 238c.
150 JORI, A., LAMESTA, L. AND VALSECCHI, A., *Boll. soc. ital. biol. sper.*, 35 (1959) 716.
150a JÜRGEN-KUSCHKE, H. AND BRAUN, H., *Strahlentherapie*, 120 (1963) 297.
151 KAHLSON, G., *Lancet*, 278 (1960) 67.
152 KAHR, H., *Naturwissenschaften*, 45 (1958) 243.
153 KAHR, H. AND FISCHER, W., *Klin. Wochschr.*, 35 (1957) 41.
154 KAMESWARAN, L. AND WEST, G. B., *J. Physiol. (London)*, 160 (1961) 13P.
155 KAMINER, B., *Federation Proc.*, 19 (1960) 258.
156 KEHL, R. AND CZYBA, J. C., *J. physiol. (Paris)*, 54 (1962) 356.
156a KEHL, R., CZYBA, J. C., ARNAUD, G. AND GUEYFFIER, H., *Compt. rend soc. biol.*, 156 (1962) 1414.
157 KEHL, R., DUMONT, L. AND CZYBA, J. C., *Compt. rend. soc. biol.*, 155 (1961) 1667.
158 KEHL, R., DUMONT, L., CZYBA, J. C. AND GERMAIN, A., *Compt. rend. soc. biol.*, 154 (1960) 1791.
159 KHAZAN, N., PRIMO, CH., DANON, A., ASSAEL, M., SULMAN, F. G. AND WINNIK, H. Z., *Arch. intern. pharmacodyn.*, 136 (1962) 291.
160 KHAZAN, N. AND SULMAN, F. G., *Proc. Soc. Exptl. Biol. Med.*, 107 (1961) 282.
161 KHAZAN, N., SULMAN, F. G. AND WINNIK, H. Z., *Proc. Soc. Exptl. Biol. Med.*, 105 (1960) 201.
162 KHAZAN, N., SULMAN, F. G. AND WINNIK, H. Z., *Proc. Soc. Exptl. Biol. Med.*, 106 (1961) 579.
163 KITAY, J. I., HOLUB, D. A. AND JAILER, J. W., *Endocrinology*, 65 (1959) 548.
164 KIZER, D. E., *Cancer Research*, 22 (1962) 196.
165 KIZER, D. E. AND CHAN, S. K., *Cancer Research*, 21 (1961) 489.
166 KONZETT, H., *Nature*, 182 (1958) 1168.
166a KOPIN, I. J., PARE, C. M. B., AXELROD, J. AND WEISSBACH, H., *J. Biol. Chem.*, 236 (1961) 3072.
167 KOVACS, K., DAVID, M. A. AND WEISZ, P., *Med. Exptl.*, 3 (1960) 113.
168 KURIATI, K. AND INOUÉ, T., *Compt. rend. soc. biol.*, 150 (1956) 1835.
169 KUSCHKE, H. J. AND GRUNER, H., *Klin. Wochschr.*, 32 (1954) 563.
170 KVEDER, S. AND MCISAAC, W. M., *J. Biol. Chem.*, 236 (1961) 3214.
171 LABORIT, H., BROUSSOLLE, B. AND PERRIMOND-TROUCHET, R., *Compt. rend. soc. biol.*, 151 (1957) 930.
172 LABORIT, H. BROUSSOLLE, B. AND PERRIMOND-TROUCHET, R., *J. physiol. (Paris)*, 49 (1957) 953.
173 LABORIT, H., JOUANY, J. H. AND NIAUSSAT, P., *Compt. rend. soc. biol.*, 153 (1959) 549.
174 LABORIT, H., NIAUSSAT, P., BROUSSOLLE, B. AND JOUANY, J. M., *Presse méd.*, 67 (1959) 927.
175 LABORIT, H., NIAUSSAT, P., BROUSSOLLE, B. AND JOUANY, J. M., *Med. Exptl.*, 1 (1959) 27.
175a LAMESTA, L., VALSECCHI, A. AND VALZELLI, L., *Boll. soc. ital. biol. sper.*, 36 (1960) 683.
176 LANGERDORFF, H. AND KOCH, R., *Strahlentherapie*, 102 (1957) 58.
177 LANGERDORFF, H. AND MELCHING, H. J., *Strahlentherapie*, 110 (1959) 505.
177a LANGERDORFF, H., MELCHING, H. J. AND LADNER, H. A., *Intern. J. Radiation Biol.*, 1 (1959) 24.
178 LANGERDORFF, H., MELCHING, H. J. AND LADNER, H. A., *Strahlentherapie*, 108 (1959) 251.
179 LANGERDORFF, H., MELCHING, H. J. AND LADNER, H. A., *Strahlentherapie*, 109 (1959) 554.
180 LANGERDORFF, H., MELCHING, H. J. AND LADNER, H. A., *Strahlentherapie*, 110 (1959) 34.
181 LANGERDORFF, H., MELCHING, H. J., LANGERDORFF, H., KOCH, R. AND JAQUES, R., *Strahlentherapie*, 104 (1957) 338.
182 LEDERER, J. AND DE MEYER, R., *Ann. endocrinol. (Paris)*, 20 (1959) 377.

183   Lefranc, G., *Compt. rend. soc. biol.*, 152 (1958) 1495.

183a   Leonard, S. L. and Day, H. T., *Proc. Soc. Exptl. Biol. Med.*, 104 (1960) 338.

184   Leonard, S. L. and Holliday, T. D., *Proc. Soc. Exptl. Biol. Med.*, 104 (1960) 338.

185   Lerner, A. B., Case, H. and Heinzelmann, R. V., *J. Am. Chem. Soc.*, 81 (1959) 6084.

186   Lerner, A. B., Case, J. D., Mori, W. and Wright, M. R. *Nature*, 183 (1959) 1821.

187   Lerner, A. B., Case, J. D. and Takahashi, Y., *J. Biol. Chem.*, 235 (1960) 1992.

188   Lerner, A. B., Case, J. D., Takahashi, Y., Lee, T. and Mori, W., *J. Am. Chem. Soc.*, 80 (1958) 2587.

189   Lindsay, R. H. and Barker, S. B., *Endocrinology*, 65 (1959) 679.

190   Lindsay, D., Poulson, E. and Robson, J. M., *J. Endocrinol.*, 23 (1961) 209.

191   Lingjaerde, P. and Skaug, O. E., *J. Biol. Chem.*, 226 (1957) 33.

191a   Linquette, M., Fossati, P., May, J. P. and Lebrun, H., *Ann. Endocrinol. (Paris)*, 23 (1962) 457.

192   Mac Donald, R. A., Schmid, R., Hakala, T. R. and Mallory, G. K., *Proc. Soc. Exptl. Biol. Med.*, 101 (1959) 83.

193   Maggiora, A., *Bull. acad. suisse sci. méd.*, 16 (1960) 315.

194   Maggiora, A., *Dermatologica*, 124 (1962) 183.

194a   Maggiora, A., *Dermatologica*, 124 (1962) 385.

194b   Maggiora, A. and Brun, R., *Dermatologica*, 126 (1963) 30.

194c   Magus, R. D., Kranse, F. W. and Riedel, B. E., *Biochem. Pharmacol.*, 13 (1964) 115.

195   Maickel, R. P. and Brodie, B. B., *Federation Proc.*, 19 (1960) 267.

196   Mansour, T. E., *Brit. J. Pharmacol.*, 12 (1957) 406.

197   Mansour, T. E., *J. Pharmacol.*, 126 (1959) 212.

198   Mansour, T. E., *Pharmacol. Revs.*, 11 (1959) 465.

199   Mansour, T. E., *J. Pharmacol.*, 135 (1962) 94.

200   Mansour, T. E., Lago, A. D. and Hawkins, J. L., *Federation Proc.*, 16 (1957) 319.

201   Mansour, T. E., Le Rouge, N. A. and Mansour, J. M., *Federation Proc.*, 20 (1961) 226f.

202   Mansour, T. E. and Mansour, J. M., *J. Biol. Chem.*, 237 (1962) 629.

203   Mansour, T. E. and Menard, J. S., *Federation Proc.*, 19 (1960) 50.

204   Mansour, T. E., Sutherland, E. W., Rall, T. W. and Bueding, E., *J. Biol. Chem.*, 235 (1960) 466.

205   Maqsood, M. and Meites, J., *Nature*, 188 (1960) 752.

206   Marois, M., *Compt. rend. soc. biol.*, 154 (1960) 1200.

207   Marshall, W. K. and Leiberman, D. M., *Lancet*, 270 (1956) 162.

208   Martel, R. and Maickel, R. P., *Pharmacologist*, 3 (1961) 56.

208a   Mayer, S. E. and Moran, N. C., *J. Pharmacol.*, 129 (1960) 271.

209   Mayer, S. W., Kelly, F. H. and Morton, M. E., *J. Pharmacol.*, 117 (1956) 197.

210   Mayer, G., Meunier, J. M. and Rouault, J., *Compt. rend.*, 247 (1958) 524.

211   Mazzone, R., Sozio, N. and Padolecchia, N., *Boll. soc. ital. biol. sper.* 35 (1959) 1050.

212   Mazzone, R., Sozio, N. and Padolecchia, N., *Boll. soc. ital. biol. sper.*, 35 (1959) 1273.

213   Mazzone, R., Sozio, N. and Padolecchia, N., *Boll. soc. ital. biol. sper.*, 35 (1959) 1275.

214   Medakovic, M. and Spuzic, I., *Nature*, 183 (1959) 1685.

214a   Meier, H., *Advances in Pharmacol.*, 2 (1963) 161.

215   Meites, J., *Proc. Soc. Exptl. Biol. Med.*, 96 (1957) 728.

216   Meites, J., *Proc. Soc. Exptl. Biol. Med.*, 100 (1959) 750.

217   Meites, J. and Hopkins, T. F., *Federation Proc.*, 19 (1960) 156e.

218   Meites, J., Nicoll, C. S. and Talwalker, P. K., *Proc. Soc. Exptl. Biol. Med.*, 101 (1959) 563.

219   Meites, J., Nicoll, C. S. and Talwalker, P. K., *Proc. Soc. Exptl. Biol. Med.*, 102 (1959) 127.

220   Meites, J., Talwalker, P. K. and Nicoll, C. S., *Proc. Soc. Exptl. Biol. Med.*, 104 (1960) 192.

221   Melching, H. J. and Langerdorff, M., *Naturwissenschaften*, 44 (1957) 377.

222   Mietkiewski, E. and Jankowska, I., *Acta Physiol. Polonica*, 12 (1961) 517.

223   Milkovic, S. and Supek, Z., *Arch. exptl. Pathol. Pharmacol., Naunyn-Schmiedeberg's*, 228 (1956) 146.

224   Milton, A. S. and Gosselin, R. E., *Pharmacologist*, 2 (1960) 68.

225   Milton, A. S. and Gosselin, R. E., *Federation Proc.*, 19 (1960) 126.

226   Mirsky, A. I., *Recent Progr. in Hormone Research, Proc. Laurention Hormone Conference*, 13 (1956) 429.

227 MIRSKY, A. I., PERISUTTI, G. AND JUSKO, R., *Endocrinology*, 60 (1957) 318.
228 MIYAWAKI, H., MICHIO, U. AND KOBAYASHI, B., *Endocrinol. Japon.*, 8 (1961) 148.
229 MONTANARI, R. AND STOCKHAM, M. A., *Brit. J. Pharmacol.*, 18 (1962) 337.
230 MOON, R. C. AND TURNER, C. W., *Proc. Soc. Exptl. Biol. Med.*, 100 (1959) 679.
231 MOON, R. C. AND TURNER, C. W., *Proc. Soc. Exptl. Biol. Med.*, 101 (1959) 332.
232 MOORE, K. E. AND GOSSELIN, R. E., *Pharmacologist*, 3 (1961) 77.
232a MOORE, K. E. AND GOSSELIN, R. E., *J. Pharmacol.*, 138 (1962) 145.
233 MOORE, K. E., MILTON, A. S. AND GOSSELIN, R. E., *Brit. J. Pharmacol.*, 17 (1961) 278.
234 MORLUNGHI, C. AND ROMEO, P., *Arch. ostet. e ginecol.*, 66 (1961) 154.
235 MORTARI, A., NANNI, E. AND PALMA, V., *Atti accad. med. lombarda*, 16 (1961) 413.
236 MOUSSATCHÉ, H. AND ALVARES-PEREIRA, N., *Naturwissenschaften*, 43 (1956) 517.
237 MOUSSATCHÉ, H. AND ALVARES-PEREIRA, N., *Acta Physiol. Latinoam.*, 7 (1957) 71.
238 MUNSON, P. L., *Recent Progr. in Hormone Research*, 13 (1957) 17.
238a NEMENOVA, N. M., MONTEIFEL, V. M. AND CHERNOV, G. A., *Biul. Exptl. Biol. Med.*, 3 (1962) 109.
239 NEWCOMER, W. S., *Am. J. Physiol.*, 202 (1962) 337.
240 NIAUSSAT, P., *Agressologie*, No. 2 (1960) 181.
241 NIAUSSAT, P., DROUET, J. AND BERTHOU, J., *Med. Exptl.*, 4 (1961) 197.
242 NIAUSSAT, P., DUBOIS, C. AND NIAUSSAT, M., *Bull. soc. études sci. Anger, Nouvelle Série*, 89e année, 2 (1959) 171.
243 NIAUSSAT, P. AND LABORIT, H., *Med. Exptl.*, 1 (1959) 207.
244 NIAUSSAT, P., LABORIT, H., DUBOIS, C. AND NIAUSSAT, M., *Compt. rend. soc. biol.*, 152 (1958) 945.
245 NIAUSSAT, P. AND NIAUSSAT, M., *Procès verbaux Soc. sci. phys. et nat. de Bordeaux*, Séance du 19 mai 1960.
246 NICOLL, C. S., *Federation Proc.*, 19 (1960) 156f.
247 NICOLL, C. S., TALWALKER, P. K. AND MEITES, J., *Am. J. Physiol.*, 198 (1960) 1103.
248 NITSCH, J. P. AND NITSCH, C. *Bull. soc. botan. France*, 105 (1958) 482.
248a OWMAN, CH., in J. ARIËNS KAPPERS AND J. P. SCHADÉ (Eds.), *Structure and Function of the Epiphysis Cerebri, Progr. in Brain Research*, Vol. 10, Elsevier, Amsterdam, 1965, p. 423.
249 PAASONEN, M. K., *Experientia*, 14 (1958) 95.
250 PAASONEN, M. K., KÄRKI, N. T. AND MOLKKA, S., *Ann. Med. Exptl. et Biol. Fenniae (Helsinki)*, 39 (1961) 405.
251 PAASONEN, M. K. AND PELTOLA, P., *Ann. Med. Exptl. et Biol. Fenniae (Helsinki)*, 38 (1960) 227.
252 PALMA, V., BUCCOMINO, B., FRESIA, P. AND LAMESTA, L., *Atti accad. med. lombarda*, 16 (1961) 415.
253 PASTAN, I. AND FIELD, J. B., *Endocrinology*, 70 (1962) 656.
254 PECZENIK, O., *Confinia Neurol.*, 21 (1961) 98.
255 PECZENIK, O. AND ZEI, M., *Confinia Neurol.*, 21 (1961) 488.
256 PEPEU, G. AND GIARMAN, N. J., *J. General Physiol.*, 45 (1962) 575.
257 PETRUCELLI, L. M., BULLE, P. H. AND OSKANI, M., *Federation Proc.*, 20 (1961) 318c.
258 PICKLES, V. R. AND SUTCHIFFE, J. F., *Biochim. Biophys. Acta*, 17 (1955) 244.
259 PILET, P. E., *Bull. soc. vaudoise sci. nat.*, 67 (1962) 93.
260 PILET, P. E., *Experientia*, 18 (1962) 153.
261 PILET, P. E., *Experientia*, 18 (1962) 169.
262 PITTMAN, J. A., WOUTERS, F. W., HILL JR., S. R., FARMER, T. A., HAMNER, D. AND ROSSER, H. F., *Acta Endocrinol.*, 35 (1960) 1175.
263 POLLICE, L. AND MAIORANO, G., *Boll. soc. ital. biol. sper.*, 34 (1958) 1491.
264 POULSON, E., BOTROS, M. AND ROBSON, J. M., *Science*, 131 (1960) 1101.
265 POULSON, E., BOTROS, M. AND ROBSON, J. M., *J. Endocrinol.*, 20 (1960) 11.
266 POULSON, E., ROBSON, J. M. AND SENIOR, J., *Summer Meeting, Brit. Pharmacol. Soc., Oxford*, 1962.
267 PUKHALSKAYA, E. C., *Problems Med. Chem.*, 8 (1962) 42.
267a PUT, T. R. AND MEDUSKI, J. W., *Acta Physiol. et Pharmacol. Neerl.*, 11 (1962) 240.
268 RADIVOJEVIC, D. V., BEAUMARIAGE, M. L. AND BACQ, Z. M., *Compt. rend. soc. biol.*, 154 (1960) 1489.
269 RAPPORT, M. M. AND VIRNO, M., *Proc. Soc. Exptl. Biol. Med.*, 81 (1952) 203.
270 REID, G., *J. Physiol. (London)*, 118 (1952) 435.

271 RENSON, J., *Arch. intern. physiol. et biochim.*, 68 (1960) 531.

272 RENSON, J. AND FISCHER, P., *Arch. intern. physiol. et biochim.*, 67 (1959) 142.

273 RESNICK, R. H., SMITH, G. T. AND GRAY, S. J., *Am. J. Physiol.*, 201 (1961) 571.

274 ROBSON, J. M. AND BOTROS, M., *J. Endocrinol.*, 22 (1961) 165.

275 ROBSON, J. M. AND SULLIVAN, F. M., *Summer Meeting, Brit. Pharmacol. Soc.*, Oxford, 1962.

275a ROBSON, J. M. AND SULLIVAN, F. M., *J. Endocrinol.*, 25 (1963) 553.

276 ROSEN, F. AND MILHOLLAND, R. J., *Federation Proc.*, 21 (1962) 237a.

277 ROSENKRANTZ, H., *Endocrinology*, 64 (1959) 355.

278 ROSENKRANTZ, H. AND LAFERTE, R. C., *Endocrinology*, 66 (1960) 832.

278a ROTHE, W. E., GRENAN, M. M. AND WILSON, S. M., *Nature*, 198 (1963) 403.

279 RUIZ GIJON, J. AND DEL RIO IBANEZ, J., *Farmacognosia (Madrid)*, 15 (1955) 221.

280 SACKLER, A. M., WELTMAN, A. S., RUSSAKOW, M., SPARBER, S. B. AND HOWENS, H., *Federation Proc.*, 21 (1962) 416c.

281 SAFFRAN, M. AND VOGT, M., *Brit. J. Pharmacol.*, 15 (1960) 165.

282 SAPEIKA, N., *Arch. intern. pharmacodyn.*, 122 (1959) 196.

283 SAWYER, C. H., *Anat. Record*, 127 (1957) 362.

284 SCALTRINI, G. C., *Haematologica (Pavia)*, 41 (1956) 681.

285 SCHEIFFARTH, F., ZICHA, L. AND FÖRSTER, W., *Arzneimittel-Forsch.*, 11 (1961) 602.

286 SCHELINE, R. R. AND SCOTT, K. G., *Cancer Research*, 18 (1958) 932.

287 SCHMID, E., SCHEIFFARTH, F., ZICHA, L. AND SCHWEMMLE, C., *Acta Endocrinol.*, 35, Suppl. 51 (1960) 843.

288 SCHWEMMLE, K. AND HEIM, F., *Med. Exptl.*, 1 (1959) 351.

289 SCHWEMMLE, K., SCHMID, E., SCHEIFFARTH, F., ZICHA, L. AND SCHICHER, E., *Arzneimittel-Forsch.*, 11 (1961) 616.

290 SCOTT, G. T., *Nature*, 193 (1962) 552.

290a SCOTT, K. G., *Proc. Western Pharmacol.*, 2 (1959) 23.

291 SCOTT, K. G., BOSTICK, W. L., SHIMKIN, M. B. AND HAMILTON, J. C., *Cancer*, 2 (1949) 692.

292 SCOTT, K. G., SCHELINE, R. R. AND STONE, R. S., *Cancer Research*, 18 (1953) 927.

293 SCOTT, K. G. AND STONE, R. S., *Cancer Research*, 19 (1959) 783.

294 SEIJI, M., *Federation Proc.*, 20 (1961) 6.

295 SELYE, H., *Allergie u. Asthma*, 7 (1961) 241.

296 SELYE, H., *Calciphylaxis*, University of Chicago Press, Chicago, Ill., 1962.

297 SELYE, H., DIENDONNE, J. M. AND GABBIANI, G., *Federation Proc.*, 21 (1962) 168d.

298 SELYE, H. AND GENTILE, G., *Naturwissenschaften*, 48 (1961) 671.

299 SETNIKAR, I., MURMANN, W. AND MAGISTRETTI, M. J., *Endocrinology*, 67 (1960) 511.

300 SHOEMAKER, W. C., KLECKNER, H., PERLOW, M. AND GALANSINO, G., *Federation Proc.*, 20 (1961) 314a.

301 SIREK, A., *Nature*, 179 (1957) 376.

302 SKANSE, B. AND HANSON, A., *Lancet*, 282 (1962) 1072.

303 SKILLEN, R. G., THIENES, C. H. AND STRAIN, L., *Endocrinology*, 70 (1962) 743.

304 SMELIK, P. G. AND DE WIED, D., *Experientia*, 14 (1958) 17.

305 SMITH, C. L., *J. Endocrinol.*, 19 (1960) 295.

306 SODEBERG, U., *Acta Physiol. Scand.*, 42 (1958) Suppl. 147.

307 SOMLYO, A. P. AND WAYE, J. D., *J. Mt. Sinai Hosp. N.Y.*, 27 (1960) 5.

308 SOULAIRAC, A. AND SOULAIRAC, M. L., *Compt. rend. soc. biol.*, 154 (1960) 510.

309 SOULAIRAC, A. AND SOULAIRAC, M. L., *Compt. rend. soc. biol.*, 155 (1961) 1010.

310 SPECTOR, W. G., *Nature*, 187 (1960) 514.

311 SPECTOR, W. G., *J. Reprod. Fertility*, 2 (1961) 362.

312 SPENCER, P. S. J. AND WEST, G. B., *Brit. J. Pharmacol.*, 17 (1961) 137.

312a SPENCER, P. S. J. AND WEST, G. B., *Intern. Arch. Allergy Appl. Immunol.*, 20 (1962) 321.

313 STEINER, F. A. AND HEDINGER, C., *Experientia*, 12 (1956) 109.

314 STEINER, F. A., SIEBENMANN, R. E., SANDRI, C. AND HEDINGER, C., *Experientia*, 13 (1957) 500.

315 STJÄRNE, L. AND SCHAPIRO, S., *Nature*, 184 (1959) 2023.

316 TAKETOMO, Y., SHORE, P. A., TOMICH, E. G., KUNTZMAN, R. AND BRODIE, B. B., *J. Pharmacol.*, 119 (1957) 188.

317 TALWALKER, P. K., MEITES, J., NICOLL, C. S. AND HOPKINS, T. F., *Am. J. Physiol.*, 199 (1960) 1073.

318 THUILLIER, J., NAKAJIMA, H., L'HUILLIER, J. AND BAGINSKI, L., *Compt. rend. soc. biol.*, 155 (1961) 2106.
319 TINDAL, J. S., *Ann. Rept. Natl. Inst. Research in Dairying*, 1956, p. 56.
320 TINDAL, J. S., *J. Endocrinol.*, 20 (1960) 78.
321 TIXIER-VIDAL, A. AND ASSENMACHER, I., *Compt. rend. soc. biol.*, 156 (1962) 37.
322 TUCHMANN-DUPLESSIS, H., *Presse méd.*, 64 (1956) 2189.
323 TUCHMANN-DUPLESSIS, H. AND MERCIER-PAROT, L., *Compt. rend.*, 252 (1961) 3882.
324 TUCHMANN-DUPLESSIS, H. AND MERCIER-PAROT, L., *Compt. rend.*, 253 (1961) 712.
325 TUCHMANN-DUPLESSIS, H. AND MERCIER-PAROT, L., *Chemotherapia*, 4 (1962) 304.
326 TURA, S., PIERAGNOLI, E., DAL MONTE, P. R. AND GIRO, C., *Boll. soc. ital. biol. sper.*, 36 (1960) 1352.
327 VALSECCHI, A., *Minerva ginecol.*, 10 (1958) 898.
328 VAUGHAN, M., *J. Biol. Chem.*, 235 (1960) 3049.
329 VEERDONK, F. C. G. VAN DE, *Nature*, 187 (1960) 948.
330 VERDESCA, A. S., WESTERMANN, C. D., CRAMPTON, R. S., BLACK, W. C., NEDELJKOVIC, R. I. AND HILTON, J. G., *Am. J. Physiol.*, 201 (1961) 1065.
331 VITTORIO, P. V., ALLEN, H. J. AND SMALL, D. L., *Radiation Research*, 15 (1961) 625.
332 VITTORIO, P. V., SMALL, D. L. AND ALLEN, H. J., *Can. J. Biochem. Physiol.*, 39 (1961) 1743.
333 WANG, R. I. H. AND BALLANTYNE, J., *Federation Proc.*, 21 (1962) 421b.
334 WAUGH, D. AND PEARL, M. J., *Am. J. Pathol.*, 36 (1960) 431.
335 WEITZEL, G., ROESTER, U., BUDDECKE, E. AND STRECKER, F. J., *Z. physiol. Chem., Hoppe-Seyler's*, 303 (1956) 161.
336 WELLS, H., BRIGGS, F. N. AND MUNSON, P. L., *Endocrinology*, 59 (1956) 571.
336a WERBOFF, J., GOTTLIEB, J. S., DEMBICKI, E. L. AND HAVLENA, J., *Exptl. Neurol.*, 3 (1961) 542.
337 WERBOFF, J., GOTTLIEB, J. S., HAVLENA, J. AND WORD, T. J., *Pediatrics*, 27 (1961) 318.
338 WEST, G. B., *Nature*, 182 (1958) 182.
339 WEST, G. B., *J. Pharm. Pharmacol.*, 12 (1960) 766.
339a WEST, G. B., *J. Pharm. Pharmacol.*, 14 (1962) 828.
340 WESTERMANN, E. O., MAICKEL, R. P. AND BRODIE, B. B., *Federation Proc.*, 19 (1960) 268.
340a WISEMAN, M. H., KALANT, N. AND HOFFMAN, M. M., *J. Lab. Clin. Med.*, 52 (1958) 27.
341 WOOLLEY, D. W., *Nature*, 180 (1957) 630.
342 YAMAZAKI, E., SLINGERLAND, D. W. AND NOGUCHI, A., *Acta Endocrinol.*, 36 (1961) 319.
343 YEH, S. D. J., SOLOMON, J. D. AND CHOW, B. F., *Federation Proc.*, 18 (1959) 357.
344 ZILBERSTEIN, R. M., personal communication.
345 ZIZINE, L., *Compt. rend. soc. biol.*, 153 (1959) 1156.

Chapter 9

SEROTONIN AND CARDIOVASCULAR SYSTEM

EFFECTS ON HEART

Serotonin exerts a marked effect on the cardiac activity of many animal species although it is present in the heart only in small concentrations (see Appendix I). The enzymes involved in the synthesis (5-hydroxytryptophan decarboxylase) or in the degradation (monoamine oxidase) of serotonin are also rather poorly represented in the cardiac tissue. For instance the monoamine oxidase activity in the heart is about 10 times less than in the liver[242].

Hearts of *molluscs* are on the whole sensitive to serotonin[10,125,169,271a,411] and they have been widely used for the bioassay of this amine (see pages 21–22). Positive chronotropic[10,125,411], inotropic[125,411] and tonotropic[125,169] effects are obtained after serotonin administration to hearts *in vitro* or *in vivo*.

The hearts of *Aplysia limacina*, *Dolium galea*, *Murex brandaris* and *Murex trunculus* are less sensitive than *Helix pomatia* or *Octopus vulgaris* hearts[122,125]. Similarly, *Modiolus modiolus* is much less sensitive than *Cyprina islandica*, *Buccinum undatum* and *Venus mercenaria*[411]. The threshold of sensitivity to serotonin for *Cyprina islandica*[411] reaches 10^{-10} g/ml. According to the views of Welsh, serotonin would be the physiological mediator of the cardioexcitatory innervation of molluscs[410,413]. A cardioexcitatory material resembling serotonin has actually been found in molluscan hearts[412]. Lysergic acid diethylamide is as stimulating as serotonin on the heart of *Venus mercenaria*[216,413], but BOL and UML inhibit the effects induced by 5-hydroxytryptamine[266]. Tryptamine[125], 5-acetyltryptamine[286] and other indoles[170] show effects similar to serotonin on *Venus mercenaria* heart.

Positive inotropic and chronotropic effects have been observed also in *crustacea* (arthropoda)[132].

Inconstant effects were observed when serotonin was placed in contact with strips of turtle ventricles[186,187], with turtle heart *in situ*[187] or with frog ventricle strips[187]. However, serotonin induces an inotropic effect in frog isolated heart[66].

In *mammals*, serotonin usually elicits tachycardia and increase of the amplitude of the contraction. These effects have been observed in cats[167,239,251,272,352,394], dogs[241,249,352], guinea pigs[251,394], rabbits[66,209,240,241,249,265,272,352,372,394], rats[70] and man[13,191,199,238]. Most of these effects have been observed working with biological preparations *in vitro* as is summarized in Table 27. Tachyphylaxis has frequently been observed[209,241,249,257,265]. *In vivo* the results are considerably influenced by the experimental conditions and often they are biphasic (for instance bradycardia followed

TABLE 27

EFFECTS OF SEROTONIN ON ISOLATED HEART

Animal species	Type of preparation	Effect observed (and references)
Cat	atria	chronotropic (394)
	auricle	inotropic (251), chronotropic (251)
	isolated heart	inotropic (251, 351), chronotropic (251, 272, 352), coronary dilatation (272, 351, 352)
	papillary muscle	inotropic (167, 239), increased excitability (167)
Dog	heart *in situ*	inotropic (241), bradycardia (241)
	heart-lung	inotropic (249), increased cardiac output (249)
	isolated heart	chronotropic (352), coronary dilatation (352)
Frog	isolated heart	inotropic (66)
	ventricular strip	no effect (187)
Guinea pig	atria	inotropic (394), chronotropic (394)
	isolated heart	chronotropic (251)
Rabbit	atria	inotropic (66), chronotropic (394)
	auricle	inotropic (240, 265), chronotropic (240, 372)
	heart-lung	inotropic (249), increased cardiac output (249)
	isolated heart	inotropic (209, 241, 272), chronotropic (209, 241, 352), coronary dilatation (272)
Rat	heart-lung	no clear effects (277)
	isolated heart	chronotropic (70), negative inotropic (70), decreased coronary flow (70)
Turtle	ventricular strip	depression (186), no effect (187)
Venus mercenaria	isolated heart	inotropic (266, 411), chronotropic (266, 411)

by tachycardia[197]). In several instances individual changes have been observed (for instance serotonin induces bradycardia only in some cats[229]). Not all the effects observed in isolated-heart preparations have been reproduced *in vivo* where the situation is much more complex. A remarkable difference has been reported in cats. While *in vitro* serotonin increases rate, contractility and excitability (papillary muscle[167,239], atria[394], auricle[251], isolated heart[251,272,352]), *in vivo* the predominant response is bradycardia[91,92,106,229,351,352]. A possible explanation for this discrepancy may be the dose used. In fact Magistretti and Valzelli report that in doses below 10 μg/kg serotonin induces tachycardia in cats, while with doses over 10 μg/kg the usual response is bradycardia[252]. In spinal cats serotonin always increases the heart rate[406].

The route of administration is not important because bradycardia has been observed when serotonin was given into the right auricle[91], into the left ventricle[91] and into the coronary artery[106,229]. It has been proposed that serotonin does not act directly but causes bradycardia through a reflex following stimulation of certain receptors[92,128,154].

In dogs, serotonin usually induces tachycardia[101,249,262,263,265] preceded by a short bradycardia[197], but when it is injected into the pulmonary artery it produces bradycardia[245,387]. Bradycardia is also observed in spinal dogs[390] after local application of serotonin to the carotid sinus area[197]. Like serotonin, 5-hydroxytryptophan induces in dogs an increase of the heart rate[33], and cardiac output[270,331].

The mechanism(s) by which serotonin acts on the cardiac activity is very complex and not yet elucidated. No simple and unifying explanation can, for the moment, account for the variety of experimental data. It is probable that some effects of serotonin are direct and some are produced through other chemical mediators.

When serotonin induces bradycardia, a cholinergic mediation seems to be predominant. Atropine prevents bradycardia in rats[252], in cats[92,352], in dogs (initial phase)[352] and in the isolated auricle of rabbits (first phase preceding tachycardia)[240]. A similar effect is obtained by vagotomy in cats[229,352] and in dogs when serotonin is injected into the pulmonary artery[387]. However, atropine does not affect the tachycardia induced by serotonin in dogs[265,352].

TABLE 28

EFFECT OF VARIOUS TREATMENTS ON ISOLATED ATRIA OF CATS (C), RABBITS (R) AND GUINEA PIGS (G) STIMULATED BY NORADRENALINE (NE), HISTAMINE (H), NICOTINE (Nic) AND SEROTONIN (5-HT)
(modified from Trendelenburg[394])

Treatment	*NE*			*H*			*Nic*			*5-HT*		
	C	*R*	*G*	*C*	*R*	*G*	*C*	*R*	*G*	*C*	*R*	*G*
Reserpine				—	=	—	+	+	+	—	+	=
Dichloroiso-proterenol	+	+	+	+	—	±	+	+	+	+	+	+
Cocaine	=	=	=	+	+	+	—	+	+	—	+	+
Morphine	—	±	—	—	—	—	+	+	+	—	+	+
TM 10	+	—	—	+	+	+	+	+	+	+	+	—
LSD	—	+	—		—					+	+	+

Parameter measured: positive chronotropic or inotropic responses.

\+ antagonism of the treatment *vs.* one of the four stimulant agents.

— no effect.

= potentiation.

An adrenergic origin has been suggested for the tachycardia induced by serotonin. However, various adrenolytic agents, including dibenamine[251], phentolamine[251,352]. dihydroergotamine[251] and *N*-ethyl-*N*-bromoethylnaphthalene methylamine[265], do not prevent tachycardia in dogs[265,352] and in the isolated auricle of cats[251]. In a very accurate study Trendelenburg compared the effect of serotonin on rabbit, cat and guinea pig atria[394]. Clear evidence for an adrenergic role was obtained only in the rabbit. In this animal species reserpine pretreatment abolished the inotropic and chronotropic effect of serotonin[209,394]. In agreement with this result are the antagonism between serotonin and choline, 2,6-xylyl ether bromide (TM 10, an agent able to prevent the release of norepinephrine from adrenergic nerve endings[126]), cocaine, morphine and dichloroisoproterenol[394]. In cat and guinea pig atria the depletion of catecholamine stores did not change the effect of serotonin. The effect of other drugs (see Table 28) suggests that while in rabbit atria the positive chronotropic action of serotonin is probably mediated through the adrenergic system, in guinea pigs it is only partially mediated and in cats it is probably independent[394]. In reserpinized dogs, serotonin produced marked bradycardia[98].

Anaesthetics prevent the effect of serotonin on the rabbit auricle[372] and on cats[351,352], but in dogs they prevent only the initial bradycardia and not the subsequent tachycardia[352].

Ganglionic blockade only partially prevents the bradycardia induced by serotonin in cats[92,352] and does not prevent serotonin-induced tachycardia (cat auricle[251], spinal cat[406], dog[352]).

Lysergic acid diethylamide[251,394], BAS[199,209] and 2-methyl-3-ethyl-5-aminoindole (ref. 92) clearly antagonize the changes of heart rate induced by serotonin in cat (refs. 92, 251, 394), rabbit[209,394], guinea pigs[394] or man[199].

Monoamine oxidase inhibitors show in some preparations an inotropic effect[237] but do not potentiate the activity of serotonin on rabbit atria[66].

Hydralazine and veratrum prevent the tachycardia induced in dogs by serotonin (ref. 265). The activities of other drugs on cardiac changes elicited by 5-hydroxytryptamine are reported in Appendix IV.

Serotonin in large doses produces disturbances of the electrocardiogram[252,257] consisting of extrasystoles, reduction of the P–Q tract and presence of the ischaemic T. These effects ultimately lead to the presence of myocardial thickening[4], ischaemia and necrosis[57,214,258,308]. However, serotonin, as 5-hydroxytryptophan, may prevent myocardial lesions induced by other treatments. For instance the myocardial necrosis elicited in rats by isoproterenol, by plasmocid (6-methoxy-8-(diethylaminopropylamino)quinoline dihydroiodide) or by ligature of coronary vessels is blocked by large doses of serotonin[12,213a]: similar results have been obtained with MAO inhibitors[431]. However, the problem is complicated by the fact that monoamine oxidase inhibitors counteract the protective effect of reserpine on myocardial lesions induced by amphetamine in mice[185]. In dogs serotonin is effective in counteracting the ectopic ventricular tachycardia resulting from acute myocardial infarction caused by ligation of the anterior descending coronary branch[218].

5-Hydroxytryptamine on the other hand is without effect on auricular fibrillation caused by vagal stimulation in thyrotoxic dogs and on ventricular tachycardia in cyclopropane- or chloroform-treated dogs[265].

EFFECTS ON THE SYSTEMIC BLOOD PRESSURE

The action of 5-HT on the vascular system cannot be defined in a simple way. According to Erspamer[119,122,124] 5-hydroxytryptamine is "neither a pure hypertensive, nor a pure hypotensive agent" and the systemic vascular action of serotonin should be considered only "as a pharmacological action". Page, who is more inclined to consider serotonin as a "physiological" factor in cardiovascular regulation, was compelled to create the term "amphibaric" to describe the multi-faceted effects of serotonin on the cardiovascular system[295]. The response to serotonin is different at different levels of systemic blood pressure, that is the response depends upon the neurogenic vasomotor tone[295]. This concept is well summarized by Haddy *et al.*[181] after a series of researches on dog foreleg[178-180]. "Serotonin antagonizes extremes of vascular tone induced by neurogenic means. It produces net dilatation when the bed is constricted and net constriction when the bed is dilated. This bidirectional response derives ultimately from the fact that changes in nervous activity change calibers of small vessels without greatly altering calibers of large vessels, and that serotonin produces small vessel dilatation at the same time that it constricts large vessels. When small vessels are already neurogenically dilated, serotonin cannot dilate them further but continues to constrict large vessels. Thus the net effect is constriction. When small vessels are highly constricted, serotonin dilates small vessels more than it constricts large vessels. The net effect is "dilatation"[181]." Most of the discrepancies reported fit into this picture. Table 29 serves to illustrate the variety of effects elicited by serotonin according to the animal species, the type of anaesthesia used, and the route of administration. The picture is further complicated by the presence of tachyphylaxis (refs. 133, 257, 363).

Rather than describe the various effects of serotonin (summarized in Table 29), an attempt will be made to classify the basic mechanisms through which serotonin affects the systemic blood pressure, keeping in mind that the effects of serotonin *in vivo* are of complex nature.

Cholinergic mediation

The vagal fibres are stimulated by serotonin in cats[278] and this may explain why this amine is mostly hypotensive in this animal species. Accordingly vagotomy (refs. 157, 229, 293, 375) or atropine[92,229,293,352] prevent the fall of blood pressure induced by serotonin in cats. At variance with this are the results obtained by Reid who found no change in the pressure response obtained by serotonin in atropinized animals[316]. Also in dogs vagotomy[23,352,390] and atropine[293,352] abolish the initial fall of blood pressure induced by serotonin, and in several instances increase the pressor phase (refs. 23, 352, 390). However, the hypotension produced in dogs by intraventricular administration of serotonin was not affected by vagotomy[26].

In rats the initial fall of blood pressure after serotonin is considered of vagal origin because it is inhibited by atropine[148,257] and by cutting of the vagi[257,340].

In rabbits vagotomy[352] or atropine[77,352] either increase or do not affect the hypotensive phase but they prevent the rise of blood pressure[352]. In chickens the hypotension induced by serotonin could not be interpreted on the basis of cholinergic mediation[61]. In mice atropine prevents the increase of blood pressure induced by large doses of serotonin[310].

TABLE 29

EFFECTS OF SEROTONIN ON SYSTEMIC BLOOD PRESSURE IN VARIOUS ANIMAL SPECIES

Animal species	Anaesthesia	Dose (in mg)	Route of administration	Effect observed (and references)
Calf	—	0.5/min	i.v.	decrease (230)
Cat	chloralose	0.1/kg	i.v.	decrease (76)
		0.1	right heart	decrease (91)
		0.1	left ventricle	decrease (91)
	chloralose (+ vagotomy)	0.009/kg	i.v.	decrease followed by rise (316)
	chloroform–chloralose	0.075	right ventricle	decrease (92)
		0.01	carotid sinus	decrease (157)
		0.01	coronary artery	decrease (229)
		0.01	cisterna magna	decrease (393)
	dial-urethane	0.05/kg	i.v.	decrease followed by rise (352)
Chicken	pentobarbital	0.001/kg	i.v.	decrease (282)
	phenobarbital	0.025	i.v.	decrease (61)
		0.2	i.v.	decrease (309)
Dog	chloralose	0.05/kg	i.v.	increase (30, 200)
	methitural	0.05/kg	i.v.	decrease (200)
	morphine-dial	0.003/min/kg	i.v.	no effect (250)
	morphine–pentobarbital	0.02/min/kg	i.v.	decrease (263)
		1.0	i.a.	increase (23)
		1.0	portal vein	decrease (23)
	pentobarbital	0.05/kg	i.v.	initial decrease followed by rise (352)
		0.02/min/kg	i.v.	increase (377)
	pentobarbitone	0.1/kg	cerebral ventricle	decrease (26)
	thiopental	0.2/kg	pulmonary artery	increase (51)
		0.2/kg	carotid sinus	decrease (197)
		0.2/kg	right heart	increase (51)
		0.2/kg	left heart	increase (51)
		0.2/kg	ascending aorta	increase (51)
Man	—	1.0	i.v.	decrease (13)
	—	0.5	i.v.	increase (376)
	—	0.004/kg	right atrium	decrease (191)
Mouse	urethan or pentobarbital	50/kg	i.v.	increase (310)
Rabbit	chloralose	0.05/kg	i.v.	decrease followed by rise (352)
		0.004/kg	i.v.	decrease (77)
Rat	chloralose	0.025/kg	i.v.	fall followed by rise and decrease (148)
	amobarbital	0.02	i.v.	increase (150)
		0.02/kg	i.v.	fall followed by rise and decrease (340)
		10.0/kg	i.p.	decrease (386)
	chloralose or urethan	0.025/kg	i.v.	fall followed by rise and decrease (257)
	urethane	0.01/kg	i.v.	decrease (141)
Sheep	thiopental	0.05/min/kg	i.v.	increase (184)

The reported inhibition or activation of cholinesterase[118,432] by serotonin does not seem adequate to explain the effects of the amine on the systemic blood pressure.

Adrenergic mediation

The interactions between serotonin and the adrenergic system are multiple and bidirectional depending on the experimental situation. To explain the depressor effect of serotonin the hypothesis has been suggested that serotonin antagonizes at the peripheral level neurogenic vasoconstriction[296] or adrenergic mediators[6]. In rats[21] and in dogs[162,204] noradrenaline antagonizes the effect of serotonin. According to Gordon *et al.*[163] serotonin and noradrenaline have an antagonistic action in arterioles and a synergistic action in large-artery segments. Small doses of serotonin potentiate and large doses inhibit the hypertension produced by adrenaline and noradrenaline in cats[347]. Noradrenaline potentiates the effect of serotonin on the vessels of rabbits (refs. 272, 273). Adrenaline has been reported as a synergist[341] but also as without effect on the serotonin responses in rabbits[272] and in cats[25].

A release of catecholamines from the adrenals was considered to be one of the mechanisms underlying the pressor responses of serotonin[235,400]. This release has been observed in cats[316,317,406] but only explains pressor effects obtained with large doses of serotonin[316,408]. In rats adrenalectomy increases the hypotensive effect of serotonin[143] and prevents the rise of blood pressure induced by serotonin[257], but a similar effect was not observed in dogs[296,338,390,403]. Several adrenolytic agents have been reported as effective in counteracting the pressor phase of serotonin activity on the systemic circulation. For instance phentolamine is effective in rats[21,257], in dogs[352] but not in cats or in rabbits[352]; dibenamine is effective in rats[257,288] and in mice[310], and in cats it potentiates the hypotensive phase[375]; dihydroergotamine in rats[288] and dogs[200]; phenoxybenzamine in rats[261] and mice[310], and *N*-ethyl-*N*-(2-chloroethyl)-9-fluorenamine (SY 21) in dogs[383]. Some of these results are reported in Table 30.

The depletion of catecholamines by reserpine in dogs[81,98,407] markedly increases the hypotensive effects of serotonin, abolishing the pressor activity. Cocaine, a known sensitizer of noradrenaline, decreases the hypertensive response to serotonin[81]. Imipramine increases both the effect of noradrenaline[337] and of serotonin[177,220,337]. It is interesting that the effect of adrenaline was inhibited under similar conditions[337]. Inhibition of the pressor responses elicited by serotonin can be achieved with drugs which cannot be considered adrenolytics. Representative examples are 2-amino-4-(*N'*-methyl-*N*-piperazino)-5-(4'-chlorophenylmercapto)pyrimidine (BW 57301) in cats[90], 1-methyllysergic acid butanolamide (UML) in dogs[79], 2-bromolysergic acid diethylamide (BOL) in rats[166,261,339] and cyproheptadine in dogs[383].

According to McCubbin *et al.*[269] an interaction between serotonin and adrenergic receptors could also explain some depressor responses induced by serotonin. These authors believe that the antagonism of serotonin toward neurogenic vasoconstriction depends upon adrenergic β-receptors rather than blockade of adrenergic constrictor receptors or release of noradrenaline at sympathetic nerve endings.

Other observations relevant to the understanding of the relationship between

TABLE 30

EFFECT OF SELECTED DRUGS OR TREATMENTS ON "AMPHIBARIC" RESPONSE OF SERO-
TONIN IN VARIOUS ANIMAL SPECIES

(Numbers in brackets indicate references)

Animal species	Change of blood pressure	Inhibited by:
Cat	initial fall	atropine (92, 229, 293, 352), vagotomy (157, 229, 293, 375)
	rise	1,2-methoxyphenylpiperazine (300), bulbocapnine (405), LSD (406)
	fall	BAS phenol (63)
Dog	initial fall	atropine (293, 352), vagotomy (23, 352, 390)
	rise	phentolamine (352), UML (127), BAS phenol (63, 424), 2-methyl-5-chlorogramine (123), LSD (207, 348), BAS (207, 208), morphine (207), hydrazinophthalazine (293), 1,2-methoxyphenylpiperazine (300), chlorpromazine (348), 2,5-dimethylserotonin (363), 1-benzyl-2,5-dimethylserotonin (363), cyproheptadine (383), bulbocapnine (405), DHE (200)
Rabbit	initial fall	cord section at C_6 (352), BAS phenol (63)
	rise	atropine (352), vagotomy (352), bulbocapnine (405)
Rat	initial fall	atropine (148, 257), vagotomy (257, 340)
	secondary rise	phentolamine (21, 257), dibenamine (257, 288), phenoxybenzamine (261), DHE (288), BOL (150, 166, 339), LSD (166, 288, 339), BAS (288)
	fall	phentolamine (257), chlorpromazine (257), ganglionic blockade (340)

serotonin and the adrenergic system are that serotonin increases the effects of excitation of adrenergic nerves (either pre- or postganglionic stimulation) in cats[11] and
facilitates the synaptic transmission in sympathetic ganglia of rat[195,195a,*]. Serotonin
was detected in the perfusate from the functioning superior cervical ganglion of the
cat *in situ*[149a].

Release of histamine

Some of the depressor effects of serotonin have been connected with the known
capacity of this amine to release histamine[129,299]. In mammals this mechanism does
not seem of great importance. Vasoconstriction induced by serotonin can be antagonized by antihistamines[272,273,383], but antihistamines are frequently also antiserotonin agents (see Chapter 6). The hypotensive action of serotonin in the dog is
inhibited by diphenhydramine[59] but other antihistamines do not show this effect in
rats[261] and cats[375].

---

* Serotonin also increases the activity of the ganglia of *Helix pomatia*[228] and stimulates the
inferior mesenteric ganglion of cats[175a].

In chickens it is more likely that the depressor effect of serotonin is mediated through histamine release[58,65]. The evidence is based upon the following data: (*a*) serotonin increases the amount of histamine in the urine[60,62]; (*b*) after depletion of tissue histamine, by means of repeated administration of Compound 48/80, serotonin does not elicit hypotension[62]; (*c*) depressor effects of serotonin are blocked by a number of antihistamine drugs[61]. Similar results have been reported for tryptamine[65], but Eble[115,116] does not agree with the hypothesis that the release of histamine is a major factor in explaining the hypotensive action of tryptamine.

Effects on peripheral receptors

The variety of cardiovascular and respiratory responses elicited by serotonin can also be considered as the net result of the many interactions produced by serotonin on the caliber of the vessels and on the specialized chemoreceptors in the aortic and carotid bodies, heart and lung, all of which induce reflexes. According to Ginzel[154] "it is not possible to determine the exact part played by these various reflex actions within the over-all effect of a given dose of serotonin injected into a given animal. Almost each single phenomenon, be it hypotension or hypertension, apnoea or hyperpnoea, that could be assigned to a particular receptor area stimulated by 5-hydroxytryptamine, and subsequently abolished through ablation of the respective afferent nerves, was found to recur when sufficiently large amounts of serotonin were used".

Aortic chemoreflex (aortic body reflex) elicits hypotension and bradycardia[106] and is stimulated by serotonin[107,157]. This effect is prevented by cutting the vagus[157].

The action on the *carotid body reflex* in dogs is proven because the pressor effect of serotonin was increased after elimination of the carotid chemoreceptors[51,382]. In man, carotid body removal abolished the hyperpnoea but not the blood pressure changes induced by serotonin[374]. Ginzel[154] suggested that the effect may be due to constriction of the vessels of the carotid body giving rise to a local anoxaemia. Reduction of oxygen tension in various tissues has been demonstrated after systemic administration of serotonin[78] (see Chapter 8).

The perfusion of isolated carotid sinus with large doses of serotonin induces a contraction[264]. *In vivo*, serotonin applied locally produced a fall of blood pressure in dogs[197] and cats[157]. 12 μg of serotonin stimulated the isolated perfused carotid sinus of dogs much more, on a weight basis, than lobeline[267].

Transection of the carotid sinus abolishes the depressor effect of serotonin[157] and enhances the pressor effect[293]. However, denervation of the carotid sinus increased the fall of blood pressure when serotonin was given intraventricularly in dogs[26]. Also the stimulation of breathing by serotonin in dogs is ascribed to an effect on the carotid sinus[110,111]. LSD and its congeners are weak antagonists of serotonin in this test[155,156].

Serotonin antagonizes some effects induced by noradrenaline (bradycardia) in carotid sinus[162]. Koella *et al.*[227] believe that in cats the stimulation of the carotid sinus area plays a role in explaining the effects of serotonin. Other authors deny such a relationship in dogs[197]. The pressor effect elicited by occlusion of the carotid arteries is blocked by serotonin[26,197,268], 5-hydroxytryptophan[33,268] and reserpine[307], a known releaser of serotonin.

Coronary chemoreflex is defined by a fall of blood pressure and heart rate on injection into the coronary arteries[106]. Serotonin (5–10 μg) elicits this reflex (refs. 92, 106, 128, 229) in cats. Vagotomy abolishes it[229]. Serotonin is not effective in dogs and man[296]. It is, however, interesting that serotonin injected intravenously in man causes substernal distress like that frequently noted by patients with coronary insufficiency (refs. 104, 296).

Pulmonary depressor chemoreflex is induced by serotonin in cats but not in dogs[106] after injection into the superior vena cava, right atrium and right ventricle[92,229]. 5-Hydroxytryptamine also causes a *pulmonary respiratory reflex* (apnoea) in cats[106] prevented by 2-methyl-3-ethyl-5-aminoindole[92]. Schneider and Yonkman[351] found that serotonin excites the slowly adapting pulmonary stretch receptors to continuous activity, an effect that Mott and Paintal[278] failed to observe. The latter authors suggested that an action on receptors situated between the great veins and the left atrium could be partly responsible for the apnoea elicited by serotonin[278].

When serotonin is injected intracarotidally, or intravenously in larger doses, there is also an interaction with the receptor sites in the brain stem[226,227].

The importance of these reflex effects of serotonin in inducing the various changes of blood pressure in different animal species is stressed by the fact that section of the vagi and of the spinal cord at C_6 produces only a uniform pressor response in dogs, cats and rabbits[352].

DRUGS INTERFERING WITH THE EFFECT OF SEROTONIN ON SYSTEMIC BLOOD PRESSURE

Many drugs which are able to alter the changes of blood pressure elicited by serotonin have already been mentioned in the various paragraphs concerning the mechanism of action of serotonin and the effect of 5-hydroxytryptamine on regional circulation. Table 30 summarizes results obtained in the animal species most frequently used for this purpose.

Ergot alkaloids

Ergotamine is a strong antagonist of the pressor effect produced by serotonin in dogs and cats[298,339]. Similarly dihydroergotamine is effective in rats[288] and dogs[200]. In chickens the depressor effect induced by serotonin is prevented by ergotamine and dihydroergotamine[64]. The antagonism is, however, not specific because the depressor effect of tryptamine and histamine as well as the pressor effect of epinephrine are prevented by these drugs[64]. Ergonovine is more specific because the inhibition of serotonin and tryptamine is not accompanied by an effect on histamine or epinephrine[64].

Lysergic acid derivatives

LSD is a good antagonist of the pressor response elicited by serotonin in rats (refs. 166, 288, 339) and dogs[207,298,348,383] but is less effective in cats[339,406] and in chickens[64]. A review on the antagonism between LSD and serotonin is given by Rothlin[329]. BOL is considered as effective as LSD in rats[150,166,261,339] but more effective than LSD in chickens[64]. UML is also a powerful inhibitor of the hypertension induced in dogs[127] and of the hypotension induced in chickens[64] by serotonin.

Indole derivatives

Many compounds have been studied with the object of finding new types of hypotensive agents. Reviews on these indole compounds are given by Erspamer[119,122], Gyermek[175], Woolley[420] and others[216,231]. Among the best-known antagonists of blood pressure changes induced by serotonin are: 2-methyl-3-ethyl-5-dimethylamino-indole (medmain)[361], active on cat hypotension[92] but not in human hypertension[376]; 2,5-dimethyltryptamine and 1-benzyl-2,5-dimethyltryptamine[363]; benzyldimethyl-bufotenin (BAB)[362], cinobufotenin[309] and 2-methyl-5-chlorogramine[123], all effective in preventing the pressor response to serotonin in dogs. 5-Acetyltryptamine is not a serotonin antagonist but it shows interesting hypotensive properties[286,360]. 5-Hydroxy-3-indoleacetamide produces a pressor effect in dogs[174]. Extensive research has been carried out on 1-benzyl-2-methyl-5-methoxytryptamine (BAS) and 1-benzyl-2-methyl-5-hydroxytryptamine (BAS phenol). BAS was found effective in rats[288] and dogs (refs. 207, 208, 383). In humans it counteracts the cardiac, respiratory and vascular effects of serotonin[199]. In hypertensive patients BAS showed hypotensive properties (refs. 417, 418).

BAS phenol[364] is an interesting compound because, according to Woolley and Shaw, it enables the pressor effects of serotonin to be differentiated from those of tryptamine[424,425]. BAS phenol blocks the pressor response of serotonin in dogs as well as the depressor response in cats, rabbits and chickens[63].

Phenothiazine and derivatives

Chlorpromazine shows antiserotonin properties and prevents the changes of blood pressure induced by 5-hydroxytryptamine. Its activity has been shown in cats (refs. 176, 419) and dogs[348] (ED_{50} = 0.12 mg/kg)[383]. In rats contradictory results have been obtained[257,261]. Promazine[419], promethazine[148,383] and trimeprazine[383] also show an effect while prochlorperazine[419] is considered less active. A recently developed compound, cyproheptadine (1-methyl-4,5-dibenzo[a,e]cycloheptatrienylidine piperidine·HCl) is a very powerful antagonist (ED_{50} = 0.05 mg/kg) of the pressor response elicited by serotonin in dogs with ganglionic blockade[383].

Adrenergic and cholinergic inhibitors

These have already been discussed (page 173–176).

Reserpine and derivatives

Besides the well-known property of releasing serotonin, reserpine also shows antiserotonin properties. It prevents the rise in blood pressure induced by serotonin in spinal-vagotomized, but not in intact dogs[350]. The schedule of treatment is important in this respect. Up to 4 h after reserpine medication there is an increase of the hypertensive properties of serotonin in dogs[349] but after a longer pretreatment serotonin becomes a powerful hypotensive agent[98]. Also, 5-hydroxytryptophan and BAS show marked hypotensive effects in dogs pretreated with reserpine[98].

When reserpine is given to animals with monoamine oxidase blockade, a marked pressor response occurs. Since this hypertension is inhibited by chlorpromazine, it was thought to be mediated by a release of serotonin[82]. However, other investigators favour the hypothesis that the hypertension is of adrenergic origin[85,141].

The hypothesis that the hypotensive properties of reserpine could be associated with a central release of serotonin has been ruled out because some derivatives such as syrosingopine[142,145,306], 10-methoxydeserpidine[144,305], diethylaminoethylreserpine and isobutylserpentinate[144] although hypotensive do not release brain serotonin.

Hypotensive agents

Because of the controversy concerning the role of serotonin in hypertension, the relationship between hypotensive agents and serotonin is of considerable interest. 1-Hydrazinophthalazine was found to be an antagonist of the pressor response induced by serotonin in dogs[279,293,389], but this antagonism was not considered to be specific[120]. 1,2-Methoxyphenylpiperazine lowers blood pressure and reduces the hypertensive effects of serotonin in cats and dogs[300].

α-Methyl-DOPA is an antihypertensive agent[52] which inhibits 5-hydroxytryptophan decarboxylase (see Chapter 4) and lowers tissue serotonin[161].

N-(3'-phenylpropyl-2')-1,1-diphenylpropyl(3)amine (segontin), a coronary dilator and hypotensive agent, releases heart serotonin[354].

Guanethidine does not affect platelet serotonin[100].

Monoamine oxidase inhibitors

Monoamine oxidase inhibitors increase the tissue serotonin and show hypotensive properties. However, no causal correlation between these two effects has been proved (refs. 373, 415, 416). The hypertensive properties of serotonin are potentiated by pretreatment with monoamine oxidase inhibitors only when serotonin is given into the lateral ventricle or cisterna magna[268,393] but not when given into the general circulation of dogs. In rats only weak effects, if any, have been obtained[21,141], but the pressor responses of 5-hydroxytryptophan[268,393] and tryptamine[67,115,116,160] are always increased.

Miscellaneous

Bulbocapnine prevents the pressor responses induced by serotonin in dogs, cats and rabbits[404,405]. A recently discovered pyrimidine derivative (2-amino-4-(N'-methyl-N-piperazine)-5-(4'-chlorophenylmercapto)pyrimidine; BW 57–301) antagonizes the rise of blood pressure caused by serotonin in cats with ganglionic blockade at doses inactive on adrenaline[90].

Acetylstrophanthidin enhances the pressor response of serotonin in dogs[113].

EFFECTS ON THE CIRCULATION OF SPECIFIC REGIONS

Coronary vessels

Serotonin dilates coronary vessels *in vitro* and *in vivo*. Coronary dilatation has been observed in the isolated heart of cat[272,351,352], of dog[352] and of rabbit[272] but not of rat where coronary constriction was found[70].

In dogs serotonin, like tryptamine[315], elicits a consistent increase of the coronary flow[101,172,263,353] with a decrease of the coronary vascular resistance[331]. As a result of this effect there is also an increase in the oxygen content of the coronary sinus (refs. 262, 263, 331) and in the cardiac oxygen consumption[262]. The coronary dilatation was also present when serotonin was injected directly into the coronary vessels[353].

A monoamine oxidase inhibitor, pheniprazine, does not increase the effects of serotonin on the coronaries[331].

Pig coronary arteries are constricted *in vitro* by serotonin[388] ($3 \cdot 10^{-7}$–10^{-5} g/ml).

Cerebral vessels

Little is known about the effects of serotonin on the cerebral circulation in spite of the fact that it is very important to evaluate the significance of the electroencephalographic changes obtained after general or local administration of serotonin. A dose of 0.2–4 μg/kg of serotonin induces in rabbits only a small decrease of cerebral flow which is probably secondary to the decrease of the systemic blood pressure[76]. In cats serotonin elicits first a short decrease and then a sustained increase of the cerebral flow probably due to a local vasodilatation[76,77]. Similar effects have been observed in dogs[77]. Atropine does not modify this effect of serotonin[76,77].

According to other authors the changes of the cerebral circulation are of little importance[86,97].

In experiments using cross-circulation in dogs, perfusion of the isolated head with serotonin caused a fall of blood pressure in the recipient[95].

Pulmonary vessels*

Reid[316] was the first to describe pulmonary vasoconstriction following systemic administration of serotonin. The occurrence of pulmonary hypertension has been confirmed by many others in dogs[113,205,244,245,335,336,382], rats[386], cats[375], rabbits[401], sheep[184], calves[230] and man[13,191] and it has been ascribed to local vasoconstriction (refs. 7, 51, 71, 91, 333, 334, 365, 403). Most of the evidence indicating the pulmonary

---

* An extensive review on the pharmacology of the pulmonary circulation has been written by Aviado[8].

vasoconstriction has come from perfusion experiments using the lungs of guinea pigs[27], cats[112,136,138,156] and dogs[114,152,365]. The constriction was deduced either from an increase in pressure during constant-flow perfusion, or a reduction in flow during constant-pressure perfusion.

Although serotonin markedly influences respiration[157,302,340,414], the pulmonary hypertension seems to be independent from changes in ventilatory mechanism[6,7]. In fact the hypothesis that pulmonary hypertension is secondary to the bronchoconstriction elicited by serotonin is apparently not valid because it has been shown that the administration of serotonin into the pulmonary artery induces hypertension without bronchoconstriction[42,288,323]. Similarly, the pulmonary hypertension seems to be dissociated from changes of the systemic pressure, because it occurs in animal species (for example cats and dogs) in which serotonin elicits opposite changes in the general blood pressure. Furthermore, the effect on pulmonary pressure is longer lasting than that on systemic pressure[9].

The pulmonary hypertension is accompanied by an increase of the pulmonary vascular resistance[270,271,336,365]. The main vasoconstriction is of the arteries, followed by the constriction of small venules (diameter 0.4 mm) and large veins (diameter 1 mm)[9]. On the contrary histamine is more effective in increasing the venous than the arterial resistance[152]. After using large doses of serotonin the effect on pulmonary venous pressure becomes more important[51,217,365,426].

Serotonin is more effective than adrenaline and noradrenaline in inducing pulmonary vasoconstriction[42]. Tryptamine[156] and 5-hydroxytryptamine[22] also produce pulmonary arterial hypertension.

The pulmonary hypertension induced by serotonin is not influenced by spinal section[280,281], vagotomy[280,281], adrenalectomy[403], ganglionic blockade[403], administration of antihistamines[323] and reserpine[323].

On the other hand the following lysergic acid derivatives are able to counteract this effect of serotonin: LSD[138,322], BOL[322] and UML[184]. Dihydroergotamine[138], BAS[322,323], promethazine[322,323], 2-ethyl-3-nitroindole[375] and heparin[375] are also antagonists of the pulmonary hypertension induced by serotonin.

The pulmonary hypertension induced by *E. coli* endotoxin is supposed to be due to a release of serotonin[105] (see Chapter 6). The inhibitory effect on pulmonary hypertension exerted by α-methyl-DOPA[225], a known decarboxylase inhibitor[161], is explained on this basis.

Repeated administrations of serotonin induce in rabbits[72,73] and, to a less extent, in rats[326] changes in the pulmonary vessels consisting of intimal hyperplasia of small arteries and arterioles (pulmonary atherosclerosis)[326-328]. This effect is increased by thyroxine and prevented by oestradiol[325].

Liver vessels

Serotonin increases portal pressure[151] and resistance to hepatic blood flow[2] in dogs. Liver vessels are constricted in guinea pigs[210] and with larger doses in rats[210,342]. Radiographic visualization of liver circulation in rats shows that the most pronounced effect of serotonin is on small venules[210]. In man serotonin increases the uptake of bromsulphalein and this effect was considered secondary to an increased hepatic flow[69].

Splenic vessels

Splenic vessels are constricted by serotonin in dogs[200].

Gastric vessels

Gastric vessels are dilated in rats by serotonin and 5-hydroxytryptophan[109].

Vessels of the mesoappendix

In rats the terminal arterioles are constricted either by intravenous injection or local application[34,94]. The topical effect of serotonin is about 200 times larger than that of general administration[34].

Mesenteric vessels

Regional intra-arterial injection of serotonin (0.5–15 μg) into the mesenteric circulation of dogs causes constriction of the arteries ranging from 0.3 to 1.2 mm of diameter. Vessels with a diameter over 0.7 mm constrict longer than vessels with a smaller diameter[269]. In rats there are small changes in intact animals[386] but strong constriction with haemorrhage in adrenalectomized animals[146].

Renal vessels

The renal vessels of dogs[297] and rats[386] are constricted after serotonin administration. Cats[147], guinea pigs[402] and rabbits[402] are less sensitive. This problem has been discussed in Chapter 7.

Skin and muscle vessels

In rats serotonin given intraperitoneally increases the muscle blood flow and reduces skin circulation[386]. The cutaneous blood flow is, however, increased when serotonin is administered intravenously[386]. Constriction of the tail artery of rats has been observed after intravenous[94] or subcutaneous[93] injection of serotonin. In man muscle blood flow is increased by 0.1–100 μg of serotonin[32]. Skin blood flow decreases after intra-arterial, intravenous[32] or intradermal[316] injection.

Human forearm vessels

The intravenous infusion of 1–4 mg/min of serotonin induces a dilatation of forearm vessels[238]. The intra-arterial administration of smaller doses (0.25–16 μg/min) reveals a vasoconstrictor effect with decrease of the blood flow, flushing and delayed cyanosis[158,182,319]. BOL and sodium salicylate antagonize the constriction of the blood vessels of the forearm and hand induced by serotonin[159].

Scrotal vessels

The effect of serotonin on scrotal vessels has been carefully analysed by Majno[255] and Palade[301]. The main effect is on vessels with a diameter of 20–30 μ and a less effect occurs on those with a diameter of 4–7 or 75–80 μ (refs. 253, 254).

Ear preparation

Rabbit's ear preparation is very sensitive to the vasoconstrictor effect of serotonin[293]. According to Ginzel and Kottegoda[156] the effect of serotonin is even stronger than that exerted by adrenaline. Serotonin reduces the relative arterio-venous anastomotic flow in this preparation[320].

The vasoconstrictor effect of serotonin is enhanced by ephedrine[137], choline-*p*-tolyl ether bromide[137] and adrenaline[341] and is antagonized by dibenamine[137], bulbocapnine[404,405], LSD[137,341], BOL[341], dihydroergotamine[137], ergonovine[341], gramine[137] and a chlorothiazide congener (6-chloro-3-dichloromethyl-3,4-dihydro-7-sulfamyl-1,2,4-benzothiazine-1,1-dioxide)[332]. Hexamethonium, tolazoline[156], piperoxan, atropine and cocaine[137] are without effect *vs.* serotonin in this test.

Perfused hind-leg preparations

Tryptamine and serotonin constrict the denervated hind-leg of dogs[293] but they show a vasodilator effect (not eliminated by atropine) in a normally innervated leg[269]. A vasoconstriction[316] abolished by LSD[80] was also observed in cats.

The circulation through the hind-limbs of the rabbit has been used to study the effect of drugs on the vasoconstrictor response elicited by serotonin[272,395]. Synergistic drugs are noradrenaline, histamine, pitressin, barium chloride and veratrine[272,273]. An antagonism was obtained with reserpine[272,397], chlorpromazine[272,273,397], hydralazine[385,397], acetylcholine[385], papaverine[272,397], promethazine[273], LSD[272,397], dihydroergotamine, ergotamine, phentolamine, yohimbine, atropine, procaine, antihistamines, pendiomid and adenosine[272].

Also in dogs[87] and in rats[88,89] serotonin shows a vasoconstrictor property which is decreased by ethylenediamine tetraacetic acid (EDTA). A similar effect was observed also in hypertensive rats[396].

In frogs the circulation through the hind-limbs shows periodical variations[427] which are potentiated by perfusion with serotonin ($1\cdot10^{-8}$g/ml)[430].

Vessels of placenta

Serotonin is vasoconstrictor on perfused human placenta[117,149] where it is about 10 times more effective than adrenaline[117]. LSD, phentolamine, yohimbine and chlorpromazine are antagonists of serotonin[5]. Trimetaphan (a ganglionic blocking agent) antagonizes serotonin but hexamethonium, tetraethylammonium, reserpine, mescaline and heparin do not inhibit serotonin[5].

Serotonin also constricts the vessels of the umbilical cord of pregnant rabbits[304].

Other vessels

A list of preparations sensitive to serotonin is shown in Table 31. It is interesting that serotonin is present in arteries and is released by injury[422]. Serotonin (20 μg/kg/min) increases the thoracic duct lymph flow in dogs[366].

TABLE 31

SYSTEMS SHOWING THE VASOCONSTRICTOR PROPERTIES OF SEROTONIN

Animal species	Biological preparation	References
Cat	hind-limbs	317
	lung	138
Dog	heart-lung	114
	hind-limbs (denervated)	293
	fore-limbs (denervated)	204
	lung	152, 323, 333, 365
	mesenterium	269
	carotid sinus	264
Frog	posterior limbs	122, 264, 427
Man	placenta	5, 117, 149
Mouse	ear	384
Pig	coronary vessels	388
Rabbit	aorta	332, 385
	aorta (chain of rings)	358
	aorta (strips)	399
	artery ring	311
	ear	311, 320, 332, 341, 404, 405
	hind-limbs	272, 273, 385, 399
	umbilical vessels	304
Rat	hind-limbs	89, 396
	mesoappendix	34
Sheep	carotid artery (spinal strip)	317

RELATION TO CLINICAL PROBLEMS

In the extensive search for the factors causing hypertension also serotonin has been considered[140,295,420,421]. One of the major difficulties in accepting a role for serotonin in the control of systemic blood pressure is the large dose of the amine required to induce a pressor change compared with the minute amounts free in blood and not bound to platelets. However, this is not a valid objection because even small amounts of serotonin released at the receptor sites may produce large effects. The idea that serotonin may be an important factor in hypertensive disease can on present evidence be considered only as an attractive working hypothesis.

Patients with essential hypertension respond to 5-hydroxytryptamine infusion with a sustained rise of blood pressure[299]. Determinations of serotonin[75,121] or 5-hydroxyindoleacetic acid[192,232,284] in the urine of hypertensive patients do not, however, reveal any difference compared with normotensive subjects. Woolley and Shaw suggested that antimetabolites of serotonin should prove useful in the therapy of hypertension[421,423]. It is interesting that with this approach[420] it was possible to discover 1-benzyl-2-methyl-5-methoxytryptamine (BAS) a drug which lowers the blood pressure in hypertensive patients[417,418]. However, it is possible that BAS produces the hypotensive effect by a mechanism unrelated to serotonin antagonism. Similar problems arise from the observation that monoamine oxidase inhibitors increase the levels of tissue and blood serotonin and show at the same time hypotensive properties[153,285,373,415,416].

The hyperserotoninaemia[318] occurring in carcinoid patients has been implicated as the cause of the circulatory changes (flushes)[391]. During the flushes there is an increase of the concentration of serotonin in the venous plasma[303]. However, in contrast to adrenaline and noradrenaline, the injection of serotonin does not elicit facial flushes in carcinoid patients[303].

The recent discovery that 1-methyl-d-lysergic acid butanolamide (UML 491 or deseril), a potent antiserotonin agent, relieves migraine and vascular headaches (refs. 24, 103, 103a, 135, 164, 183, 196, 243, 321, 369, 370) suggests the possibility that serotonin may be involved in the vasomotor alteration of cerebral vessels. The search for the presence of serotonin in the subcutaneous fluid during migraine attack was, however, negative[102]. The infusion of 5-hydroxytryptamine or 5-hydroxytryptophan in susceptible subjects does not initiate headache[223]. In another investigation, where 5-HT induced headache, constriction of the superficial veins of the scalp was observed, while in migraine a dilatation is always observed[287].

The possible participation of serotonin in the onset of portal* hypertension, suggested on the basis of an increased serotoninaemia in this disease[248], was rejected because infusion of serotonin (30 μg/kg/min) in normal or cirrhotic subjects did not change the venous portal pressure[83,84]. An increased excretion of urinary 5-hydroxyindoleacetic acid was reported in subjects with acute coronary insufficiency[3].

The pulmonary arterial hypertension induced by serotonin (see page 181) has been discussed as a cause of the general effects seen after pulmonary embolism[91,92,375].

SEROTONIN AND BLOOD COAGULATION

Clotting and release of platelet serotonin

For many years it has been known that blood after it has been shed[54] or after the clotting process[53] releases a factor which constricts arteries[292]. This vasoconstrictor factor was purified from beef serum and identified as 5-hydroxytryptamine[312,313]. All the blood 5-hydroxytryptamine is present in the platelets in concentrated form (refs. 44, 45, 47, 202, 433, 434) (see Appendix I). The platelets must be considered as a store of 5-HT, they do not synthesize or metabolize the amine (see Chapter 4), although recently it has been suggested that a low level of monoamine oxidase is present in platelets[289–291]. Serotonin is taken up by platelets both by absorption[36,37] and by an active process[40,201,275]. Various agents (reserpine[68,74,188,201], imipramine[259,378], nicotine[343], tryptamine[378], heparin[215], dinitrophenol[39,330], see Chapter 4) inhibit the 5-HT uptake. A quantitative relationship exists between 5-HT uptake and the concentration of ATP present in platelets[35,38,41].

---

* Portal blood contains more serotonin than arterial blood[392].

The fact that during clotting there is a release of 5-HT[189,203,433] suggested a role of serotonin in various aspects of coagulation. The release of 5-HT during clotting is caused by an action of thrombin on platelets[56,171,433]. Trypsin shows the same effect[56,139]. The effect of thrombin is distinguishable from that of reserpine in two respects; it is more rapid (82% release in 5 min with thrombin while with reserpine there is 30% release in 60 min[139]) and it is blocked by heparin[139]. After clotting human blood loses the ability to take up serotonin[359]. In stored blood the serotonin content is decreased[355].

Effect of serotonin on clotting

The direct effects of serotonin on the processes of coagulation are ill-defined and contradictory.

Several authors report that serotonin does not affect *in vitro* the blood-clotting systems[94,247,294,409]. However, according to Keller[221] serotonin reduces the anti-thrombin effect exerted by heparin and similar conclusions are shared by Milne and Cohn[276]. Serotonin has also been considered as a "clot retraction factor" (on bovine or human platelet-poor plasma)[14,15,130,131] but this effect has not been observed in other animal species including man[247,294,409]. However, the finding that antiserotonin agents prevent clot retraction reopens the problem[108,173].

Fibrinolytic activity *in vitro* is inhibited in man[398] and increased in rabbits[233] also *in vivo*[234]. The conversion of fibrinogen to fibrin is accelerated[46,276] and the inhibitory effect of thrombin on this process is prevented by serotonin[219,276]. However, 5-HT does not affect thrombolysis in rabbits[198].

Although *in vitro* serotonin does not seem to affect coagulation, *in vivo* there are a number of effects that indirectly may act on coagulation. High doses of 5-HT injected in rats induce an increase of circulating platelets[18,55,194,371,380]. The effect was found in adrenalectomized rats[381] but was not observed after a single dose of serotonin in rabbits[43].

Bleeding time is considerably reduced by serotonin in mice[94], rats[93,94], rabbits (refs. 94, 236), guinea pigs[283], chickens[94] and man[47,48,357]. This activity is present when serotonin is given either by a general route[47,93,94,236] or by local application (refs. 48, 283, 357). The dose required in rabbits[236] is 10 μg/kg and in rats[93] 0.4 mg/kg where the effect was observed even in heparinized and eviscerated rats[94]. The simple hypothesis that the reduction of bleeding time is the result of the vasoconstrictor activity of serotonin is not supported by experiments which show that there is no direct relation between the vascular and haemostatic effects[93]. The results of the haemostatic and the vasoconstrictor effects following the administration of large doses of serotonin are the formation of intravascular thrombi[317,324], embolization[224], cyanosis[319] and necrosis[246] at the peripheral level.

Thrombosis after serotonin is particularly severe in conditions of atherosclerosis and blood hypercoagulability[96].

According to Bracco *et al.* the haemostatic effects of serotonin are not due to the amine but to some products of its metabolic transformation[46,48-50].

References p. 190

Serotonin and haematological diseases

Considerable efforts have been devoted to find if diseases showing various types of disorders in the coagulation process are correlated with changes in platelet serotonin*. In cases of thrombocytopenia there is frequently a decrease of 5-HT in platelets[17,29,99] which seems related to a decreased 5-HT uptake[344]. It is interesting that in patients with idiopathic thrombocytopenia the bone marrow contains megakaryocytes which produce platelets with a reduced capacity to take up serotonin[206]. Newborn infants show a tendency to haemorrhage and have a low concentration of 5-HT in platelets[165]. However, other authors do not think that patients with haemorrhagic defects show a causal change of 5-HT in platelets[190,356,409]. Patients with pseudo-haemophilia (v. Willebrand-Jürgens syndrome) do not show major changes in serotonin metabolism[29,256].

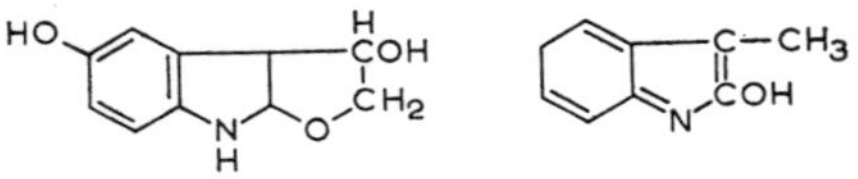

Fig. 24. Serotonin metabolites.

It is interesting that serotonin has been used as a therapeutic agent. Favourable results have been reported after local treatment of haemorrhage[357]. The haemorrhagic symptoms and the reduced capillary resistance present in thrombocytopenia are controlled by general administration of serotonin[1,19,20,47,260,357,371,379], although 5-HT load does not modify the level of serotonin in platelets[28,**]. Also Werlhof's disease is favourably influenced by serotonin[314]. The doses used range from 5 mg (refs. 19, 357) to 10 mg (ref. 260) intravenously 2–4 times a day. According to Ballerini[16] the favourable results depend upon the presence of a critical level of platelets.

Miscellaneous

One method of investigating the importance of serotonin in the control of haemostasis has been the use of reserpine, a depletor of tissue and platelet serotonin[68,74,168,188,201,367,368]. The results have been negative in the sense that animals and patients treated with reserpine, although depleted of their stores of serotonin, do not show apparent changes in coagulation[193], bleeding time[367,368] and capillary resistance of skin vessels[428]. However, in some patients reserpine prolongs the clotting time[222]. A combination of reserpine and anticoagulants induces occult haemorrhage (ref. 213), alteration in haemostatic efficiency[211] and prolongation of bleeding time[212].

After excision of the intestine when 5-HIA excretion is reduced and platelet 5-HT is lowered, there are no changes in the capillary resistance (skin petechiae produced

---

* A review on this problem was recently written by Schmid and Witte[345] and Ballerini *et al.*[17].
** However, recent findings show that an oral repeated load of serotonin (about 20 mg every 6 h for 18 days) results in a marked increase of platelet serotonin[274].

by vacuum)[429]. However, an antagonist of 5-HT, LSD, decreases capillary resistance and haemostasis[31]. Experimental embolism induced in rats does not change the urinary excretion of 5-hydroxyindoleacetic acid[134].

Prednisolone induces after chronic treatment haemorrhagic effects and retards the loss of serotonin in stored platelets[346].

REFERENCES

1   ALLEGRI, A. AND FERRARI, V., *Minerva med.*, 45 (1954) 1660.
2   ANDREWS, W. H. M. AND BUTTERWORTH, K. R., *J. Physiol. (London)*, 141 (1958) 38P.
3   ANGELINO, P. F., CROLLE, G., PELLEGRINI, A., TARTARA, D., PASSAGGIO, A. M. AND PAGANO, P. G., *Presse méd.*, 69 (1961) 728.
4   ANGRIST, A., OKA, M. AND NAKAO, K., *Federation Proc.*, 21 (1962) 95.
5   ASTRÖM, A. AND SAMELIUS, U., *Brit. J. Pharmacol.*, 12 (1957) 410.
6   ATTINGER, E. O., *Am. Heart J.*, 54 (1957) 837.
7   ATTINGER, E. O., *Arch. intern. pharmacodyn.*, 125 (1960) 463.
8   AVIADO, D. M., *Pharmacol. Revs.*, 12 (1960) 159.
9   AVIADO, D. M., *Am. J. Physiol.*, 198 (1960) 1032.
10 BACQ, Z. M., FISCHER, P. AND GHIRETTI, F., *Arch. intern. physiol.*, 60 (1951) 165.
11 BACQ, Z. M. AND RENSON, J., *Arch. intern. pharmacodyn.*, 130 (1961) 464.
12 BAJUSZ, E. AND JASMIN, G., *Rev. can. biol.*, 21 (1962) 51.
13 BALDRIGHI, V. AND FERRARI, V., *Folia Cardiol.*, 14 (1955) 7.
14 BALLERINI, G., *Atti accad. sci. Ferrara*, 32 (1954–55) 63.
15 BALLERINI, G., *Progr. med.*, 11 (1955) 455.
16 BALLERINI, G., *Boll. soc. ital. ematol.*, 4 (1956) 200.
17 BALLERINI, G., BERETTA, P. AND LA PAGLIA, S., *Ann. univ. Ferrara (med. interna)*, 2 (1958) 115.
18 BALLERINI, G. AND CANTELLI, T., *Boll. soc. ital. ematol.*, 3 (1955) 405.
19 BARONI, G., *Gazz. med. ital.*, 115 (1956) 51.
20 BASERGA, A. AND BALLERINI, G., *Schweiz. med. Wochschr.*, 85 (1955) 914.
21 BELESLIN, D. AND VARAGIC, V., *Arch. intern. pharmacodyn.*, 128 (1960) 100.
22 BEN, M., BOXILL, G. C. AND WINBURY, M. M., *Federation Proc.*, 21 (1962) 113.
23 BERGAMASCO, G., BINDA, G. AND AGOSTONI, A., *Atti soc. lombarda sci. med. e biol.*, 11 (1956) 215.
24 BERGOUIGNAN, M. AND SEILHEAN, A., *Presse méd.*, 68 (1960) 2176.
25 BHARGAVA, K. P. AND BORISON, H. L., *J. Pharmacol. Exptl. Therap.*, 119 (1957) 395.
26 BHARGAVA, K. P. AND TANGRI, K. K., *Brit. J. Pharmacol.*, 14 (1959) 411.
27 BHATTACHARYA, B. K., *Arch. intern. pharmacodyn.*, 103 (1955) 357.
28 BIANCHI, P. A., CROSTI, P. F. AND LUCCHELLI, P. E., *Progr. med.*, 16 (1960) 174.
29 BIGELOW, F. S., *J. Lab. Clin. Med.*, 43 (1954) 759.
30 BINET, L. AND BURSTEIN, M., *Compt. rend. soc. biol.*, 149 (1955) 1860.
31 BLAIR, E. L., INGRAM, G. I. C., WAKEFIELD, M. AND ARMITAGE, P., *Nature*, 176 (1955) 563.
32 BOCK, K. D., DENGLER, H., KUHN, H. M. AND MATTHES, K., *Arch. exptl. Pathol. Pharmakol., Naunyn-Schmiedeberg's*, 230 (1957) 257.
33 BOGDANSKI, D. F., WEISSBACH, H. AND UDENFRIEND, S., *J. Pharmacol. Exptl. Therap.*, 122 (1958) 182.
34 BOHR, D. F., WOLF, M. E. AND RONDELL, P. A., *Am. J. Physiol.*, 182 (1955) 311.
35 BORN, G. V. R., *Biochem. J.*, 68 (1958) 695.
36 BORN, G. V. R. AND BRICKNELL, J., *J. Physiol. (London)*, 145 (1958) 8P.
37 BORN, G. V. R. AND BRICKNELL, J., *J. Physiol. (London)*, 147 (1959) 153.
38 BORN, G. V. R. AND GILLSON, R. E., *J. Physiol. (London)*, 137 (1957) 82P.
39 BORN, G. V. R. AND GILLSON, R. E., *J. Physiol. (London)*, 141 (1958) 39P.
40 BORN, G. V. R. AND GILLSON, R. E., *J. Physiol. (London)*, 146 (1959) 472.
41 BORN, G. V. R., INGRAM, G. I. C. AND STACEY, R. S., *Brit. J. Pharmacol.*, 13 (1958) 62.
42 BORST, H. G., BERGLUND, E. AND McGREGOR, M., *J. Clin. Invest.*, 36 (1957) 669.
43 BOUNAMEAUX, Y. AND LECOMTE, J., *Arch. intern. physiol.*, 62 (1954) 543.
44 BRACCO, M. AND CURTI, P. C., *Haematologica (Pavia)*, 37 (1953) 721.
45 BRACCO, M. AND CURTI, P. C., *Experientia*, 10 (1954) 71.
46 BRACCO, M. AND CURTI, P. C., *Boll. soc. ital. ematol.*, 6 (1958) 31.
47 BRACCO, M., CURTI, P. C. AND BALLERINI, G., *Boll. e mem. soc. piemont. chir.*, 24 (1954) 634.
48 BRACCO, M., CURTI, P. C. AND BALLERINI, G., *Farmaco (Pavia), Ed. sci.*, 9 (1954) 318.
49 BRACCO, M., CURTI, P. C., GIULIANO, V. AND PESSINA, G., *Boll. soc. ital. ematol.*, 6 (1958) 26.
50 BRACCO, M., CURTI, P. C. AND PESSINA, G., *Boll. soc. ital. ematol.*, 6 (1958) 27.
51 BRAUN, K. AND STERN, S., *Am. J. Physiol.*, 201 (1961) 369.
52 BREST, A. N. AND MOYER, J. H., *Am. J. Cardiol.*, 9 (1962) 116.

53 BRUN, G. C., *Acta Pharmacol. Toxicol.*, 4 (1948) 251.
54 BRUN, G. C., *Acta Pharmacol. Toxicol.*, 5 (1949) 53.
55 BRUNI, G. AND SERIA, U., *Atti soc. lombarda sci. med. e biol.*, 11 (1956) 2.
56 BUCKINGHAM, S. AND MAYNERT, E. W., *Pharmacologist*, 4 (1962) 169.
57 BULLE, P. H., *Science*, 126 (1957) 24.
58 BUNAG, R. D. AND WALASZEK, E. J., *Federation Proc.*, 20 (1961) 111.
59 BUNAG, R. D. AND WALASZEK, E. J., *Experientia*, 17 (1961) 503.
60 BUNAG, R. D. AND WALASZEK, E. J., *Pharmacologist*, 3 (1961) 63.
61 BUNAG, R. D. AND WALASZEK, E. J., *J. Pharmacol. Exptl. Therap.*, 133 (1961) 52.
62 BUNAG, R. D. AND WALASZEK, E. J., *J. Pharmacol. Exptl. Therap.*, 135 (1962) 151.
63 BUNAG, R. D. AND WALASZEK, E. J., *J. Pharmacol. Exptl. Therap.*, 136 (1962) 59.
64 BUNAG, R. D. AND WALASZEK, E. J., *Arch. intern. pharmacodyn.*, 135 (1962) 142.
65 BUNAG, R. D. AND WALASZEK, E. J., *Japan. J. Pharmacol.*, 11 (1962) 171.
66 BURFORD, H., LEICK, J. AND WALASZEK, E. J., *Arch. intern. pharmacodyn.*, 128 (1960) 39.
67 BURFORD, H. AND WALASZEK, E. J., *Federation Proc.*, 19 (1960) 124.
68 BURKHALTER, A., COHN JR., V. H. AND SHORE, P. A., *Biochem. Pharmacol.*, 3 (1960) 328.
69 CAMPANACCI, D., *Fegato*, 6 (1960) 267.
70 CARBONIN, P. U., SENSI, S., PIFFANELLI, A., GRECO, A. V. AND GAMBASSI, G., *Biochim. e biol. sper.*, 2 (1962–63) 90.
71 CARLIER, J., LEJEUNE-LEDANT, G. AND BARAC, G., *Compt. rend. soc. biol.*, 152 (1958) 1034.
72 CARLIER, J., LEJEUNE-LEDANT, G. AND KEIL, CH., *Med. Exptl.*, 1 (1959) 73.
73 CARLIER, J., LEJEUNE-LEDANT, G. AND KEIL, CH., *Bull. soc. intern. chir.*, 1 (1960) 1.
74 CARLSSON, A., SHORE, P. A. AND BRODIE, B. B., *J. Pharmacol. Exptl. Therap.*, 120 (1957) 334.
75 CARTIER, P., *Semaine hôp.*, 34 (1958) 665.
76 CASELLA, C., FUMAGALLI, B. AND NOLI, S., *Boll. soc. ital. biol. sper.*, 33 (1957) 1550.
77 CASELLA, C., MAGGI, G. C. AND NOLI, S., *Arch. fisiol.*, 59 (1959) 182.
78 CATER, D. B., GARATTINI, S., MARINA, F. AND SILVER, I. A., *Proc. Roy. Soc. (London)*, B, 155 (1961) 136.
79 CERLETTI, A., *II Symposio centro cefalee clin. med. gen. univ. Firenze*, (1960) 27.
80 CERLETTI, A. AND KONZETT, H., *Arch. exptl. Pathol. Pharmakol., Naunyn-Schmiedeberg's*, 228 (1956) 146.
81 CERLETTI, A. AND WEIDMANN, H., *Arch. intern. pharmacodyn.*, 139 (1962) 177.
82 CHESSIN, M., KRAMER, E. R. AND SCOTT, C. C., *J. Pharmacol. Exptl. Therap.*, 119 (1957) 453.
83 CHIANDUSSI, L., GRECO, F., CESANO, L., MURATORI, F., VACCARINO, A., INDOVINA, D. AND CAMPI, L., *Minerva med.*, 53 (1962) 3422.
84 CHIANDUSSI, L., GRECO, F., INDOVINA, D., CESANO, L., VACCARINO, A. AND MURATORI, F., *Proc. Soc. Exptl. Biol. Med.*, 112 (1963) 326.
85 CHIESARA, E., MORTARI, A. AND SIOLI, G., *Boll. soc. ital. biol. sper.*, 37 (1961) 1253.
86 CLINI, V., FERRARA, G. B., GLÄSSER, A. AND MANTEGAZZINI, P., *Arch. ital. sci. farmacol.*, 10 (1960) 110.
87 COHEN, Y. AND BLANCHARD, P., *J. physiol. (Paris)*, 52 (1960) 569.
88 COHEN, Y. AND BLANCHARD, P., *J. physiol. (Paris)*, 53 (1961) 301.
89 COHEN, Y. AND BLANCHARD, P., *J. physiol. (Paris)*, 54 (1962) 316.
90 COLVILLE, K. I. AND LINDSAY, L. A., *Federation Proc.*, 21 (1962) 174.
91 COMROE JR., J. H., *Am. J. Physiol.*, 171 (1952) 715.
92 COMROE JR., J. H., LINGEN, B. V., STROUD, R. C. AND RONCORONI, A., *Am. J. Physiol.*, 173 (1953) 379.
93 CORREALE, P., *Arch. intern. pharmacodyn.*, 97 (1954) 106.
94 CORRELL, J. T., LYTH, L. F., LONG, S. AND VANDERPOEL, J. C., *Am. J. Physiol.*, 169 (1952) 537.
95 COSTA, E. AND APRISON, M. H., *Federation Proc.*, 16 (1957) 25.
96 COSTANTINIDES, P. AND LAWDER, J., *Federation Proc.*, 22 (1963) 251.
97 CREPAX, P. AND INFANTELLINA, F., *Arch. sci. biol. (Bologna)*, 41 (1957) 207.
98 CRONHEIM, G. E. AND GOURZIS, J. T., *J. Pharmacol. Exptl. Therap.*, 130 (1960) 444.
99 CROSTI, P. F., BIANCHI, P. A. AND LUCCHELLI, P. E., *Haematol. Lat.*, 3 (1960) 161.
100 CROSTI, P. F. AND LUCCHELLI, L., *Atti accad. med. lombarda*, 16 (1961) 481.
101 CRUMPTON, C. W., CASTILLO, C. A., ROWE, G. G. AND MAXWELL, G. M., *Ann. N.Y. Acad. Sci.*, 80 (1959) 960.

102 DALESSIO, D. J., *World Neurol.*, 3 (1962) 66.
103 DALESSIO, D. J., *J. Am. Med. Assoc.*, 181 (1962) 318.
103a DALSGAARD-NIELSEN, T., *Praxis*, 49 (1960) 867.
104 DAVIES, D. F., GROPPER, A. L. AND SCHROEDER, H. A., *Circulation*, 3 (1951) 543.
105 DAVIS, R. B., MEEKER, W. R. AND McQUARRIE, D. G., *Surg. Forum*, 10 (1960) 401.
106 DAWES, G. S. AND COMROE JR., J. H., *Physiol. Revs.*, 34 (1954) 167.
107 DAWES, G. S., MOTT, J. C. AND WIDDICOMBE, J. G., *Arch. intern. pharmacodyn.*, 90 (1952) 203.
108 DEUTSCH, E. AND MARTINY, K., *Thromb. Diath. Haematol.*, 2 (1958) 111.
109 DOLCINI, H. A., ZAIDMAN, I. AND GRAY, S. J., *Am. J. Physiol.*, 199 (1960) 1157.
110 DOUGLAS, W. W. AND TOH, C. C., *J. Physiol. (London)*, 117 (1952) 71P.
111 DOUGLAS, W. W. AND TOH, C. C., *J. Physiol. (London)*, 120 (1953) 311.
112 DUKE, H. N., *J. Physiol. (London)*, 135 (1957) 45.
113 DUTEIL, J., *Federation Proc.*, 21 (1962) 128.
114 DUTEIL, J. J. AND AVIADO, D. M., *Circulation Research*, 11 (1962) 466.
115 EBLE, J. N., *Federation Proc.*, 21 (1962) 330.
116 EBLE, J. N., *J. Pharmacol. Exptl. Therap.*, 140 (1963) 243.
117 ELIASSON, R. AND ÅSTRÖM, A., *Acta Pharmacol. Toxicol.*, 11 (1955) 254.
118 ERDÖS, E. G., BAART, N., FOLDES, F. F. AND ZSIGMOND, E. K., *Science*, 126 (1957) 1176.
119 ERSPAMER, V., *Ricerca sci.*, 22 (1952) 694.
120 ERSPAMER, V., *Arch. Internal Med.*, 90 (1952) 505.
121 ERSPAMER, V., *Ciba Foundation Symposium on Hypertension*, Churchill, London, 1954, p. 90.
122 ERSPAMER, V., *Pharmacol. Revs.*, 6 (1954) 425.
123 ERSPAMER, V., *Science*, 121 (1955) 369.
124 ERSPAMER, V., *Arzneimittel-Forsch.*, 8 (1958) 571.
125 ERSPAMER, V. AND GHIRETTI, F., *J. Physiol. (London)*, 115 (1951) 470.
126 EXLEY, K. A., *Brit. J. Pharmacol.*, 12 (1957) 297.
127 FANCHAMPS, A., DOEPFNER, W., WEIDMANN, H. AND CERLETTI, A., *Schweiz. med. Wochschr.*, 90 (1960) 1040.
128 FASTIER, F. N., McDOWALL, M. A. AND WAAL, H., *Brit. J. Pharmacol.*, 14 (1959) 527.
129 FELDBERG, W. AND SMITH, A. N., *Brit. J. Pharmacol.*, 8 (1953) 406.
130 FENICHEL, R. L. AND SEEGERS, W. H., *J. Appl. Physiol.*, 10 (1957) 71.
131 FENICHEL, R. L. AND SEEGERS, W. H., *Am. J. Physiol.*, 181 (1955) 19.
132 FLOREY, E. AND FLOREY, E., *Naturwissenschaften*, 40 (1953) 413.
133 FREYBURGER, W. A., GRAHAM, B. E., RAPPORT, M. M., SEAY, P. H., GOVIER, W. M., SWOAP, O. F. AND VANDER BROOK, M. J., *J. Pharmacol. Exptl. Therap.*, 105 (1952) 80.
134 FRICK, H., *Ann. Med. Exptl. et Biol. Fenniae (Helsinki)*, 36 (1958) 407.
135 FRIEDMAN, A. P. AND LOSIN, S., *Arch. Neurol.*, 4 (1961) 241.
136 GADDUM, J. H., *Pharmacol. Revs.*, 9 (1957) 211.
137 GADDUM, J. H. AND HAMEED, K. A., *Brit. J. Pharmacol.*, 9 (1954) 240.
138 GADDUM, J. H., HEBB, C. O., SILVER, A. AND SWAN, A. A. B., *Quart. J. Exptl. Physiol.*, 38 (1953) 255.
139 GAINTNER, J. R., JACKSON, D. P. AND MAYNERT, E. W., *Bull. Johns Hopkins Hosp.*, 111 (1962) 185.
140 GALLINI, R., *Sett. med.*, 45 (1957) 349.
141 GARATTINI, S., FRESIA, P., MORTARI, A. AND PALMA, V., *Med. Exptl.*, 2 (1960) 252.
142 GARATTINI, S., KATO, R. AND VALZELLI, L., *Experientia*, 16 (1960) 120.
143 GARATTINI, S., LAMESTA, L., MORTARI, A., PALMA, V. AND VALZELLI, L., *J. Pharm. and Pharmacol.*, 13 (1961) 385.
144 GARATTINI, S., LAMESTA, L., MORTARI, A. AND VALZELLI, L., *J. Pharm. and Pharmacol.*, 13 (1961) 548.
145 GARATTINI, S., MORTARI, A., VALSECCHI, A. AND VALZELLI, L., *Nature*, 183 (1959) 1273.
146 GARATTINI, S., NANNI, E. AND PALMA, V., unpublished results.
147 GARCIA DE JALON, P. D., *Arch. inst. farmacol. exptl. (Madrid)*, 11 (1959) 79.
148 GARCIA DE JALON, P. D., GOMEZ ALONZO, B. AND LASTRA SANTOS, L. A., *Arch. inst. farmacol. exptl. (Madrid)*, 11 (1959) 60.
149 GAUTIERI, R. F. AND CIUCHTA, H. P., *J. Pharm. Sci.*, 51 (1962) 55.

149a GERTNER, S. B., PAASONEN, M. K. AND GIARMAN, N. J., *J. Pharmacol. Exptl. Therap.*, 127 (1959) 268.

150 GESSNER, P. K., KHAIRRALLAH, P. A., MCISAAC, W. M. AND PAGE, I. H., *J. Pharmacol. Exptl. Therap.*, 130 (1960) 126.

151 GIBERTINI, G. AND LODI, R., *Chir. e patol. sper.*, 8 (1960) 990.

152 GILBERT, R. P., HINSHAW, L. B., KUIDA, H. AND VISSCHER, M. B., *Am. J. Physiol.*, 194 (1958) 165.

153 GILLESPIE JR., L., TERRY, L. L. AND SJOERDSMA, A., *Am. Heart J.*, 58 (1959) 1.

154 GINZEL, K. H., in G. P. LEWIS (Ed.), *5-Hydroxytryptamine*, Pergamon, London, 1958, p. 131.

155 GINZEL, K. H., *Brit. J. Pharmacol.*, 13 (1958) 250.

156 GINZEL, K. H. AND KOTTEGODA, S. R., *Quart. J. Exptl. Physiol.*, 38 (1953) 225.

157 GINZEL, K. H. AND KOTTEGODA, S. R., *J. Physiol. (London)*, 123 (1954) 277.

158 GLOVER, W. E., GREENFIELD, A. D. M., KIDD, B. S. L. AND WHELAN, R. F., *J. Physiol. (London)*, 140 (1958) 113.

159 GLOVER, W. E., MARSHALL, R. J. AND WHELAN, R. F., *Brit. J. Pharmacol.*, 12 (1957) 498.

160 GOLDBERG, L. I., *Ann. N.Y. Acad. Sci.*, 80 (1959) 639.

161 GOLDBERG, L. I., DA COSTA, F. M. AND OZAKI, M., *Nature*, 188 (1960) 502.

162 GORDON, P., HADDY, F. J. AND LIPTON, M. A., *Science*, 128 (1958) 531.

163 GORDON, P., HADDY, F. AND LIPTON, M., *Federation Proc.*, 18 (1959) 397.

164 GRAHAM, J. R., *New Engl. J. Med.*, 263 (1960) 1273.

165 GRASSO, E., BIANCHI, P., CROSTI, P. F. AND ERMACORA, E., *Panminerva med.*, 3 (1961) 49.

166 GRECO, F. DEL, MASSON, G. M. C. AND CORCORAN, A. C., *Am. J. Physiol.*, 187 (1956) 509.

167 GREEN, J. P. AND NAHUM, L. H., *Circulation Research*, 5 (1957) 634.

168 GREEN, J. P., PAASONEN, M. K. AND GIARMAN, N. J., *Proc. Soc. Exptl. Biol. Med.*, 94 (1957) 428.

169 GREENBERG, M. J., *Brit. J. Pharmacol.*, 15 (1960) 365.

170 GREENBERG, M. J., *Brit. J. Pharmacol.*, 15 (1960) 375.

171 GRETTE, K., *Scand. J. Clin. Lab. Invest.*, 11 (1959) 50.

172 GRIGGS JR., D. M. AND CASE, R. B., *Circulation*, 22 (1960) 758.

173 GROSS, R. AND STANFENBERG, E., *Thromb. Diath. Haemorrhag.*, 2 (1958) 125.

174 GYERMEK, L., *Nature*, 192 (1961) 465.

175 GYERMEK, L., *Pharmacol. Revs.*, 13 (1961) 399.

175a GYERMEK, L. AND BINDLER, E., *J. Pharmacol. Exptl. Therap.*, 135 (1962) 344.

176 GYERMEK, L., LAZAR, G. AND CSAK, Z., *Arch. intern. pharmacodyn.*, 107 (1956) 62.

177 GYERMEK, L. AND POSSEMATO, C., *Med. Exptl.*, 3 (1960) 225.

178 HADDY, F. J., *Minn. Med.*, 41 (1958) 162.

179 HADDY, F. J., FLEISHMAN, M. AND EMANUEL, D. A., *Circulation Research*, 5 (1957) 247.

180 HADDY, F. J., *Angiology*, 11 (1960) 21.

181 HADDY, F. J., GORDON, P. AND EMANUEL, D. A., *Circulation Research*, 7 (1959) 123.

182 HADDY, F. J., MOLNAR, J. I., BORDEN, C. W. AND TEXTER JR., E. C., *Circulation*, 25 (1962) 239.

183 HALE, A. R. AND REED, A. F., *Am. J. Med. Sci.*, 243 (1962) 92.

184 HALMAGY, D. F. J. AND COLEBATCH, H. J. H., *J. Pharmacol. Exptl. Therap.*, 134 (1961) 47.

185 HALPERN, B. N., MORARD, J. C. AND DRUDI-BARACCO, C., *Compt. rend. soc. biol.*, 156 (1962) 773.

186 HANSON, A. S. AND MAGILL, T., *Federation Proc.*, 20 (1961) 122.

187 HANSON, A. S. AND MAGILL, T., *Proc. Soc. Exptl. Biol. Med.*, 109 (1962) 329.

188 HARDISTY, R. M., INGRAM, G. I. C. AND STACEY, R. S., *Experientia*, 12 (1956) 424.

189 HARDISTY, R. M. AND STACEY, R. S., *J. Physiol. (London)*, 130 (1955) 711.

190 HARDISTY, R. M. AND STACEY, R. S., *Brit. J. Haematol.*, 3 (1957) 292.

191 HARRIS, P., FRITTS, H. W. AND COURNAND, A., *Circulation*, 21 (1960) 1134.

192 HAVERBACK, B. J., SJOERDSMA, A. AND TERRY, L. L., *New Engl. J. Med.*, 255 (1956) 270.

193 HAVERBACK, B. J., DUTCHER, T. F., SHORE, P. A., TOMICH, M. S., TERRY, L. L. AND BRODIE, B. B., *New Engl. J. Med.*, 256 (1957) 343.

194 HEDINGER, C. AND LANGEMANN, H., *Schweiz. med. Wochschr.*, 85 (1955) 368.

195 HERTZLER, E. C., *Federation Proc.*, 20 (1961) 317.

195a HERTZLER, E. C., *Brit. J. Pharmacol.*, 17 (1961) 406.

196 HEYCK, H., *Schweiz. med. Wochschr.*, 90 (1960) 203.

197 HEYMANS, C. AND HEUVEL-HEYMANS, G. VAN DEN, *Arch. intern. pharmacodyn.*, 93 (1953) 95.

198 HOLEMANS, R. AND CHO, M. H., *Federation Proc.*, 21 (1962) 63.

199 HOLLANDER, W., MICHAELSON, A. L. AND WILKINS, R. W., *Circulation*, 16 (1957) 246.

200 HOTOVY, R. AND ROESCH, E., *Arch. exptl. Pathol. Pharmakol., Naunyn-Schmiedeberg's*, 232 (1958) 369.

201 HUGHES, F. B. AND BRODIE, B. B., *J. Pharmacol. Exptl. Therap.*, 127 (1959) 96.

202 HUMPHREY, J. H. AND JAQUES, R., *J. Physiol. (London)*, 124 (1954) 305.

203 HUMPHREY, J. H. AND JAQUES, R., *J. Physiol. (London)*, 128 (1955) 9.

204 HURWITZ, R., CAMPBELL, R. W., GORDON, P. AND HADDY, F. J., *J. Pharmacol. Exptl. Therap.*, 133 (1961) 57.

205 HYLAND, J. W., PIEMME, T. E., ALEXANDER, S., HAYNES, F. W., SMITH, G. T. AND DEXTER, L., *Federation Proc.*, 21 (1962) 109.

206 IZAK, G., NELKEN, D. AND GUREVITCH, J., *Blood*, 12 (1957) 507.

207 JACOB, J., in E. ROTHLIN (Ed.), *Neuro-Psychopharmacology*, Vol. 2, Elsevier, Amsterdam, 1961 p. 53.

208 JACOB, J. AND CUGURRA, F., *Arch. intern. pharmacodyn.*, 123 (1960) 362.

209 JACOB, J. AND POITE-BEVIERRE, M., *Arch. intern. pharmacodyn.*, 127 (1960) 11.

210 JÄNKÄLÄ, E. O. AND VIRTAMA, P., *Ann. Med. Exptl. et Biol. Fenniae (Helsinki)*, 40 (1962) 1.

211 JAQUES, L. B., in S. JOHNSON (Ed.), *Blood Platelets*, Little, Brown and Co., Boston, Mass., 1961, p. 135.

212 JAQUES, L. B., *Thromb. Diath. Haemorrhag.*, 7 (1962) 507.

213 JAQUES, L. B. AND FISHER, L. M., *Arch. intern. pharmacodyn.*, 123 (1960) 325.

213a JASMIN, G. AND BAJUSZ, E., *Rev. can. biol.*, 21 (1962) 135.

214 JASMIN, G. AND BOIS, P., *Lab. Invest.*, 9 (1960) 503.

215 JOHANSSON, S. A., *Acta Physiol. Scand.*, 50 (1960) 95.

216 JULIA, M., IGOLEN, J., FELIX, M. AND JACOB, J., *Compt. rend.*, 250 (1960) 1741.

217 KABINS, S. A., MOLINA, C. AND KATZ, L. N., *Am. J. Physiol.*, 197 (1959) 955.

218 KAPILA, K. AND ARORA, R. B., *J. Pharm. Pharmacol.*, 14 (1962) 831.

219 KATO, L. AND GÖZSY, B., *Am. J. Physiol.*, 195 (1958) 66.

220 KAMMANN, A. J., COUSSIO, J. D. AND IZQUIERDO, J. A., *Med. Exptl.*, 6 (1962) 1.

221 KELLER, R., *Experientia*, 14 (1958) 181.

222 KHIRWADKAR, M. G., *Indian J. Med. Sci.*, 9 (1955) 162.

223 KIMBALL, R. W., FRIEDMAN, A. P. AND VALLEJO, E., *Neurology*, 10 (1960) 107.

224 KNISELY, W. H., WALLACE, J. M. AND ADDISON, W., *Federation Proc.*, 17 (1958) 88.

225 KOEHLER, J. A., TSAGARIS, T. J., KUIDA, H. AND HECHT, H. H., *Am. J. Physiol.*, 204 (1963) 987.

226 KOELLA, W. P. AND CZICMAN, J., *Federation Proc.*, 20 (1961) 305.

227 KOELLA, W. P., SMYTHIES, J. R., BULL, D. M. AND LEVY, C. K., *Am. J. Physiol.*, 198 (1960) 205.

228 KOSHTOIANTS, KH. S. AND ROZHA, K., *Acta Physiol. Acad. Sci. Hung.*, 19 (1961) 189.

229 KOTTEGODA, S. R. AND MOTT, J. C., *Brit. J. Pharmacol.*, 10 (1955) 66.

230 KUIDA, H., BROWN, A. M., TORNE, J. L., LANGE, R. L. AND HECHT, H. H., *Federation Proc.*, 20 (1961) 106.

231 KÜNG, H. L. AND SCHINDLER, W., *Experientia*, 15 (1959) 66.

232 KUSCHKE, H. J., IGATA, A. AND FERNEDING, B., *Klin. Wochschr.*, 40 (1962) 322.

233 KWAAN, H. C., LO, R. AND MCFADZEAN, A. J. S., *Clin. Sci.*, 16 (1957) 255.

234 KWAAN, H. C., LO, R. AND MCFADZEAN, A. J. S., *Brit. J. Haematol.*, 4 (1958) 51.

235 LECOMTE, J., *Arch. intern. pharmacodyn.*, 100 (1955) 457.

236 LECOMTE, J., BOUNAMEAUX, Y., FISCHER, P. AND OSTERRIETH, P., *Arch. intern. pharmacodyn.*, 97 (1954) 389.

237 LEE, W. C., SHIN, Y. H. AND SHIDEMAN, F. E., *J. Pharmacol. Exptl. Therap.*, 133 (1961) 180.

238 LEMESSURIER, D. H., SCHWARTZ, C. J. AND WHELAN, R. F., *Brit. J. Pharmacol.*, 14 (1959) 246.

239 LEUSEN, I. AND LACROIX, E., *Arch. intern. physiol. et biochim.*, 67 (1959) 93.

240 LEVY, J. AND MICHEL-BER, E., *Compt. rend.*, 243 (1956) 326.

241 LOUBATIÈRES, A., SASSINE, A. AND MAUCHE, J., *Compt. rend. soc. biol.*, 149 (1953) 1634.

242 LOVENBERG, W., LEVINE, R. J. AND SJOERDSMA, A., *J. Pharmacol. Exptl. Therap.*, 135 (1962) 7.

243 LOVSHIN, L. L., *Diseases of Nervous System*, 24 (1963) 3.

244 LUISADA, A. A., LIU, C. K., JONA, E. AND POLLI, J. F., *Angiology*, 6 (1955) 503.

245 MCCANON, D. M. AND HORVATH, S. M., *Am. J. Physiol.*, 179 (1954) 131.

246 MCDONALD, R. A., ROBBINS, S. L. AND MALLORY, G. K., *A.M.A. Arch. Pathol.*, 65 (1958) 369.

247 MAGALINI, S. I. AND STEFANINI, M., *Proc. Soc. Exptl. Biol. Med.*, 90 (1955) 615.
248 MAGDELAINE, M., DREUX, C., BONVARLET, A. AND LEGER, L., *Presse méd.*, 70 (1962) 7.
249 MAGGI, G. C. AND NOLI, S., *Boll. soc. ital. biol. sper.*, 34 (1958) 1290.
250 MAGISTRETTI, M. AND MANTEGAZZA, P., *Rend. ist. lombardo sci.*, *Pt. I*, 88 (1955) 774.
251 MAGISTRETTI, M. AND VALZELLI, L., *Boll. soc. ital. biol. sper.*, 31 (1955) 1035.
252 MAGISTRETTI, M. AND VALZELLI, L., *Boll. soc. ital. biol. sper.*, 31 (1955) 1038.
253 MAJNO, G. AND PALADE, G. E., *J. Biophys. Biochem. Cytol.*, 11 (1961) 571.
254 MAJNO, G., PALADE, G. E. AND SCHOEFF, G. I., *J. Biophys. Biochem. Cytol.*, 11 (1961) 607.
255 MAJNO, G., SCHOEFF, G. I. AND PALADE, G. E., *Federation Proc.*, 20 (1961) 119.
256 MARCACCI, M., *Sintesi*, 1 (1957) 163.
257 MARIANI, L. AND MORTARI, A., *Atti accad. med. lombarda*, 15 (1960) 210.
258 MARIANI, L., MORTARI, A. AND SIOLI, G., *Mal. cardiovasc.*, 2 (1961) 577.
259 MARSHALL, E. F., STIRLING, G. S., TEVIT, A. C. AND TODRICK, A., *Brit. J. Pharmacol.*, 15 (1960) 35.
260 MARTINETTI, L. AND MARAGNANI, U., *Farmaco (Ed. prat.)*, 10 (1955) 122.
261 MASSON, G. M. C., CORCORA, A. C. AND FRANCO-BROWDER, S., *Am. J. Physiol.*, 195 (1958) 407.
262 MAXWELL, G. M., CASTILLO, C. A., CLIFFORD, J. E., CRUMPTON, C. W. AND ROWE, G. G., *Am. J. Physiol.*, 197 (1959) 736.
263 MAXWELL, G. M., CASTILLO, C. A., WHITE JR., D. H. AND CRUMPTON, C. W., *J. Lab. Clin. Med.*, 50 (1957) 930.
264 MAZZELLA, H., *Arch. intern. pharmacodyn.*, 116 (1958) 37.
265 MCCAWLEY, E. L., LEVEQUE, P. E. AND DICK, H. L. H., *J. Pharmacol. Exptl. Therap.*, 106 (1952) 406.
266 MCCOY WRIGHT, A., MOORHEAD, M. AND WELSH, J. H., *Brit. J. Pharmacol.*, 18 (1962) 440.
267 MCCUBBIN, J. W., GREEN, J. H., SALMOIRAGHI, G. C. AND PAGE, I. H., *J. Pharmacol. Exptl. Therap.*, 116 (1956) 191.
268 MCCUBBIN, J. W., KANEKO, Y. AND PAGE, I. H., *Circulation Research*, 8 (1960) 849.
269 MCCUBBIN, J., KANEKO, Y. AND PAGE, I. H., *Circulation Research*, 11 (1962) 74.
270 MCGAFF, C. J. AND MILNOR, W. R., *Federation Proc.*, 19 (1960) 96.
271 MCGAFF, C. J. AND MILNOR, W. R., *Am. J. Physiol.*, 202 (1962) 957.
271a MCINTYRE, A. R., WILLIAMS, D. E. AND HUMOLLER, F. L., *Federation Proc.*, 19 (1960) No. 1, part 1.
272 MEIER, R., TRIPOD, J. AND STUDER, A., *Arch. intern. pharmacodyn.*, 117 (1958) 185.
273 MEIER, R., TRIPOD, J. AND WIRZ, E., *Arch. intern. pharmacodyn.*, 109 (1957) 55.
274 MELMON, K. AND SJOERDSMA, A., *Lancet*, 285 (1963) 316.
275 MEYTHALER, CHR., SCHMID, E., BLUMBERG, J., ZICHA, L. AND WITTE, S., *Med. Exptl.*, 7 (1962) 232.
276 MILNE, W. L. AND COHN, S. H., *Am. J. Physiol.*, 189 (1957) 470.
277 MINELLI, R., *Boll. soc. ital. biol. sper.*, 37 (1961) 345.
278 MOTT, J. C. AND PAINTAL, A. S., *Brit. J. Pharmacol.*, 8 (1953) 238.
279 NAESS, K. AND SKRAMSTAD, K. H., *Acta Pharmacol. Toxicol.*, 10 (1954) 178.
280 NAHAS, G. G., *J. physiol. (Paris)*, 50 (1958) 425.
281 NAHAS, G. G. AND MCDONALD, I., *Am. J. Physiol.*, 196 (1959) 1045.
282 NATOFF, I. L. AND LOCKETT, M. F., *J. Pharm. and Pharmacol.*, 9 (1957) 464.
283 NIAUSSAT, P., LABORIT, H. AND BARON, C., *Compt. rend. soc. biol.*, 152 (1958) 1100.
284 OKA, M. AND LEPPÄNEN, V. V. E., *Acta Med. Scand.*, 166 (1960) 297.
285 ORVIS, H. H., TAMAGNA, I. G. AND THOMAS, R. E., *Am. J. Med. Sci.*, 238 (1959) 118.
286 OSBORNE, M., DETAR, R. AND GABEL, L., *Federation Proc.*, 21 (1962) 324.
287 OSTFELD, A. M. AND WOLFF, H. G., *Med. Clin. N. Am.*, 42 (1958) 1497.
288 OUTSCHOORN, A. S. AND JACOB, J., *Brit. J. Pharmacol.*, 15 (1960) 131.
289 PAASONEN, M. K., *Biochem. Pharmacol.*, 8 (1961) 301.
290 PAASONEN, M. K., *Biochem. Pharmacol.*, 8 (1961) 241.
291 PAASONEN, M. K., AND PLETSCHER, A., *Experientia*, 16 (1960) 30.
292 PAGE, I. H., *Am. Heart J.*, 28 (1948) 161.
293 PAGE, I. H., *J. Pharmacol. Exptl. Therap.*, 105 (1952) 58.
294 PAGE, I. H., *Physiol. Revs.*, 38 (1958) 277.
295 PAGE, I. H., in G. P. LEWIS (Ed.), *5-Hydroxytryptamine*, Pergamon, London, 1958, p. 93.

[296] PAGE, I. H. AND McCUBBIN, J. W., *Circulation Research*, 1 (1953) 354.

[297] PAGE, I. H. AND McCUBBIN, J. W., *Am. J. Physiol.*, 173 (1953) 411.

[298] PAGE, I. H. AND McCUBBIN, J. W., *Am. J. Physiol.*, 174 (1953) 436.

[299] PAGE, I. H. AND McCUBBIN, J. W., *Am. J. Physiol.*, 184 (1956) 265.

[300] PAGE, I. H., WOLFORD, R. W. AND CORCORAN, A. C., *Arch. intern. pharmacodyn.*, 119 (1959) 214.

[301] PALADE, G. E., *Circulation*, 24 (1961) 368.

[302] PARKS, V. J., SANDISON, A. G., SKINNER, S. L. AND WHELAN, R. F., *J. Physiol. (London)*, 151 (1960) 342.

[303] PEART, W. S., ANDREWS, T. M. AND ROBERTSON, J. I. S., *Lancet*, 280 (1961) 577.

[304] PEPEY, G. AND GIARMAN, N. J., *J. Gen. Physiol.*, 45 (1962) 575.

[305] PETERFALVI, M. AND JEQUIER, R., *Arch. intern. pharmacodyn.*, 124 (1960) 237.

[306] PLUMMER, A. J., BARRETT, W. E., MAXWELL, R. A., FINOCCHIARO, D., LUCAS, R. A. AND EARL, A. E., *Arch. intern. pharmacodyn.*, 119 (1959) 245.

[307] PLUMMER, A. J., EARL, A., SCHNEIDER, J. A., TRAPOLD, J. AND BARRETT, W., *Ann. N.Y. Acad. Sci.*, 59 (1955) 8.

[308] POUMAILLOUX, M., *Arch. Maladies Coeur Vaisseaux*, 52 (1959) 721.

[309] POWELL, C. E., SWANSON, E. E. AND CHEN, K. K., *J. Am. Pharm. Assoc.*, 44 (1955) 399.

[310] PRADHAM, S. N., ACHINSTEIN, B. AND SHEAR, M. J., *Arch. intern. pharmacodyn.*, 120 (1959) 1.

[311] RAPPORT, M. M., GREEN, A. AND PAGE, I. H., *Science*, 108 (1948) 329.

[312] RAPPORT, M. M., GREEN, A. AND PAGE, I. H., *J. Biol. Chem.*, 174 (1948) 735.

[313] RAPPORT, M. M., GREEN, A. AND PAGE, I. H., *J. Biol. Chem.*, 176 (1948) 1243.

[314] RAVETTA, A., *Minn. Med.*, 46 (1955) 1193.

[315] REID, G., *Australian J. Exptl. Biol. Med. Sci.*, 29 (1952) 101.

[316] REID, G., *J. Physiol. (London)*, 118 (1952) 435.

[317] REID, G. AND RAND, M., *Nature*, 169 (1952) 801.

[318] ROBERTSON, J. I. S. AND ANDREWS, T. M., *Lancet*, 280 (1961) 570.

[319] RODDIE, I. C., SHEPHERD, J. T. AND WHELAN, R. F., *Brit. J. Pharmacol.*, 10 (1955) 445.

[320] RONDELL, P. A., PALMER, L. E. AND BOHR, D. F., *Federation Proc.*, 16 (1957) 109.

[321] ROOKE, E. D., RUSHTON, J. G. AND PETERS, G. A., *Proc. Staff Meetings Mayo Clinic*, 37 (1962) 433.

[322] ROSE, J. C., *Circulation*, 16 (1957) 931.

[323] ROSE, J. C. AND LAZAR, E. J., *Circulation Research*, 6 (1958) 283.

[324] ROSSI, P., MASCAGNI, M. T. AND SPANÒ, G., *Boll. soc. ital. cardiol.*, 2 (1957) 249.

[325] ROSSI, P., MOTOLESE, M. AND GIACOMELLI, F., *Acta Tertii Europaei Cordis Scientia Conventus, Roma*, A (1960) 27.

[326] ROSSI, P., MOTOLESE, M. AND ZAMBONI, L., *Am. Heart J.*, 58 (1959) 715.

[327] ROSSI, P., MOTOLESE, M. AND ZAMBONI, L., *Arch. ital. anat. istol. patol.*, 33 (1959) 374.

[328] ROSSI, P. AND ZAMBONI, L., *Nature*, 181 (1958) 1216.

[329] ROTHLIN, E., in S. GARATTINI AND V. GHETTI (Eds.), *Psychotropic Drugs*, Elsevier, Amsterdam, 1957, p. 36.

[330] ROTHSCHILD, A. M. AND GARCIA LIMA, E., *Experientia*, 19 (1963) 19.

[331] ROWE, G. G., SKODA, A., CASTILLO, C. A., KYLE, J. C., LEICHT, T. R. AND CRUMPTON, C. W., *Clin. Pharmacol. Therap.*, 4 (1963) 467.

[332] RUBIN, A. A., BEAUREGARD, S. C. AND FREEMAN, B. G., *Federation Proc.*, 19 (1960) 101.

[333] RUDOLPH, A. M. AND AULD, P. A. M., *Am. J. Physiol.*, 198 (1960) 864.

[334] RUDOLPH, A. M., KURLAND, M. D., AULD, P. A. M. AND PAUL, M. H., *Am. J. Physiol.*, 197 (1959) 617.

[335] RUDOLPH, A. M. AND PAUL, M. H., *Am. J. Physiol.*, 189 (1957) 263.

[336] RUDOLPH, A. M. AND PAUL, M. H., *Federation Proc.*, 16 (1957) 110.

[337] SABELLI, H. C. AND SINAY, J., *Arzneimittel-Forsch.*, 10 (1960) 935.

[338] SALMOIRAGHI, G. C. AND McCUBBIN, J. W., *Circulation Research*, 2 (1954) 280.

[339] SALMOIRAGHI, G. C., McCUBBIN, J. W. AND PAGE, I. H., *J. Pharmacol. Exptl. Therap.*, 119 (1957) 240.

[340] SALMOIRAGHI, G. C., PAGE, I. H. AND McCUBBIN, J. W., *J. Pharmacol. Exptl. Therap.*, 118 (1956) 477.

[341] SAVINI, E. C., *Brit. J. Pharmacol.*, 11 (1956) 313.

342   SCHEIFFARTH, F. VON, HESSE, R., ZICHA, L. AND SCHMID, E., *Intern. Arch. Allergy Appl. Immunol.*, 17 (1960) 289.
343   SCHIEVELBEIN, H. AND WERLE, E., *Psychopharmacologia*, 3 (1962) 35.
344   SCHMID, E., ORDNUNG, W., SCHÖN, H., WITTE, S. AND MEYTHALER, CH., *Blut*, 7 (1961) 129.
345   SCHMID, E. AND WITTE, S., *Med. Welt*, 17 (1962) 918.
346   SCHMID, E., WITTE, S., PFANNMÜLLER, B. AND SCHRICHKER, K. TH., *Thromb. Diath. Haemorrhag.*, 7 (1962) 258.
347   SCHMITT, H. AND SCHMITT, H., *Compt. rend. soc. biol.*, 153 (1959) 917.
348   SCHMITT, H. AND SCHMITT, H., *Arch. intern. pharmacodyn.*, 120 (1959) 251.
349   SCHMITT, H. AND SCHMITT, H., *Arch. intern. pharmacodyn.*, 125 (1960) 30.
350   SCHNEIDER, J. A. AND RINEHART, R. K., *Arch. intern. pharmacodyn.*, 105 (1956) 253.
351   SCHNEIDER, J. A. AND YONKMAN, F. F., *Am. J. Physiol.*, 174 (1953) 127.
352   SCHNEIDER, J. A. AND YONKMAN, F. F., *J. Pharmacol. Exptl. Therap.*, 111 (1954) 84.
353   SCHOFIELD, B. M. AND WALKER, J. M., *J. Physiol. (London)*, 122 (1953) 489.
354   SCHÖNE, H. H. AND LINDNER, E., *Arzneimittel-Forsch.*, 10 (1960) 583.
355   SCHRICKER, K. TH., SCHMID, E., MEYTHALER, CH. AND WITTE, S., *Bibliotheca Haematol.*, 12 (1961) 351.
356   SCHULLMAN, N. R., PLITMAN, G. I. AND UDENFRIEND, S., *Clin. Research Proc.*, 4 (1956) 27.
357   SERRAVALLI, S., *Boll. soc. ital. ematol.*, 2 (1954) 335.
358   SETNIKAR, I. AND RAVASI, M., *Nature*, 183 (1959) 898.
359   SHARMAN, D. F. AND SULLIVAN, F. M., *Nature*, 177 (1956) 332.
360   SHAVEL JR., J., STRANDTMANN, M. VON, AND COHEN, M. P., *J. Am. Chem. Soc.*, 84 (1962) 881.
361   SHAW, E. AND WOOLLEY, D. W., *J. Pharmacol. Exptl. Therap.*, 111 (1954) 43.
362   SHAW, E. AND WOOLLEY, D. W., *Proc. Soc. Exptl. Biol. Med.*, 93 (1956) 217.
363   SHAW, E. AND WOOLLEY, D. W., *J. Pharmacol. Exptl. Therap.*, 116 (1956) 164.
364   SHAW, E. AND WOOLLEY, D. W., *Proc. Soc. Exptl. Biol. Med.*, 96 (1957) 439.
365   SHEPHERD, J. T., DONALD, D. E., LINDER, E. AND SWAN, H. J. C., *Am. J. Physiol.*, 197 (1959) 963.
366   SHIM, W. K. T., POLLACK, E. L. AND DRAPANAS, T., *Am. J. Physiol.*, 201 (1961) 81.
367   SHORE, P. A., PLETSCHER, A. AND BRODIE, B. B., *J. Pharmacol. Exptl. Therap.*, 116 (1956) 51.
368   SHORE, P. A., PLETSCHER, A., TOMICH, E. G., KUNTZMAN, R. AND BRODIE, B. B., *J. Pharmacol. Exptl. Therap.*, 117 (1956) 232.
369   SICUTERI, F., *Intern. Arch. Allergy Appl. Immunol.*, 15 (1959) 300.
370   SICUTERI, F., MICHELACCI, S. AND FRANCHI, G., *Intern. Arch. Allergy Appl. Immunol.*, 15 (1959) 291.
371   SIEGENTHALER, W., *Schweiz. Med. Wochschr.*, 86 (1956) 1443.
372   SINHA, Y. K. AND WEST, G. B., *J. Pharm. and Pharmacol.*, 5 (1953) 370.
373   SJOERDSMA, A., GILLESPIE, L. AND UDENFRIEND, S., *Ann. N.Y. Acad. Sci.*, 80 (1959) 969.
374   SKINNER, S. L. AND WHELAN, R. F., *J. Physiol. (London)*, 162 (1962) 35.
375   SMITH, G. AND SMITH, A. N., *Surg. Gynecol. Obstet.*, 101 (1955) 691.
376   SPIES, T. D. AND STONE, R. E., *J. Am. Med. Assoc.*, 150 (1952) 1599.
377   SPINAZZOLA, A. J. AND SHERROD, T. R., *J. Pharmacol. Exptl. Therap.*, 113 (1955) 49.
378   STACEY, R. S., *Brit. J. Pharmacol.*, 16 (1961) 284.
379   STEFANINI, M. AND MAGALINI, S. I., *Arch. Internal Med.*, 98 (1956) 23.
380   STEINER, F. A. AND HEDINGER, C., *Experientia*, 12 (1956) 109.
381   STEINER, F. A., SIEBENMANN, R. E., SANDRI, C. AND HEDINGER, C., *Experientia*, 13 (1957) 500.
382   STERN, S. AND BRAUN, K., *Bull. Research Council Israel, Sect. E*, (1961) 175.
383   STONE, C. A., WENGER, H. C., LUDDEN, C. T., STAVORSKI, J. M. AND RISS, A. C., *J. Pharmacol. Exptl. Therap.*, 131 (1961) 73.
384   STROBEL, G. E., MCINTOSH, J. AND KULKA, J. P., *Federation Proc.*, 20 (1961) 118.
385   STUDER, A. AND TRIPOD, J., *Helv. Physiol. et Pharmacol. Acta*, 18 (1960) 384.
386   TAKACS, L. AND VAJDA, V., *Am. J. Physiol.*, 204 (1963) 301.
387   TAKASAKI, K., *Federation Proc.*, 20 (1961) 111.
388   TAKENAKA, F., *Japan. J. Pharmacol.*, 9 (1959) 55.
389   TAYLOR, R. D., PAGE, I. H. AND CORCORAN, A. C., *Arch. Internal Med.*, 88 (1951) 1.
390   THIEBLOT, L., DUCHÊNE-MARULLAZ, P. AND BERTHELAY, J., *Compt. rend. soc. biol.*, 151 (1957) 34.
391   THORSON, A., BIÖRCK, G., BJÖRKMAN, G. AND WALDENSTRÖM, J., *Am. Heart J.*, 47 (1954) 795.
392   TOH, C. C., *J. Physiol. (London)*, 126 (1954) 248.
393   TRABUCCHI, E., *Atti accad. med. lombarda*, 12 (1957) 18.

394   TRENDELENBURG, U., *J. Pharmacol. Exptl. Therap.*, 130 (1960) 450.
395   TRIPOD, J., in S. GARATTINI AND V. GHETTI (Eds.), *Psychotropic Drugs*, Elsevier, Amsterdam, 1959, p. 437.
396   TRIPOD, J. AND BEIN, H. J., *Helv. Physiol. et Pharmacol. Acta*, 18 (1960) 394.
397   TRIPOD, J., STUDER, A. AND MEIER, R., *Arch. intern. pharmacodyn.*, 126 (1960) 126.
398   TSITOURIS, G., SANDBERG, H., BELLET, S., FEINBERG, L. J. AND SCHRAEDER, J., *J. Lab. Clin. Med.*, 59 (1962) 25.
399   VACEK, L., *Arch. intern. pharmacodyn.*, 124 (1960) 461.
400   VANE, J. R., COLLIER, H. O. J., CORNE, S. J., MARLEY, E. AND BRADLEY, P. B., *Nature*, 194 (1962) 486.
401   VIRTAMA, P. AND JÄNKÄLÄ, E., *Angiology*, 12 (1961) 77.
402   VIRTAMA, P. AND JÄNKÄLÄ, E., *Angiology*, 12 (1961) 372.
403   VITOLO, E., OBBIASSI, M., ZOCCHE, G. P. AND SCOTTI, G. C., *Brit. Heart J.*, 24 (1962) 22.
404   WALASZEK, E. J. AND CHAPMAN, J. E., *Federation Proc.*, 20 (1961) 314.
405   WALASZEK, E. J. AND CHAPMAN, J. E., *J. Pharmacol. Exptl. Therap.*, 137 (1962) 285.
406   WEIDMANN, H. AND CERLETTI, A., *Arch. intern. pharmacodyn.*, 111 (1957) 98.
407   WEIDMANN, H. AND CERLETTI, A., *Helv. Physiol. et Pharmacol. Acta*, 19 (1961) C 119.
408   WEIDMANN, H. AND CERLETTI, A., *Biochem. Pharmacol.*, 8 (1961) 165.
409   WEINER, M. AND UDENFRIEND, S., *Circulation*, 15 (1957) 353.
410   WELSH, J. H., *Arch. exptl. Pathol. Pharmakol., Naunyn-Schmiedeberg's*, 219 (1953) 23.
411   WELSH, J. H., *Nature*, 173 (1954) 955.
412   WELSH, J. H., *J. Marine Biol. Assoc. United Kingdom*, 35 (1956) 193.
413   WELSH, J. AND MCCOY, A. C., *Science*, 125 (1957) 348.
414   WESTERMANN, E., *Arch. exptl. Pathol. Pharmakol., Naunyn-Schmiedeberg's*, 233 (1958) 518.
415   WHIPPLE, H. E. (Ed.), *New Reflections on Monoamineoxidase Inhibition. Ann. N.Y. Acad. Sci.*, 107 (1963) 809–1158.
416   WHITELOCK, O. V. S. (Ed.), *Amine Oxidase Inhibitors. Ann. N.Y. Acad. Sci.*, 80 (1959) 551–1045.
417   WILKINS, R. W., *New Engl. J. Med.*, 255 (1956) 115.
418   WILKINS, R. W. AND HOLLANDER, W., *Circulation*, 14 (1956) 1018.
419   WIRTH, W., GÖSSWALD, R. AND VATER, W., *Arch. intern. pharmacodyn.*, 123 (1959) 78.
420   WOOLLEY, D. W., *The Biochemical Bases of Psychoses*, Wiley, New York, 1962, p. 96.
421   WOOLLEY, D. W. AND SHAW, E., *J. Pharmacol. Exptl. Therap.*, 108 (1953) 87.
422   WOOLLEY, D. W. AND SHAW, E., *J. Biol. Chem.*, 203 (1953) 69.
423   WOOLLEY, D. W. AND SHAW, E., *J. Am. Chem. Soc.*, 74 (1953) 2948.
424   WOOLLEY, D. W. AND SHAW, E., *J. Pharmacol. Exptl. Therap.*, 121 (1957) 13.
425   WOOLLEY, D. W. AND SHAW, E., *Nature*, 194 (1962) 486.
426   YOUNG JR., R. C., NAGANO, H., VAUGHAN, T. R. AND STAUB, N. C., *Federation Proc.*, 21 (1962) 442.
427   ZAMBONI, P., *Arch. sci. biol. (Bologna)*, 40 (1956) 317.
428   ZAMBONI, P., *Arch. intern. pharmacodyn.*, 113 (1957) 203.
429   ZAMBONI, P., BERTACCINI, G. AND NOBILI, M. B., *Arch. ital. sci. farmacol.*, 9 (1959) 394.
430   ZAMBONI, P. AND CARLEVARO, G. C., *Boll. soc. ital. biol. sper.*, 30 (1954) 1011.
431   ZBINDEN, G., *Am. Heart J.*, 60 (1960) 450.
432   ZSIGMOND, E. K., FOLDES, F. F. AND FOLDES, V. M., *J. Neurochem.*, 8 (1961) 72.
433   ZUCKER, M. B. AND BORRELLI, J., *Appl. Physiol.*, 7 (1954–55) 425.
434   ZUCKER, M. B., FRIEDMAN, B. K. AND RAPPORT, M. M., *Proc. Soc. Exptl. Biol. Med.*, 85 (1954) 282.

Chapter 10

SEROTONIN AND
CENTRAL NERVOUS SYSTEM

This is a fascinating and controversial field, in which there are sharp contrasts between enthusiastic support and negative criticism. Certainly the discovery of the presence of serotonin in the brain, and particularly the knowledge about the effects of psychotropic drugs on serotonin, gave a strong impetus to the investigation of the chemistry of the brain. Furthermore, studies about serotonin synthesis and metabolism in the brain have been the source of many new drugs, some of which have proved to have therapeutic significance.

It is interesting that the original ideas about the importance of serotonin for the central nervous functions arose from the observation that a potent hallucinogenic agent, namely lysergic acid diethylamide (LSD), was an antagonist[222] or an antimetabolite[604] of serotonin, whereas nowadays the causal relationship between hallucinations by LSD and serotonin is strongly doubted.

It would be impossible to describe here all the problems of biochemistry, physiology, pathology and pharmacology of the central nervous system in which serotonin has been implicated in some way. The reader will find additional information in specialized reviews[81,116,191,219,244,256,492,585,588].

BIOCHEMISTRY OF SEROTONIN IN THE BRAIN

Distribution

The presence of serotonin in the brain was established about ten years ago by various investigators using biological and chemical methods of determination (refs. 9, 64, 148, 564, 574).

Serotonin is unevenly distributed in the mammalian brain, and is concentrated mainly in certain areas[65,224]. Thus, portions of the brain such as the hypothalamus and midbrain contain relatively large amounts of serotonin, while low concentrations are present in the cortex and cerebellum. A localization in the glial cells has been suggested[190,354], but glioma and oligodendroglioma do not contain serotonin[321]. The microscopic localization of serotonin in the brain has been studied by means of a histochemical technique[122]. Its distribution does not always closely follow the presence of other amines[9,120,148,158,344,488,580]. A typical example is the high content of serotonin[408] and the low level of noradrenaline[344] in the limbic system.

A map (with references to the pertinent literature) of the concentration of 5-HT and other substances in various parts of the brain of several animal species is reported in Appendix V (see page 356).

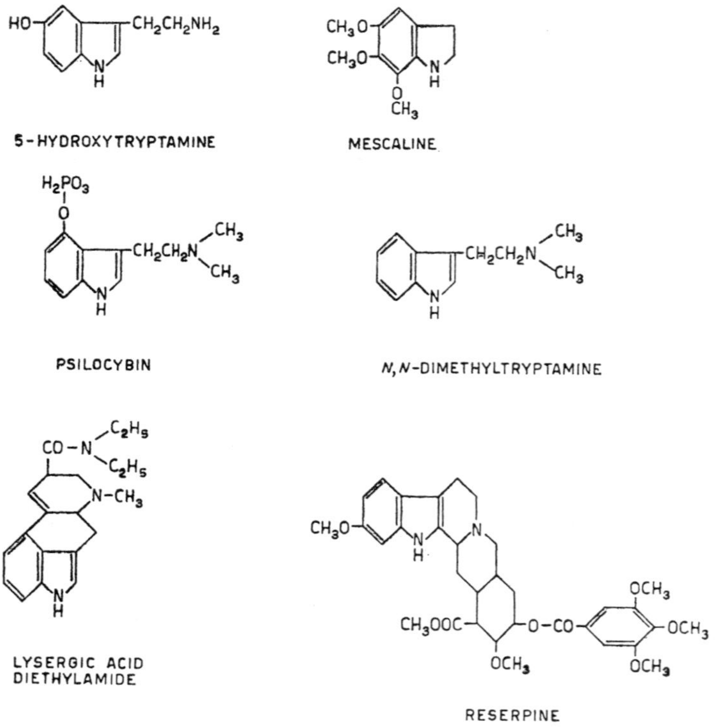

5-HYDROXYTRYPTAMINE

MESCALINE

PSILOCYBIN

N,N-DIMETHYLTRYPTAMINE

LYSERGIC ACID DIETHYLAMIDE

RESERPINE

Fig. 26. 5-Hydroxytryptamine and other structurally related compounds showing central effects

TABLE 32

DISTRIBUTION OF SEROTONIN IN SOME AREAS OF THE HUMAN BRAIN (IN $\mu g/g$)

Human brain	Bertler (ref. 55)*	Costa and Aprison (ref. 151)**	Garattini and Valzelli (ref. 244)*
Corpora quadrigemina	—	0.63	0.70
Hypothalamus	0.37	0.81	0.78
Nucleus caudatus	0.27	0.44	0.30
Pons	0.19	0.7	0.29
Putamen	0.23	0.55	0.42
Thalamus	0.23	0.62	0.37
Substantia nigra	—	1.42	0.86
Number of determinations	4	6	8
Time after death (h)	2	1–6	24

* Spectrofluorimetric method.
** Bioassay.

It is interesting that the pattern of distribution of serotonin in the human brain is similar to that of other mammals. The values for some selected areas are reported in Table 32, which compares figures obtained by bioassay and by spectrofluorimetric determinations[55,151,244]. Daily rhythmical changes in the mouse[7] and hamster [376a] brain, but not in rats[454], have been described.

In newborn animals there is a low concentration of brain serotonin[324,326].

It should be emphasized, however, that the concentration of serotonin in the brain is only a static aspect of the problem. It is more important to know the dynamic aspects involving synthesis, transport, storage, inactivation and turnover of the amine.

Synthesis

General information about the enzymes involved in the synthesis of serotonin is given in Chapter 3. An important question is to establish whether serotonin is completely synthesized in the brain, or does the brain depend for its serotonin supply

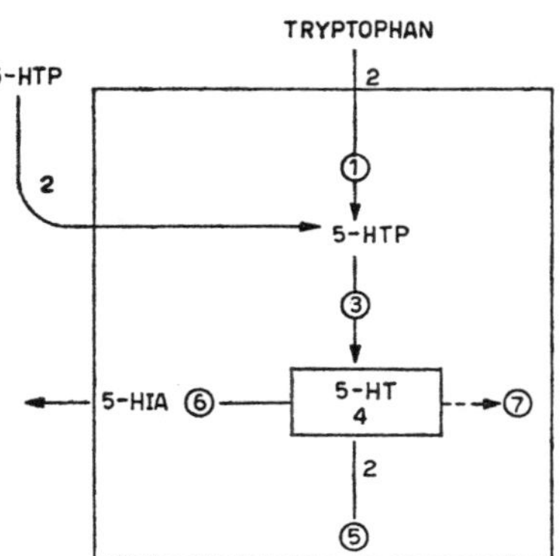

Fig. 27. Schematic diagram of serotonin synthesis and metabolism in the nervous system.

1. Tryptophan-5-hydroxylase (particulate fraction)
2. Active transport
3. 5-HTP decarboxylase (cytoplasm)
4. Mobile pool of serotonin
5. Reserve pool (granules)
6. Monoamine oxidase (mitochondria)
7. Possible other metabolic routes

upon a precursor formed in other parts of the body. There is little doubt that the amount of serotonin in the brain depends on the availability of tryptophan[569]. Increasing the intake of tryptophan in the diet tends to be followed by an increase in the level of brain serotonin[270,587], whereas animals with a diet deficient in this amino acid show a reduced level of brain serotonin[227–229,612]. A good correlation between the level of plasma tryptophan and the concentration of brain serotonin has been demonstrated[162,*]. Tryptophan enters the brain as a result of active transport[278]. Other amino acids compete for the same mechanism of transport[397,496]. Tryptophan has to be hydroxylated in the 5-position through the activity of an enzyme which has not yet been well characterized. It should be recalled that tryptamine is not hydroxylated in the 5-position[570].

---

* Inanition does not decrease brain serotonin[226].

TRYPTOPHAN-5-HYDROXYLASE

This enzyme has recently been found by Cooper[146] in a particulate fraction of the intestinal mucosa and kidney.

Also, Freedland et al.[216] found a liver enzyme which hydroxylates tryptophan and which requires DPN and oxygen. The liver enzyme seems to be different from that of the intestine, because the latter is active anaerobically and does not require DPN, but does require Cu^{2+} and ascorbic acid[146,*]. However, other authors[457] questioned the physiological significance of this enzyme and have shown that the liver enzyme is in fact similar to phenylalanine hydroxylase[217,457]. Further support for these findings was added by Chari-Briton[139], although this author found that in certain experimental conditions the two enzymes behave differently.

Tryptophan-5-hydroxylase has not yet been detected in brain tissue. Indirect evidence, however, suggests that a complete synthesis of serotonin from tryptophan occurs in the brain. For instance, an intracerebral injection of labelled tryptophan permits the presence of labelled 5-hydroxyindoles to be demonstrated when mono-amine oxidase is blocked. No labelled 5-HT appears in the brain when labelled tryptophan is given intraperitoneally[230,231]. Also, unpublished data from our laboratories show an increase of 5-hydroxyindoles after intracerebral administration of tryptophan.

The complete removal of the gastrointestinal tract decreases 5-HT in serum, spleen and lung, but has little effect on rat brain 5-HT[53]. A good correlation exists between the concentrations of tryptophan and serotonin in various parts of the brain[446].

5-HYDROXYTRYPTOPHAN DECARBOXYLASE

There is now little doubt that 5-HTP and DOPA are decarboxylated by the same enzyme[57,204,595,609].

This conclusion is strongly supported by work showing that the ratio 5-HTP/DOPA decarboxylase activity in various brain regions is constant[344]. The decarboxylase is localized mostly in the non-particulate fraction of the cells[255], and since the enzyme is widely present, 5-hydroxytryptophan has never been detected in the nervous tissue**.

5-Hydroxytryptophan decarboxylase can be found in the brain of various animal species including rats[250,325,445], guinea pigs[306,325,445], pigeons[21a,22], mice[445], rabbits (refs. 57, 445), hogs[57], dogs[158,567], and man[2,481].

The distribution of 5-HTP decarboxylase in the various parts of the brain is reported in Appendix V. A direct correlation between serotonin content and decarboxylase activity is quite apparent[158,567]. It is seen that the highest values of the enzyme are found in such areas as the nucleus caudatus and hypothalamus, while the lowest values are those in the cerebral hemispheres and cerebellum[57].

---

* Recent observations suggest that α-methyldopa and α-propyldopacetamide are strong inhibitors of tryptophan hydroxylase[106a,467a].

** In mouse mastocytoma 5-HTP is stored in special granules[283].

The administration of 5-hydroxytryptophan results in an increase of brain serotonin[67,156,220,574,575]. The decarboxylase activity is found to continue in the brain after death[250,435].

Transport

As already reported, tryptophan is taken up in the brain through active transport (ref. 278). For 5-hydroxytryptophan also, a similar transport mechanism has been described[497,499]. This could explain the rapid formation of brain serotonin when 5-hydroxytryptophan is given by general route[574]. The 5-HTP transport in rat brain slices is enhanced by oxygen and glucose, but it is reduced by dinitrophenol and it requires optimal pH and temperature conditions[496]. The various parts of dog brain show different capacities for concentrating 5-HTP, the caudate nucleus, amygdala and hippocampus being the most active[499,534]. However, there is no direct relationship between 5-HTP uptake and 5-HTP decarboxylase or serotonin level in the various parts of the brain. If serotonin is not completely synthesized in the brain, the transport of 5-HTP could be a mechanism of physiological importance in maintaining the levels of brain serotonin.

Such a transport mechanism does not exist for serotonin because, as is well known, the amine does not cross the blood-brain barrier[33,152,157,236,514,531,574,605]. However, a system to concentrate serotonin in specialized granules inside the cells has been demonstrated.

Brain slices, therefore, differ from platelets or mast cells in their ability to concentrate serotonin[496].

Storage

Serotonin is stored in the central nervous system in a form which is proof against the effect of monoamine oxidase, an enzyme present in all portions of the brain[284]. This situation is usually referred to in terms of "bound" and "free" serotonin. The serotonin in the "bound" form does not act on the receptor sites, while the "free" serotonin is subjected to uptake from specialized granules, to inactivation by monoamine oxidase and to interaction with the receptor sites. The term "bound" means only an association with a particulate fraction, although some types of complexes have been described between serotonin and ATP[72,73] or heparin[328] or lipoid substances[178a,206,274,311a]. The ratio between the "bound" and the "free" form* is not known and may be different in the various brain areas. The reported[258,498] value of 20 or 30% of serotonin existing as "free" in the rat brain was probably affected by a release occurring during the manipulations. In any case, the greatest percentage of serotonin in the brain exists in the particulate fraction in rats[258,498], mice[272] and guinea pigs[478].

The granules** storing serotonin are morphologically[42] and biochemically

---

* Brodie and Costa prefer the terms "reserve" and "mobile" pools[81].
** For a review on granules see Hagen[285] and Green[271].

(refs. 35, 119, 596) distinct from mitochondria and are referred to as pinched-off nerve endings[119,174,268,269] and synaptic vesicles[174]. Serotonin is present also in the microsomal fraction[325a,383]. The particles of guinea pig brain[598] have a diameter of 0.02–0.3 μ. The specific gravity of the granules is higher than that of the mitochondria of the argentaffin cells[450] but lower than that of the mitochondria of the brain (refs. 443, 478, 597), and they are probably different from the granules carrying acetylcholine[173,479,596] or histamine[119].

Inactivation

When serotonin is formed or released from granules, it is subjected to the action of monoamine oxidase* (see Chapter 3). This enzyme, which is present in excess in relation with the substrate available, is rather uniformly distributed throughout the brain[393,572,574], although there is a threefold difference between the highest (hypothalamus) and the lowest (corpus callosum) monoamine oxidase activity[66]. Hagen and Weiner have studied the activity of monoamine oxidase in the brain of various animal species; they found that the highest activity, with serotonin as a substrate, is present in the rat followed by the mouse, dog, guinea pig, rabbit and cat[284]. Monoamine oxidase is present in the mitochondria[23,511,530,568] located around the granules[166a]. The enzyme is present only in nerve cells and not in glia cells[23]. It seems to be practically the only enzyme able to metabolize serotonin in the brain. For this reason, a marked increase of brain serotonin takes place after the administration of a monoamine oxidase inhibitor. That the monoamine oxidase is active in the brain even in normal conditions is suggested by the presence of 5-hydroxyindoleacetic acid in considerable concentrations[466]. This metabolite increases after administration of 5-HTP, disappears after blocking monoamine oxidase, but does not change after administration of exogenous 5-HIA[466].

TURNOVER RATE

Studies on the metabolism of serotonin are important in order to establish the rate at which the amine can be formed in the brain. The monoamine oxidase inhibitors are particularly useful in this respect. It has been calculated, by using harmaline, that the half-life of 5-HT in the brain is of the order of 10 to 30 minutes[573] though it may be less[90].

Fig. 28 illustrates the increase of brain serotonin after intravenous administration of a rapid-acting monoamine oxidase inhibitor, *trans*-phenylcyclopropylamine (tranilcypromine)[244]. The half-life measured as the time required to increase rat brain 5-HT by 50% (0.235 μg/g) is calculated to be around 12.5 minutes. The true time will be less than this because minor alternative metabolic pathways for 5-HT could be present in the brain and phenylcyclopropylamine must take some time to completely inhibit monoamine oxidase.

---

* In reference to the importance of monoamine oxidase for the central nervous system, see the reviews of Zeller[614] and Pletscher *et al.*[436].

This means that the rat brain should be able to synthesize something like 50 μg of serotonin per day. Considering that the cortex and cerebellum do not synthesize 5-HT, it must be concluded that the brain stem may make and destroy serotonin at a very rapid rate.

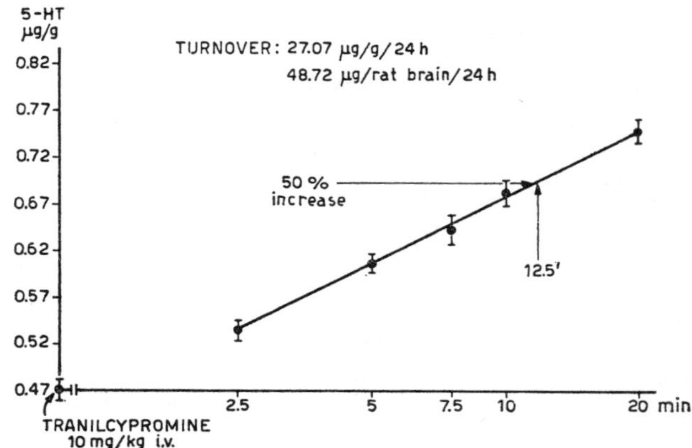

Fig. 28. Increase of brain serotonin in rats after administration of a fast-acting monoamine oxidase inhibitor. Serotonin was measured by a spectrofluorimetric method according to Bogdanski[64].

CENTRAL EFFECTS OF SEROTONIN

The presence in the brain of enzymes for the synthesis and inactivation of serotonin, and its uneven distribution, storage and release, and high turnover all indicate a role for serotonin in the central nervous tissue as a so-called neurohormone*. However, the function of serotonin in the brain is not yet known. The difficulties in assessing the central effects of serotonin depend primarily on the impermeability of the blood-brain barrier to the amine. The use of the precursor 5-hydroxytryptophan must always be considered critically because this compound is also decarboxylated in parts of the brain where serotonin is not present. Furthermore, the presence of noradrenaline in the brain, and the many interactions between serotonin and catecholamines, make it invariably difficult to establish whether serotonin is acting *per se* or whether its activity is influenced by other mediators.

Another difficult point to establish with regard to the so-called "central effects" of serotonin, is the significance of the peripheral and aspecific effects. Usually, the doses employed induce changes in blood pressure (see Chapter 9), reduce cerebral flow[131,132], decrease body temperature[236,351], potentiate the effect of anaesthesia (see below), and induce cerebral oedema[263,367] and pain[24,25,281,327].

---

* "By neurohormone is meant either a transmitter substance released from nerve endings and activating the post-synaptic membrane, or a substance enhancing or inhibiting (modulating) the action of a transmitter" (ref. 81). A model of a neurochemical transducer system is given by Brodie and Beaven[80].

Furthermore, serotonin is used in pharmacological doses many times higher than the amounts present in the body. It must, however, be recalled that the turnover of 5-HT in the central nervous system is high and that there is a possibility of high concentration of the neurohormone in certain specialized parts of the brain.

BIOCHEMICAL EFFECTS ON NERVOUS TISSUE

Very little is known about the effect of serotonin on brain metabolism. Serotonin in high concentrations, but not 5-HTP, inhibits *in vitro* cerebral acetylcholinesterase in a reversible way[16,17,558,*]. *In vivo*, however, there is evidence of activation of acetylcholinesterase (histochemical method) in rat hypothalamus[19]. In the same experiments, acid phosphatase, but not succinic dehydrogenase, was found to be increased[334].

Perfusion of cat brain cortex with 5-HT or 5-HTP results in an increase of glucose content without change of glycogen or lactate[365]. Similar results, *i.e.* decreased glucose consumption, have been observed after intravenous administration of 5-HT in rabbits[112,114]. Increased oxygen uptake by rat brain homogenate has also been described[525] (for other data concerning the effect of serotonin on glucose metabolism, see page 157). Serotonin decreases the uptake of α-aminoisobutyric acid[264]. In guinea pigs, even large doses of 5-HT do not affect the permeability of the brain stem to dyes[50].

EFFECTS ON GLIA AND NEURONS

Chick brain stem, the thalamus of newborn kittens and puppies, and human foetal brain contain oligodendroglia cells that are characterized when cultivated *in vitro* by a pulsating, rhythmical movement (about 10 per hour)[362,396]. Serotonin increases the number of pulsations[387,396] and the contraction[47,48,246,392,605] in a manner reminiscent of its effect on the smooth muscles. The effect of serotonin is antagonized by chlorpromazine[396], l-methylmedmain[605] and LSD[605].

In adult brain cells serotonin (1–5 μg/ml) "causes slow pumping of the perinuclear cytoplasm (not observed with any other drug) which forces cytoplasm down the axon. The nucleus enlarges at the beginning of contraction, and aggregation of the cytoplasmic granules is most strikingly seen in dendrites. Small undulating membranes appear on the fibres. The optical density of the nucleus progressively increases during the first hour, but subsequently it becomes more transparent" (ref. 247).

PHARMACOLOGICAL EFFECTS

Behaviour

The administration of serotonin results in a type of sedation which has been described in various animal species. It is impossible to summarize all the observations made on this subject. Therefore only a few examples will be reported.

------

* Serotonin is antagonistic to acetylcholine in the sympathetic superior cervical ganglion[171].

In pigeons, 5-hydroxytryptophan shows inhibition in an operative conditioning situation[19,20]. This effect is heightened by prior administration of a monoamine oxidase inhibitor[21] and is correlated with the increase of serotonin in the telencephalon and diencephalon[22].

In mice, after administration of serotonin (from 5 mg/kg), there is sedation[523] with reduced spontaneous locomotion[67,94,95,*]. An example of this sedative effect is also the antagonism *versus* stimulation induced by morphine[382], mescaline[205,560], methylphenethydate[560] and caffeine[210]. 5-Hydroxytryptophan in small doses tends to decrease spontaneous motility, but in large doses[144,233,574] or after pretreatment with monoamine oxidase inhibitors[350], it induces typical tremors, ataxia and piloerection.

In rats, serotonin has a sedative effect[424], but the administration of 5-hydroxytryptophan induces tremor and agitation with a symptomatology recalling that exhibited by LSD[574]. This effect of 5-HTP has been used to evaluate central antiserotonin activity[555]. However, relatively small doses of 5-HTP reduce spontaneous activity[321b]. Here also, the action of 5-HTP is enhanced by prior administration of monoamine oxidase inhibitors. With rats trained to a conditioned escape response, it has been observed that serotonin produces prolongation of the reaction time and partly abolishes the escape reaction. The effect is present with a dose of 0.5 mg/kg and with higher doses may last up to 7 hours[267]. Similar results were obtained with 5-HTP[321a]. Serotonin inhibits audiogenic reflexes[560]. Olds developed a technique for "self-injection" of drugs in various parts of the brain[403]. Serotonin by itself has a sedative effect on rats[404] but it is able to restore normal activity in animals pretreated with LSD[402]. This effect is, however, present only when the drug is injected into certain parts of the brain, *i.e.* the telencephalon[402].

In rabbits, opposite effects were observed according to the dose used. Small doses of 5-HT[388,390,424] and 5-HTP[389,390] induce a condition of sedation, while large doses of both compounds result in a form of excitation[388,390,574].

In cats, the effect of serotonin has been carefully studied after injecting the amine into a cannula implanted in the lateral ventricle. The most evident reactions were sedation[199,200], the disappearance of spontaneous activity[494], lethargy[225], and prolonged tremors[179,180,**]. These effects, which are present also when serotonin is given by general route[74], are prolonged by a monoamine oxidase inhibitor[493] and counteracted by LSD[74,225] and morphine[225]. Haley injected 5-HT into the lateral ventricle and described such main effects as salivation, tachypnoea, vomiting, scratching and convulsions[286].

In dogs, a careful description of the symptomatology following intravenous administration of 5-HTP has been given by Udenfriend *et al.*[574] and includes signs of tremors, postural incoordination and loss of placing reactions.

In humans, only a few data are available because the autonomic effects induced

---

* In mice, 5-HT is supposed to partially cross the blood-brain barrier[574].
** Tremors have also been observed after intraventricular injection of tryptamine[493].

by serotonin do not allow the dose to be increased. Without going into details, it may be concluded that no significant behavioral changes have been observed[492]. 5-HTP in doses up to 120 mg did not affect behaviour[168] or induce a feeling of sedation[391]. Borges *et al.*[71] found an electroencephalographic improvement in subjects with hepatic coma after administration of 10–40 mg of 5-HTP.

CHANGES OF BODY TEMPERATURE

The hypothermia induced by 5-HT in various animal species is still considered mainly of peripheral origin. In mice, the decrease of body temperature is already evident after a dose of 5 mg/kg, being of rapid onset (about 20 minutes) and of short duration (up to 3 hours according to the actual dose)[197,351,352]. The hypothermia is still present at thermoneutrality (32°C), suggesting that the thermoregulatory centres are not involved, but that either a fall in heat production or an increase in heat loss is operative[352]. However, it has also been reported[515] that serotonin becomes hyperthermic when the room temperature is more than 30°C.

In rats 5-HT also produces hypothermia[236]. The decrease of body temperature is enhanced by prior treatment with a monoamine oxidase inhibitor, but with a time effect different from the heightened effect of tryptamine or the antagonism toward reserpine, suggesting that the hypothermia is not of central origin[243]. The hypothermia is also considerably increased in adrenalectomized rats[236]. The decrease of body temperature by serotonin is antagonized both by antiserotonin agents (*e.g.* UML and cyproheptadine[244]) and by central stimulants (*e.g.* LSD, amphetamine and tetrahydro-β-naphthylamine[352]).

Different effects have been described in other animal species. For instance, in rabbits[428] and in cats[267] serotonin may be hyperthermic, this effect being enhanced by monoamine oxidase inhibitors[428]. In various animal species, though not in rats, 5-HTP is often hyperthermic, particularly when there is blockade of monoamine oxidase[155,303,304,476]. However, at low doses 5-HTP also produces hypothermia[323].

It is interesting that antiserotonin agents such as UML and cyproheptadine also antagonize this hyperthermia[244].

Feldberg and Myers[198] ascribe to serotonin in the hypothalamus a role in regulating body temperature. 5-HT would be an effective hyperthermic agent when the temperature has already been elevated as, for example, after administration of pyrogens[198].

ENHANCEMENT OF NARCOSIS

The observation that the toxicity of barbiturates or ether anaesthesia increases when serotonin is injected is not new[149,187]. Quantitative measurements show that the duration of narcosis is prolonged in proportion to the dose of serotonin used (refs. 14, 369, 533, 556). This effect has been observed in various animal species, including mice[94,213,483], rats[13,213,240] and humans[461].

The effects of various types of narcotic are enhanced by serotonin, including hexobarbital[13,94,104,196,213,266], hexobarbitone[223], pentobarbital[240,425], thiopental[425], ethanol[58], chloral hydrate[101,102,104,353], ethyl ether[214,425,607], urethane[425], and chloroform[607], but not halothane[607]. 5-HTP also shares the effect of serotonin[104,353,366a]. Various hypotheses have been put forward to explain this effect of serotonin, among which are vascular effects, decreased reabsorption of barbiturates[477], retardation of barbiturate detoxification[366a], decreased body temperature[197], and relative brain anoxia. None of these explanations, however, adequately accounts for the observed effect. For instance, the heightened effect of chloral hydrate, though somewhat reduced, still persists when animals are prevented from losing body temperature by high room temperature[197].

The influence of hormonal factors has also been claimed because the effect of hexobarbital was enhanced more in female than in male rats[13].

It is interesting that serotonin injected into the anterior brain of mice (40 μg), but not into the posterior brain, prolongs hexobarbital sleeping time[505].

Buchel and Levy[102,104], in studying the enhanced effects of narcotics due to serotonin and 5-HTP, found that 5-HTP induces sleep in animals when the effects of barbiturates have worn off. They consider these results as evidence that serotonin enhances the effects of narcotics, at least partially, in the central nervous system.

The heightening effect of serotonin is further enhanced by a monoamine oxidase inhibitor when the sleep is induced by chloral hydrate or hexobarbital[104], but not by ethanol[58].

This effect of serotonin is inhibited by various drugs including chlorpromazine[240], reserpine[240], β-tetrahydronaphthylamine[223], DOPA[234], isopropylnoradrenaline[240] and lysergic acid derivatives[178,484,551]. LSD and BOL, in small doses, enhance the effect of serotonin[483], but are antagonistic in large doses[483,484,551].

EFFECTS ON ANALGESIA

Serotonin increases the effect of some analgesic agents[103,331,524], including methadone[103] and morphine[381], but not pethidine and dihydrocodeine[381].

Local anaesthesia is not affected at the level of isolated nerves[345], but is antagonized by serotonin *in vivo*[347,473].

Ethanol

It is known that alcohol induces changes in serotonin metabolism *in vivo*, including an increase of blood serotonin[259,415]. This is particularly the case in carcinoid patients[259,290] where it is possibly related to a decreased degradation of the amine (refs. 377, 471). The central depressant effect of ethanol has been explained on the basis of a reduced level of brain serotonin and noradrenaline[279,280]. This effect was partially confirmed by Duritz and Truitt who considered that the lowering of brain amines was brought about by the formation of acetaldehyde[183]. However, other authors have found that alcohol has no effect on the brain serotonin and noradrenaline of rats and rabbits[185,452].

Much work has been devoted to the effects of serotonin on spontaneous or evoked electrical activity in the brain or nerves. Again there is disagreement between those who interpret these effects as centrally acting and those who consider them only as artifacts.

It is beyond the scope of this book to review in detail the various effects exerted by serotonin on brain electrical activity. A good summary can be found in the paper by Domer and Longo[181].

In general, serotonin has been described as a "depressant" (refs. 163, 426, 582). Thus in rabbits a decreased excitability of the reticular ascending system has been found[232,388,390]. Similarly, serotonin has a depressant effect on the preparation known as "cortex isolé" (ref. 177). On the other hand, Crepax and Infantellina[159] found that the EEG remained unaffected after intracarotid administration of serotonin. Pineda and Snyder[426] consider the depressant action of 5-HT on evoked potentials in the brain following peripheral nerve stimulation to be a "general nonspecific effect". Vascular responses[364] and cortical oedema[367] have been suggested as explanations; cerebral decrease of oxygen tension[137] should also be considered.

A number of authors who have treated cats with physiological doses of serotonin, injected intracarotidally, have found a clear cortical activation[260,370,474] which was inhibited by pentobarbital[370]. A decreased threshold of cortical excitability was also observed by Scarinci[491]. In preparations designed to produce a state of sleep (mesencephalic transection) or a state of wakefulness (midpontine-pretrigeminal transection), the content of 5-HT in the mesencephalic-hypothalamic area of the cat was not significantly changed[15].

To avoid the blood-brain barrier, 5-hydroxytryptophan was used in some experiments. Flattening of the electrocorticogram and development of rhinencephalic discharge[181,390] were observed.

Monnier found common features between the effect of reserpine and serotonin or 5-HTP[388,389]. According to the size of the dose of 5-HT or 5-HTP used, there were signs of depression (small doses) or of activation (large doses) of the brain electrical activity. In rabbits, 5-HTP evokes alerting which is inhibited by sections above the midbrain[509a]. Kimishima observed that reserpine decreased hippocampal seizure discharges, whilst 5-HT or 5-HTP had no effect[331].

Serotonin counteracts the effect of LSD on the rabbit EEG[422,423], however, some authors working with small doses of serotonin have found a synergism[75]. Vogt concluded that antagonism or synergism between LSD and 5-HT depends on the sites at which the electrodes are placed for recording[581].

Marrazzi devoted much work to characterize the inhibition exerted by intracarotid injection of serotonin on the evoked potentials from the terminal cortical synapses of the transcallosal pathway joining symmetrical points in the association optic cortices[261,372,373,472,*].

These results have been confirmed, but evidence has also been presented in favour of the possibility that the inhibition observed was mostly of peripheral origin (ref. 340). Using a different preparation, and recording from the hippocampus, other authors have concluded that serotonin does not exert a direct synaptic action[459,**]. Koella *et al.*[341] also agree about the lack of direct synaptic inhibition and consider serotonin to be a "modulator".

Serotonin has been applied electrophoretically to the extracellular fluid surrounding neurons in the brain stem of cats. No effect on the response of neurons in the medullary and mesencephalic reticular formation to peripheral stimuli were observed[165]. However, inhibition was noted in single neurons of the lateral geniculate nucleus after stimulation of optical nerves[164].

EFFECT ON NERVE CONDUCTION

In vitro, serotonin, even in very large doses (500 μg/ml), does not affect the excitability and conduction of the sciatic nerves of the rat and frog[133]. However, there is an increase of rat nerve conduction when serotonin is injected *in vivo* (refs. 134, 140, 141). This effect requires the presence of intact adrenals[135] and pituitary[136], but is not activated through a release of ACTH or adrenaline[135,136]. The threshold of the nerve of the leg (crab *Carcinus maenas*) is increased by serotonin[329].

SEROTONINERGIC FIBRES

The electrolytic destruction of the medial forebrain bundle within the lateral hypothalamus produces a considerable fall in brain serotonin[292]. Other lesions in the septal region, dorsomedial midbrain tegumentum and ventral midbrain tegumentum induce a moderate decrease of the brain amine, while other types of lesions not related to the medial forebrain bundle are without effect[292]. The decrease of serotonin following the destruction of the medial forebrain bundle starts the second day after operation, reaching the maximum after 12 days, and cannot be explained on the basis of the loss of cerebral material[288]. That the fall of serotonin is related to the degeneration of nerve fibres is shown by the observation that a unilateral lesion produces a decrease of brain serotonin only in the hemibrain containing the lesion and not in the contralateral side[288]. These results may be interpreted as an indication of the presence of medial forebrain bundle fibres producing serotonin[288]. Further experiments indicate that in rat brain the distribution of serotonin-producing fibres is different from the distribution of noradrenaline-producing fibres[291]. A large decrease of serotonin in the

---

* Serotonin depresses monosynaptic transmission in the isolated spinal cord of the frog[118].
** Tryptamine inhibits the firing of pyramidal and subicular hippocampal cells and increases that of granule cells[169].

spinal cord of rabbits has been found after transection at the level of the second thoracic segment[129]; a similar situation obtains in the case of noradrenaline[366]. Also histochemical[122] and electrophysiological[9b] findings indicate that serotonin may serve in the spinal cord as a synaptic transmitter in descending pathways. This suggestion is corroborated by the observation that the nerve stimulation of frog or mouse spinal cord results in a release of serotonin[9a]. This approach may lead to a mapping of the "serotoninergic" fibres in the central nervous system.

INTERACTION WITH CATECHOLAMINES

The fact that serotonin and catecholamines act as antagonists in many experimental conditions has already been mentioned several times; it is emphasized again at this point because of the possible physiological significance of this antagonism in the central nervous system.

The precursors of serotonin and catecholamines (5-HTP and DOPA) compete with each other for the transport mechanism in the brain[496]. Furthermore, 5-HTP and DOPA compete for the same decarboxylase *in vivo* as is shown by the fact that large doses of DOPA decrease the level of serotonin in the brain[57]. 5-HTP may antagonize the stimulant effect of DOPA in reserpinized animals[3].

Serotonin and noradrenaline have been observed to have opposite effects on the electrical activity of the brain[390] and on the processes of temperature regulation[198]. Furthermore, catecholamines inhibit the enhanced effect of barbiturates induced by 5-HT[234,240] and prevent the tremors following intracisternal administration of serotonin[180]. The serotonin–noradrenaline antagonism towards the effect of barbiturates occurs also when the drugs are injected intracerebrally[503,504]. Whether these pharmacological effects are also indicative of physiological importance in the regulation of certain central functions of the nervous system cannot be said at present. It is interesting that the theory of Hess[296] concerning the existence of "trophotropic" and "ergotropic" systems with opposite functions has found pharmacological support in the hypothesis of Brodie[91] which suggests that serotonin and noradrenaline could be involved as neurohormones in such systems.

Contrasting changes of 5-HT and noradrenaline in the brain have been reported during hibernation[576].

CONVULSIONS AND ANTICONVULSANT DRUGS

Serotonin itself after a single dose does not affect convulsant treatments. This has been established in various animal species by evaluating convulsions induced by camphor[59], metrazol[59,69,560], strychnine[113] and picrotoxin[113]. Even when serotonin is injected intracerebrally in mice, either into the anterior or posterior brain, there is no modification of the metrazol convulsant effect[506]. However, when serotonin is given for several days by general route, an antagonism towards metrazol and picrotoxin develops[113]. In epileptic patients serotonin tends to be an antagonist of the anomalous electrical activity[117]. In cats and monkeys, however, the administration of 5-HTP is considered to favour the onset of epileptogenic activity[584]. Several investigations have been devoted to the study of the effect of convulsant treatments on the level of brain serotonin. By measuring serotonin with bioassay or spectrofluorimetry, an increase of brain 5-HT was observed after electroshock[241] without changes in brain monoamine oxidase activity[540]. Further studies demonstrated that the increase was present in dogs and rats, particularly in the mesencephalic and hypothalamic areas[234]. Metrazol and, to a lesser extent, strychnine also, increased brain serotonin in rats[234]. The increase was not a function of the convulsions, because after preventing convulsion with anticonvulsant drugs, electroshock was still effective in elevating the level of brain serotonin[234]. The finding of increased brain serotonin after electroshock has not been confirmed by other authors[69] and has been severely criticized by Bertaccini (ref. 52). The latter, using bioassay, regularly observed a small increase of brain serotonin after electroshock or metrazol, but did not consider the results obtained as significant. More recently, Breitner et al.[76] found a small but significant increase in brain serotonin in various animal species after general, but not after local, electrostimulation.

Electroshock in patients induced either no change[273] or a decrease[43] in blood serotonin, while an increase in blood noradrenaline was observed[557]. After electroshock therapy, a significant decrease of the urinary excretion of 5-HIA was observed (ref. 557). When the intensity of electroshock was increased, a considerable increase of urinary 5-HIA was measured[579].

Another type of stress has been shown to be capable of increasing brain serotonin (ref. 384). Insulin coma in rabbits causes an elevation of serotonin in various parts of the brain, including the telencephalon, hippocampus and midbrain, and the diencephalon, but not in the medulla oblongata and pons[154].

Anticonvulsant drugs are known to affect brain serotonin levels. A specific increase has been found after administration of diphenylhydantoin*, N-methylphenylsuccinimide, phenylacetylurea, phenobarbitone, sodium bromide[69] and several barbiturates[68,234]. Diphenylhydantoin counteracts the depletion of brain serotonin induced by reserpine, and enhances the increase following 5-HTP[70]. The protective effect of diphenylhydantoin against electroshock was for some time related to the increase of brain serotonin[11]. Such correlation seems to exist also for some barbiturates[10,**], and for sensitivity to the convulsant effect of hexafluorodiethyl ether[561]. Many studies have been carried out to establish a relationship between changes in brain serotonin and changes in the threshold for convulsions. Reserpine, which depletes brain serotonin, increases the intensity and decreases the threshold of various types of convulsant treatments[59,142,313,339,359,506]. This effect is prevented by the intracerebral injection of serotonin or noradrenaline[506]. On the other hand, several monoamine oxidase inhibitors exhibit an anticonvulsant effect[409,410,442,447,448]

---

* Other authors have not been able to confirm these results[185,409,448].

** Other authors have failed to confirm these results[185,409,448].

and in addition they counteract the activity of reserpine[339] and enhance the activity of other anticonvulsant agents[409,410,608],*.

By contrast, several studies have demonstrated that there is no relation between changes in brain serotonin levels and sensitivity to convulsions[175,409,420,421].

The problem of the relationship between serotonin and convulsions is not yet resolved, but requires further research, particularly in regard to sensitivity to convulsions in relation to changes in serotonin levels in specific areas of the brain.

RELEASERS OF BRAIN SEROTONIN

The prototype of this class of drug is represented by reserpine which has been shown to decrease the level of brain serotonin[85,437]. The effect has been related to the sedative action of reserpine, because only the alkaloids of *Rauwolfia serpentina*, having sedative action, are known to decrease brain serotonin[89,188,189,438,521]. It must be added at once that reserpine releases not only serotonin**, but also noradrenaline (refs. 56, 84, 123, 130, 275, 298, 517, 519, 589), 3-hydroxytyramine or dopamine[124], and probably histamine as well[583].

One of the first problems to be resolved was whether the decrease of brain serotonin by reserpine was due to a capacity of the alkaloid to block serotonin biosynthesis. The level of brain serotonin and noradrenaline is certainly very low after the administration of reserpine. The decrease and return to normal levels are closely related in time, though in varying degrees in different animal species, to the typical sedative effect[295,520] while there is no relation between the level of brain reserpine and changed levels of brain amines[399,440]. However, during long-term treatment in rabbits, a low level of brain monoamines has been observed when the gross behaviour was apparently normal[285a]. Data collected by several workers indicate that serotonin biosynthesis is not in fact impaired by reserpine[93,192,289]. At all events, 5-hydroxytryptophan decarboxylase remains normal or is even increased (refs. 54, 150)***. The brain 5-HIA[468] and the excretion in the urine of 5-HIA are increased after administration of reserpine[189,192,212,487,522] suggesting that serotonin is released from tissues. Furthermore, when serotonin is released by reserpine, the administration of a monoamine oxidase inhibitor results in a prompt increase of brain serotonin[90]. This fact can be explained only by assuming that serotonin is continuously synthesized and metabolized even when the effect of reserpine is operative. Concerning the mechanisms involved in the depletion of amines, it is likely that reserpine blocks the "pump" responsible for concentrating serotonin inside the granules (see page 70).

If the decrease of brain serotonin and noradrenaline induced by reserpine is prevented by prior treatment with a monoamine oxidase inhibitor, then the central

---

* This effect may, however, be related to the blocking effect of MAO inhibitors on the enzymes able to metabolize several drugs[29,344].
** It is noteworthy that the brain contains a factor able to release serotonin from tissues[257].
*** It can, however, be decreased with large doses of reserpine[594].

pharmacological effects of reserpine are counteracted. The sedation is often replaced by excitation[58,87,427,518], the deconditioning effect is inhibited[248,300,356], and the decrease of the threshold for convulsion is prevented[339,448].

A further problem of importance is to decide whether the decrease of brain amines after the administration of reserpine should be regarded in terms of a deficiency of serotonin and noradrenaline or, by contrast, as an increased availability of the amines for the receptors, due to the fact that the newly synthesized amines cannot be stored. The question is still unresolved and requires further investigation. It must also be admitted that the study of the relationship between the biochemical and pharmacological effects of reserpine is made particularly difficult by the number of factors involved. In fact, all attempts to determine the central effects of reserpine are complicated by the influence of peripheral activities (such as hypotension) and by hormonal effects (for instance, pituitary stimulation). The research work of several laboratories has recently served to clarify a number of points.

(*1*) By working with reserpine derivatives, it has been possible to show that the sedative and the hypotensive effects of reserpine may be distinguished. This is not unexpected, since reserpine itself in small doses shows hypotensive activity without any sedative effect. However, the autonomy of the two types of effect has been shown more clearly by compounds like syrosingopine (Su 3118)[235,238,310,439], β-diethylaminoethylreserpine (DL 152)[110,237], isobutylserpentinate (IBS)[237,449] and 10-methoxydeserpidine[237,419,578], which are hypotensive like reserpine but not sedative.

On the other hand, compounds like dimethylaminobenzoylmethylreserpate (Su 5171)[82,235,238], methyl-18-epireserpate ethyl ester (Su 9064)[160,463], 10-chlorodeserpidine[578] and tetrabenazine[455,508] have a clear sedative action in doses which do not affect blood pressure.

(*2*) The hypotensive properties of reserpine derivatives are not dependent upon the lowering of brain amines[237,310]. On the other hand, all the reserpine derivatives showing sedative activity are able to decrease brain amines[89,237,238,521] (see Table 33). To date, there are no exceptions to this rule, and this would indicate that the sedative activity exerted by reserpine could be strictly related to changes in brain amines. However, there is as yet no clear-cut evidence in favour of the hypothesis that changes in the levels of brain amines are the cause of the sedation.

(*3*) Reserpine sedation can be dissociated from the effect of reserpine on ACTH hypersecretion[83,379]. This distinction has been made possible with the aid of selective inhibitors of reserpine. While a monoamine oxidase inhibitor prevents not only depletion of brain amines, but also sedation and the rise of plasma corticosterone levels after reserpine administration, desmethylimipramine prevents only sedation without affecting brain amines and the hormonal response[374].

Considerable interest is linked to the possibility that the effect of reserpine may

TABLE 33

BRAIN AMINES AND SEDATION AFTER ADMINISTRATION OF RESERPINE
DERIVATIVES

Compound	Dose (mg/kg)	Depletion (> 65%) brain		Sedation (ptosis, hypothermia)
		5-HT	NOR	
Reserpine	1.25– 2.5 i.p.	+	+	present
Syrosingopine (Su 3118)	2.5 – 5.0 i.p.	—	—	absent
Dimethylaminobenzoyl-methylreserpate (Su 5171)	5.0 –10.0 i.p.	+	+	present
Methyl-18-epireserpate ethyl ester (Su 9064)	5.0 i.v.	+	+	present
Isobutylserpentinate (IBS)	2.5 –10.0 i.p.	—	—	absent
β-Diethylaminoethylreserpine (DL 152)	2.5 –10.0 i.p.	—	—	absent
10-Methoxydeserpidine	2.5 –10.0 i.p.	—	—	absent
10-Chlorodeserpidine	5.0 –10.0 i.v.	+	+	present

i.p. = intraperitoneal;   i.v. = intravenous

be ascribable to changes either in serotonin or in noradrenaline. Brodie and co-workers have put forward the hypothesis of serotonin mediation as a cause of the sedation induced by reserpine[81,88,92], while Carlsson[120,121] and Paasonen[406] point to the importance of noradrenaline in explaining the pharmacological effects of reserpine. The matter is still undecided and the arguments for and against the two points of view will be presented here.

Rats kept in a cold room for at least 4 hours do not show the usual symptomatology when injected with reserpine[242,547]. Under these conditions, brain serotonin is only slightly, if at all, decreased[242,547], but the brain noradrenaline is lowered[82,547]. However, this virtual lack of effect on brain serotonin of reserpine in a low environmental temperature has not been confirmed[349,593], nor has the absence of sedative effect[337,349,593].

Reserpinized rabbits treated with 1-phenylisopropyl-2-isopropylhydrazine recover from sedation and show a rise of brain serotonin, whilst low levels of noradrenaline and dopamine still persist[538]. However, this result could be ascribed to the presence of sympathomimetic properties in the monoamine oxidase used (it is well known that amphetamine and other central stimulants antagonize reserpine[545,559]).

As a matter of fact, isopropylhydrazine, given after reserpine, does increase brain serotonin, but does not affect the sedation[115]. Su 5171, a reserpine derivative, decreases noradrenaline to a greater extent than serotonin. The sedation parallels the decrease of serotonin, but not the decrease of noradrenaline[82].

Two DOPA decarboxylase inhibitors (α-methyl-m-tyrosine and α-methyl-DOPA)

TABLE 34

SIMILARITIES BETWEEN THE PHARMACOLOGICAL EFFECTS OF RESERPINE
AND SEROTONIN

Effect	Reference for serotonin	Reference for reserpine
Sedation	145, 225, 475	107, 205, 225
antagonized by LSD	225, 475	107, 225
Hypothermia	351, 352	88, 172, 351
antagonized by LSD, amphetamine and tetra-hydro-β-naphthylamine	352	352
Hyperthermia after MAO inhibitors	428	428
Barbiturate potentiation	94, 214, 293, 424	94
antagonized by LSD (large doses)	94, 522	94, 522
antagonized by BOL (large doses)	138, 484	138, 484
Potentiation of hypnotics not barbiturates	197	88, 197
Antagonism *vs.* central stimulants	205, 559, 560	205, 559, 560
Antagonism *vs.* audiogenic convulsions	559, 560	559, 560
Deconditioning effect	248, 300, 356	245, 312, 320, 590, 591

show the unique property of decreasing brain noradrenaline with a slight effect on brain 5-HT[469,512]. These compounds do not exhibit reserpine-like properties, but when reserpine is given and serotonin is lowered the typical sedation appears[81,153]. Recently, Carlsson[125,126] has presented evidence that these two amino acid analogues yield the corresponding α-methylamines and β-hydroxylated derivatives in the brain, these being stored instead of dopamine and noradrenaline. However, with α-methyl-*m*-tyrosine a depletion of brain serotonin is certainly present, and this without substitution by α-methyl-*m*-tyramine[81].

High concentrations of *N*-(3'-phenylpropyl-2')-1,1-diphenylpropyl-3-amine (segontin), a coronary dilator agent[357], decrease brain noradrenaline (about 35%), but not brain serotonin[244]. Segontin does not induce sedation, but a similar decrease of brain noradrenaline accompanied by a decrease of serotonin, as after administration of a reserpine derivative (Su 9064), does result in sedation [244].

Other experiments carried out in rats receiving a large dose of reserpine after blockade of monoamine oxidase with phenelzine, have indicated a decrease of brain noradrenaline with normal values of brain serotonin. These animals were not sedated (ref. 244).

It is also suggestive that a number of pharmacological effects induced by reserpine are shared by serotonin (see Table 34).

One of the strongest points in support of the "noradrenaline theory" is the fact that 5-HTP is not active in antagonizing reserpine sedation, while DOPA is very effective when given to animals fully reserpinized[127]. That DOPA antagonizes various effects of reserpine is well established[3,194,398,510]. However, it remains to evaluate the significance and the specificity of this result, particularly in relation to the fact that the antagonism could be a physiological one due to the fact that the amines formed by DOPA may be stimulants.

In concluding discussion of this problem, it is interesting to note that 5-HTP and DOPA given together in certain doses produce a better antagonism to reserpine sedation than does DOPA alone[127].

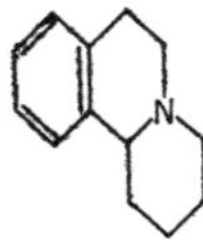

Fig. 29. Basic structure of benzoquinolizine.

Another class of drugs able to release brain amines is represented by the benzo-quinolizine derivatives.

An extensive review of the biochemical and pharmacological properties of these drugs has recently been published[431].

MONOAMINE OXIDASE INHIBITORS

Since 5-HT is metabolized in the brain almost completely through monoamine oxidase, the administration of a monoamine oxidase inhibitor results in a considerable increase of brain serotonin[87,182]. It is beyond the scope of this book to review this subject. It has been well summarized recently by Pletscher *et al.*[436]. There are monoamine oxidase inhibitors with hydrazine[61,62,432,509,616], amine[363,490], imine[287], aldoxime[380] and propinyl-amine[538,553] groups. Some inhibitors are short-acting (harmaline[429], phenylcyclopropylamine[490], α-methyltryptamine[276]), others long-acting (iproniazid[86,88], phenelzine[143] and phenylisopropylhydrazine[301,302,304]). In brain as in other organs, these compounds increase the level of several amines, including serotonin[88,451,518], dopamine[124,262,568], noradrenaline[32,128,568] (probably only in certain animal species), and possibly tryptamine[294,401,529,592]. An increase in brain serotonin after iproniazid treatment has been observed also in man[232a,232b].

The central pharmacological effects of the monoamine oxidase inhibitors have not yet been completely analysed.

Acute effects are practically absent. The behaviour of the animal does not change even if the level of brain amines is considerably increased.

However, pharmacological analysis reveals a number of effects which are related partly to the block of monoamine oxidase and partly to a direct pharmacological action of the compound *per se*.

For instance, the influence of phenylisopropylhydrazine on barbiturate narcosis is due to at least four factors:

(*a*) block of the microsome enzymes, which is present until the molecule as such is present in the liver[29,346];

(*b*) direct antagonism due to the amphetamine-like activity[302];

(*c*) increase of brain amines, which has a different pattern according to the time and the species. In the first few hours serotonin is increased in some species without changes in noradrenaline, which increases only after several hours[128,518];

(*d*) hypothermia, which is present, at least in rats, for several hours[368].

Other central activities of monoamine oxidase inhibitors, such as the symptomatology occurring after the administration of DOPA[120,170,368,456], 5-HTP (refs. 67, 155, 156, 209, 304, 430, 476), tryptophan[297a] or tryptamine[243,554], parallel the block of brain monoamine oxidase.

Antagonism toward reserpine is common to all monoamine oxidase inhibitors. To date, all the activities of reserpine are found to be prevented by the inhibitors of monoamine oxidase.

Chronic treatment with monoamine oxidase inhibitors induces in some species hyperactivity and signs of central sympathetic predominance. According to Brodie, this type of reaction is closely correlated with an increase of catecholamines rather than with an increase of serotonin in brain[81,539].

Table 35 summarizes the principal activities common to all the monoamine oxidase inhibitors.

TABLE 35

SOME PHARMACOLOGICAL EFFECTS EXERTED BY MONOAMINE OXIDASE INHIBITORS

Acute treatment:

1. Enhancement of central symptomatology by: DOPA, 5-HTP, tryptamine

2. Increase of hypothermia by serotonin

3. Prevention of reserpine sedation

4. Interference with barbiturate narcosis

5. Anticonvulsant activity

Chronic treatment:

1. Excitation

2. Hypotension

DRUGS ALTERING PERMEABILITY TO SEROTONIN

Chlorpromazine and many phenothiazine and thioxanthene derivatives are known to prevent several of the pharmacological effects of serotonin[239,240,282]. Most of the studies concerned have been carried out to evaluate the peripheral effects of serotonin, which have been dealt with in previous chapters.

Chlorpromazine and chlorprothixene interfere also with the metabolism of serotonin (and noradrenaline). Recently this has been interpreted as an inhibition of the "penetration of monoamines to the receptor sites" (refs. 251, 434).

Phenothiazines decrease the rise of serotonin induced by administration of 5-HTP[253,254] or by monoamine oxidase inhibitors[253], and they counteract the release induced by reserpine[252] and benzoquinolizine[433] without directly affecting 5-HTP decarboxylase or monoamine oxidase[249,434], or the metabolism of the drugs used[507].

However, according to Morpurgo[390a], these effects are related to the hypothermia induced by phenothiazines.

OTHER TREATMENTS AFFECTING BRAIN SEROTONIN

A decrease of brain serotonin has been reported after administration of vincamin (an alkaloid posessing hypotensive activity)[265], guanethidine[208], and nicotine[501,502], and after exposure to irradiation[537] and ozone[532]. Also other treatments like adrenalectomy[236,453] and administration of *Shigella shigae* toxin[371] have resulted in a decrease of brain serotonin.

Administration of methylene blue increases the level of brain serotonin, probably through a block of monoamine oxidase[186]. LSD increases brain serotonin levels and counteracts the depletion induced by reserpine[218,219]. Hepatectomy in rats also increases brain serotonin[565].

SEROTONIN AND MENTAL DISEASES

A hypothesis concerning the possible role of serotonin in brain functions stimulated considerable interest in searching for abnormalities in the metabolism of this amine in patients with mental illness. Furthermore, the clinical improvement obtained by the use of drugs able to release, accumulate or antagonize serotonin stressed from a practical point of view the importance of serotonin for brain function (refs. 79, 601). It is beyond the scope of this book to discuss the many problems posed by the present classification of mental diseases, particularly when one tries to make correlations between biochemical changes and the occurrence of a given disease[330].

It must also be realized that even a serious defect in the metabolism of serotonin in a localized area of the brain would hardly be recognized by investigating the metabolism of the amine in the blood or urine. The results summarized below must be necessarily regarded as indications of abnormalities which have meaning only when detected also in the brain[527,528].

URINARY EXCRETION OF HYDROXYINDOLES *

Serotonin is excreted in small quantities in the urine of normal subjects, but is decreased in the urine of selected schizophrenic subjects[211]. However, the average excretion remains unchanged[464]. Several papers report the excretion of *5-hydroxyindoleacetic acid* in the urine of mental patients. The majority of the observations indicate that there is no difference in the excretion of 5-HIA in the urine of schizophrenic as compared to other psychotic or normal subjects (refs. 108, 109, 201,

---

* Reviews on this subject have been published by Benda[46], Smythies[535], Kety[330], Price *et al.*[444], Sprince[541] and Tissot[557].

202a, 221, 289, 311, 342, 462, 464, 465, 489). Similar results have been obtained when the excretion of 5-HIA is measured from hour to hour[202,311]. However, there is a tendency for schizophrenics to show larger variations than normal subjects[201,311]. Several authors report that in selected patients, and also in the average values, there can be a slight decrease[44,215,355,465] or an increase[37,376] in the urinary elimination of 5-HIA. Buscaino and Stefanachi[108] were able to detect an increase in the excretion of 5-HIA only in patients with catatonic schizophrenia. Brune and Himwich[97], studying changes in 5-HIA excretion in selected patients, found that aggravation of psychotic symptoms was associated with a marked rise of urinary 5-HIA, while tranquillization was associated with a moderate increase or a reduction of 5-HIA. A similar pattern is followed in the excretion of indoleacetic acid[98,100] and tryptamine (refs. 98, 544). It must be remembered, however, that this type of study requires very careful control of the dietetic and environmental conditions[305]. Anderson *et al.*[12] reported the increased excretion of 5-HIA after ingestion of bananas (see also page 59), while recently the periodical enhancement of urinary 5-HIA in a psychiatric patient was related to ingestion of walnuts, which have a high concentration of serotonin (refs. 332, 333).

A way to establish possible anomalies of serotonin metabolism in mental patients is to study 5-HIA excretion after administration of serotonin precursors.

Loading with tryptophan resulted in a consistent increase of 5-HIA excretion in normal subjects, but not in schizophrenics[348,613,615]. Some authors have found opposite results, schizophrenic patients doubling their excretion of 5-HIA, while controls did not[38]. More extensive investigations, however, have led to the conclusion that there are no differences in the two groups after tryptophan loading (refs. 44, 311, 342, 513)*.

The administration of 5-hydroxytryptophan increases urinary excretion of 5-HIA (refs. 335, 458, 552). Feldstein *et al.*[203] administered 5-HTP labelled with ^{14}C in position 3 or in the carboxyl group to schizophrenic and to normal subjects, but were unable to detect any difference in the excretion of labelled 5-HIA. Furthermore, serotonin loading did not result in major changes in schizophrenics as against normal subjects as regards 5-HIA excretion[108,202a,203a].

The release of endogenous serotonin, induced by reserpine, did not reveal any difference in the enhanced 5-HIA excretion observed in mentally defective patients or in schizophrenics[577].

On the whole, these results suggest that in schizophrenia there are no systemic changes in the metabolism of serotonin.

Another line of research is the study of abnormal indoles present in the urine of psychiatric patients. A number of observations are suggestive, but require further confirmation.

More indoles are usually detected in the urine of schizophrenics than in normals

---

* Low tryptophan levels are found in the urine of schizophrenics[184]. Loading with tryptophan did not induce any major change in the excretion of tryptophan metabolites[44,96].

(refs. 355, 460, 542). Sherwood[516] reports the presence of a tryptophan peroxide with a hydroxyl group in the β-position. Particular interest attaches to the reported presence of *N*-dimethylserotonin (bufotenine) in the urine of schizophrenics[106,211]. Bufotenine is also present in a poisonous mushroom (*Amanita mappa*)[599] and is able to impair the passage of nerve impulses[193] and to induce mental changes and hallucinations when administered to normal subjects[195,470,482,548,563]. Some authors, however, have not been able to confirm the presence of bufotenine in the urine of schizophrenics[464,535,544]. Similar disagreements also exist with regard to the excretion of *N*-methyl serotonin[106,464,535,544].

Dimethyltryptamine is transformed in the 6-hydroxy derivative[549,550], a metabolite without psychotomimetic activity in humans[470].

Indoles with a hydroxyl group in the 6 position are also more abundant in schizophrenics than in normal subjects[542]. Similarly, 6-hydroxyskatole sulphate is excreted in the urine of schizophrenics[395,543], but it disappears after treatment with antibiotics[395]. Furthermore, no differences have been found in the excretion of this metabolite in schizophrenics, manic depressives, psychotic and normal subjects[343].

The absence or the decrease of some indolic substances in the urine of schizophrenic children has been reported[526]. A hydroxylate indole (not serotonin or 5-HIA) is also deficient or in abnormally low concentrations in the urine of schizophrenics[111].

BLOOD SEROTONIN

The levels of serotonin in the blood of schizophrenic patients are within the normal limits[43,201,322]. On the other hand, more variations are generally found in the extreme values in schizophrenics than in normal subjects[201,322]. However, in selected patients, high values of serotonin in platelets have been observed[407,414]. High values have also been detected in autistic children[495].

The finding that serotonin is oxidized in serum by caeruloplasmin[375] has heightened interest in the reported high level of caeruloplasmin in mental diseases[1,6]. More recent results, however, have failed to confirm these reports[4,18,45,546,610].

SEROTONIN IN CEREBROSPINAL FLUID

Since serotonin does not cross the blood-brain barrier[514,566,574,600,603,604], any change in the serotonin content of the cerebrospinal fluid may be of significance in assessing the role of the amine in mental diseases.

The human spinal fluid contains, at most, very small concentrations of serotonin (ref. 480). Some authors[562] report values around 0.006 μg/ml; others[536,602] believe that the concentration of serotonin is less than this figure. The total 5-hydroxyindoles are of the order of 0.03–0.09 μg/ml (ref. 30) and, according to Jensen[314] and Roos[467], are mostly represented by 5-hydroxyindoleacetic acid (tryptamine and indoleacetic acid are also present).

In children the concentration of serotonin in cerebrospinal fluid is higher than in adults and may reach 0.1–0.3 μg/ml (ref. 5).

An increase has been found in hydrocephalus[30], tubercular meningitis[5] and certain brain tumours[602], while a decrease was present in depressive psychosis[30]. No changes were found in mongoloids, phenylketonurics[418] and schizophrenics[602]. Several authors reported an increase of the serotonin concentration in cerebrospinal fluid, but no quantitative data were given[105].

EFFECT OF SEROTONIN ON HUMAN BEHAVIOUR

Since serotonin does not cross the blood-brain barrier, any central effect after injection or administration of serotonin is unexpected. It is therefore surprising to read a report about therapeutic effects obtained in paranoids, but not in hebephrenics, after oral administration of 200–600 mg of serotonin per day[34].

5-Hydroxytryptophan has been given with the aim of increasing the level of serotonin in the brain. No alleviation of schizophrenic psychosis has been observed after 5-hydroxytryptophan administered alone[78,335], or in combination with a monoamine oxidase inhibitor[441], or with a serotonin inhibitor to minimize the peripheral effects of the serotonin formed[297,603,606]. However, tryptophan* and methionine induce mood elevation when given to schizophrenics pretreated with a monoamine oxidase inhibitor[8,99,147,441]. Very rapid relief of severe depressions has been observed after the administration of a combination of a monoamine oxidase inhibitor with 5-hydroxytryptophan[336]. The observation that 5-hydroxytryptophan alleviates the psychological effects induced by LSD is interesting[77,410a]. On the other hand, olfactory hallucinations[1,2,3] have been observed[335] in one schizophrenic patient following the administration of 5-HTP and in other patients the disease became worse[601]. Carcinoid patients do not show any sign of mental disturbance[531]. Although there are no uniform results, further research in this field does seem to be justified.

Miscellaneous

In the brains of patients with *Parkinson's syndrome* there is a decrease in the serotonin content at the level of the nucleus caudatus, putamen, pallidum, thalamus, hypothalamus and substantia nigra[51]. This decrease parallels the observations with dopamine[39]. Similarly, the urinary excretion of 5-hydroxyindoleacetic acid[40] and dopamine[41] is decreased in patients with Parkinson's disease. DOPA[63], but not 5-HTP[458], ameliorates the symptoms.

In *mongoloids* the excretion of 5-HIA in the urine is reduced[316,317], although the blood level of serotonin[414] and the serum serotonin[416] are normal. Differences in 5-HTP decarboxylase or in tryptophan absorption have been excluded[316].

---

* Administration of tryptophan following monoamine oxidase inhibition produces in man neurological abnormalities, which are chiefly excitatory in nature[297a].

Two mentally retarded sisters showed an increase of 5-HIA excretion, both in the basal state, and after loading with tryptophan or 5-hydroxytryptophan[299].

SEROTONIN AND PHENYLKETONURIA

This hereditary metabolic disease is characterized by the presence of high levels of phenylalanine in the plasma, by the occurrence in the urine of phenylpyruvic acid[166], and by the development of idiocy early in childhood or infancy[176,319,358] which can, however, be partially cured by reducing the level of phenylalanine in the diet[28,60,338].

A disturbance of indole metabolism, both in the blood and the urine, is characteristic of this disease.

In *blood* a decrease of serotonin[36,49,411,413] has been observed in phenylketonuric children as compared with other mentally defective controls. In *urines* there is an increased elimination of indoleacetic acid[26,27,413], indolelactic acid[26], with a concomitant marked reduction in the excretion of serotonin[417], 5-hydroxyindoleacetic acid[26,27,207,315,413,416,417,485] and of tryptamine[416]. These metabolic disturbances were reduced during a "low phenylalanine" diet[412,417]. The administration of a monoamine oxidase inhibitor results in an increase of serotonin and tryptamine excretion in the urine, which is, however, less marked than in normal children[416]. The same pattern has been observed after tryptophan loading[417]. No serotonin has been detected in the cerebrospinal fluid of phenylketonurics and mongoloids, even after monoamine oxidase blockade[418].

The disturbance of indole metabolism should not, however, be considered specific for phenylketonurics, because in these patients a decrease of dopamine, noradrenaline and epinephrine in serum and urine has also been reported[49,392,*].

Several investigations have been carried out with the aim of elucidating the biochemical reasons for the decreased level of serotonin in phenylketonurics. The data obtained have been integrated with the experimental observation that a diet rich in phenylalanine reduces learning in rats, and induces a pattern of metabolic disturbances similar to those observed in phenylketonurics[31].

A first hypothetical approach was to consider the decreased formation of serotonin as the consequence of a deficiency in phenylalanine hydroxylase[318,386]. This deficiency would then be responsible for a decreased hydroxylation of tryptophan[417] and, furthermore, the presence of high levels of phenylalanine in the serum would be another negative factor in competing for the hydroxylase[217]. However, recent studies have not supported the hypothesis that phenylalanine hydroxylase has physiological significance for the hydroxylation of tryptophan[457].

Some authors have demonstrated that the high levels of phenylalanine in serum (ref. 571) and the aromatic acid metabolites of phenylalanine (phenylacetic, phenyllactic and phenylpyruvic acid)[167] reduce the decarboxylation of 5-hydroxytryptophan decarboxylase.

---

* In phenylketonurics there is production of phenylethylamine[400].

The fact that a reduced decarboxylase activity is present in phenylketonurics was demonstrated by loading with 5-HTP. In this condition, when the hydroxylation of tryptophan was not a limiting factor, there was still a reduced excretion of serotonin and of 5-hydroxyindoleacetic acid in the urine[412].

Another indirect demonstration of the inhibition of 5-HTP decarboxylase is the report that phenylacetic acid temporarily reduces the excretion of 5-HIA in carcinoidosis[486] and in an atypical carcinoid secreting 5-hydroxytryptophan[485]. This interpretation is, however, contradicted by the finding that phenylacetic acid blocks the renal excretion of 5-hydroxyindoleacetic acid[385].

Furthermore, experiments carried out in animals tend to support the conclusion that the inhibition of 5-HTP decarboxylase is not the cause of the low level of serotonin in phenylketonurics.

In rats, a diet rich in phenylalanine reduces serotonin in the serum[308,586] and the brain[161,270,361,457,586,587,611], decreases 5-HIA in the urine[310] and also reduces learning capacity[360,361,605a]. However, phenylpyruvic and phenylacetic acid do not affect the level of serotonin[161]. The transformation of 5-HTP to serotonin is diminished in the kidney[307] and brain[457], but this has been ascribed not to inhibition of the decarboxylase[587], but to the fact that phenylalanine inhibits the transport of 5-HTP across the blood-brain barrier[278,378,454a,500]. In guinea pigs, the reduction of brain 5-HT in the course of a diet rich in phenylalanine was likewise unaccompanied by a change in the brain 5-hydroxytryptophan decarboxylase[306]. This situation is different from the one present in newborn animals where the low level of brain serotonin[306,309,325] is sustained by a low level of decarboxylase[306,309].

In conclusion, it seems that the low level of serotonin formation in phenylketonurics can be explained by several possible mechanisms, including decreased hydroxylation of tryptophan and reduced utilization of 5-HTP, with a consequently diminished possibility of transport inside the cells and the probability of reduced decarboxylase activity. Apart from the reasons mentioned above, the decreased excretion of 5-hydroxyindoleacetic acid in the urine can also be explained on the basis of reduced renal excretion. The presence of phenylalanine and its metabolites in high concentrations seems to be responsible for the biochemical disturbances in the indole metabolism in phenylketonurics.

REFERENCES

1 ABOOD, L., *Brain Research Foundation Rept.*, Jan. 12 (1957) 12.
2 ACKERMANN, H. AND LANGEMANN, H., *Helv. Physiol. et Pharmacol. Acta*, 18 (1960) C5.
3 AIRAKSINEN, M. M. AND MATTILA, M., *Acta Pharmacol. et Toxicol.*, 19 (1962) 199.
4 AIVAZIAN, G. H. AND GRIFFITH, J. D., *A.M.A. Arch. Gen. Psychiat.*, 2 (1960) 7.
5 AKCASU, A., AKCASU, M. AND TUMAY, S. B., *Nature*, 187 (1960) 324.
6 AKERFELDT, S., *Science*, 125 (1957) 117.
7 ALBRECHT, P., VISSCHER, M. B., BITTNER, J. J. AND HALBERG, F., *Proc. Soc. Exptl. Biol. Med.*, 92 (1956) 703.
8 ALEXANDER, F., CURTIS, G. C., SPRINCE, H. AND CROSLEY, A. P., *J. Nervous Mental Disease*, 137 (1963) 135.
9 AMIN, A. H., CRAWFORD, T. B. B. AND GADDUM, J. H., *J. Physiol. (London)*, 126 (1954) 596.
9a ANDÉN, N. E., CARLSSON, A., HILLARP, N. A. AND MAGNUSSON, T., *Life Sci.*, 3 (1964) 473.
9b ANDÉN, N. E., LUNDBERG, A., ROSENGREN, E. AND VYKLICKY, L., *Experientia*, 19 (1963) 654.
10 ANDERSON, E. G. AND BONNYCASTLE, D. D., *J. Pharmacol. Exptl. Therap.*, 130 (1960) 138.
11 ANDERSON, E. G., MARKOWITZ, S. D. AND BONNYCASTLE, D. D., *J. Pharmacol. Exptl. Therap.*, 136 (1962) 179.
12 ANDERSON, J. A., ZIEGLER, M. R. AND DOEDEN, D., *Science*, 127 (1958) 236.
13 ANGIBEAUD, P., BUCHEL, L. AND LEVY, J., *Anesthésie et analgésie*, 16 (1959) 1.
14 ANTONA, G., *Farmaco (Pavia), Ed. pract.*, 10 (1955) 86.
15 ANTONELLI, A. R., BERTACCINI, G. AND MANTEGAZZINI, P., *J. Neurochem.*, 8 (1961) 157.
16 APRISON, M. H., *Recent Advances in Biological Psychiatry*, Grune and Stratton, New York, 1962, p. 4.
17 APRISON, M. H., *Federation Proc.*, 19 (1960) 275.
18 APRISON, M. H. AND DREW, A. L., *Science*, 127 (1958) 758.
19 APRISON, M. H. AND FERSTER, C. B., *Experientia*, 16 (1960) 159.
20 APRISON, M. H. AND FERSTER, C. B., *J. Pharmacol. Extpl. Therap.*, 131 (1961) 100.
21 APRISON, M. H. AND FERSTER, C. B., *J. Neurochem.*, 6 (1961) 350.
21a APRISON, M. H., TAKAHASHI, R. AND FOLKERTH, T. L., *J. Neurochem.*, 11 (1964) 341.
22 APRISON, M. H., WOLF, M. A., POULOS, G. L. AND FOLKERTH, T. L., *J. Neurochem.*, 9 (1962) 575.
23 ARIOKA, J. AND TANIMUKAI, H., *J. Neurochem.*, 1 (1957) 311.
24 ARMSTRONG, D., in G. P. LEWIS (Ed.), *5-Hydroxytryptamine*, Pergamon, London, 1958, p. 140.
25 ARMSTRONG, D., DRY, R. M. L., KEELE, C. A. AND MARKHAM, J. W., *J. Physiol. (London)*, 120 (1953) 326.
26 ARMSTRONG, M. D. AND ROBINSON, K. S., *Arch. Biochem. Biophys.*, 52 (1954) 287.
27 ARMSTRONG, M. D. AND ROBINSON, K. S., *Federation Proc.*, 13 (1954) 175.
28 ARMSTRONG, M. D. AND TYLER, F. H., *J. Clin. Invest.*, 34 (1955) 565.
29 ARRIGONI-MARTELLI, E. AND KRAMER, M., *Med. Exptl.*, 1 (1959) 45.
30 ASHCROFT, G. W. AND SHARMAN, D. F., *Nature*, 186 (1960) 1050.
31 AUERBACH, V. H., WAISMAN, H. A. AND WYCKOFF JR., L. B., *Nature*, 182 (1958) 871.
32 AXELROD, J., *Science*, 127 (1958) 754.
33 AXELROD, J. AND INSCOE, J. K., *J. Pharmacol. Exptl. Therap.*, 141 (1963) 161.
34 BACIOCCHI, M., GRANDMONTAGNE, O. AND ROUAUX, J., *Gaz. méd. France*, (1961) 1545.
35 BAKER, R. V., *J. Physiol. (London)*, 145 (1959) 473.
36 BALDRIDGE, R. C., BOROFSKY, L., BAIRD, H., REICHLE, F. AND BULLOCK, D., *Proc. Soc. Exptl. Biol. Med.*, 100 (1959) 529.
37 BANERJEE, S. AND AGARWAL, P. S., *Proc. Soc. Exptl. Biol. Med.*, 97 (1958) 65.
38 BANERJEE, S. AND AGARWAL, P. S., *Proc. Soc. Exptl. Biol. Med.*, 97 (1958) 657.
39 BARBEAU, A., *A.M.A. Arch. Neurol.*, 4 (1961) 97.
40 BARBEAU, A. AND JASMIN, G., *Rev. can. biol.*, 20 (1961) 837.
41 BARBEAU, A., MURPHY, G. F. AND SOURKES, T. L., *Science*, 133 (1961) 1706.
42 BARNETT, R. J., HAGEN, P. AND LEE, F. L., *Biochem. J.*, 69 (1958) 36P.
43 BAUDIŠ, P., VÁŇA, J., CERHOVÁ, M. AND ŠÍDLOVÁ, A., *Československ. psychiat.*, 57 (1961) 164.
44 BENASSI, C. A., BENASSI, P., ALLEGRI, G. AND BALLARIN, P., *J. Neurochem.*, 7 (1961) 264.
45 BENDA, P., *Presse méd.*, 67 (1959) 1818.

46 BENDA, P., *Presse méd.*, 67 (1959) 1887.

47 BENITEZ, H., MURRAY, M. AND WOOLLEY, D. W., *Proc. Intern. Congr. Neuropathol.*, 2nd Congr., *London*, 1955, Part II, p. 423.

48 BENITEZ, H., MURRAY, M. AND WOOLLEY, D. W., *Anat. Record*, 121 (1955) 446.

49 BERENDES, H., ANDERSON, J. A., ZIEGLER, M. R. AND RUTTENBERG, D., *A.M.A. J. Diseases Children*, 96 (1958) 430.

50 BERKINSHAW-SMITH, E. M. I., MORGAN, R. S. AND PAYLING WRIGHT, G., *Nature*, 196 (1962) 173.

51 BERNHEIMER, H., BIRKMAYER, W. AND HORNYKIEWICZ, O., *Klin. Wochschr.*, 39 (1961) 1056.

52 BERTACCINI, G., *J. Neurochem.*, 4 (1959) 217.

53 BERTACCINI, G., *J. Physiol. (London)*, 153 (1960) 239.

54 BERTLER, Å., *Acta Physiol. Scand.*, 51 (1961) 75.

55 BERTLER, Å., *Acta Physiol. Scand.*, 51 (1961) 97.

56 BERTLER, Å., CARLSSON, A. AND ROSENGREN, E., *Naturwissenschaften*, 43 (1956) 521.

57 BERTLER, Å. AND ROSENGREN, E., *Experientia*, 15 (1959) 382.

58 BESENDORF, H. AND PLETSCHER, A., *Helv. Physiol. et Pharmacol. Acta*, 14 (1956) 383.

59 BIANCHI, C., *Nature*, 179 (1957) 202.

60 BICKEL, H., GERRARD, J. AND HICKMANS, E. M., *Lancet*, ii (1953) 812.

61 BIEL, J. H., DRUKKER, A. E., SHORE, P. A., SPECTOR, S. AND BRODIE, B. B., *J. Am. Chem. Soc.*, 80 (1958) 1519.

62 BIEL, J. H., NUHFER, P. A. AND CONWAY, A. C., *Ann. N.Y. Acad. Sci.*, 80 (1959) 568.

63 BIRKMAYER, W. AND HORNYKIEWICZ, O., *Wien. klin. Wochschr.*, 73 (1961) 787.

64 BOGDANSKI, D. F., PLETSCHER, A., BRODIE, B. B. AND UDENFRIEND, S., *J. Pharmacol. Exptl. Therap.*, 117 (1956) 82.

65 BOGDANSKI, D. F. AND UDENFRIEND, S., *J. Pharmacol. Exptl. Therap.*, 116 (1956) 7.

66 BOGDANSKI, D. F., WEISSBACH, H. AND UDENFRIEND, S., *J. Neurochem.*, 1 (1957) 272.

67 BOGDANSKI, D. F., WEISSBACH, H. AND UDENFRIEND, S., *J. Pharmacol. Exptl. Therap.*, 122 (1958) 182.

68 BONNYCASTLE, D. D., BONNYCASTLE, H. F. AND ANDERSON, E. G., *J. Pharmacol. Exptl. Therap.*, 135 (1962) 17.

69 BONNYCASTLE, D. D., GIARMAN, N. J. AND PAASONEN, M. K., *Brit. J. Pharmacol.*, 12 (1957) 228.

70 BONNYCASTLE, D. D., PAASONEN, M. K. AND GIARMAN, N. J., *Nature*, 178 (1956) 990.

71 BORGES, F. J., MERLIS, J. K. AND BESSMAN, S. P., *J. Clin. Invest.*, 38 (1959) 715.

72 BORN, G. V. R. AND GILLSON, R. E., *J. Physiol. (London)*, 137 (1957) 82P.

73 BORN, G. V. R., INGRAM, G. I. C. AND STACEY, R. S., *Brit. J. Pharmacol.*, 13 (1958) 62.

74 BRADLEY, P. B., in G. P. LEWIS (Ed.), *5-Hydroxytryptamine*, Pergamon, London, 1958, p. 214.

75 BRADLEY, P. B. AND HAUCE, A. J., *J. Physiol. (London)*, 132 (1956) 50P.

76 BREITNER, C., PICCHIONI, A., CHIN, L. AND BURTON, L. E., *Diseases of Nervous System*, 22, Monograph Suppl. (1961) 1.

77 BRENGELMANN, J. C., PARE, C. M. B. AND SANDLER, M., *J. Mental Sci.*, 104 (1958) 1237.

78 BRENGELMANN, J. C., PARE, C. M. B. AND SANDLER, M., *J. Mental Sci.*, 105 (1959) 770.

79 BRODIE, B. B., in G. P. LEWIS (Ed.), *5-Hydroxytryptamine*, Pergamon, London, 1958, p. 64.

80 BRODIE, B. B. AND BEAVEN, M. A., *Med. Exptl.*, 8 (1963) 320.

81 BRODIE, B. B. AND COSTA, E., in J. DE AJURIAGUERRA (Ed.), *Monoamines et Système Nerveux Central*, Symposium Bel-Air, Genève, 1961, p. 13.

82 BRODIE, B. B., FINGER, K. F., ORLANS, F. B., QUINN, G. P. AND SULSER, F., *J. Pharmacol. Exptl. Therap.*, 129 (1960) 250.

83 BRODIE, B. B., MAICKEL, R. P. AND WESTERMANN, E. O., in S. S. KETY AND J. ELKES (Eds.), *Regional Neurochemistry*, Pergamon, Oxford, 1961, p. 351.

84 BRODIE, B. B., OLIN, J. S., KUNTZMAN, R. G. AND SHORE, P. A., *Science*, 125 (1957) 1293.

85 BRODIE, B. B., PLETSCHER, A. AND SHORE, P. A., *Science*, 122 (1955) 968.

86 BRODIE, B. B., PLETSCHER, A. AND SHORE, P. A., *J. Pharmacol. Exptl. Therap.*, 116 (1956) 9.

87 BRODIE, B. B., PLETSCHER, A. AND SHORE, P. A., *J. Pharmacol. Exptl. Therap.*, 116 (1956) 9.

88 BRODIE, B. B. AND SHORE, P. A., *Ann. N.Y. Acad. Sci.*, 66 (1957) 631.

89 BRODIE, B. B., SHORE, P. A. AND PLETSCHER, A., *Science*, 123 (1956) 992.

90 BRODIE, B. B., SPECTOR, S., KUNTZMAN, R. G. AND SHORE, P. A., *Naturwissenschaften*, 10 (1958) 243.
91 BRODIE, B. B., SPECTOR, S. AND SHORE, P. A., *Pharmacol. Revs.*, 11 (1959) 548.
92 BRODIE, B. B., SPECTOR, S. AND SHORE, P. A., *Ann. N.Y. Acad. Sci.*, 80 (1959) 609.
93 BRODIE, B. B., TOMICH, E. G., KUNTZMAN, R. AND SHORE, P. A., *J. Pharmacol. Exptl. Therap.*, 119 (1957) 461.
94 BROWN, B. B., *Ann. N.Y. Acad. Sci.*, 66 (1957) 677.
95 BROWN, B. B., *Arch. intern. pharmacodyn.*, 128 (1960) 391.
96 BROWN, F. C., WHITE, J. B. AND KENNEDY, J. K., *Am. J. Psychiat.*, 117 (1960) 63.
97 BRUNE, G. G. AND HIMWICH, H. E., *Science*, 133 (1961) 190.
98 BRUNE, G. G. AND HIMWICH, H. E., *A.M.A. Arch. Gen. Psychiat.*, 6 (1962) 324.
99 BRUNE, G. G. AND HIMWICH, H. E., *J. Nervous Mental Disease*, 134 (1962) 447.
100 BRUNE, G. G. AND PSCHEIDT, G. R., *Federation Proc.*, 20 (1961) 889.
101 BUCHEL, L., LEVY, J. AND TANGUY, O., *Rev. agressol.*, 1 (1960) 389.
102 BUCHEL, L. AND LEVY, J., *Anesthésie et analgésie*, 17 (1960) 314.
103 BUCHEL, L., LEVY, J. AND TANGUY, O., *Compt. rend. acad. sci.*, 246 (1958) 2947.
104 BUCHEL, L., LEVY, J. AND TANGUY, O., *Rev. agressol.*, 1 (1960) 389.
105 BULLE, P. H. AND KONCHEGUL, L., *Quart. Rev. Psychiat. Neurol.*, 18 (1957) 287.
106 BUMPUS, F. M. AND PAGE, I. H., *J. Biol. Chem.*, 212 (1955) 111.
106a BURKARD, W. P., GEY, K. F. AND PLETSCHER, A., *Life Sci.*, 3 (1964) 27.
107 BURTON, R. M., *Ann. N.Y. Acad. Sci.*, 66 (1957) 695.
108 BUSCAINO, G. A. AND STEFANACHI, L., *A.M.A. Arch. Neurol. Psychiat.*, 80 (1958) 78.
109 BUSCAINO, G. A. AND STEFANACHI, L., *Confinia Neurol.*, 18 (1958) 188.
110 BUZAS, A. AND RÉGNIER, G., *Compt. rend. acad. sci.*, 250 (1960) 1340.
111 CAFRUNY, E. J. AND DOMINO, E. F., *A.M.A. Arch. Neurol. Psychiat.*, 79 (1958) 336.
112 CAHN, J., *Thérapie*, 13 (1958) 62.
113 CAHN, J., GEORGES, G. AND HEROLD, H., in S. GARATTINI AND V. GHETTI (Eds.), *Psychotropic Drugs*, Elsevier, Amsterdam, 1957, p. 481.
114 CAHN, J., HEROLD, M., DUBRASQUET, M. AND BURET, J. P., *Compt. rend. soc. biol.*, 151 (1957) 82.
115 CANAL, N., MAFFEI-FACCIOLI, A. AND SCAMAZZO, T., *Boll. soc. ital. biol. sper.*, 35 (1959) 436.
116 CANAL, N. AND MANZINI, B., *II Simposio centro cefalee, clin. med. univ. Firenze*, (1960) 1.
117 CANALI, G. AND GRISONI, R., in S. GARATTINI AND V. GHETTI (Eds.), *Psychotropic Drugs*, Elsevier, Amsterdam, 1957, p. 289.
118 CARELS, G., *Arch. intern. pharmacodyn.*, 138 (1962) 326.
119 CARLINI, E. A. AND GREEN, J. P., *Brit. J. Pharmacol.*, 20 (1963) 264.
120 CARLSSON, A., *Pharmacol. Revs.*, 11 (1959) 490.
121 CARLSSON, A., in E. ROTHLIN (Ed.), *Neuropsychopharmacology*, Vol. 2, Elsevier, Amsterdam, 1961, p. 417.
122 CARLSSON, A., FALCK, B. AND HILLARP, N. Å., *Acta Physiol. Scand.*, 56, Suppl. 196 (1962).
123 CARLSSON, A. AND HILLARP, N. Å., *Kgl. Fysiograf. Sällskap. Lund, Förh.*, 26 (1956) n. 8.
124 CARLSSON, A., LINDQVIST, M., MAGNUSSON, T. AND WALDECK, B., *Science*, 127 (1958) 471.
125 CARLSSON, A. AND LINDQVIST, M., in J. DE AJURIAGUERRA (Ed.), *Monoamines et Système Nerveux Central*, Symposium Bel-Air, Genève, 1961, p. 89.
126 CARLSSON, A. AND LINDQVIST, M., *Acta Physiol. Scand.*, 54 (1962) 87.
127 CARLSSON, A., LINDQVIST, M. AND MAGNUSSON, T., *Nature*, 180 (1957) 1200.
128 CARLSSON, A., LINDQVIST, M. AND MAGNUSSON, T., in J. R. VANE, G. E. W. WOLSTENHOLME AND M. O'CONNOR (Eds.), *Adrenergic Mechanisms*, Churchill, London, 1960, p. 432.
129 CARLSSON, A., MAGNUSSON, T. AND ROSENGREN, E., *Experientia*, 19 (1963) 359.
130 CARLSSON, A., ROSENGREN, E., BERTLER, Å. AND NILSSON, J., in S. GARATTINI AND V. GHETTI (Eds.), *Psychotropic Drugs*, Elsevier, Amsterdam, 1957, p. 363.
131 CASELLA, C., FUMAGALLI, B. AND NOLI, S., *Boll. soc. ital. biol. sper.*, 33 (1957) 1550.
132 CASELLA, C., MAGGI, G. C. AND NOLI, S., *Arch. fisiol.*, 59 (1959) 182.
133 CASELLA, G. AND RAPUZZI, G., *Boll. soc. ital. biol. sper.*, 33 (1957) 719.
134 CASELLA, C. AND RAPUZZI, G., *Boll. soc. ital. biol. sper.*, 33 (1957) 721.
135 CASELLA, C. AND RAPUZZI, R., *Boll. soc. ital. biol. sper.*, 33 (1957) 1569.
136 CASELLA, C. AND RAPUZZI, G., *Rend. ist. lombardo sci.*, B 92 (1958) 265.

137   CATER, D., GARATTINI, S., SILVER, J. AND MARINA, F., *Proc. Royal Soc. (London)*, B, 155 (1961) 136.
138   CERLETTI, A. AND ROTHLIN, E., *Nature*, 176 (1955) 785.
139   CHARI-BRITON, A., *Biochem. Biophys. Research Communs.*, 12 (1963) 310.
140   CHAUCHARD, P. AND MAZOUÉ, H., *Compt. rend. soc. biol.*, 151 (1957) 463.
141   CHAUCHARD, P. AND MAZOUÉ, H., *Compt. rend. soc. biol.*, 157 (1963) 525.
142   CHEN, G., ENSOR, G. R. AND BOHNER, B., *Proc. Soc. Exptl. Biol. Med.*, 86 (1954) 507.
143   CHESSIN, M., DUBNICK, B. LEESON, G. AND SCOTT, C. C., *Ann. N.Y. Acad. Sci.*, 80 (1959) 597.
144   CHESSIN, M., KRAMER, E. R. AND SCOTT, C. C., *J. Pharmacol. Exptl. Therap.*, 119 (1957) 453.
145   COOK, L. AND WEIDLEY, E., *Ann. N.Y. Acad. Sci.*, 66 (1957) 740.
146   COOPER, J. R. AND MELCER, I., *J. Pharmacol. Exptl. Therap.*, 132 (1961) 265.
147   COPPEN, A. AND SHAW, D. M., *Lancet*, i (1963) 79.
148   CORREALE, P., *J. Neurochem.*, 1 (1956) 22.
149   CORREL, J. T., LYTH, L. F., LONG, S. AND VANDERPOEL, J. C., *Am. J. Physiol.*, 169 (1952) 537.
150   COSTA, E., personal communication.
151   COSTA, E. AND APRISON, M. H., *J. Nervous Mental Disease*, 126 (1958) 289.
152   COSTA, E. AND APRISON, M. H., *Am. J. Physiol.*, 192 (1958) 95.
153   COSTA, E., GESSA, G. L., HISCH, C., KUNTZMAN, R. AND BRODIE, B. B., *Ann. N.Y. Acad. Sci.*, 96 (1962) 118.
154   COSTA, E. AND HIMWICH, H. E., in F. BRÜCKE (Ed.), *Biochemistry of the Central Nervous System*, Pergamon, London, 1959, p. 283.
155   COSTA, E., HIMWICH, W. A., GOLDSTEIN, S. G., CANHAM, R. AND HIMWICH, H. E., *Federation Proc.*, 18 (1959) 379.
156   COSTA, E. AND RINALDI, F., *Am. J. Physiol.*, 194 (1958) 214.
157   COSTA, E., RINALDI, F. AND HIMWICH, H. E., in S. GARATTINI AND V. GHETTI (Eds.), *Psychotropic Drugs*, Elsevier, Amsterdam, 1957, p. 21.
158   CRAWFORD, T. B. B., in G. P. LEWIS (Ed.), *5-Hydroxytryptamine*, Pergamon, London, 1958, p. 20.
159   CREPAX, P. AND INFANTELLINA, F., *Arch. sci. biol. (Bologna)*, 41 (1957) 207.
160   CUENCA, E., KUNTZMAN, R., COSTA, E. AND BRODIE, B. B., *Federation Proc.*, 20 (1961) 322.
161   CULLEY, W. J., SAUNDERS, R. N., MERTZ, E. T. AND JOLLY, D. H., *Proc. Soc. Exptl. Biol. Med.*, 111 (1962) 444.
162   CULLEY, W. J., SAUNDERS, R. N., MERTZ, E. T. AND JOLLY, D. H., *Proc. Soc. Exptl. Biol. Med.*, 113 (1963) 645.
163   CURTIS, D. R., *J. Physiol. (London)*, 145 (1959) 175.
164   CURTIS, D. R. AND DAVIS, R., *Nature*, 192 (1961) 1083.
165   CURTIS, D. R. AND KOIZUMI, K., *J. Neurophysiol.*, 24 (1961) 80.
166   DALGLIESH, C. E., *Advances in Protein Chem.*, 10 (1955) 31.
166a DARIN DE LORENZO, A. J., *Bull. Johns Hopkins Hosp.*, 108 (1961) 258.
167   DAVIDSON, J. AND SANDLER, M., *Nature*, 181 (1958) 186.
168   DAVIDSON, J., SJOERDSMA, A., LOOMIS, L. N. AND UDENFRIEND, S., *J. Clin. Invest.*, 36 (1957) 1594.
169   DE BARAN, L., GOGOLAK, G., LONGO, V. G. AND STUMPF, CH., *J. Pharmacol. Exptl. Therap.*, 139 (1963) 337.
170   DEGKWITZ, R., FROWEIN, R., KULENKAMPFF, C. AND MOHS, U., *Klin. Wochschr.*, 38 (1960) 120.
171   DE KALBERMATTEN, J. P., *Thesis*, University of Lausanne, 1962.
172   DEMING, Q., BOGDANSKI, D. F., UDENFRIEND, S., SHORE, P. A. AND BRODIE, B. B., *Federation Proc.*, 15 (1956) 416.
173   DE ROBERTIS, E., *Biochem. Pharmacol.*, 8 (1961) 29.
174   DE ROBERTIS, E., PELLEGRINO DE IRALDI, A., RODRIGUEZ, G. AND GOMEZ, C. J., *J. Biophys. Biochem. Cytol.*, 9 (1961) 229.
175   DE SCHAEPDRYVER, A. F., PIETTE, Y. AND DELAUNOIS, A. L., *Arch. intern. pharmacodyn.*, 140 (1962) 358.
176   DI GEORGE, A. M., *Pediat. Clin. N. Am.*, 10 (1963) 723.
177   DI STEFANO, V., LEARY, D. E. AND FELDMAN, S., *Federation Proc.*, 15 (1956) 417.
178   DOBKIN, A., HARLAND, J. H. AND FEDORUK, S., *Anesthesiology*, 21 (1960) 13.
178a DOMBRO, R. S., BRADHAM, L. S., CAMPBELL, N. K. AND WOOLLEY, D. W., *Biochim. Biophys. Acta*, 54 (1961) 516.

179 DOMER, F. R. AND FELDBERG, W., *J. Physiol. (London)*, 154 (1960) 34P.
180 DOMER, F. R. AND FELDBERG, W., *Brit. J. Pharmacol.*, 15 (1960) 578.
181 DOMER, F. R. AND LONGO, V. G., *Arch. intern. pharmacodyn.*, 136 (1962) 204.
182 DUBNICK, B., LEESON, G. A. AND PHILLIPS, G. E., *Biochem. Pharmacol.*, 11 (1962) 45.
183 DURITZ, G. AND TRUITT JR., E. B., *Federation Proc.*, 22 (1963) 272.
184 EDELSTEIN, E. L. AND KRASSILOWSKY, D., *J. Nervous Mental Disease*, 128 (1959) 459.
185 EFFRON, D. H. AND GESSA, G. L., *Arch. intern. pharmacodyn.*, 142 (1963) 111.
186 EHRINGER, H., HORNYKIEWICZ, O. AND LECHNER, K., *Arch. exptl. Pathol. Pharmakol., Naunyn-Schmiedeberg's*, 241 (1961) 568.
187 ERSPAMER, V., *Pharmacol. Revs.*, 6 (1954) 425.
188 ERSPAMER, V., *Lancet*, i (1956) 511.
189 ERSPAMER, V., *Naturwissenschaften*, 43 (1956) 61.
190 ERSPAMER, V., *XXI Congr. intern. cienc. fisiol.*, Buenos Aires, 1959, Conf. espec., p. 216.
191 ERSPAMER, V., *Progr. in Drug Research*, 3 (1961) 151.
192 ERSPAMER, V. AND CICERI, C., *Experientia*, 13 (1957) 87.
193 EVARTS, E. V., *Ann. N.Y. Acad. Sci.*, 66 (1957) 479.
194 EVERETT, G. M. AND WIEGAND, R. G., *Biochem. Pharmacol., Proc. 1st Intern. Pharmacol. Meeting*, 8 (1962) 85.
195 FABING, H. B. AND HAWKINS, J. R., *Science*, 123 (1956) 886.
196 FASTIER, F. N., *Experientia*, 12 (1956) 351.
197 FASTIER, F. N., SPEDEN, R. N. AND WAAL, H., *Brit. J. Pharmacol.*, 12 (1957) 251.
198 FELDBERG, W. AND MEYERS, R. D., *Nature*, 200 (1963) 1325.
199 FELDBERG, W. AND SHERWOOD, S. L., *J. Physiol. (London)*, 120 (1953) 3P.
200 FELDBERG, W. AND SHERWOOD, S. L., *J. Physiol. (London)*, 123 (1954) 148.
201 FELDSTEIN, A., HOAGLAND, H. AND FREEMAN, H., *Science*, 128 (1958) 358.
202 FELDSTEIN, A., HOAGLAND, H. AND FREEMAN, H., *J. Nervous Mental Disease*, 129 (1959) 62.
202a FELDSTEIN, A., HOAGLAND, H. AND FREEMAN, H., *Arch. Gen. Psychiat.*, 5 (1961) 246.
203 FELDSTEIN, A., WONG, K. K. AND FREEMAN, H., *Federation Proc.*, 20 (1961) 340.
203a FELDSTEIN, A., WONG, K. K. AND FREEMAN, H., *J. Psychiat. Research*, 2 (1964) 41.
204 FELLMAN, J. H., *Enzymologia*, 20 (1959) 366.
205 FELLOWS, E. J. AND COOK, L., in S. GARATTINI AND V. GHETTI (Eds.,) *Psychotropic Drugs*, Elsevier, Amsterdam, 1957, p. 397.
206 FENSTER, E. AND TOWNE, J. C., *Federation Proc.*, 22 (1963) 624.
207 FERRARI, V., CAMPAGNARI, F. AND GUIDA, A., *Minerva med.*, 46 (1955) 119.
208 FERRINI, R. AND GLÄSSER, A., *J. Pharm. Pharmacol.*, 15 (1963) 772.
209 FERSTER, C. B. AND APRISON, M. H., *Pharmacologist*, 1 (1959) 75.
210 FIORETTI-LUCARELLI, L., *Boll. chim. farm.*, 7 (1958) 16.
211 FISCHER, E., VAZQUEZ, F. A., FERNANDEZ, T. A. AND LISKOWSKI, L., *Lancet*, i (1961) 890.
212 FISCHER, P. AND LECOMTE, J. C., *Compt. rend. soc. biol.*, 150 (1956) 1026.
213 FORNAROLI, P. AND KOLLER, M., *Farmaco (Pavia), Ed. Pract.*, 10 (1954) 546.
214 FORNAROLI, P. AND KOLLER, M., *Farmaco (Pavia), Ed. Pract.*, 12 (1955) 984.
215 FORREST, A. D., *J. Mental Sci.*, 103 (1957) 614.
216 FREEDLAND, R. A., WADZINSKI, I. M. AND WAISMAN, H. A., *Biochem. Biophys. Research Communs.*, 5 (1961) 94.
217 FREEDLAND, R. A., WADZINSKI, I. M. AND WAISMAN, H. A., *Biochem. Biophys. Research Communs.*, 6 (1961) 227.
218 FREEDMAN, D. X., *Federation Proc.*, 19 (1960) 266.
219 FREEDMAN, D. X. AND GIARMAN, N. J., in G. H. GLASER (Ed.), *EEG and Behaviour*, Basic Books, New York, 1963, p. 198.
220 FRETER, K., WEISSBACH, H., UDENFRIEND, S. AND WITKOP, B., *Proc. Soc. Exptl. Biol. Med.*, 94 (1957) 725.
221 FROHMAN, C., GOODMAN, M., LUBY, E., BECKETT, P. AND SEUF, R., *A.M.A. Arch. Neurol. Psychiat.*, 79 (1958) 730.
222 GADDUM, J. H., *Ciba Foundation Symposium on Hypertension*, Little, Brown and Co., 1954, Boston, p. 75.
223 GADDUM, J. H., *Ann. N.Y. Acad. Sci.*, 66 (1957) 643.
224 GADDUM, J. H. AND GIARMAN, N. J., *Brit. J. Pharmacol.*, 11 (1956) 88.

225 GADDUM, J. H. AND VOGT, M., *Brit. J. Pharmacol.*, 11 (1956) 175.

226 GAL, E. M. AND DREWES, P. A., *Proc. Soc. Exptl. Biol. Med.*, 106 (1961) 295.

227 GAL, E. M. AND DREWES, P. A., *Proc. Soc. Exptl. Biol. Med.*, 110 (1962) 368.

228 GAL, E. M., DREWES, P. A. AND BARRACLOUGH, C. A., *Biochem. Pharmacol.*, 8 (1961) 32.

229 GAL, E. M., DREWES, P. A. AND BARRACLOUGH, C. A., *Biochem. Pharmacol., Proc. 1st Intern. Pharmacol. Meeting*, 8 (1962) 107.

230 GAL, E. M. AND MARSHALL JR., F. D., in H. E. HIMWICH AND W. A. HIMWICH (Eds.), *Binding Sites of Brain Biogenic Amines*, Elsevier, Amsterdam, 1964 (in press).

231 GAL, E. M., POCZIK, M. AND MARSHALL JR., F. D., *Biochem. Biophys. Research Communs.*, 12 (1963) 39.

232 GANGLOFF, H. AND MONNIER, H., *Helv. Physiol. et Pharmacol. Acta*, 15 (1957) 83.

232a GANROT, P. O., GOTTFRIES, C. G. AND ROSENGREN, E., *Acta Psychiat. Scand.*, 39, Suppl. 169 (1963) 3.

232b GANROT, P. O., ROSENGREN, E. AND GOTTFRIES, C. G., *Experientia*, 18 (1962) 260.

233 GARATTINI, S., *Schweiz. Arch. Neurol. Neurochir. Psychiat.*, 84 (1959) 269.

234 GARATTINI, S., KATO, R. AND VALZELLI, L., *Psychiat. et Neurol.*, 140 (1960) 190.

235 GARATTINI, S., KATO R. AND VALZELLI, L., *Experientia*, 16 (1960) 120.

236 GARATTINI, S., LAMESTA, L., MORTARI, A., PALMA, V. AND VALZELLI, L., *J. Pharm. Pharmacol.*, 13 (1961) 385.

237 GARATTINI, S., LAMESTA, L., MORTARI, A. AND VALZELLI, L., *J. Pharm. Pharmacol.*, 13 (1961) 548.

238 GARATTINI, S., MORTARI, A., VALSECCHI, A. AND VALZELLI, L., *Nature*, 183 (1959) 1273.

239 GARATTINI, S. AND VALZELLI, L., *Boll. soc. ital. biol. sper.*, 31 (1955) 1648.

240 GARATTINI, S. AND VALZELLI, L., *Boll. soc. ital. biol. sper.*, 32 (1956) 292.

241 GARATTINI, S. AND VALZELLI, L., in S. GARATTINI AND V. GHETTI (Eds.), *Psychotropic Drugs*, Elsevier, Amsterdam, 1957, p. 428.

242 GARATTINI, S. AND VALZELLI, L., *Science*, 128 (1958) 1278.

243 GARATTINI, S. AND VALZELLI, L., *Symposium "Le Sindromi Depressive"*, Ed. Minerva Medica, Torino, 1960, p. 7.

244 GARATTINI, S. AND VALZELLI, L., in J. DE AJURIAGUERRA (Ed.), *Monoamines et Système Nerveux Central*, Symposium Bel-Air, Geneve, 1961, p. 59.

245 GATTI, G. L., in S. GARATTINI AND V. GHETTI (Eds.), *Psychotropic Drugs*, Elsevier, Amsterdam, 1957, p. 125.

246 GEIGER, R. S., *Federation Proc.*, 16 (1957) 44.

247 GEIGER, R. S., *Federation Proc.*, 17 (1958) 52.

248 GENOVESE, E. AND GOLDWURM, G. F., *Boll. soc. ital. biol. sper.*, 34 (1958) 1399.

249 GEY, K. F., BURKARD, W. P. AND PLETSCHER, A., *Biochem. Pharmacol.*, 8 (1961) 383.

250 GEY, K. F. AND PLETSCHER, A., *Experientia*, 16 (1960) 372.

251 GEY, K. F. AND PLETSCHER, A., in J. M. BORDELEAU (Ed.), *Extrapyramidal System and Neuroleptics*, Éditions Psychiatriques, Montreal, 1961, p. 175.

252 GEY, K. F. AND PLETSCHER, A., *Helv. Physiol. et Pharmacol. Acta*, 19 (1961) C22.

253 GEY, K. F. AND PLETSCHER, A., *J. Pharmacol. Exptl. Therap.*, 133 (1961) 18.

254 GEY, K. F. AND PLETSCHER, A., *Nature*, 194 (1962) 387.

255 GIARMAN, N. J., *Federation Proc.*, 15 (1956) 428.

256 GIARMAN, N. J., *Yale J. Biol. Med.*, 32 (1959) 73.

257 GIARMAN, N. J. AND POTTER, L. T., *Federation Proc.*, 17 (1958) 371.

258 GIARMAN, N. J. AND SCHANBERG, S. M., *Biochem. Pharmacol.*, 1 (1958) 301.

259 GIRARD, J. P., *Med. Exptl.*, 7 (1962) 287.

260 GLÄSSER, A. AND MANTEGAZZINI, P., *Arch. ital. biol.*, 98 (1960) 351.

261 GLUCKMAN, M. I., ROSS HART E. AND MARRAZZI, A. S., *Science*, 126 (1957) 448.

262 GOLDSTEIN, M., FRIEDHOFF, A. J., SIMMONS, C. AND PROCHOROFF, N. N., *Experientia*, 15 (1959) 254.

263 GOMIRATO, G. AND ZANALDA, A., *Arch. sci. med.*, 106 (1958) 323.

264 GORDON, M. W., SIMS, J. A., HANSON, R. K. AND KUTTNER, R. E., *J. Neurochem.*, 9 (1962) 477.

265 GÖRÖG, P. AND SZPORNY, L., *Biochem. Pharmacol.*, 11 (1961) 165.

266 GRANA, E. AND LILLA, L., *Farmaco (Pavia), Ed. sci.*, 12 (1957) 1025.

267 GRANDJEAN, R. AND BÄTTIG, K., *Helv. Physiol. et Pharmacol. Acta*, 15 (1957) 366.

268   GRAY, E. G. AND WHITTAKER, V. P., *J. Physiol. (London)*, 153 (1960) 35P.
269   GRAY, E. G. AND WHITTAKER, V. P., *J. Anat.*, 96 (1962) 79.
270   GREEN, H., GREENBERG, S. M., ERICKSON, R. W., SAWYER, J. L. AND ELLISON, T., *J. Pharmacol. Exptl. Therap.*, 136 (1962) 174.
271   GREEN, J. P., *Advances in Pharmacol.*, 1 (1962) 349.
272   GREEN, J. P., FURANO, A. V. AND CARLINI, E. A., *Federation Proc.*, 21 (1962) 365.
273   GREEN, J. P., PAASONEN, M. K. AND GIARMAN, N. J., *Proc. Soc. Exptl. Biol. Med.*, 94 (1957) 428.
274   GREEN, J. P., ROBINSON, J. D. AND DAY, M., *J. Pharmacol. Exptl. Therap.*, 131 (1961) 12.
275   GREENBERG, R. AND TOMAN, J. E. P., *Federation Proc.*, 17 (1958) 60.
276   GREIG, M. E., WALK, R. A. AND GIBBONS, A. J., *J. Pharmacol. Exptl. Therap.*, 127 (1959) 110.
277   GRISEMER, E. C., BARSKY, J., DRAGSTELDT, C. A., WELLS, J. A. AND ZELLER, E. A., *Proc. Soc. Exptl. Biol. Med.*, 84 (1953) 699.
278   GUROFF, G. AND UDENFRIEND, S., *J. Biol. Chem.*, 237 (1962) 803.
279   GURSEY, D. AND OLSON, R. E., *Proc. Soc. Exptl. Biol. Med.*, 104 (1960) 280.
280   GURSEY, D., VESTER, J. W. AND OLSON, R. E., *J. Clin. Invest.*, 38 (1959) 1008.
281   GUZMAN, F., BRAUN, C. AND LINN, R. K. S., *Arch. intern. pharmacodyn.*, 136 (1962) 353.
282   GYERMEK, L., LAZAR, I. AND CSAK, A. Zs., *Arch. intern. pharmacodyn.*, 107 (1956) 62.
283   HAGEN, P., BARNETT, R. J. AND FU-LI, L., *J. Pharmacol. Exptl. Therap.*, 126 (1959) 91.
284   HAGEN, P. AND WEINER, N., *Federation Proc.*, 18 (1959) 1005.
285   HAGEN, P. B., *J. Neuropsychiat.*, 4 (1962) 107.
285a HÄGGENDAL, J. AND LUNDQVIST, M., *Intern. J. Neuropharmacol.*, 3 (1964) 59.
286   HALEY, T. J., in S. GARATTINI AND V. GHETTI (Eds.), *Psychotropic Drugs*, Elsevier, Amsterdam, 1957, p. 313.
287   HARRIS, E., ALBURN, H. E. AND SEIFTER, J., *Abstracts Fall Meeting American Society Pharmacology*, Ann Arbor, Mich., 1958, p. 17.
288   HARVEY, J. A., HELLER, A. AND MOORE, R. Y., *J. Pharmacol. Exptl. Therap.*, 140 (1963) 103.
289   HAVERBACK, B. J., SJOERDSMA, A. AND TERRY, L. L., *New Engl. J. Med.*, 255 (1956) 270.
290   HEGGLIN, R. AND LANGEMANN, H., *Helv. Med. Acta*, 22 (1955) 463.
291   HELLER, A. AND HARVEY, J. A., *Pharmacologist*, 5 (1963) 264.
292   HELLER, A., HARVEY, J. A. AND MOORE, R. Y., *Biochem. Pharm.*, 11 (1962) 859.
293   HEROLD, M., GEORGES, G. AND CAHN, J., in S. GARATTINI AND V. GHETTI (Eds.), *Psychotropic Drugs*, Elsevier, Amsterdam, 1957, p. 58.
294   HESS, S. M., REDFIELD, B. G. AND UDENFRIEND, S., *J. Pharmacol. Exptl. Therap.*, 127 (1959) 178.
295   HESS, S. M., SHORE, P. A. AND BRODIE, B. B., *J. Pharmacol. Exptl. Therap.*, 118 (1956) 84.
296   HESS, W. R., *Das Zwischenhirn*, Schwabe, Basel, 1954.
297   HOAGLAND, H., *J. Nervous Mental Disease*, 126 (1958) 211.
297a HODGE, J. V., OATES, J. A. AND SJOERDSMA, A., *Clin. Pharmacol. Therap.*, 5 (1964) 149.
298   HOLZBAUER, M. AND VOGT, M., *J. Neurochem.*, 1 (1956) 8.
299   HOOFT, C., DE LAEY, P., DE LOORE, F. AND VERBEECK, J., *Helv. Paediat. Acta*, 17 (1962) 1.
300   HORISBERGER, B. AND GRANDJEAN, E., *Helv. Physiol. et Pharmacol. Acta*, 16 (1958) 146.
301   HORITA, A., *J. Pharmacol. Exptl. Therap.*, 122 (1958) 176.
302   HORITA, A., *Ann. N.Y. Acad. Sci.*, 80 (1959) 590.
303   HORITA, A. AND GOGERTY, J. H., *Federation Proc.*, 16 (1957) 308.
304   HORITA, A. AND GOGERTY, J. H., *J. Pharmacol. Exptl. Therap.*, 122 (1958) 195.
305   HORWITT, M. K., *Science*, 124 (1956) 429.
306   HSIA, D. Y.-Y., NISHIMURA, K. AND BRENCHLEY, Y., *Nature*, 200 (1963) 578.
307   HUANG, I. AND HSIA, D. Y.-Y., *Proc. Soc. Exptl. Biol. Med.*, 112 (1963) 81.
308   HUANG, I., TANNENBAUM, S., BLUME, L. AND HSIA, D. Y.-Y., *Proc. Soc. Exptl. Biol. Med.*, 106 (1961) 533.
309   HUANG, I., TANNENBAUM, S. AND HSIA, D. Y.-Y., *Nature*, 186 (1960) 717.
310   HUGHES, F. B., FINGER, K. F. AND BRODIE, B. B., *J. Pharmacol. Exptl. Therap.*, 128 (1960) 131.
311   HUSZAK, I. AND DURKO, I., *Psychiat. et Neurol.*, 143 (1962) 407.
311a INOUYE, A., KATAOKA, K. AND SHINAGAWA, J., *Nature*, 198 (1963) 291.
312   JACOBSEN, E., in S. GARATTINI AND V. GHETTI (Eds.), *Psychotropic Drugs*, Elsevier, Amsterdam, 1957, p. 119.

313 JENNEY, E. H., *Federation Proc.*, 13 (1954) 370.
314 JENSEN, K., *Acta Neurol. Scand.*, 38 (1962) 278.
315 JEPSON, J. B., *Lancet*, ii (1955) 1009.
316 JÉRÔME, H., *Bull. mém. soc. méd. Hôp. Paris*, 113 (1962) 168.
317 JÉRÔME, H., LEJEUNE, J. AND TURPIN, R., *Compt. rend. acad. sci.*, 251 (1960) 2075.
318 JERVIS, G. A., *Proc. Soc. Exptl. Biol. Med.*, 82 (1953) 514.
319 JERVIS, G. A., *Diseases of Nervous System*, 18, Monograph Suppl. 93 (1957).
320 JOHN, E. R., WENZEL, B. M. AND TSCHIRGI, R. D., *Science*, 127 (1958) 25.
321 JOYCE, D., *Experientia*, 19 (1963) 187.
321a JOYCE, D. AND HURWITZ, H. M. B., *Psychopharmacologia*, 5 (1964) 424.
321b JOYCE, D. AND MROSOVSKY, N., *Psychopharmacologia*, 5 (1964) 417.
322 JUS, A. D., LASKOWSKA, D. AND ZIMMY, S., *Ann. méd. psychiat.*, 116 (1958) 898.
323 KÄRJÄ, J., KÄRKI, N. T. AND TALA, E., *Acta Pharmacol. et Toxicol.*, 18 (1961) 253.
324 KÄRKI, N. T., KUNTZMAN, R. AND BRODIE, B. B., *Federation Proc.*, 19 (1960) 184.
325 KÄRKI, N. T., KUNTZMAN, R. AND BRODIE, B. B., *J. Neurochem.*, 9 (1962) 53.
325a KATAOKA, K., *Japan. J. Physiol.*, 12 (1962) 623.
326 KATO, R., *J. Neurochem.*, 5 (1960) 202.
327 KEELE, C. A. AND ARMSTRONG, D., *Substances Producing Pain and Itch*, E. Arnold, London, 1964.
328 KELLER, R., *Arzneimittel-Forsch.*, 8 (1958) 390.
329 KERKUT, G. A. AND PRICE, M. A., *Life Sci.*, (1963) 722
330 KETY, S. S., *Science*, 129 (1959) 1590.
331 KIMISHIMA, K., *Yonago Acta Med.*, 7 (1963) 1.
332 KIRBERGER, E., *Deut. med. Wochschr.*, 87 (1962) 929.
333 KIRBERGER, E. AND BRAUN, L., *Biochim. Biophys. Acta*, 49 (1961) 391.
334 KIVALO, E., RINNE, I. K. AND MÄKELÄ, S., *Experientia*, 14 (1958) 408.
335 KLEE, G. D., BERTINO, J., GOODMAN, A. AND ARONSON, H., *J. Mental Sci.*, 106 (1960) 309.
336 KLINE, N. S. AND SACKS, W., *Am. J. Psychiat.*, 120 (1963) 274.
337 KNOLL, J. AND KNOLL, B., *Arch. intern. pharmacodyn.*, 133 (1961) 310.
338 KNOW, W. E., *Pediatrics*, 26 (1960) 1.
339 KOBINGER, W., *Arch. exptl. Pathol. Pharmakol., Naunyn-Schmiedeberg's*, 233 (1958) 559.
340 KOELLA, W. P., SMYTHIES, J. R. AND BULL, W. P., *Science*, 129 (1959) 1231.
341 KOELLA, W. P., SMYTHIES, J. R., LEVY, C. K. AND CZICMAN, J. S., *Am. J. Physiol.*, 199 (1960) 381.
342 KOPIN, I. J., *Science*, 129 (1959) 835.
343 KRALL, A. R., LEVER, P. G., VILLAVERDE R. AND BILLETT, B., *J. Am. Med. Assoc.*, 184 (1963) 280.
344 KUNTZMAN, R., SHORE. P A., BOGDANSKI, D. F. AND BRODIE, B. B., *J. Neurochem.*, 6 (1961) 226.
345 LAROCHE, M. J. AND BATHELLIER, C., *Compt. rend. soc. biol.*, 154 (1960) 716.
346 LAROCHE, M. J. AND BRODIE, B. B., *J. Pharmacol. Exptl. Therap.*, 130 (1960) 134.
347 LAROCHE, M. J. AND BRODIE, B. B., *Compt. rend. soc. biol.*, 154 (1960) 713.
348 LAUER, J. W., INSKIP, W. M., BERNSOHN, J. AND ZELLER, E. A., *A.M.A. Arch. Neurol. Psychiat.*, 80 (1958) 122.
349 LAVERTY, R., *J. Neurochem.*, 10 (1963) 151.
350 LESSIN, A. W., *Biochem. Pharmacol.*, 2 (1959) 290.
351 LESSIN, A. W. AND PARKES, M. W., *Brit. J. Pharmacol.*, 12 (1957) 245.
352 LESSIN, A. W. AND PARKES, M. W., *J. Pharm. and Pharmacol.*, 9 (1957) 657.
353 LEVY, J. AND MICHEL-BER, E., *J. Physiol. (Paris)*, 52 (1960) 789.
354 LEWIS, G. P., *J. Pharm. and Pharmacol.*, 10 (1958) 529.
355 LEYTON, G., *Brit. Med. J.*, 2 (1958) 1136.
356 LIBRO, E., *Boll. soc. ital. biol. sper.*, 23 (1957) 890.
357 LINDNER, E., *Arzneimittel-Forsch.*, 10 (1960) 568.
358 LIPPMAN, R. W., PERRY, T. L. AND WRIGHT, S. W., *Metabolism*, 7 (1958) 274.
359 LITTLE, M. J. AND CONRAD, E., *J. Pharmacol. Exptl. Therap.*, 129 (1960) 454.
360 LOO, Y. H., DILLER, E. AND OWEN JR., J. E., *Nature*, 194 (1962) 1286.
361 LOUTTIT, R. T., *J. Comp. Physiol. Psychol.*, 55 (1962) 425.

362 LUMSDEM, C. E. AND POMERAT, C. M., *Exptl. Cell Research*, 2 (1951) 103.
363 MAASS, A. R. AND NIMMO, H. J., *Nature*, 184 (1959) 547.
364 MAGGI, G. G. AND SAVOLDI, F., *Sett. med.*, 48 (1960) 881.
365 MAGNES, J. AND HESTRIN-LERNER, S., *J. Neurochem.*, 5 (1960) 128.
366 MAGNUSSON, T. AND ROSENGREN, E., *Experientia*, 19 (1963) 229.
366a MAHLER, J. D. AND HUMOLLER, F. L., *J. Neuropsychiat.*, 5 (1964) 252.
367 MALCOM, J. L., in G. P. LEWIS (Ed.), *5-Hydroxytryptamine*, Pergamon, London, 1958, p. 221.
368 MANTEGAZZA, P. AND RIVA, M., *Med. Exptl.*, 4 (1961) 367.
369 MANTEGAZZINI, P., *Boll. soc. ital. biol. sper.*, 32 (1956) 839.
370 MANTEGAZZINI, P., *Arch. intern. pharmacodyn.*, 112 (1957) 199.
371 MAREK, K., SMETANA, R. AND RASKOVA, H., *Intern. J. Neuropharmacol.*, 1 (1962) 71.
372 MARRAZZI, A. S., *Ann. N. Y. Acad. Sci.*, 66 (1957) 496.
373 MARRAZZI, A. S., *Ann. N.Y. Acad. Sci.*, 92 (1961) 990.
374 MARTEL, R. R., WESTERMANN, E. O. AND MAICKEL, R. P., *Life Sci.*, (1962) 151.
375 MARTIN, G. M., BENDITT, E. P. AND ERIKSEN, N., *Arch. Biochem. Biophys.*, 90 (1960) 208.
376 MASUDA, M., SLONECKER, J. S. AND DORPAT, T. L., *J. Nervous Mental Disease*, 130 (1960) 125.
376a MATUSSEK, N. AND PATSCHKE, U., *Med. Exptl.*, 11 (1964) 81.
377 MAYNARD, L. S. AND SCHENKER, V. J., *Nature*, 196 (1962) 575.
378 MCKEAN, C. M., SCHANBERG, S. M. AND GIARMAN, N. J., *Science*, 137 (1962) 604.
379 MEICKEL, R. P., WESTERMANN, E. O. AND BRODIE, B. B., *J. Pharmacol. Exptl. Therap.*, 134 (1961) 167.
380 MELTZER, H. Y., *Federation Proc.*, 20 (1961) 227b.
381 MERCIER, F., ERTZENSPERGER, P. AND MERCIER, J., *Anesthésie et analgésie*, 16 (1959) 70.
382 MERCIER, J. AND LACREUSETTE, J., *Compt. rend. soc. biol.*, 157 (1963) 1036.
383 MICHAELSON, I. A. AND WHITTAKER, V. P., *Biochem. Pharmacol.*, 11 (1962) 505.
384 MILINE, R., STERN, P. AND HUKOVIC, S., *Experientia*, 14 (1958) 415.
385 MILNE, M. D., *Lancet*, ii (1959) 467.
386 MITOMA, C., AULD, R. M. AND UDENFRIEND, S., *Proc. Soc. Exptl. Biol. Med.*, 94 (1957) 634.
387 MIURA, T., TSUJIYAMA, Y., MAKITA, K., NAKAZAWA, T., SATO, K. AND NAKAHARA, M., in S. GARATTINI AND V. GHETTI (Eds.), *Psychotropic Drugs*, Elsevier, Amsterdam, 1957, p. 478.
388 MONNIER, M., *Arch. intern. pharmacodyn.*, 124 (1960) 281.
389 MONNIER, M. AND GANGLOFF, H., *Helv. Physiol. et Pharmacol. Acta*, 15 (1957) 83.
390 MONNIER, H. AND TISSOT, R., *Helv. Physiol. et Pharmacol. Acta*, 16 (1958) 255.
390a MORPURGO, C., *Biochem. Pharmacol.*, 11 (1962) 967.
391 MURPHREE, H. B., JENNEY, E. H. AND PFEIFFER, C. C., *Pharmacologist*, 2 (1960) 64.
392 MURRAY, M. R., in W. F. WINDLE (Ed.), *Biology of Neuroglia*, C. C. Thomas, Springfield, Ill., 1958, p. 176.
393 MUSTAKALLIO, K. K., LEVONEN, E. AND RAEKALLIO, J., *Science*, 134 (1961) 344.
394 NADLER, H. L. AND HSIA, D. Y.-Y., *Proc. Soc. Exptl. Biol. Med.*, 107 (1961) 721.
395 NAKAO, A. AND BALL, M., *J. Nervous Mental Diseases*, 130 (1960) 417.
396 NAKAZAWA, T., *Texas Repts. Biol. Med.*, 18 (1960) 52.
397 NEAME, K. D., *J. Neurochem.*, 6 (1961) 358.
398 NEČINA, J. AND KREJČI, I., *Biochem. Pharmacol.*, 8 (1961) 35.
399 NUMEROF, P., GORDON, M. AND KELLY, J. M., *J. Pharmacol. Exptl. Therap.*, 115 (1955) 427.
400 OATES, J. A., NIRENBERG, P. Z., JEPSON, J. B., SJOERDSMA, A. AND UDENFRIEND, S., *Proc. Soc. Exptl. Biol. Med.*, 112 (1963) 1078.
401 OATES, J. A. AND ZALTMAN, P., *Ann. N.Y. Acad. Sci.*, 80 (1959) 977.
402 OLDS, J., *Science*, 127 (1958) 315.
403 OLDS, J., *Physiol. Revs.*, 42 (1962) 554.
404 OLDS, J. AND OLDS, M. E., *Science*, 127 (1958) 1175.
405 OLSON, R. E., GURSEY, D. AND VESTER, J. W., *New Engl. J. Med.*, 263 (1960) 1169.
406 PAASONEN, M. K., *Biochem. Pharmacol.*, 5 (1961) 389.
407 PAASONEN, M. K. AND KIVALO, E., *Psychopharmacologia*, 3 (1962) 188.
408 PAASONEN, M. K., MACLEAN, P. D. AND GIARMAN, N. J., *J. Neurochem.*, 1 (1957) 326.
409 P'AN, S. Y., FUNDERBURK, W. H. AND FINGER, K. F., *Proc. Soc. Exptl. Biol. Med.*, 108 (1961) 680.
410 P'AN, S. Y., FUNDERBURK, W. H. AND FINGER, K. F., *Federation Proc.*, 20 (1961) 323.

410a PARE, C. M. B. AND LaBROSSE, E. H., *J. Psychiat. Research*, 1 (1963) 271.

411 PARE, C. M. B., SANDLER, M. AND STACEY, R. S., *Lancet*, i (1957) 551.

412 PARE, C. M. B., SANDLER, M. AND STACEY, R. S., *Lancet*, ii (1958) 1099.

413 PARE, C. M. B., SANDLER, M. AND STACEY, R. S., *Arch. Disease Childhood*, 34 (1959) 422.

414 PARE, C. M. B., SANDLER, M. AND STACEY, R. S., *J. Neurol. Neurosurg. Psychiat.*, 23 (1960) 341.

415 PERMAN, E. S., *Acta Physiol. Scand.*, 51 (1961) 62.

416 PERRY, T. L., *Science*, 136 (1962) 879.

417 PERRY, T. L., HANSEN, S., TISCHLER, B. AND HESTRIN, M., *Proc. Soc. Exptl. Biol. Med.*, 115 (1964) 118.

418 PERRY, T. L., SHAW, K. N. F. AND WALKER, D., *Nature*, 189 (1961) 926.

419 PETERFALVI, M. AND JEQUIER, R., *Arch. intern. pharmacodyn.*, 124 (1960) 237.

420 PICCHIONI, A. L., CHIN, L. AND BREITNER, C., *Federation Proc.*, 21 (1962) 416.

421 PICCHIONI, A. L., CHIN, L., CHOISSER, D. AND BREITNER, C., *Pharmacologist*, 5 (1963) 238.

422 PIERRE, R., *Compt. rend. soc. biol.*, 151 (1957) 890.

423 PIERRE, R., *Compt. rend. soc. biol.*, 151 (1957) 1135.

424 PIERRE, R. AND CAHN, J., *Compt. rend. soc. biol.*, 149 (1955) 1406.

425 PIERRE, R. AND CAHN, J., *Anesthésie et analgésie*, 4 (1956) 723.

426 PINEDA, A. AND SNYDER, R. S., *Neurology*, 13 (1963) 166.

427 PLETSCHER, A., *Experientia*, 12 (1956) 479.

428 PLETSCHER, A., in S. GARATTINI AND V. GHETTI (Eds.), *Psychotropic Drugs*, Elsevier, Amsterdam, 1957, p. 468.

429 PLETSCHER, A. AND BESENDORF, H., *Experientia*, 15 (1959) 25.

430 PLETSCHER, A., BESENDORF, H., BÄCHTOLF, H. P. AND GEY, K. F., *Helv. Physiol. et Pharmacol. Acta*, 17 (1959) 202.

431 PLETSCHER, A., BROSSI, A. AND GEY, K. F., *Intern. Rev. Neurobiol.*, 4 (1962) 275.

432 PLETSCHER, A. AND GEY, K. F., *Science*, 128 (1958) 900.

433 PLETSCHER, A. AND GEY, K. F., *Med. Exptl.*, 2 (1960) 259.

434 PLETSCHER, A. AND GEY, K. F., in J. DE AJURIAGUERRA (Ed.), *Monoamines et Système Nerveux Central*, Symposium Bel-Air, Genève, 1961, p. 105.

435 PLETSCHER, A. AND GEY, K. F., *Nature*, 190 (1961) 918.

436 PLETSCHER, A., GEY, K. F. AND ZELLER, P., *Progr. in Drug Research*, 2 (1961) 417.

437 PLETSCHER, A., SHORE, P. A. AND BRODIE, B. B., *Science*, 122 (1955) 374.

438 PLETSCHER, A., SHORE, P. A. AND BRODIE, B. B., *J. Pharmacol. Exptl. Therap.*, 116 (1956) 84.

439 PLUMMER, A. J., BARRETT, W. R., MAXWELL, R. A., FINOCCHI, D., LUCAS, R. A. AND EARL, A. E., *Arch. intern. pharmacodyn.*, 119 (1959) 245.

440 PLUMMER, A. J., SHEPPARD, H. AND SHULERT, A. R., in S. GARATTINI AND V. GHETTI (Eds.), *Psychotropic Drugs*, Elsevier, Amsterdam, 1957, p. 350.

441 POLLIN, W., CARDON JR., P. V. AND KETY, S. S., *Science*, 133 (1961) 104.

442 PORTILLA, J. L., *Arch. inst. farmacol. exptl. (Madrid)*, 12 (1960) 21.

443 POTTER, L. T. AND AXELROD, J., *Nature*, 194 (1962) 581.

444 PRICE, J. M., BROWN, R. R. AND PETERS, H. A., *Neurology*, 9 (1959) 456.

445 PRICE, S. A. P. AND WEST, G. B., *J. Pharm. and Pharmacol.*, 12 (1960) 617.

446 PRICE, S. A. P. AND WEST, G. B., *Nature*, 185 (1960) 470.

447 PROCKOP, D. J., SHORE, P. A. AND BRODIE, B. B., *Experientia*, 15 (1959) 145.

448 PROCKOP, D. J., SHORE, P. A. AND BRODIE, B. B., *Ann. N.Y. Acad. Sci.*, 80 (1959) 643.

449 PROOSDIJ-HARTZEMA, E. G. VAN, AKKERMAN, A. M. AND DE JONGH, D. K., *Arch. intern. pharmacodyn.*, 123 (1959) 168.

450 PRUSOFF, W. H., *Brit. J. Pharmacol.*, 15 (1960) 520.

451 PSCHEIDT, G. R. AND HIMWICH, H. E., *Biochem. Pharmacol.*, 12 (1963) 65.

452 PSCHEIDT, G. R., ISSEKUTZ, B. AND HIMWICH, H. E., *Quart. J. Studies Alc.*, 22 (1961) 550.

453 PUT, T. R. AND MEDUSKI, J. W., *Acta Physiol. et Pharmacol. Neerl.*, 11 (1962) 240.

454 QUAY, W. B., *Gen. Comp. Endocrinol.*, 3 (1963) 473.

454a QUAY, W. B., *Proc. Soc. Exptl. Biol. Med.*, 114 (1963) 718.

455 QUINN, G. P., SHORE, P. A. AND BRODIE, B. B., *J. Pharmacol. Exptl. Therap.*, 127 (1959) 103.

456 RANDALL, L. O. AND BADGON, R. E., *Ann. N.Y. Acad. Sci.*, 80 (1959) 626.

457 RENSON, J., WEISSBACH, H. AND UDENFRIEND, S., *J. Biol. Chem.*, 237 (1962) 2261.

458 RESNICK, R. H., GRAY, S. J., KOCH, J. P. AND TIMBERLAKE, W. H., *Proc. Soc. Exptl. Biol. Med.*, 110 (1962) 77.

459 REVZIN, A. M. AND COSTA, E., *Am. J. Physiol.*, 198 (1960) 959.

460 RIEGELHAUPT, L., *J. Nervous Mental Disease*, 127 (1958) 228.

461 RIGHINI, E. AND VOLTA, A., *Boll. soc. med. chir. Pisa*, 26 (1958) 369.

462 ROBINS, E., LOWE, I. P. AND HAVNER, N. M., *Clin. Research Proc.*, 4 (1956) 149.

463 ROBINSON, M. H., LUCAS, R. A., MAC PHILLAMY, H. B., BARRETT, W. AND PLUMMER, A. J., *Experientia*, 17 (1961) 14.

464 RODNIGHT, R., *Biochem. J.*, 64 (1956) 621.

465 RODNIGHT, R. AND AVES, E. K., *J. Mental Sci.*, 104 (1958) 1149.

466 ROOS, B. E., *Life Sci.*, (1962) 25.

467 ROOS, B. E., *Life Sci.*, (1963) 1.

467a ROOS, B. E., ANDÉN, N. E. AND WERDINIUS, B., *J. Neuropharmacol.*, 3 (1964) 117.

468 ROOS, B. E. AND WERDINIUS, B., *Life Sci.*, (1962) 105.

469 ROOS, B. E. AND WERDINIUS, B., *Life Sci.*, (1963) 92.

470 ROSENBERG, D. E., ISBELL, H. AND MINER, E. J., *Psychopharmacologia*, 4 (1963) 39.

471 ROSENFELD, G., *Proc. Soc. Exptl. Biol. Med.*, 103 (1960) 144.

472 ROSS HART, E., RODRIGUEZ, J. M. AND MARRAZZI, A. S., *Science*, 134 (1961) 1696.

473 ROSSIGNOL, P. AND BOULU, R., *Compt. rend. soc. biol.*, 150 (1956) 2126.

474 ROTHBALLER, A. B., *Electroencephalog. Clin. Neurophysiol.*, 9 (1957) 409.

475 ROTHLIN, E., *Ann. N.Y. Acad. Sci.*, 66 (1957) 668.

476 ROWE, R. P., BLOOM, B. M., P'AN, S. Y. AND FINGER, K., *Federation Proc.*, 18 (1959) 441.

477 RÜMKE, CHR. L., *Arch. exptl. Pathol. Pharmakol., Naunyn-Schmiedeberg's*, 243 (1962) 298.

478 RYALL, R. W., *Nature*, 196 (1962) 680.

479 RYALL, R. W., *Biochem. Pharmacol.*, 11 (1962) 1233.

480 SACHS, E., *J. Neurosurg.*, 14 (1957) 22.

481 SACKS, W., *J. Appl. Physiol.*, 16 (1961) 1050.

482 SAI-HALASZ, A., BRUNECKER, G. AND SZARA, S., *Psychiat. Neurol.*, 135 (1958) 285.

483 SALMOIRAGHI, G. C. AND PAGE, I. H., *J. Pharmacol. Exptl. Therap.*, 120 (1957) 20.

484 SALMOIRAGHI, G. C., SOLLERO, L. AND PAGE, I. H., *J. Pharmacol. Exptl. Therap.*, 117 (1956) **166.**

485 SANDLER, M. AND CLOSE, H. G., *Lancet*, ii (1959) 316.

486 SANDLER, M., DABIES, A. AND RIMINGTON, C., *Lancet*, ii (1959) 318.

487 SANDLER, M. AND WEST, G. B., *J. Physiol. (London)*, 140 (1958) 99.

488 SANO, I., GAMO, T., KAKIMOTO, Y., TANIGUCHI, K., TAKESADA, M. AND NISHINUMA, K., *Biochim. Biophys. Acta*, 32 (1959) 586.

489 SANO, I., KAKIMOTO, Y., OKAMOTO, T., NAKAIMA, H. AND KUDO, Y., *Schweiz. med. Wochschr.*, 87 (1957) 214.

490 SARKAR, S., BANERJEC, R., ISE, M. S. AND ZELLER, E. A., *Helv. Chim. Acta*, 43 (1960) 439.

491 SCARINCI, V., *Boll. soc. ital. biol. sper.*, 31 (1955) 779.

492 SCARLATO, G., MANZINI, B. AND CANAL, N., *Riv. sper. freniat.*, 83, Suppl. III (1959) 1087.

493 SCHAIN, R. J., *Brit. J. Pharmacol.*, 17 (1961) 261.

494 SCHAIN, R. J., *World Neurol.*, 3 (1962) 706.

495 SCHAIN, R. J. AND FREEDMAN, D. X., *J. Pediat.*, 58 (1961) 315.

496 SCHANBERG, S. M., *J. Pharmacol. Exptl. Therap.*, 139 (1963) 191.

497 SCHANBERG, S. M. AND GIARMAN, N. J., *Biochim. Biophys. Acta*, 41 (1960) 556.

498 SCHANBERG, S. M. AND GIARMAN, N. J., *Biochem. Pharmacol.*, 12 (1962) 187.

499 SCHANBERG, S. M., MCILROY, C. A. AND GIARMAN, N. J., *Federation Proc.*, 20 (1961) 141.

500 SCHANBERG, S. M., MCKEAN, C. M. AND GIARMAN, N. J., *Federation Proc.*, 21 (1962) 269.

501 SCHIEVELBEIN, H. AND WERLE, E., *Psychopharmacologia*, 3 (1962) 35.

502 SCHIEVELBEIN, H., WERLE, E. AND JACOBY, W., *Naturwissenschaften*, 18 (1961) 602.

503 SCHMIDT, J., *Arch. exptl. Pathol. Pharmakol., Naunyn-Schmiedeberg's*, 241 (1961) 509.

504 SCHMIDT, J. AND MATTHIES, H., *Acta Biol. Med. Germ.*, 7 (1961) 107.

505 SCHMIDT, J. AND MATTHIES, H., *Acta Biol. Med. Germ.*, 7 (1961) 443.

506 SCHMIDT, J. AND MATTHIES, H., *Acta Biol. Med. Germ.*, 8 (1962) 426.

507 SCHWARTZ, D. E., BURKARD, W. P., ROTH, M., GEY, K. F. AND PLETSCHER, A., *Arch. intern. pharmacodyn.*, 141 (1963) 135.

508 SCHWARTZ, D. E., PLETSCHER, A., GEY, K. F. AND RIEDER, J., *Helv. Physiol. et Pharmacol. Acta*, 18 (1960) 10.

509 SCHWARTZ, M. S., *Federation Proc.*, 19 (1960) 277.

509a SCHWEIGERDT, A. K. AND HIMWICH, H. E., *J. Pharmacol. Exptl. Therap.*, 144 (1964) 253.

510 SEIDEN, L. S. AND CARLSSON, A., *Psychopharmacologia*, 4 (1963) 418.

511 SEIDEN, L. S. AND WESTLEY, J., *Biochim. Biophys. Acta*, 58 (1962) 363.

512 SHARMAN, D. F. AND SMITH, S. E., *J. Neurochem.*, 9 (1962) 403.

513 SHAW, C. R., LUCAS, J. AND RABINOVITCH, R. D., *A.M.A. Arch. Gen. Psychiat.*, 1 (1959) 366.

514 SHAW, E. AND WOOLLEY, D. W., *J. Pharmacol. Exptl. Therap.*, 111 (1954) 43.

515 SHEMANO, I. AND NICKERSON, M., *Can. J. Biochem. Physiol.*, 36 (1958) 1243.

516 SHERWOOD, W., *J. Nervous Mental Disease*, 125 (1957) 490.

517 SHORE, P. A. AND BRODIE, B. B., in S. GARATTINI AND V. GHETTI (Eds.), *Psychotropic Drugs*, Elsevier, Amsterdam, 1957, p. 423.

518 SHORE, P. A. AND BRODIE, B. B., *Science*, 127 (1958) 704.

519 SHORE, P. A., MEAD, J. A. R., KUNTZMAN, R. G., SPECTOR, S. AND BRODIE, B. B., *Science*, 126 (1957) 1063.

520 SHORE, P. A., PLETSCHER, A. AND BRODIE, B. B., *J. Pharmacol. Exptl. Therap.*, 116 (1956) 51.

521 SHORE, P. A., PLETSCHER, A., TOMICH, E. G., CARLSSON, A., KUNTZMAN, R. AND BRODIE, B. B., *Ann. N.Y. Acad. Sci.*, 66 (1957) 609.

522 SHORE, P. A., SILVER, S. L. AND BRODIE, B. B., *Science*, 122 (1955) 284.

523 SHORE, P. A., SILVER, S. L. AND BRODIE, B. B., *Experientia*, 11 (1955) 272.

524 SIGG, E. B., CAPRIO, G. AND SCHNEIDER, J. A., *Proc. Soc. Exptl. Biol. Med.*, 97 (1958) 97.

525 SIVA SANKAR, D. V., *J. Neuropsychiat.*, 3 (1961) 123.

526 SIVA SANKAR, D. V., GOLD, E., PHIPPS, E. AND SANKAR, D. B., *Ann. N.Y. Acad. Sci.*, 96 (1962) 392.

527 SJOERDSMA, A., *New Engl. J. Med.*, 261 (1959) 181.

528 SJOERDSMA, A., *New Engl. J. Med.*, 261 (1959) 231.

529 SJOERDSMA, A., OATES, J. A., ZALTZMAN, P. AND UDENFRIEND, S., *J. Pharmacol. Exptl. Therap.*, 126 (1959) 217.

530 SJOERDSMA, A., SMITH, T. E., STEFENSON, T. D. AND UDENFRIEND, S., *Proc. Soc. Exptl. Biol. Med.*, 89 (1955) 36.

531 SJOERDSMA, A., WEISSBACH, H. AND UDENFRIEND, S., *Am. J. Med.*, 20 (1956) 520.

532 SKILLEN, R. G., THIENES, C. H., CANGELOSI, J. AND STRAIN, L., *Proc. Soc. Exptl. Biol. Med.*, 108 (1961) 121.

533 SLOCOMBE, A. G., HOAGLAND, H. AND TORIAN, L. S., *Am. J. Physiol.*, 185 (1956) 601.

534 SMITH, S. E., *Brit. J. Pharmacol.*, 20 (1963) 178.

535 SMYTHIES, J. R., *Lancet*, i (1960) 1287.

536 SNOW, P. J. D., LENNARD-JONES, J., CURSON, G. AND STACEY, R., *Lancet*, ii (1955) 1004.

537 SPECK, L., *J. Neurochem.*, 9 (1962) 573.

538 SPECTOR, S., KUNTZMAN, R., HIRSCH, C. AND BRODIE, B. B., *Federation Proc.*, 19 (1960) 279.

539 SPECTOR, S., SHORE, P. A. AND BRODIE, B. B., *J. Pharmacol. Exptl. Therap.*, 128 (1960) 15.

540 SPILMAN, E., *A.M.A. Arch. Gen. Psychiat.*, 2 (1960) 545.

541 SPRINCE, H., *Clin. Chem.*, 7 (1961) 203.

542 SPRINCE, H., *Ann. N.Y. Acad. Sci.*, 96 (1962) 399.

543 SPRINCE, H., HOUSER, E. AND JAMESON, D., *A.M.A. Arch. Gen. Psychiat.*, 2 (1960) 268.

544 SPRINCE, H., PARKER, C. M., JAMESON, D. AND ALEXANDER, F., *J. Nervous Mental Disease*, 137 (1963) 246.

545 STEIN, L. AND RAY, O. S., *Nature*, 188 (1960) 1199.

546 STENNETT, R. G. AND CALLOWHILL, C. R., *J. Nervous Mental Disease*, 128 (1959) 453.

547 SULSER, F. AND BRODIE, B. B., *Science*, 131 (1960) 1440.

548 SZARA, S., *Experientia*, 12 (1956) 441.

549 SZARA, S. AND HEARST, E., *Ann. N.Y. Acad. Sci.*, 96 (1962) 134.

550 SZARA, S., HEARST, E. AND PUTNEY, F., *Federation Proc.*, 19 (1960) 23.

551 TAESCHLER, M. AND CERLETTI, A., *J. Pharmacol. Exptl. Therap.*, 120 (1957) 179.

552 TAÜTZ, N. A., BENTE, D., SCHMID, E. AND SCHMID-CLAUDY, E., *Med. Exptl.*, 9 (1963) 313.

553 TAYLOR, J. D., WYKES, A. A., GLADISH, J. C., MARTIN, W. B. AND EVERETT, G. M., *Federation Proc.*, 19 (1960) 278.

554   TEDESCHI, D. H., TEDESCHI, R. E. AND FELLOWS, E. J., *Federation Proc.*, 18 (1959) 450.
555   TEDESCHI, D. H., TEDESCHI, R. E., FOWLER, P. J., GREEN, H. AND FELLOWS, E. J., *Biochem. Pharmacol.*, 11 (1962) 481.
556   THORSEN, A., BIÖRCK, G., BJORKMAN, G. AND WALDENSTRÖM, J., *Am. Heart J.*, 47 (1954) 795.
557   TISSOT, R., in J. DE AJURIAGUERRA (Ed.), *Monoamines et Système Nerveux Central*, Symposium Bel-Air, Genève, 1961 p. 169.
558   TONINI, G., *Boll. soc. ital. biol. sper.*, 31 (1955) 766.
559   TRIPOD, J., in S. GARATTINI AND V. GHETTI (Eds.), *Psychotropic Drugs*, Elsevier, Amsterdam, 1957, p. 437.
560   TRIPOD, J., STUDER, A. AND MEIER, R., *Arch. intern. pharmacodyn.*, 112, (1957) 319.
561   TRUITT JR., E. B. AND EBERSBERGER, E. M., *Federation Proc.*, 20 (1961) 320.
562   TURNER, W. J. AND MAUSS, E., *A.M.A. Arch. Gen. Psychiat.*, 1 (1959) 646.
563   TURNER, W. J. AND MERLIS, S., *A.M.A. Arch. Neurol. Psychiat.*, 81 (1959) 121.
564   TWAROG, B. M. AND PAGE, I. H., *Am. J. Physiol.*, 175 (1953) 157.
565   TYCE, G. M., FLOCK, E. V. AND OWEN, C. A., *Federation Proc.*, 22 (1963) 633.
566   UDENFRIEND, S., *Federation Proc.*, 15 (1956) 493.
567   UDENFRIEND, S., in G. P. LEWIS (Ed.), *5-Hydroxytryptamine*, Pergamon, London, 1958, p. 43.
568   UDENFRIEND, S., in F. BRÜCKE (Ed.), *Biochemistry of Central Nervous System*, Pergamon, London, 1959, p. 301.
569   UDENFRIEND, S., CLARK, C. AND TITUS, E., *J. Am. Chem. Soc.*, 75 (1953) 501.
570   UDENFRIEND, S., CREVELING, C. R., POSNER, H., REDFIELD, B. G., DALY, J. AND WITKOP, B., *Arch. Biochem. Biophys.*, 83 (1959) 501.
571   UDENFRIEND, S., LOVENBERG, W. M. AND WEISSBACH, H., *Federation Proc.*, 19 (1960) 7.
572   UDENFRIEND, S., SHORE, P. A., BOGDANSKI, D. F., WEISSBACH, H. AND BRODIE, B. B., *Recent Progr. in Hormone Research*, 13 (1957) 1.
573   UDENFRIEND, S. AND WEISSBACH, H., *Proc. Soc. Exptl. Biol. Med.*, 97 (1958) 748.
574   UDENFRIEND, S., WEISSBACH, H. AND BOGDANSKI, D. F., *Ann. N.Y. Acad. Sci.*, 66 (1957) 602.
575   UDENFRIEND, S., WEISSBACH, H. AND BOGDANSKI, D. F., *J. Biol. Chem.*, 224 (1957) 803.
576   UUSPÄÄ, V. J., *Experientia*, 19 (1963) 156.
577   VALCOURT, A. J., *A.M.A. Arch. Neurol. Psychiat.*, 81 (1959) 292.
578   VELLUZ, L., PETERFALVI, M. AND JEQUIER, R., *Compt. rend. acad. sci.*, 24 (1958) 1905.
579   VOELKEL, A., *Naturwissenschaften*, 20 (1957) 541.
580   VOGT, M., *J. Physiol. (London)*, 123 (1954) 451.
581   VOGT, M., in G. P. LEWIS (Ed.), *5-Hydroxytryptamine*, Pergamon, London, 1958, p. 209.
582   VOGT, M., GUNN JR., C. J. AND SAWYER, C. H., *Neurology*, 7 (1957) 559.
583   WAALKES, T. P. AND WEISSBACH, H., *Proc. Soc. Exptl. Biol. Med.*, 93 (1956) 394.
584   WADA, J. A., *Science*, 134 (1961) 1688.
585   WALASZEK, E. J., *World Neurol.*, 2 (1961) 602.
586   WANG, H. L., HARWALKAR, V. R. AND WAISMAN, H. A., *Federation Proc.*, 20 (1961) 6.
587   WANG, H. L., HARWALKAR, V. R. AND WAISMAN, H. A., *Arch. Biochem. Biophys.*, 96 (1962) 181.
588   WAY, E. L. AND SUTHERLAND, V. C., *Anesthesiology*, 24 (1963) 543.
589   WEIL-MALHERBE, H. AND BONE, A. D., *Nature*, 181 (1958) 1474.
590   WEISKRANTZ, L., in S. GARATTINI AND V. GHETTI (Eds.), *Psychotropic Drugs*, Elsevier, Amsterdam, 1957, p. 67.
591   WEISKRANTZ, L. AND WILSON JR., W. A., *Science*, 123 (1956) 1116.
592   WEISSBACH, H., KING, W., SJOERDSMA, A. AND UDENFRIEND, S., *J. Biol. Chem.*, 234 (1959) 81.
593   WERDINIUS, B., *Acta Pharmacol. Toxicol.*, 19 (1962) 43.
594   WEST, G. B., *J. Pharm. and Pharmacol.*, 10 (1958) 92T.
595   WESTERMANN, E., BALZER, H. AND KNELL, J., *Arch. exptl. Pathol. Pharmakol., Naunyn-Schmiedeberg's*, 234 (1958) 194.
596   WHITTAKER, V. P., *Biochem. Pharmacol.*, 1 (1958) 351.
597   WHITTAKER, V. P., *Biochem. J.*, 72 (1959) 694.
598   WHITTAKER, V. P., *Biochem. J.*, 73 (1959) 37P.
599   WIELAND, T. AND MOTZEL, W., *Ann. Chem., Liebigs*, 581 (1953) 10.
600   WOOLLEY, D. W., *Research Publs. Assoc. Research Nervous Mental Disease*, 36 (1958) 381.
601   WOOLLEY, D. W., *The Biochemical Bases of Psychoses*, Wiley, New York, 1962.
602   WOOLLEY, D. W. AND CAMPBELL, N. K., *Ann. N.Y. Acad. Sci.*, 96 (1962) 108.

603   WOOLLEY. D. W. AND SHAW, E., *Proc. Natl. Acad. Sci.*, 40 (1954) 228.
604   WOOLLEY, D. W. AND SHAW, E., *Brit. Med. J.*, 2 (1954) 122.
605   WOOLLEY, D. W. AND SHAW, E. N., *Ann. N.Y. Acad. Sci.*, 66 (1957) 649.
605a   WOOLLEY, D. W. AND VAN DER HOEVEN, TH., *Science*, 144 (1964) 883.
606   WOOLLEY, D. W., VAN WINKLE, E. AND SHAW, E., *Proc. Natl. Acad. Sci.*, 43 (1957) 128.
607   WULFSOHN, N. L. AND POLITZER, W. M., *Anaesthesia*, 17 (1962) 64.
608   YEN, H. C. Y., SALVATORE, A. T., SILVERMAN, A. J. AND KING, T. O., *Arch. intern. pharmacodyn.*,
       140 (1962) 631.
609   YUWILER, A., GELLER, E. AND EIDUSON, S., *Arch. Biochem. Biophys.*, 80 (1959) 162.
610   YUWILER, A., JENKINS, I. M. AND DUKAY, A., *A.M.A. Arch. Gen. Psychiat.*, 4 (1961) 395.
611   YUWILER, A. AND LOUTTIT, R. T., *Science*, 134 (1961) 831.
612   ZBINDEN, G., PLETSCHER, A. AND STUDER, A., *Z. ges. exptl. Med.*, 129 (1958) 615.
613   ZELLER, E. A., *J. Clin. Exptl. Psychopathol.*, 19, Suppl. 1 (1958) 106.
614   ZELLER, E. A., in J. DE AJURIAGUERRA (Ed.), *Monoamines et Système Nerveux Central*, Sym-
       posium Bel-Air, Genève, 1961, p. 51.
615   ZELLER, E. A., BERNSOHN, J., INSKIP, W. M. AND LAUER, J. W., *Naturwissenschaften*, 44 (1957)
       427.
616   ZELLER, P., PLETSCHER, A., GEY, K. F., GUTMANN, H., HEGEDÜS, B. AND STRAUB, O., *Ann.
       N.Y. Acad. Sci.*, 80 (1959) 555.

DISTRIBUTION OF 5-HT IN SOME ANIMAL AND VEGETAL SPECIES

LEGEND

Acetone   = Amin's method[4]
Acetone (G) = Garven's method[49]
Butanol (1) = Udenfriend's method[114]
Butanol (2) = Bogdanski's method[13]
Butanol (3) = Weissbach's method[128]
Butanol (4) = Shore's method[102]
Butanol (5) = Waalkes's method[119]
Chr.   = Chromatographic method
Col.   = Colorimetric method
Sph.   = Spectrophotometric method
Sphf.   = Spectrophotofluorimetric method
G. P. Ileum = Guinea pig ileum
R. Colon   = Rat colon
R. Stomach = Rat stomach
R. Uterus   = Rat uterus
S. S. Heart   = *Spisula solida* heart
V. M. Heart = *Venus mercenaria* heart
Rb. Ear   = Isolated perfused rabbit ear
(+)   = Mean of authors' data
(m)   = Male
(f)   = Female
n.m.   = Non-measurable amounts

Animal or vegetal species	Tissue or fluid	5-HT µg/g or ml content	Method of extraction	Method of measurement	Reference
Agkistrodon piscivorus (Reptiles)	Venom	0.89	Chr.	Sphf.	133a
Amphitrite ornata (Polychaeta)	Body wall	0.19	Butanol (2)	Sphf.	129
	Gills	0.26			129
	Nerve cord	5.4			129
	Tentacles	0.08			129
Apis mellifera (Insecta)	Heads	0.12(+)	Butanol (2)	Sphf.	129
	Tips of abdomen	2.15(+)			129
	Venom apparatus	21.0			129
Arenicola marina (Polychaeta)	Body wall	2.3	Butanol (2)	Sphf.	129
	Nerve cord	3.1			129
Artica islandica (Pelecypoda)	Ganglia (cerebro-pleural, pedal and visceral g.)	20.0	Butanol (2)	Sphf.	129
Ass	Blood (serum)	0.41	Acetone	R. Uterus	38
	Spleen	3.1			35
Asterias forbesi (Asteroidea)	Without gonads	0.016	Butanol (2)	Sphf.	129
Avocado		10	Butanol (1)	Sphf.	113
Banana	Pulp	24	Butanol (2)	Sphf.	75
		28			120
	Peel	65			120
		50	Butanol (1)		113
		150			113
	Hard green outer peel	74			113
		0.1	Acetone	R. Uterus	130
	inner peel	13	Butanol (1)	Sphf.	113
		0.2	Acetone	R. Uterus	130
	pulp	24	Butanol (1)	Sphf.	113
		25	Acetone	R. Uterus	130
	Ripe outer peel	96	Butanol (1)	Sphf.	113
		52	Acetone	R. Uterus	130

Animal or vegetal species	Tissue or fluid		5-HT µg/g or ml content	Method of extraction	Method of measurement	Reference
Banana	Ripe	inner peel	38	Butanol (1)	Sphf.	113
			40	Acetone	R. Uterus	130
		pulp	36	Butanol (1)	Sphf.	113
			19	Acetone	R. Uterus	130
	Overripe	outer peel	161	Butanol (1)	Sphf.	113
			39	Acetone	R. Uterus	130
		inner peel	170	Butanol (1)	Sphf.	113
			30	Acetone	R. Uterus	130
		pulp	35	Butanol (1)	Sphf.	113
			22	Acetone	R. Uterus	130
Bat	Blood	(serum)	3.56	Acetone	R. Uterus	35
	Spleen		18.9			35
Blue-red plum			8	Butanol (1)	Sphf.	113
Bombinator pachypus	Intestine	(small + large int.)	0.75	Acetone	R. Uterus	38
	Skin	(dry weight)	700			35
			1000			38
	Stomach		1			38
Buccinum undatum (Gastropoda)	Ganglia		7.7	Butanol (2)	Sphf.	129
Bufo americanus (Amphibia)	Brain		9.1	Butanol (2)	Sphf.	12a
	Small intestine		5.1			12a
	Stomach		2.6			12a
Bufo bufo bufo (Amphibia)	Intestine		0.87(+)	Acetone	R. Uterus	84
	Brain		1.5	Butanol (2)	Sphf.	12a
	Small intestine		7.1			
	Stomach		2.7			
Bufo marinus (Amphibia)	Dry venom		1000	Acetone	R. Uterus	26
	Ganglia and nerves		6.1			129

Animal or vegetal species	Tissue or fluid		5-HT µg/g or ml content	Method of extraction	Method of measurement	Reference
Busycon canaliculatum (Gastropoda)	Ganglia		9.2	Butanol (2)	Sphf.	129
	Ganglia and nerves		6.1			129
Buthotus minax	Dry venom		40.0	Acetone	R. Stomach	3
Calf	Abomasus		1.2	Acetone	R. Uterus	35
	Intestine	(large)	4.1(+)			35
		(small)	3.3(+)			35
	Brain		0.32	Acetone	R. Uterus	28
Calliactis parasitica	Body wall		27(+)	Acetone	R. Uterus	70
	Coelenteric tissue		550(+)			70
	Tentacle		9(+)			70
Cancer borealis (Crustacea)	Pericardial organs		3.3	Butanol (2)	Sphf.	70a
	Thoracic ganglion		0.02			70a
	Supra-oesophageal ganglia		0.08			70a
Cancer irroratus (Crustacea)	Brain		0.11	Butanol (2)	Sphf.	129
Carcinus maenas (Crustacea)	Green glands		<0.12	Butanol (2)	Sphf.	129
	Leg nerves		<0.03			129
	Pericardial organs		2.8			129
	Ventral ganglia		<0.04			129
Carica papaya	Pulp		1.1–2.1	Acetone	R. Uterus and Colon	42
Cat	Blood	(total)	0.68	Acetone	R. Uterus	31
	Brain	(total)	0.20			28
			0.30		R. Uterus	131
		cere-bellum	0.30	Butanol (2)	Sphf.	14
			0.27	Butanol (1)		116
		cerebral gyri	0.24	Butanol (2)		14
		hypo-thalamus	1.78			14
		medulla	0.55			14

Animal or vegetal species	Tissue or fluid		5-HT µg/g or ml content	Method of extraction	Method of measurement	Reference
Cat	Brain	mesen-cephalon	1.23	Butanol (2)	Sphf.	14
		pons	0.33			14
		white matter	0.07			14
	Duodenum		0.9	Acetone	R. Uterus	131
	Ear		0.13			83
	Ileum		0.5			131
	Kidney		0.1			131
	Intestine	(large)	1.19			35
		(small)	0.7(+)			35
	Liver		0.56			83
			0.6			131
	Lung		0.62			83
	Platelets		0.9			
			(μg/10⁹ plat.)	Acetone	R. Uterus or Colon	57
	Skin		0.1		R. Uterus	131
		(abdomi-nal)	0.08			83
	Spleen		8.4			35
			8.5			131
			8.5			83
	Stomach		0.45			35
			0.5			131
Cerebratulus lacteus (Trematoda)	Head		2.9	Butanol (2)	Sphf.	129
Chicken	Blood	(total)	2.5(+)	Butanol (3)	Sphf.	128
		(serum)	2.75	Acetone	R. Uterus	35
	Brain		0.8	Butanol (1)	Sphf.	125
	Liver		1.6			125
	Spleen		12.54	Acetone	R. Uterus	35
Chiton tuberculatus (Amphineura)	Nerve tissue		17.5	Butanol (2)	Sphf.	129
Ciprinus auratus (Fish)	Brain		0.15	Butanol (2)	Sphf.	12a
	Small intestine		1.1			12a
	Stomach		1.1			12a

Animal or vegetal species	Tissue or fluid		5-HT µg/g or ml content	Method of extraction	Method of measurement	Reference
Cow	Brain	basal ganglia + dien- cephalon + mesen- cephalon	0.10	Acetone	R. Uterus	28
		cortex	0.03			28
Cowage, see *Mucuna pruriens*						
Crab, see *Cancer borealis, Cancer irroratus*						
Crotalus ada- manteus (Reptiles)	Venom		0.24	Chr.	Sphf.	133a
Crotalus atrox	Venom		0.55(+)	Chr.	Sphf.	133a
Discoglossus pictus	Skin	(dry weight)	430(+)	Acetone	R. Uterus	29
Dog	Blood	(total)	0.43	Butanol (1)	Sphf.	114
			0.49	Acetone	R. Uterus or Colon	57
			0.4	Butanol (2)	Sphf.	126
		(plasma)	0.02	Butanol (2)	Sphf.	126
		(serum)	0.21	Acetone	R. Uterus	35
			0.18			39
			0.21			38
	Brain	(total)	1.04	Butanol (2)	Sphf.	48
			0.23	Acetone	V. M. Heart	111
			0.1	Acetone	R. Uterus	28
			0.2		R. Uterus	131
		amygdala	2.1	Butanol (1)	Sphf.	112
			0.489	Acetone	V. M. Heart	79
		area 4 (motoria)	0.078	Butanol (2)		43
			0.021	Acetone	R. Uterus	4
		area 51 (olfactive)	0.016			4
		area postrema	0.262		S. S. Heart	81
			0.215		R. Uterus	4

Animal or vegetal species	Tissue or fluid		5-HT µg/g or ml content	Method of extraction	Method of measurement	Reference
Dog	Brain	area 17 (visiva)	n.m.	Acetone	R. Uterus	4
			0.057	Butanol (2)	Sphf.	43
		basal ganglia + dien- cephalon + mesen- cephalon	0.5	Acetone	R. Uterus	28
		caudate nucleus	n.m.	Butanol (2)	Sphf.	4
			0.116			45
			0.062	Acetone	S. S. Heart	45
			0.229		V. M. Heart	79
			0.274			81
			0.72	Butanol (1)	Sphf.	112
			1.03	Butanol (2)	Sphf.	52
		cere- bellum	n.m.	Acetone	R. Uterus	4
			0.23(+)		V. M. Heart	111
			0.07	Butanol (1)	Sphf.	14
			0.057		Sph.	44
			0.09		Sphf.	112
		+ mesen- cephalon	0.14(+)	Acetone	V. M. Heart	111
		cortex	0.09		R. Uterus and	
					S. S. Heart	45
			0.09		S. S. Heart	81
		central white matter	0.28		R. Uterus	4
		cerebral gyri	0.17	Butanol (1)	Sphf.	14
		cerebro- spinal fluid	0.004	Acetone	R. Uterus	4
		cervical ganglia	n.m.		S. S. Heart	45
			n.m.		R. Uterus	4

Animal or vegetal species	Tissue or fluid		5-HT µg/g or ml content	Method of extraction	Method of measurement	Reference
Dog	Brain	corpus callosum	n.m.	Acetone	R. Uterus	4
		grey matter	0.029			4
			0.14(+)		V. M. Heart	111
			0.34	Butanol (1)	Sphf.	14
			0.20			114
			0.27			112
		hypo-thalamus	0.28	Acetone	R. Uterus	4
			0.37			45
			0.28		S. S. Heart	45
			0.37			81
			0.375		V. M. Heart	79
			0.99	Butanol (2)	Sphf.	23
			1.5	Butanol (1)		21
			1.7			22
			1.75			18
		(anterior)	0.22	Acetone	R. Uterus	4
		(posterior)	0.225			4
		hippo-campus	0.045			4
			0.260		V. M. Heart	79
			0.64	Butanol (1)	Sphf.	15
		medulla oblongata	0.033	Acetone	R. Uterus	4
		mesen-cephalon	0.205			4
		(midbrain)	0.97	Butanol (1)	Sphf.	14
			0.72	Butanol (2)		43
			1	Butanol (1)		112
		(collicula)	n.m.	Acetone	R. Uterus	4
		(pedun-culus)	n.m.			4
		olfactory bulb	0.38	Butanol (1)	Sphf.	14
			0.35			112
			0.048	Acetone	R. Uterus	4
			0.039			45
			0.039			81
			0.078		V. M. Heart	79

Animal or vegetal species	Tissue or fluid		5-HT $\mu g/g$ or ml content	Method of extraction	Method of measurement	Reference
Dog	Brain	optic nerve	n.m.	Acetone	R. Uterus	4
		optic tract	n.m.	Butanol (1)	Sphf.	112
		pons	0.38			112
			0.41			14
		putamen	0.115	Acetone	V. M. Heart	79
		septal area	1.5	Butanol (1)	Sphf.	112
		septum pellucidum	1.5			112
			0.46	Acetone	V. M. Heart	79
		thalamus	0.018	Acetone	R. Uterus	4
			0.65	Butanol (1)	Sphf.	14
			0.47			43
		(medial) 4th ventr.	0.067	Acetone	R. Uterus	4
		floor	0.301		S. S. Heart	81
			0.098		R. Uterus	4
		white matter	n.m.	Butanol (1)	Sphf.	14
	Duodenum		3.7	Acetone	R. Uterus	131
		(mucosa)	8.5			40
		(muscularis externa)	0.15			40
		(submucosa and m. mucae)	0.2			40
	Ear		0.03			83
	Ejaculate		0.15	Butanol (2)	Sphf.	66a
	Ganglia	paravertebral	0.2(+)			4
			n.m.		S. S. Heart	45
		stellate	n.m.		R. Uterus	4
	Heart		0.18(+)		V. M. Heart	111
	Kidney		0.34(+)			111
			0.1		R. Uterus	131
	Intestine	(large)	2.8			35
		(small)	4(+)			35
			4.3			131
	Liver		0.54			83
			0.5			131
			0.1	Butanol (1)	Sphf.	114

Animal or vegetal species	Tissue or fluid		5-HT $\mu g/g$ or ml content	Method of extraction	Method of measurement	Reference
Dog	Lung		0.26	Acetone	R. Uterus	83
	Nerve	femoral	0.01			4
		sciatic	0.11(+)			4
		vagus	0.49(+)			4
	Parotid gland		0.012(+)	Acetone	V. M. Heart	77
	Platelets		1.7			
			($\mu g/10^9$ plat.)		R. Uterus or Colon	57
			0.59			
			($\mu g/mg$ prot.)	Butanol (2)	Sphf.	126
	Skin		0.1	Acetone	R. Uterus	131
		(abdominal)	0.03			83
	Spinal cord (*in toto*)		n.m.			4
			0.55	Butanol (1)	Sphf.	14
			0.62			112
	Spinal cord fasc.					
		cuneif.	n.m.	Acetone	R. Uterus	4
		fasc. gracil.	n.m.			4
		grey matter	0.082			4
		n. cuneif.	0.17			4
		n. gracil.	0.17			4
		white matter	n.m.			4
	Spleen		4.6			83
			1.4			35
			1.4			38
			4.6			131
	Stomach		5.2			35
			0.42			38
			5.2			131
	Submaxillary gland		0.017(+)	Acetone	V. M. Heart	77
	Thyroid		0.025		V. M. Heart	77
	Urine		0.21(+)			111
Dogfish, see *Squalus acanthias*						
Duck	Intestine	(large)	3.6	Acetone	R. Uterus	35
		(small)	3.6(+)			35
	Serum		1.16			35
	Spleen		4.1			35

Animal or vegetal species	Tissue or fluid		5-HT µg/g or ml content	Method of extraction	Method of measurement	Reference
Dugesia dorotocephala (Turbellaria)	Whole		2.5(+)	Butanol (2)	Sphf.	129
Dugesia tigrina (Turbellaria)	Whole		2.0	Butanol (2)	Sphf.	129
Eastern diamond-black rattlesnake, see *Crotalus adamanteus*						
Eggplant			2	Butanol (1)	Sphf.	113
Eledone moschata	Blood		0.04	Acetone	R. Uterus	35
	Intestinal tract		0.2			35
	Optic ganglia		4			84
	Posterior salivary glands		125			84
Ensis directus (Pelecypoda)	Ganglia		39.0	Butanol (2)	Sphf.	129
Fasciolaria tulipa (Gastropoda)	Ganglia and nerves		9.4	Butanol (2)	Sphf.	129
	Kidneys		1.7			129
Flatworm, see *Cerebratulus lacteus, Dugesia dorotocephala, Dugesia tigrina, Lineus ruber, Pneumonoecis similiplexus*						
Frog, see *Hyla arborea, Hyla cinerea, Rana esculenta, Rana pipiens*						
Giant toad, see *Bufo marinus*						
Gluta renghas	Leaves		0.1	Water	R. Stomach	65
Glycera dibranchiata (Polychaeta)	Nerve cord		4.6	Butanol (2)	Sphf.	129
Goat	Blood	(total)	3	Butanol (2)	Sphf.	126
	Brain	basal ganglia	0.17	Acetone	V.M.Heart	83a

Animal or vegetal species	Tissue or fluid		5-HT µg/g or ml content	Method of extraction	Method of measurement	Reference
Goat	Brain	hippo-campus	0.15	Acetone	V.M. Heart	83a
		occipital lobe	0.09			83a
		parietal lobe	0.10			83a
		pineal body	3.20			83a
	Lung		1.56			83a
Goat foetus	Allantoic fluid		0.31	Acetone	V.M. Heart	83a
	Amniotic fluid		0.02			83a
	Blood		3.35			83a
	Brain	basal ganglia	0.27			83a
		hippocam-pus	0.25			83a
		occipital lobe	0.29			83a
		parietal lobe	0.36			83a
	Lung		0.65			83a
	Placenta		0.26			83a
	Platelets		4.3 (µg/10^9 plat.)		R. Colon or Uterus	57
	Serum		2.18		R. Uterus	35
			2.2			39
			2.18	Butanol (2)	Sphf.	126
	Spleen		4.8	Acetone	R. Uterus	35

Goldfish, see *Ciprinus auratus*

Animal or vegetal species	Tissue or fluid		5-HT µg/g or ml content	Method of extraction	Method of measurement	Reference
Goose	Platelets		350	Ethanol	Chr.	21
Guinea pig	Blood	(total)	0.15(+)	Butanol (2)	Sphf.	14
			0.18	Acetone	R. Uterus or Colon	57
		(serum)	0.21		Sphf.	35
	Brain		0.3		R. Uterus	28
			0.46	Butanol (2)	Sphf.	14

Animal or vegetal species	Tissue or fluid		5-HT $\mu g/g$ or ml content	Method of extraction	Method of measurement	Reference
Guinea pig	Brain		1.21	Butanol (2)	Sph.	48
			0.3	Acetone	R. Uterus	131
	Duodenum		5			131
	Ear		0.18			83
	Ileum		3.4			131
	Intestine		5.6	Butanol (3)	Sphf.	128
			6.2	Butanol (1)		128
		(large)	0.7	Acetone	R. Uterus	35
		(small)	4.2(+)			35
	Kidney		0.1			131
	Leucocytes polymorphonuclear		<0.01 $\mu g/10^9$		R. Uterus or Colon	57
	Liver		0.06		R. Uterus	83
			0.1			131
	Lung		0.2	Butanol (2)	Sphf.	127
			0.06	Acetone	R. Uterus	83
			<0.2	Butanol (3)	Sphf.	128
			<0.3	Butanol (1)		128
			0.15	Butanol (2)		62
			0.59			68
	Platelets		0.21 $\mu g/10^9$	Acetone	R. Uterus or Colon	57
	Skin		0.02		R. Uterus	83
		(abdominal)	0.1			131
	Spleen		1.06			35
			1.1			83
			1.1			131
	Stomach		1.4			35
			1.4			131
Hamster	Brain		0.2	Acetone	R. Uterus	131
	Duodenum		0.9			131
	Ear		0.08			83
	Ileum		1.3			131
	Kidney		0.1			131
	Liver		0.22			83
			0.2			131
	Lung		2.75			83
	Skin		0.1			131
		(abdominal)	0.08			83

Animal or vegetal species	Tissue or fluid		5-HT μg/g or ml content	Method of extraction	Method of measurement	Reference
Hamster	Spleen		20.5	Acetone	R. Uterus	83
			20.5			131
	Stomach		1.2			131
Hare (Leporid)	Brain		0.075	Acetone	R. Uterus	72
Helix aspersa (Gasteropoda)	Brain		0.5–4	Chr.	Chr.	62a
	Mantle		1			62a
	Heart		3			62a
Heloderma horridum (Reptiles)	Venom		5.22	Chr.	Sphf.	133a
Hirudo medicinalis (Hirudinea)	Nerve cord and sinus sheaths		6.9	Butanol (2)	Sphf.	129
Hog	Brain	basal ganglia + diencephalon + mesencephalon	0.5	Acetone	R. Uterus	28
Honeybee, see *Apis mellifera*						
Horse	Blood	(serum)	0.41	Acetone	R. Uterus	35
	Brain	basal ganglia + diencephalon + mesencephalon	0.3			28
		cortex	0.03			28
	Intestine	(large)	0.75(+)			35
		(small)	0.52(+)			35
	Spleen		1.76			35
Horseshoe crab, see *Limulus poliphemus*						
Human	Appendix		0.9	Butanol (2)	Sphf.	90a
	Brain	amygdaloid nucleus	0.32	Acetone	R. Uterus	30

Animal or vegetal species	Tissue or fluid		5-HT µg/g or ml content	Method of extraction	Method of measurement	Reference
Human	Brain	anterior perforated subst. area	0.98	Acetone	R. Uterus	30
		postrema	0.57			30
		cere-bellum	0.03			30
		cervical spinal cord	0.14			30
		corpus callosum	0.25			30
		cortex (frontal)	0.02			30
		(grey matter)	0.048			30
		fornix (ant. pars)	0.36			30
		globus pallidus	0.27			30
		gyrus (cingulate)	0.042			30
		(hippo-campal)	0.16			30
		(sub-callosum)	0.31			30
		hypo-thalamus	0.81			30
			0.29	Butanol (2)	Sphf.	8a
		hippo-campus	0.088	Acetone	R. Uterus	30
		mammil-lary bodies	0.39			30
		medulla oblongata	0.34			30
		nucleus caudatus	0.44			30
		pole (frontal)	0.014			30
		(occipital)	0.02			30
		(temporal)	0.037			30

Animal or vegetal species	Tissue or fluid		5-HT µg/g or ml content	Method of extraction	Method of measurement	Reference
Human	Brain	pons	0.7	Acetone	R. Uterus	30
		putamen	0.55			30
			0.32	Butanol (2)	Sphf.	8a
		putamen + nucleus caud.	0.51	Acetone	R. Uterus	30
		quadrig. (anterior bodies)	0.81			30
		(posterior bodies)	0.46			30
		substantia nigra	1.42			30
		thalamus	0.62			30
		uncus	0.22			30
	Blood	(total)	0.12	Acetone	R. Uterus	6
			0.0055(+)		V. M. Heart	53
			0.2			117
			0.19		R. Uterus or Colon	57
			0.13(+)		Chr.	20
			0.12(+)	Butanol (5)	Sphf.	119
			0.15(+)	Butanol (1)		104
			0.117	Acetone	R. Uterus	94
			0.20	Butanol (2)	Sphf.	69
		(m)	0.19	Butanol (1)	Sphf.	41
		(f)	0.15			41
		plasma	0.019		Sphf.	31a
		(serum)	0.12	Acetone	R. Uterus	35
			0.16			39
			0.21			92
			0.12			34
			0.062		Rb. Ear	135
			0.15		R. Uterus	95
			0.07			106
			0.098			7
			0.21	Butanol (2)	Sphf.	33
	Cerebrospinal fluid		0.006			105
			n.m.			4
			n.m.			5

Animal or vegetal species	Tissue or fluid	5-HT μg/g or ml content	Method of extraction	Method of measurement	Reference
Human	Colon	1.7	Butanol (1)	Sphf.	90a
	Duodenum	3.7			90a
	Ejaculate	0.15			66a
		0.5–0.0	Acetone	Sphf.	33a
		135		Chr. or Sphf.	62
	Placenta	0.13	Acetone–ethanol	S. S. Heart	64
	Platelets	0.036 (μg/10^8 plat.)	Acetone	R. Uterus	55
		0.25(+) (μg/10^9 plat.)		R. Uterus or Colon	57
		0.06 (μg/10^8 plat.)		R. Uterus	18
		0.057 (μg/10^8 plat.)			56
		0.76 (μg/10^9 plat.)		Rb. Ear	136
		0.87 (μg/10^9 plat.)	Butanol (1)	Sphf.	124
		0.51 (μg/mg plat. prot.)	Butanol (3)	Sphf.	100
		0.44(+) (μg/mg prot.) (newborn infants)	Butanol		105
		0.053 (μg/10^9 plat.) (infants 7–29 weeks)	Acetone	R. Uterus	73
		0.197 (μg/10^9 plat.) (children $1\frac{1}{2}$–12 years)			73
		0.423 (μg/10^9 plat.) (man 25–50 years)			73
		1.52 (μg/10^9 plat.)	Butanol (1)	Sph.	99
		0.52		Sphf.	99

Animal or vegetal species	Tissue or fluid		5-HT µg/g or ml content	Method of extraction	Method of measurement	Reference
Human	Platelets	(man 70 years)	0.82		Sph.	99
	Skin		0.7	Butanol (1)	Sphf.	103
		(ab-domi-nal)	0.03	Acetone	R. Uterus	57
	Spleen		2.37	Butanol (3)	Sphf.	71a
	Stomach	antrum	0.9	Butanol (2)		90a
		body	0.7			90a
		pylorus	0.6			90a
	Urine		0.23			64
			0.095		S.S. Heart	45
			0.06		Sph. or. Chr.	91
Human foetus (6th month)	Brain	basal ganglia + diencephalon + mesencephalon	0.05	Acetone	R. Uterus	28
Hydra oligactis (Hydrozoa)			1.5	Butanol (2)	Sphf.	129
Hyla arborea (Amphibia)	Skin	(dry weight)	95(+)	Acetone	R. Uterus	35
Hyla cinerea (Amphibia)	Brain		2.0	Butanol (2)	Sphf.	12a
	Small intestine		4.8			12a
	Stomach		3.6			12a
Leiurus quinquestriatus	Dry venom		2000–4000	Acetone	R. Uterus	2
			3500	Water	Chr.	1
			3000			26
Limulus poliphemus (Arachnoidea)	Coxal glands		0.80	Butanol (2)	Sphf.	129
	Heart and cardiac ganglion		0.09			129
	Intestine		0.09			129
	Leg nerves		0.22(+)			129
	Nerve cords		0.15(+)			129

Animal or vegetal species	Tissue or fluid		5-HT μg/g or ml content	Method of extraction	Method of measurement	Reference
Lineus ruber (Trematoda)	Head		0.43	Butanol (2)	Sphf.	129
Lizard, see *S. cyanogenys*						
Loligo pealii (Cephalopoda)	Brain		0.7	Butanol (2)	Sphf.	129
	Optic ganglia		1.15			129
	Median salivary glands		0.38			129
	Skin		0.04			129
Lumbricus terrestris (Oligochaeta)	Nerve cord		10.4	Butanol (2)	Sphf.	129
Meal worm, see *Tenebrio molitor*						
Metridium senile (Coelenterata)	Acontia		0.96(+)	Butanol (2)	Sphf.	128a
	Body wall		0.05(+)			128a
	Oral disk and tentacles		0.12(+)			128a
	Tentacles		0.34(+)			128a
	Whole animal		0.05(+)			128a
Mexican beaded lizard, see *Heloderma horridum*						
Molgula manhattensis (Ascidiacea)	Intestine		<0.17	Butanol (2)	Sphf.	129
	Kidney		<0.43			129
	Part of body wall, including some nervous tissue		<0.11			129
Mouse	Animal	*in toto*	2.4	Butanol (1)	Sphf.	125
			1.2			114
	Blood	(serum)	1.48	Acetone	R. Uterus	35
	Brain		0.9	Butanol (2)	Sphf.	85
			0.9		Sph.	48
			0.85	Butanol (1)	Sphf.	115
			0.6	Butanol (2)		118
			1			86
			0.3	Acetone	R. Uterus	131
		(without cerebellum)	0.82	Butanol (2)	Sphf.	118

Animal or vegetal species	Tissue or fluid		5-HT μg/g or ml content	Method of extraction	Method of measurement	Reference
Mouse	Carcass		0.7	Butanol (1)	Sphf.	115
	Duodenum		1.2	Acetone	R. Uterus	131
	Ear		1.12			83
	Ileum		1			131
	Intestine		5.4	Butanol (1)	Sphf.	128
			5.2	Butanol (3)		128
		(large)	3.1	Acetone	R. Uterus	35
		(small)	1.6			35
	Liver		0.7			131
			0.68			83
			0.9	Butanol (1)	Sphf.	114
	Lung		2.95	Acetone	R. Uterus	83
			1.9	Butanol (1)	Sphf.	127
	Mast cells		130			103
	Skin		0.4	Acetone	R. Uterus	131
		(abdominal)	0.37			83
	Spleen		1.64(+)			35
			2.7			131
			2.7			83
	Stomach		1			131
			8.85			35
Mucuna pruriens	Trichomes		150	Ethanol–water	G. P. Ileum and R. Uterus	19

Mud-dauber wasp, see *Sceliphron coementarium*

Animal or vegetal species	Tissue or fluid		5-HT μg/g or ml content	Method of extraction	Method of measurement	Reference
Musa sapientum	Pulp		45	Butanol (1)	Sphf.	113
var. *paradisiaca*	Pulp	(unripe)	19.9	Acetone	R. Uterus and Colon	42
		(ripe)	56.7			42
		(overripe)	12			42
	Peel	(unripe)	17.6			42
		(ripe)	41.3			42
		(overripe)	5.6			42
Mya arenaria (Pelecypoda)	Ganglia		22	Butanol (2)	Sphf.	129
Mytilus edulis (Pelecypoda)	Ganglia	(cerebro-pleural)	15.0	Butanol (2)	Sphf.	129
		(pedal)	15.0			129
		(visce-ral)	10.0			129

Animal or vegetal species	Tissue or fluid		5-HT μg/g or ml content	Method of extraction	Method of measurement	Reference
Nettle, see *Urtica dioica*						
Octopus macropus	Blood	(serum)	0.04	Acetone	R. Uterus	35
Octopus vulgaris	Optic ganglia		4	Acetone	R. Uterus	84
(Cephalopoda)	Post. salivary glands		125			84
(Mediterraneus)						
(Florida)	Brain	optic and stellate ganglia	3.2	Butanol (2)	Sphf.	129
(Bermuda)	Brain		0.8			129
		optic ganglia	2.3			129
		optic and stellate ganglia	0.56			129
	Ant. salivary glands		1.44(+)			129
	Post. salivary glands		70.0			129
Ox	Abomasus		1.6	Acetone	R. Uterus	35
	Intestine	(large)	3.05(+)			35
		(small)	3.35(+)			35
	Platelets	(dry weight)	0.025			127
	Pineal body		0.4		V. M. Heart	50
	Serum		1.48			35
	Spleen		7.8			35
Palinurus argus	Nerve cord		<0.10	Butanol (2)	Sphf.	129
(Crustacea)						
Papaw, see *Carica papaya*						
Passiflora foetida	Pulp		1.4–3.5	Acetone	R. Uterus and Colon	42
Passion fruit, see *Passiflora foetida*						
Pecten magellanicus	Visceral ganglia		36.0	Butanol (2)	Sphf.	129
(Pelecypoda)						
Pig	Blood	(serum)	0.26	Acetone	R. Uterus	35

Animal or vegetal species	Tissue or fluid		5-HT µg/g or ml content	Method of extraction	Method of measurement	Reference
Pig	Intestine	(large)	0.64(+)	Acetone	R. Uterus	35
		(small)	2.25(+)			35
	Spleen		1.23			35
	Stomach		0.5			35
Pigeon	Blood	(serum)	0.33	Acetone	R. Uterus	35
	Brain		0.15			28
			0.7	Butanol (2)	Sphf.	12a
		cerebellum	0.25(+)	Butanol (2)	Sphf.	4a
		diencepha-lon + optic				
		lobes	0.77(+)			4a
	Heart		0.27(+)			4a
	Intestine		1.1	Acetone	R. Uterus	35
	Liver		0.43(+)	Butanol (2)	Sphf.	4a
	Lung		0.15(+)			4a
	Pons-medulla		1.41(+)			4a
	Telencephalon		1.29(+)			4a
Pineapple	Fresh juice		12.000		Col.	24
	Canned juice		24.300			24
Plantain, see *Musa sapientum*						
Pneumonoecis similiplexus (Trematoda)			<0.19	Butanol (2)	Sphf.	129
Polistes gallica (Insecta)	Sting apparatus		300	Acetone	R. Uterus	26
Polistes sp. (Insecta)	Abdomen		34.4	Butanol (2)	Sphf.	129
Rabbit	Blood	(total)	3.5(+)	Butanol (2)	Sphf.	13
			4.5(+)	Butanol (1)		38
			4			122
			3.5			102
			3.75			97
			3.8			102
			5.2	Acetone	R. Uterus or Colon	57

Animal or vegetal species	Tissue or fluid		5-HT µg/g or ml content	Method of extraction	Method of measurement	Reference
Rabbit	Blood	(total)	5	Butanol (3)	Sphf.	128
			9.4	Butanol (1)	Col.	132
			4.6			121
			4.5	Butanol (2)	Sphf.	89
		(plasma)	0.9			123
		(serum)	3.8	Acetone	R. Uterus	35
			4.4	Acetone (G)		49
			4.35	Acetone		76
			3.53			38
			4.3			39
			0.35(+)	Butanol (3)	Sphf.	123
	Brain	(total)	0.46	Butanol (2)		15
			0.3	Acetone	R. Uterus	28
					Chr.	28
			0.65(+)	Butanol (2)	Sphf.	87
			0.57			101
			0.57			22
			0.54	Butanol (1)		114
			0.50			115
			0.55	Butanol (2)		97
			0.67	Acetone	R. Uterus	30
			0.3			131
		brain stem	0.87(+)	Butanol (2)	Sphf.	87
			0.51(+)			98
			0.7			99
			0.62(+)			23
			0.65			107
		cerebellum	0.112	Acetone	R. Uterus	31
			0.035	Acetone (G)		49
		hippo-campus	0.272	Acetone		31
		hypo-thalamus	0.39			49
		medulla + pons	0.582			31
		mesen-cephalon	0.647			31
			0.31	Acetone (G)		49
		olfactory bulb	0.17			49

Animal or vegetal species	Tissue or fluid		5-HT µg/g or ml content	Method of extraction	Method of measurement	Reference
Rabbit	Brain	telen-cephalon	0.282	Acetone	R. Uterus	31
	Diaphragm		0.09	Acetone (G)		49
	Duodenum		3.3	Acetone		131
	Ear		0.11			83
	Heart		0.023	Acetone (G)		49
	Kidney		0.020	Acetone		49
			0.055(+)	Acetone (G)		49
			0.1	Acetone		131
	Ileum		3.7			131
	Intestine	(large)	3.5			49
			2.7			35
		(small)	20	Butanol (1)	Sphf.	86
			11.25(+)			131
			6	Butanol (2)	Sphf.	89
			10.0			12a
			3.8	Acetone	R. Uterus	49
			10	Butanol (2)	Sphf.	128
			10.7	Butanol (3)		128
			3.5(+)	Acetone	R. Uterus	35
	Liver		0.56			49
			0.92(+)	Acetone (G)		49
			0.55	Acetone		83
			0.6			131
	Lung		0.127(+)	Acetone (G)		49
			0.06	Acetone		49
			1.61			83
			2.1	Butanol (2)	Sphf.	127
	Pancreas		0.21	Acetone	R. Uterus	49
			0.145	Acetone (G)		49
	Parotid gland		0.16	Acetone	V. M. Heart	77
	Platelets		7.5 (μg/10^9 plat.)	Acetone	R. Uterus or Colon	57
			2.7 (μg/10^8 plat.)	Butanol (1)	Col.	132
			16.1 (μg/mg plat. prot.)			132
			6.8 (μg/mg plat. prot.)			121

Animal or vegetal species	Tissue or fluid		5-HT µg/g or ml content	Method of extraction	Method of measurement	Reference
Rabbit	Platelets		5.5 (µg/mg plat. prot.)	Butanol (3)	Sphf.	100
	Salivary glands		0.07	Acetone (G)	R. Uterus	49
	Skin		0.1	Acetone		131
		(abdominal)	0.04			83
	Spleen		24.3			83
			11.7			49
			19.6			35
			24.3			131
	Stomach	(total)	6.2			49
			4.9			131
		(body mucosa)	8.75			81
		(fundus)	4.9			35
		(pyloric mucosa)	1.25			40
		(pylorus)	0.85			35
	Submaxillary gland		0.25	Acetone	V. M. Heart	77
	Thymus		0.05	Acetone (G)		49
	Thyroid		0.09			49
			0.19	Acetone	V. M. Heart	77
Rana esculenta (Amphibia)	Blood	(serum)	0.18	Acetone	R. Uterus	35
	Skin	(dry weight)	19(+)			35
	Spleen		0.08			35
Rana pipiens (Amphibia)	Brain		3.7	Butanol (2)	Sphf.	12a
	Small intestine		5.4			12a
	Stomach		3.0			12a
Rat	Adipous tissue		n.m.			82
	Blood	(plasma)	0.04(+)	Acetone	R. Uterus	109
		(serum)	0.64(+)			36
			0.97			35
			1.05			39
			0.57			38
			0.82			12
			0.75(+)			10
		(rat 35 g)	0.46(f)			9
		(rat 60 g)	0.46(f)			9

Animal or vegetal species	Tissue or fluid		5-HT µg/g or ml content	Method of extraction	Method of measurement	Reference
Rat	Blood	(rat 100 g)	0.73(m)	Acetone	R. Uterus	9
			0.62(f)			9
		(rat 200 g)	0.83(m)			9
			0.75(f)			9
		(rat 300 g)	1.16(m)			9
		(total)	0.15(+)	Butanol (2)	Sphf.	13
			0.34	Acetone	R. Uterus or Colon	57
			0.2		R. Stomach	59
	Brain	(total)	0.317(+)		R. Uterus	35
			0.4			28
			0.33		V. M. Heart	16
			0.395(+)			17
			0.24			111
			0.40	Butanol (2)	Sphf.	13
			0.55			85
			0.37	Butanol (1)		115
			0.51(+)			112
			0.86		Sph.	74
			0.97			48
			0.47	Butanol (2)	Sphf.	46
			0.43(+)	Acetone	R. Uterus	10
			0.2			131
			0.49		R. Stomach	59
			0.326			78
			0.216			60
			0.98(+)		R. Uterus	110
			0.27			11
			0.40	Butanol (1)	Sphf.	32
			0.44	Butanol (2)		25
		(free)	0.063			51
		(bound)	0.038			51
		(2 days)	0.17	Butanol (2)		61
		(5 days)	0.21			61
		(120–200 days)	0.40(m)			61
			0.54(f)			61
			0.33			12

Animal or vegetal species	Tissue or fluid		5-HT µg/g or ml content	Method of extraction	Method of measurement	Reference
Rat	Brain	(rat 35 g)	0.18(m)	Acetone	R. Uterus	9
			0.18(f)			9
		(rat 60 g)	0.33(f)			9
		(rat 100 g)	0.49(m)			9
			0.24(f)			9
		(rat 200 g)	0.30(m)			9
			0.24(f)			9
		(rat 300 g)	0.30(m)			9
		cerebellum	0.117	Butanol (2)	Sphf.	46
		dien- cephalic area	0.81			46
		hemi- spheres	0.37			46
		hypo- thalamus	2.75	Acetone	R. Uterus	90
		mesen- cephalic area	0.56			90
	Duodenum		1.2			131
			5.6			108
	Ear		1.01			83
			0.46(+)			10
			0.69(+)			11
		(rat 35 g)	0.10(m)			9
			0.26(f)			9
		(rat 60 g)	0.38(f)			9
		(rat 100 g)	0.13(m)			9
			0.43(f)			9
			0.35			12
			1.4			108
		(rat 200 g)	0.19(m)			9
			0.47(f)			9
		(rat 300 g)	0.25(m)			9
	Foetus		0.1	Butanol (1)	Sphf.	32
		brain (total)	0.1	Butanol (2)	Sphf.	61
	Gastrointestinal tract		2.17(+)	Acetone	R. Uterus	35
			1.6(+)			11

Animal or vegetal species	Tissue or fluid		5-HT μg/g or ml content	Method of extraction	Method of measurement	Reference
Rat	Heart		0.1	Butanol (1)	Sphf.	114
			0.1			32
	Ileum		1.2	Acetone	R. Uterus	131
			1.02		R. Stomach	71
			5.5		R. Uterus	108
			0.66	Butanol (2)	Sphf.	62a
	Intestine	(jejunum)	4.6	Acetone	R. Uterus	93
			3.2			108
		(large)	3.9			35
			2.3		R. Colon	47
			4.9			109
			2.39	Butanol (1)	Sph.	48
			2.95	Acetone	R. Uterus	10
			1.1	Butanol (1)	Sphf.	128
			1.7	Butanol (3)		128
			4.38(+)	Acetone	R. Uterus	11
			3.8			108
		(rat 35 g)	1.45(m)			9
			0.75(f)			9
		(rat 50 g)	0.86(f)			9
		(rat 100 g)	1.90(m)			9
			1.16(f)			9
		(rat 200 g)	2.3(m)			9
			2.12(f)			9
		(rat 300 g)	2.8(m)			9
			1.92(+)			11
		(small)	1.7			10
			1.2			35
			2.2		R. Colon	47
			3.95(+)		V. M. Heart	17
		(small + large)	4.76	Butanol (2)	Sphf.	118
			3.95	Acetone	R. Stomach	78
			2.79			60
			3.02	Butanol (2)	Sphf.	75
	Kidney		0.1	Acetone	R. Uterus	131
			0.23	Butanol (2)	Sphf.	75
			0.34			118
			0.44		Sph.	67
			0.1	Butanol (1)	Sphf.	32

Animal or vegetal species	Tissue or fluid		5-HT $\mu g/g$ or ml content	Method of extraction	Method of measurement	Reference
Rat	Liver		0.1	Acetone	R. Uterus	131
			0.25(+)			10
			0.14			83
			<0.1	Butanol (1)	Sphf.	114
			0.3			114
			<0.1			32
		(rat 35 g)	0.03(m)	Acetone	R. Uterus	9
			0.04(f)			9
		(rat 60 g)	0.05(f)			9
		(rat 100 g)	0.06(m)			9
			0.06(f)			9
		(rat 200 g)	0.12(m)			9
			0.14(f)			9
		(rat 300 g)	0.14(m)			9
	Lung		2.3	Butanol (2)	Sphf.	127
			2.7	Butanol (3)		128
			3.5	Butanol (1)		128
			2.86	Butanol (2)		118
			1.49(+)	Acetone	R. Uterus	10
			3.9			108
			2			93
			3		R. Stomach	12
			1.19(+)		R. Uterus	11
		(rat 35 g)	0.52(m)			9
			0.36(f)			9
		(rat 60 g)	0.36(f)			9
		(rat 100 g)	0.64(m)			9
			0.51(f)			9
		(rat 200 g)	0.97(m)			9
			1.22(f)			9
		(rat 300 g)	2.02(m)			9
			1.16			83
	Mast cells (mesenteric)		800			37
			0.54	Butanol (2)	Sphf.	62a
	Parotid glands		0.56(+)	Acetone	V. M. Heart	77
	Paws		0.25	Acetone	R. Uterus	12
			3.5	Butanol (2)	Sphf.	73a
	Platelets		0.4 ($\mu g/10^9$ plat.)		R. Uterus or Colon	57

Animal or vegetal species	Tissue or fluid		5-HT µg/g or ml content	Method of extraction	Method of measurement	Reference
Rat	Salivary glands		1.74	Butanol (4)	Sphf.	118
	Skin		1.48	Acetone	R. Uterus	54
			1.3			131
			1.087			71
		(abdominal)	1.34			83
			1.9			108
		(back area lat.)	1.65(+)			8
		(foot area lat.)	6.4(+)			8
		(subcutaneous)	1			93
			2.14			133
			1.35(+)			11
	Spleen		1.58(+)			36
			2.8			35
			2.5			83
			5.9(+)		V. M. Heart	17
			4			93
			2.37(+)			10
			3.4		R. Colon	109
			3.51	Butanol (2)	Sphf.	118
			2.5	Acetone	R. Uterus	131
			2.40			12
		(rat 35 g)	1.37(m)			9
			1.41(f)			9
		(rat 60 g)	1.8(f)			9
		(rat 100 g)	1.31(f)			9
		(rat 200 g)	1.29(m)			9
			1.89(f)			9
		(rat 300 g)	1.16(m)			9
	Stomach	(fundus)	1.8	Acetone	R. Uterus	108
		(pyloric part)	2.1			108
			1.4			35
			6		R. Colon	109
			1.4		R. Uterus	131
			1.73			10
	Sublingual glands		0.85(+)	Acetone	V. M. Heart	77

Animal or vegetal species	Tissue or fluid		5-HT µg/g or ml content	Method of extraction	Method of measurement	Reference
Rat	Submaxillary glands		0.25(+)			77
	Thyroid		3.7			77
			4.32			80
	Uterus		0.2	Butanol (2)	Sphf.	114
			0.1	Butanol (1)		32
	Whole rat		24.3	Acetone	R. Uterus	108
Red plum			10	Butanol (1)	Sphf.	113
Sagartia luciae (Anthozoa)			1.30	Butanol (2)	Sphf.	129
Salamander	Brain		2.8	Butanol (2)	Sphf.	12a
	Small intestine		5.0			12a
	Stomach		3.0			12a
Sceliphron coementarium (Insecta)	Heads		0.47	Butanol (2)	Sphf.	129
	Thorax		0.19			129
	Abdomen		29.0			129

Scorpion, see *Buthotus minax, Leiurus quinquestriatus, Vejovis spiniger, Vejovis* sp.

Animal or vegetal species	Tissue or fluid		5-HT µg/g or ml content	Method of extraction	Method of measurement	Reference
S. cyanogenys (Reptiles)	Brain		3.1	Butanol (2)	Sphf.	12a
	Stomach		2.2			12a
	Small intestine		6.9			12a
Scylliorhinus canicula	Intestine		1.5	Acetone	R. Uterus	84
	Brain		0.2			28

Sea anemone, see *Calliactis parasitica, Sagartia luciae*

Animal or vegetal species	Tissue or fluid		5-HT µg/g or ml content	Method of extraction	Method of measurement	Reference
Sea gull	Blood	(serum)	0.69	Acetone	R. Uterus	35
Sea urchin	Blood	(serum)	1.84	Acetone	R. Uterus	35
	Spleen		1.8			35
Sheep	Abomasus		3.2	Acetone	R. Uterus	35
	Blood	(serum)	0.85			35

Animal or vegetal species	Tissue or fluid		5-HT µg/g or ml content	Method of extraction	Method of measurement	Reference
Sheep	Brain	basal ganglia + dien-cephalon + mesen-cephalon	0.04	Acetone	R. Uterus	28
		cortex	0.02			28
	Intestine	(large)	2.7(+)			35
		(small)	3.05(+)			35
	Spleen		3.8			35
	Thyroid		4.6		V. M. Heart	77
Snail, see *Helix aspersa*						
Spisula solidissima (Pelecypoda)	Connective		2.2	Butanol (2)	Sphf.	129
	Ganglia		14.3			129
Squalus acanthias (Spiny dogfish)	Ejaculate		437			66
Stork	Blood	(serum)	0.03	Acetone	R. Uterus	35
Tench, see *Tinca vulgaris*						
Tenebrio molitor (Insecta)	Whole larvae		0.05	Butanol (2)	Sphf.	129
Testudo graeca	Blood	(serum)	0.01	Acetone	R. Uterus	35
	Spleen		0.01			35
Thyone briareus (Holothuroidea)	Region of nerve ring		0.04	Butanol (2)	Sphf.	129
Tinca vulgaris	Blood	(serum)	0.04	Acetone	R. Uterus	35
	Spleen		0.05			35
Toad, see *Bufo americanus, Bufo bufo bufo, Bufo marinus*						
Tomato			12	Butanol (1)	Sphf.	113

Animal or vegetal species	Tissue or fluid		5-HT $\mu g/g$ or ml content	Method of extraction	Method of measurement	Reference
Torpedo marmorata	Blood	(serum)	0.02	Acetone	R. Uterus	35
	Spleen		0.06			35
Tropidonotus natrix	Blood	(serum)	0.41	Acetone	R. Uterus	35
	Brain		0.2			28
	Spleen		0.16			35
Turkey cock	Blood	(serum)	0.11	Acetone	R. Uterus	35
Turtle, see *Testudo graeca*						
Urtica dioica	Sting		0.0048/ sting	Acetone	R. Uterus	27
			0.004/ sting		G. P. Ileum	27
			0.0034/ sting		R. Colon	27
	Sting fluid		200			26
Vejovis sp. (Arachnoidea)	Sting segment		94.0(+)	Butanol (2)	Sphf.	129
Vejovis spiniger (Arachnoidea)	Sting segment		n.m.	Butanol (2)	Sphf.	129
Venus mercenaria (Pelecypoda)	Ganglia		40.0	Butanol (2)	Sphf.	129
Vespa vulgaris	Dry venom		320		Chr. G. P. Ileum	58

Wasp, see *Polistes gallica*, *Polistes* sp., *Sceliphron coementarium*, *Vespa vulgaris*

Water moccasin, see *Agkistrodon piscivorus*

Western diamond-black rattlesnake, see *Crotalus atrox*

REFERENCES

1   ADAM, K. R. AND WEISS, C., *Nature*, 178 (1956) 421.
2   ADAM, K. R. AND WEISS, C., *J. Exptl. Biol.*, 35 (1958) 39.
3   ADAM, K. R. AND WEISS, C., *Nature*, 183 (1959) 1398.
4   AMIN, A. H., CRAWFORD, T. B. B. AND GADDUM, J. H., *J. Physiol. (London)*, 126 (1954) 596.
4a APRISON, M. H., WOLF, M. N., POULOS, G. L. AND FOLKERTH, T. L., *J. Neurochem.*, 9 (1962) 575.
5   ASHROFT, G. W. AND SHARMAN, D. F., *Nature*, 186 (1960) 1050.
6   BALLERINI, G., BERETTA, P. AND LA PAGLIA, S., *Ann. univ. Ferrara*, 2 (1958) 115.
7   BARKHAN, P., *S. African J. Med. Sci.*, 20 (1955) 49.
8   BENDITT, E. P., in G. P. LEWIS (Ed.), *5-Hydroxytryptamine*, Pergamon, London, 1957, p. 32.
8a BERNHEIMER, H., BIRKMEYER, W. AND HORNYKIEWICZ, O., *Klin. Wochschr.*, 20 (1961) 1056.
9   BERTACCINI, G., *Ricerca sci.*, 28 (1958) 2261.
10 BERTACCINI, G., *Naturwissenschaften*, 45 (1958) 548.
11 BERTACCINI, G., *J. Neurochem.*, 4 (1959) 217.
12 BERTACCINI, G., *J. Physiol. (London)*, 153 (1960) 239.
12a BOGDANSKI, D. F., BONOMI, L. AND BRODIE, B. B., *Life Sci.*, 1 (1963) 80.
13 BOGDANSKI, D. F., PLETSCHER, A., BRODIE, B. B. AND UDENFRIEND, S., *J. Pharmacol. Exptl. Therap.*, 117 (1956) 82.
14 BOGDANSKI, D. F. AND UDENFRIEND, S., *J. Pharmacol. Exptl. Therap.*, 116 (1956) 7.
15 BOGDANSKI, D. F., WEISSBACH, H. AND UDENFRIEND, S., *J. Neurochem.*, 1 (1957) 272.
16 BONNYCASTLE, D. D., PAASONEN, M. K. AND GIARMAN, N. J., *Nature*, 178 (1956) 991.
17 BONNYCASTLE, D. D., PAASONEN, M. K. AND GIARMAN, N. J., *Brit. J. Pharmacol.*, 12 (1957) 228.
18 BORN, G. V. R., INGRAM, G. I. C. AND STACEY, R. S., *Brit. J. Pharmacol.*, 13 (1958) 62.
19 BOWDEN, K., BROWN, B. AND BATTY, J. E., *Nature*, 174 (1954) 925.
20 BRACCO, M., CURTI, P. C. AND GIULIANO, M., *Farmaco (Pavia), Ed. sci.*, 10 (1955) 595.
21 BRACCO, M., CURTI, P. C. AND GIULIANO, M., *Experientia*, 12 (1955) 31.
22 BRODIE, B. B., SHORE, P. A. AND PLETSCHER, A., *Science*, 123 (1956) 992.
23 BRODIE, B. B., SPECTOR, S., KUNTZMAN, K. G. AND SHORE, P. A., *Naturwissenschaften*, 45 (1958) 243.
24 BRUCE, D. W., *Nature*, 188 (1960) 147.
25 CANAL, N., MAFFEI-FACCIOLI, A. AND MANZINI, B., *Boll. soc. ital. biol. sper.*, 34 (1958) 1525.
26 COLLIER, H. O. J., in G. P. LEWIS (Ed.), *5-Hydroxytryptamine*, Pergamon, London, 1957, p. 5.
27 COLLIER, H. O. J. AND CHESHER, G. B., *Brit. J. Pharmacol.*, 11 (1956) 186.
28 CORREALE, P., *J. Neurochem.*, 1 (1956) 22.
29 COSTA, E., personal communication.
30 COSTA, E. AND APRISON, M. H., *J. Nervous Mental Disease*, 126 (1958) 289.
31 COSTA, E. AND RINALDI, F., *Am. J. Physiol.*, 194 (1958) 289.
31a CRAWFORD, N., *Clin. Chim. Acta*, 8 (1963) 39.
32 DAVIDSON, J., SJOERDSMA, A., LOOMIS, L. N. AND UDENFRIEND, S., *J. Clin. Invest.*, 36 (1957) 1594.
33 DAVIS, R. B., *J. Lab. Clin. Med.*, 54 (1959) 344.
33a ELIASSON, R., *J. Urol.*, 86 (1961) 676.
34 ERSPAMER, V., *Ciba Foundation Symposium on Hypertension*, Little, Brown and Co., Boston, 1954, p. 7–8.
35 ERSPAMER, V., *Rend. sci. Farmitalia*, 1 (1954) 1.
36 ERSPAMER, V., *Experientia*, 12 (1956) 63.
37 ERSPAMER, V., *Z. Vitamin-, Hormon- u. Fermentforsch.*, 9 (1957) 74.
38 ERSPAMER, V. AND FAUSTINI, B., *Naturwissenschaften*, 40 (1953) 317.
39 ERSPAMER, V. AND SALA, G., *Brit. J. Pharmacol.*, 9 (1954) 31.
40 FELDBERG, W. AND TOH, C. C., *J. Physiol. (London)*, 119 (1953) 352.
41 FELDSTEIN, A., HOAGLAND, H. AND FREEMAN, H., *J. Nervous Mental Disease*, 192 (1959) 62.
42 FOY, J. M. AND PARRATT, J. R., *J. Pharm. Pharmacol.*, 12 (1960) 360.
43 FRESIA, P., GENOVESE, E., KATO, R. AND VALZELLI, L., *Boll. soc. ital. biol. sper.*, 34 (1958) 1397.
44 FRESIA, P., GENOVESE, E., VALSECCHI, A. AND VALZELLI, L., *Boll. soc. ital. biol. sper.*, 33 (1957) 888.
45 GADDUM, J. H. AND PAASONEN, M. K., *Brit. J. Pharmacol.*, 10 (1955) 474.

46   GARATTINI, S., KATO, R. AND VALZELLI, L., *Atti soc. lombarda sci. med. biol.*, 13 (1958) 300.
47   GARATTINI, S., VALSECCHI, A. AND VALZELLI, L., *Experientia*, 13 (1957) 330.
48   GARATTINI, S. AND VALZELLI, L., *Psychotropic Drugs*, Elsevier, Amsterdam, 1957, p. 428.
49   GARVEN, J. D., *Brit. J. Pharmacol.*, 11 (1956) 66.
50   GIARMAN, N. J. AND DAY, M., *Biochem. Pharmacol.*, 1 (1958) 235.
51   GLUCKMAN, M. I., *Federation Proc.*, 19 (1960) 265.
52   GOLDBERG, L. I., DA COSTA, F. M. AND OZAKI, M., *Nature*, 188 (1960) 502.
53   GREEN, J. P., PAASONEN, M. K. AND GIARMAN, N. J., *Proc. Soc. Exptl. Biol. Med.*, 94 (1957) 428.
54   HALPERN, B. N., LIACOPOULOS, P. AND LIACOPOULOS-BRIOT, M., *Compt. rend. soc. biol.*, 151 (1957) 1692.
55   HARDISTY, R. M., INGRAM, G. I. C. AND STACEY, R. S., *Experientia*, 12 (1956) 424.
56   HARDISTY, R. M. AND STACEY, R. S., *J. Physiol. (London)*, 130 (1955) 711.
57   HUMPHREY, J. H. AND JAQUES, R., *J. Physiol. (London)*, 124 (1954) 305.
58   JAQUES, R. AND SCHACHTER, M., *Brit. J. Pharmacol.*, 9 (1954) 53.
59   KÄRKI, N. T. AND PAASONEN, M. K., *Acta Pharmacol. Toxicol.*, 16 (1959) 20.
60   KÄRKI, N. T. AND PAASONEN, M. K., *J. Neurochem.*, 3 (1959) 352.
61   KATO, R., *J. Neurochem.*, 5 (1960) 202.
62   KATO, R., MARIANI, L. AND VALZELLI, L., *Atti soc. lombarda sci. med. biol.*, 13 (1958) 297.
62a KERKUT, G. A. AND COTTRELL, G. A., *Comp. Biochem. Physiol.*, 8 (1963) 53.
62b KINDWALL, P. A., BORÉUS, L. O. AND WESTERHOLM, B., *Am. J. Physiol.*, 203 (1962) 389.
63   KATSH, S., *J. Urology*, 81 (1959) 570.
64   KURIAKI, K. AND INOUE, T., *Compt. rend. soc. biol.*, 150 (1956) 1835.
65   LIN, R. C. Y. AND WHITTOW, G. C., *Brit. J. Pharmacol.*, 15 (1960) 440.
66   MANN, T., *Nature*, 188 (1960) 941.
66a MANN, T., SEAMARK, R. F. AND SHARMAN, D. F., *Brit. J. Pharmacol.*, 17 (1961) 208.
67   MARIANI, L. AND VALZELLI, L., *Boll. soc. ital. biol. sper.*, 34 (1958) 1167.
68   MARIANI, L. AND VALZELLI, L., *Boll. soc. ital. biol. sper.*, 34 (1958) 1169.
69   MARSHALL, E. F., STIRLING, G. S., TAIT, A. C. AND TODRICK, A., *Brit. J. Pharmacol.*, 15 (1960) 35.
70   MATHIAS, A. P., ROSS, D. M. AND SCHACHTER, M., *Nature*, 180 (1957) 658.
70a MAYNARD, D. M. AND WELSH, J. H., *J. Physiol (London)*, 149 (1959) 215.
71   MEDAKOVIC, M. AND RADMANOVIC, B., *J. Pharm. Pharmacol.*, 12 (1960) 695.
71a MELLINKOFF, S., CRADDOCK, C., FRANKLAND, M., KENDRICKS, F. AND GREIPEL, M., *Am. J. Digest. Dis.*, 7 (1962) 347.
72   MILINE, R., STERN, P. AND HUKOVIC, S., *Experientia*, 14 (1958) 415.
73   MITCHELL, R. G. AND CASS, R., *J. Clin. Invest.*, 38 (1959) 595.
73a MÖRSDORF, K., *Med. Exptl.*, 1 (1959) 87.
74   MURELLI, B., VALSECCHI, A. AND VALZELLI, L., *Boll. soc. ital. biol. sper.*, 33 (1957) 257.
75   MUSSINI, E. AND VALZELLI, L., *Atti soc. lombarda sci. med. biol.*, 13 (1958) 310.
76   NAESS, K. AND SCHANCHE, S., *Nature*, 177 (1956) 1130.
77   PAASONEN, M. K., *Experientia*, 14 (1958) 95.
78   PAASONEN, M. K. AND KÄRKI, N. T., *Brit. J. Pharmacol.*, 14 (1959) 164.
79   PAASONEN, M. K., MCLEAN, P. D. AND GIARMAN, N. J., *J. Neurochem.*, 1 (1957) 326.
80   PAASONEN, M. K. AND PELTOLA, P., *Ann. Med. Exptl. et Biol. Fenniae (Helsinki)*, 38 (1960) 228.
81   PAASONEN, M. K. AND VOGT, M., *J. Physiol. (London)*, 132 (1956) 617.
82   PAOLETTI, R., personal communication (1961).
83   PARRATT, J. R. AND WEST, G. B., *J. Physiol. (London)*, 137 (1957) 617.
83a PEPEU, G. AND GIARMAN, N. J., *J. Gen. Physiol.*, 45 (1962) 575.
84   PICCINELLI, D., *Arch. intern. pharmacodyn.*, 117 (1958) 452.
85   PLETSCHER, A., *Experientia*, 12 (1956) 479.
86   PLETSCHER, A., SHORE, P. A. AND BRODIE, B. B., *Science*, 122 (1955) 968.
87   PLETSCHER, A., SHORE, P. A. AND BRODIE, B. B., *J. Pharmacol. Exptl. Therap.*, 116 (1956) 84.
88   POLLI, E. E., BIANCHI, P. A. AND CROSTI, R. S., *Haematol. Latina (Milan)*, 1 (1958) 250.
89   QUINN, G. P., SHORE, P. A. AND BRODIE, B. B., *J. Pharmacol. Exptl. Therap.*, 127 (1959) 103.
90   RENSON, J. AND FISCHER, P., *Arch. int. physiol. et biochem.*, 67 (1959) 142.
90a RESNICK, R. H. AND SEYMOUR, J. G., *Gastroenterology*, 41 (1961) 119.
91   RODNIGHT, R., *Biochem. J.*, 64 (1956) 621.

92   ROSEMBERG, J. C., DAVIS, R., MORAN, W. H. AND ZIMMERMAN, B.. *Federation Proc.*, 18 (1959) 503.
93   SANYAL, R. K. AND WEST, G. B., *J. Physiol. (London)*, 142 (1958) 571.
94   SCHMID, E., SCHEIFFARTH, F. AND ZICHA, L., *Lancet*, ii (1959) 354.
95   SHARMAN, D. F. AND SULLIVAN, F. M., *Nature*, 177 (1956) 332.
96   SHORE, P. A., *Pharmacol. Revs.*, 11 (1959) 276.
97   SHORE, P. A. AND BRODIE, B. B., in S. GARATTINI AND V. GHETTI (Eds.), *Psychotropic Drugs*, Elsevier, Amsterdam, 1957, p. 423.
98   SHORE, P. A. AND BRODIE, B. B., *Science*, 127 (1958) 704.
99   SHORE, P. A. AND BRODIE, B. B., *J. Clin. Exptl. Psychopath. and Quart. Rev. Psychiat. Neurol.*, 19 (1958) 56.
100 SHORE, P. A., GILLESPIE JR., L., SPECTOR, S. AND PROCKOP, D., *Naturwissenschaften*, 45 (1958) 340.
101 SHORE, P. A., PLETSCHER, A., TOMICH, E. G., CARLSSON, A., KUNTZMAN, R. AND BRODIE, B. B.. *Ann. N. Y. Acad. Sci.*, 66 (1957) 609.
102 SHORE, P. A., PLETSCHER, A., TOMICH, E. G., KUNTZMAN, R. AND BRODIE, B. B., *J. Pharmacol. Exptl. Therap.*, 117 (1956) 232.
103 SJOERDSMA, A., WAALKES, T. P. AND WEISSBACH, H., *Science*, 125 (1957) 1202.
104 SJOERDSMA, A., WEISSBACH, H. AND UDENFRIEND, S., *Am. J. Med.*, 20 (1956) 520.
105 SJOERDSMA, A., WEISSBACH, H., TERRY, L. L. AND UDENFRIEND, S., *Am. J. Med.*, 23 (1957) 5.
106 SNOW, P. J. D., LENNARD-JONES, J. E., CURZON, G. AND STACEY, R. S., *Lancet*, ii (1955) 1004.
107 SPECTOR, S., SHORE, P. A. AND BRODIE, B. B., *J. Pharmacol. Exptl. Therap.*, 128 (1960) 15.
108 TELFORD, J. M. AND WEST, G. B., *Brit. J. Pharmacol.*, 15 (1960) 532.
109 TOH, C. C., *J. Physiol. (London)*, 138 (1957) 488.
110 TOWNE, J., PUT, T. AND SCHWARTZ, N., *Federation Proc.*, 18 (1959) 452.
111 TWAROG, B. M. AND PAGE, I. H., *Am. J. Physiol.*, 175 (1953) 157.
112 UDENFRIEND, S., in G. P. LEWIS (Ed.), *5-Hydroxytryptamine*, Pergamon, London, 1957, p. 43.
113 UDENFRIEND, S., LOVENBERG, W. AND SJOERDSMA, A., *Arch. Biochem. Biophys.*, 85 (1959) 487.
114 UDENFRIEND, S., WEISSBACH, H. AND BOGDANSKI, D. F., *J. Biol. Chem.*, 224 (1957) 803.
115 UDENFRIEND, S., WEISSBACH, H. AND BOGDANSKI, D. F., *J. Pharmacol. Exptl. Therap.*, 120 (1957) 255.
116 UDENFRIEND, S., WEISSBACH, H. AND BOGDANSKI, D. F., *Ann. N. Y. Acad. Sci.*, 66 (1957) 602.
117 UDENFRIEND, S., WEISSBACH, H. AND SJOERDSMA, A., *Science*, 123 (1956) 669.
118 VALZELLI, L., unpublished observations.
119 WAALKES, T. P., *J. Lab. Clin. Med.*, 53 (1959) 824.
120 WAALKES, T. P., SJOERDSMA, A., CREVELING, C. R., WEISSBACH, H. AND UDENFRIEND, S., *Science*, 127 (1958) 648.
121 WAALKES, T. P. AND WEISSBACH, H., *Proc. Soc. Exptl. Biol. Med.*, 93 (1957) 602.
122 WAALKES, T. P., WEISSBACH, H., BORICEVICH, J. AND UDENFRIEND, S., *Proc. Soc. Exptl. Biol. Med.*, 95 (1957) 479.
123 WAALKES, T. P., WEISSBACH, H., BORICEVICH, J. AND UDENFRIEND, S., *J. Clin. Invest.*, 36 (1957) 1115.
124 WEINER, M. AND UDENFRIEND, S., *Circulation*, 15 (1957) 353.
125 WEISSBACH, H., BOGDANSKI, D. F., REDFIELD, B. G. AND UDENFRIEND, S., *J. Biol. Chem.*, 227 (1957) 617.
126 WEISSBACH, H., BOGDANSKI, D. F. AND UDENFRIEND, S., *Arch. Biochem. Biophys.*, 73 (1958) 492.
127 WEISSBACH, H., WAALKES, T. P. AND UDENFRIEND, S., *Science*, 125 (1957) 235.
128 WEISSBACH, H., WAALKES, T. P. AND UDENFRIEND, S., *J. Biol. Chem.*, 230 (1958) 865.
128a WELSH, J. H., *Nature*, 186 (1960) 811.
129 WELSH, J. H. AND MOORHEAD, M. *J. Neurochem.*, 6 (1960) 146.
130 WEST, G. B., *J. Pharm. and Pharmacol.*, 10 (1958) 589.
131 WEST, G. B., *J. Pharm. and Pharmacol.*, 10 (1958) 92 T.
132 WOOLLEY, D. W. AND EDELMAN, P. M., *Science*, 127 (1958) 281.
133 YEH, S. D. J., SOLOMON, J. D. AND CHOW, B. W., *Federation Proc.*, 18 (1959) 357.
133a ZARAFONETIS, C. J. D. AND KALAS, J. P., *Am. J. Med. Sci.*, 240 (1960) 764.
134 ZUCKER, M. B. AND BORRELLI, J., *J. Appl. Physiol.*, 7 (1955) 425.
135 ZUCKER, M. B., LEWIS, J. H. AND BORRELLI, J., *Proc. Sixth Intern. Congr. of Intern. Soc. Haematology*, Grune and Stratton, New York, 1958, p. 540.
136 ZUCKER, M. B. AND RAPPORT, P. P., *Federation Proc.*, 13 (1954) 170.

Appendix II

ACTION OF SOME DRUGS AND TREATMENTS ON 5-HT CONTENT OF VARIOUS TISSUES OF DIFFERENT ANIMAL SPECIES

LEGEND

I.A. = intra-arterially
I.P. = intraperitoneally
I.V. = intravenously
O.   = orally
S.C. = subcutaneously
U.   = units
(+) = mean of authors' data

References p. 317

Substance	Dose mg/kg	Route of administration	Time between drug adm. and killing	Animal	tissue	5-HT µg/g or ml content		Reference
						controls	treated	
ACTH, see Adrenocorticotropic hormone								
Adrenalectomy	—	—	—	Rat	liver	0.41	0.71	92
	—	—	—		blood	0.276	0.095	46
			72 h		brain	0.38	0.28	30
					spleen	3.16	2.45	30
					lung	0.59	0.87	30
					kidney	0.20	0.26	30
			24 h		hemispheres	0.85	1.95	23
					base	0.97	1.15	23
					medulla oblon-gata	0.37	1.40	23
					serum	0.50	0.12	23
					brain	0.65	0.67	82
Adrenocorticotropic hormone	20	—	18 h	Mouse	brain	0.47	0.43	86
Ajmalacine	2	I.V.	4 h	Rabbit	brain	0.57	0.52	15
							0.58	15
Ajmaline	2	I.V.	4 h	Rabbit	brain	0.57	0.44	15
Amitriptyline	20	I.P.	1 h	Rat	brain	0.60	0.68	61a
			4 h				0.55	61a
	50	I.P.	1 h				0.65	61a
			4 h				0.63	61a

		doses)							
		15	S.C.	4½ h	Dog	hypothalamus	0.354	0.347	54
		20					0.526	0.217	54
							0.521	0.294	54
		20	S.C.	5 h		hypothalamus	0.317	0.095	54
		30		2 h			0.317	0.105	54
		15		4½ h		caudate nucleus	0.280	0.303	54
							0.310	0.047	54
							0.298	0.102	54
		20	S.C.	5 h		caudate nucleus	0.480	0.100	54
		30		2 h			0.480	0.260	54
		20	S.C.	4½ h	Puppy	hypothalamus	0.443	0.250	54
						caudate nucleus	0.150	0.142	54
		30				hypothalamus	0.443	0.197	54
						area postrema	0.858	0.515	54
		35 (fractionated doses)	S.C.	4 h	Rat	brain	0.331	0.349	50
		35	I.P.	15 min		hemispheres	0.37	0.36	27
						cerebellum	0.117	0.101	27
						diencephalic area	0.81	0.79	27
						mesencephalic area	0.56	0.67	27
						total brain	0.47	0.43	27
Anaphylactic shock	Horse serum				Rat	spleen	4	4.1	67
						lung	2	2	67
						jejunum	4.6	3.3	67
						subcutaneous	1	0.8	67
	Egg albumin				Guinea pig	lung	0.15	0.53	42

Substance	Dose mg/kg	Route of administration	Time between drug adm. and killing	Animal	tissue	5-HT µg/g or ml content		Reference
						controls	treated	
Anoxia			15 min	Rat	hemispheres	0.37	0.42	27
					cerebellum	0.117	0.184	27
					diencephalic area	0.81	0.84	27
					mesencephalic area	0.56	0.70	27
					total brain	0.47	0.52	27
Apomorphine	100	S.C.	4 h	Rat	brain	0.331	0.258	50
Azacyclonol	400 (fractionated doses)	I.P.	4 h	Rat	brain	0.331	0.299	50
Banana feeding				Rat	brain	0.48	0.49	48
					kidney	0.23	0.21	48
					large intestine	3.02	4.39	48
Barbital	200 (fractionated doses)	I.V.	4 h	Rabbit	brain	0.57	0.51	11
							0.48	11
BAS, see 1-Benzyl-2-methyl-5-methoxytryptamine								
Benzoquinolizine derivative RO 1–9569	40	I.V.	3 h	Rabbit	brain	0.49	0.18	61
	50		4 h		brain	0.75	0.40	64

			25 h	Rabbit	brain	0.50	1	59
				Mouse	brain	1	0.30	59
1-Benzyl-2-methyl-5-methoxytryptamine (BAS)	15 (daily for 3 days)	I.P.	1 h	Rabbit	μg/10^8 platelets	2.7	3.1	91
					μg/mg plat. proteins	6.1	20.6	91
					μg/ml blood	9.4	8.2	91
			2 days		μg/10^8 platelets	2.7	1	91
					μg/mg plat. proteins	6.1	10	91
Bulbocapnine	350 (fractionated doses)	I.P.	4 h	Rat	brain	0.33	0.68	50
					duodenum	2.77	1.90	50
					blood	0.48	0.68	50
Calcium aspartate	200 (daily for 6 days)	O.		Rat	brain	0.38	0.39	32
					spleen	2.89	3.21	32
	400 (daily for 6 days)	O.			brain	0.38	0.39	32
					spleen	2.89	3.42	32
Calcium chloride	295.4	I.P.	1 h	Rat	brain	0.39	0.46	32
					spleen	3.06	3.35	32
			4 h		brain	0.39	0.48	32
					spleen	3.06	3.54	32
Carbethoxysyringoyl-methylreserpate (Syrosingopine or SU 3118)	2.5	I.P.	4 h	Rat	brain	0.47	0.49	31
	10						0.45	31
	20						0.46	31

Substance	Dose mg/kg	Route of administration	Time between drug adm. and killing	Animal	tissue	5-HT µg/g or ml content		Reference
						controls	treated	
Carbethoxysyringoyl-	40	I.P.	4 h	Rat	brain	0.47	0.20	31
methylreserpate	50						0.13	31
(Syrosingopine or SU 3118)	20				small intestine	4.76	3.37	28
	50						2.64	28
	20				kidney	0.34	0.29	28
					lung	2.86	1.64	28
					spleen	3.51	1.04	28
Carbon dioxide	5% in O_2	Inhalation	Immed.	Rat	brain	0.331	0.361	50
Cardiazol	70	I.P.	15 min	Rat	hemispheres	0.37	0.56	27
(Metrazol, Leptazol)					cerebellum	0.117	0.139	27
					diencephalic area	0.81	0.99	27
					mesencephalic area	0.56	0.83	27
					total brain	0.47	0.63	27
	60		10 min		brain	0.27(+)	0.33(+)	3
					small intestine	1.92(+)	1.93(+)	3
					large intestine	4.38(+)	5 (+)	3
					spleen	1.35(+)	1.40(+)	3
					lung	1.19(+)	2.05(+)	3
					ear	0.69(+)	1.21(+)	3
					serum	0.57(+)	0.82(+)	3
			15–20 min		brain	0.27(+)	0.38	3
					gastroint. tract	1.6 (+)	2.7	3

			60 min		brain	0.27(+)	0.29(+)	3
					small intestine	1.92(+)	2.10(+)	3
					large intestine	4.38(+)	6.75(+)	3
					spleen	1.35(+)	1.40(+)	3
					lung	1.19(+)	2.65(+)	3
					ear	0.69(+)	1.30	3
					serum	0.57(+)	0.80(+)	3
Catechol	5	I.P.	4 h	Rat	brain	0.38	0.38	32
Chloral hydrate	300	I.P.	1 h	Rat	brain	0.47	0.58	6a
						0.36	0.56	6a
Chloralose	100	I.P.	1 h	Rat	brain	0.36	0.55	6a
Chlorpromazine	10	I.P.	5 h	Rat	brain	0.241	0.278	17
	10	I.V.	4 h	Rabbit	brain	0.57	0.52	17
	15	I.P.	5 h	Rat	brain	0.47	0.54	29
	30	I.V.	3 h	Rabbit	telencephalon	0.282	0.352	21
					hippocampus	0.272	0.347	21
					mesencephalon	0.647	0.683	21
					medulla + pons	0.582	0.627	21
					cerebellum	0.112	0.125	21
	75 (fractionated doses)	S.C.	4 h	Rat	brain	0.331	0.276	50
	25 (for 3 days)	I.P.			brain	0.2	0.08	90
					spleen	2.5	0.62	90

Substance	Dose mg/kg	Route of administration	Time between drug adm. and killing	Animal	tissue	5-HT µg/g or ml content		Reference
						controls	treated	
Chlorpromazine	25 (for 3 days)	I.P.	4 h	Rat	skin	1.3	0.22	90
					stomach	1.4	0.98	90
					duodenum	1.2	0.60	90
					ileum	1.2	1.08	90
Cortisone	26 (fractionated doses)	I.V.	4 h	Rabbit	brain	0.57	0.49	11
	50 (every 24 h per 15 days)	I.P.	6 h	Rat	hemispheres	0.37	0.50	43
					cerebellum	0.127	0.103	43
					diencephalic area	0.82	0.91	43
					mesen- cephalic area	0.53	0.74	43
					total brain	0.48	0.57	43
					liver	0.41	0.27	92
	100		30 min	Mouse	brain	0.47	0.39	86
	50 (daily for 14 days)	I.M.		Rat	brain	0.2	0.18	90
					spleen	2.5	0.18	90
					skin	1.3	0.45	90
					stomach	1.4	1.54	90
					duodenum	1.2	1.2	90
					ileum	1.2	1.31	90
	50	I.M.		Rat	subcutaneous	0.70	0.048	16a

Compound 48/80	1000–3000 (in 6 doses, 1 every 12 h)	I.P.		Rat	derma	1.48	0.7	36
	3 (in two doses)		24 h		abdominal skin	1.68	0.22	55
	4 (fract.dos.)		3 h		brain	0.331	0.378	50
Cyproheptadine (Periactin)	10 (daily for 9 days)	I.P.		Rat	brain	0.41	0.38	32
					spleen	3.84	4.53	32
Decompression				Rat	brain	0.26	0.29	43a
					lung	0.71	0.70	43a
					ileum	0.66	0.74	43a
					skin	0.41	0.58	43a
					mesenteric mast cells	5.2	6.0	43a
11-Demethoxyreserpine	2	I.V.	4 h	Rabbit	brain	0.57	0.06	74
Deoxycorticosterone	100		30 min	Mouse	brain	0.47	0.43	86
Deserpidine	2	I.V.	4 h	Rabbit	brain	0.57	0.12	50
							0.09	50
							0.06	11
							0.13	11
Dexamethasone	10 (daily for 4 days)	I.M.		Rat	abdom. skin	1.9	0.83	80
					pyloric stomach	2.1	2.14	80
					jejunum	3.2	1.89	80

Substance	Dose mg/kg	Route of administration	Time between drug adm. and killing	Animal	tissue	5-HT $\mu g/g$ or ml content		Reference
						controls	treated	
Dextran	1800 (in 5 doses 1 every 12 h)	S.C.		Rat	derma	1.48	0.78	36
Dibenamine	100 (fractionated doses)	S.C.	14 h	Rat	brain	0.331	0.293	50
Diethylaminoethylreserpine (DL 152)	20	I.P.	4 h	Rat	brain	0.38	0.31	29
					spleen	2.52	1.25	29
	10	I.P.	12 h		brain	0.38	0.38	29
Diethyl ether		inhalation	20 min	Rat	brain	0.36	0.61	6a
Dihydroxyphenylalanine (DOPA)	200	I.P.	30 min	Mouse	brain	0.82	0.72	90a
Diphenhydramine	35 (fractionated doses)	I.V.	4 h	Rabbit	brain	0.57	0.54	11
5,5-Diphenyloxazolidine-2,4-dione	400	I.P.	12 h	Rat	brain	0.341	0.581	7
Diphenylhydantoin	12.5	I.P.	30 min	Rat	brain	0.38	0.475	1a

	doses)					0.392	0.580	8
					spleen	5	3.2	8
					small intestine	4.6	2.9	8
					caval blood	0.5	0.4	8
	100	I.P.	35 min		hemispheres	0.37	0.59	6
					cerebellum	0.117	0.142	6
					diencephalic area	0.81	1.1	6
					mesencephalic area	0.56	0.92	6
					total brain	0.47	0.66	6
DOPA, see Dihydroxyphenylalanine								
Electroshock	110 V (0.2 sec)		15 min	Rat	hemispheres	0.37	0.53	6
					cerebellum	0.117	0.171	6
					diencephalic area	0.81	1.28	6
					mesencephalic area	0.56	1.08	6
					total brain	0.46	0.78	6
			73 h		hemispheres	0.37	0.50	32
					cerebellum	0.117	0.18	32
					diencephalic area	0.81	1.11	32
					mesencephalic area	0.56	0.83	32
					total brain	0.47	0.56	32
	125 V (0.2 sec)		10 min		brain	0.27(+)	0.35(+)	3
					gastrointestinal tract	1.6(+)	1.2	3

Substance	Dose mg/kg	Route of administration	Time between drug adm. and killing	Animal	tissue	5-HT µg/g or ml content		Reference
						controls	treated	
Electroshock	125 V (0.2 sec)		10 min	Rat	small intestine	1.92(+)	1.77(+)	3
					large intestine	4.38(+)	2.99(+)	3
					spleen	1.35(+)	1.89(+)	3
					lung	1.19(+)	1.2(+)	3
					ear	0.69(+)	0.36(+)	3
					serum	0.57(+)	0.52(+)	3
			60 min		brain	0.27(+)	0.34(+)	3
					gastrointestinal tract		1.2	3
					small intestine		1.99(+)	3
					large intestine		2.2 (+)	3
					spleen		1.19(+)	3
					lung		1.53(+)	3
					ear		0.48(+)	3
					serum		0.39(+)	3
	150 mA (0.2 sec)		10 min		brain	0.56	0.62	9a
	80 V (0.15 sec)		15 min	Mouse	brain	0.60	0.65	32
	120 V (0.6 sec)		15 min	Dog	area 17 (motor)	0.078	0.118	25
					area 21 (visual)	0.057	0.081	25
					cerebellum	0.037	0.049	25
					hypothalamus	0.99	1.37	25
					thalamus	0.47	0.59	25
					mesencephalon	0.72	1.15	25
	150 mA		10 min	Cat	brain	0.48	0.50	9a

Substance	Dose	Route	Time	Animal	Region			Ref.
			10 min	Rat	brain	0.252	0.368	18
Ephedrine	100	S.C.	30 h	Dog	hypothalamus	0.370	0.270	54
					caudate nucleus	0.274	0.170	54
	250			Puppy	hypothalamus	0.324	0.366	54
							0.234	54
					caudate nucleus	0.145	0.168	54
							0.095	54
	300				hypothalamus	0.324	0.315	54
					caudate nucleus	0.145	0.166	54
Ethanol	15% in H$_2$O	drinking	Immediat.	Rat	brain	0.331	0.407	50
	2000	I.V.	1 h			0.80	0.68	23a
	2000 (daily for 7 days)					0.68	0.80	23a
	4475		70 min			0.36	0.56	6a
	2000	I.V.	1 h			0.67	0.33	35a
Ether				Puppy	hypothalamus	0.325	0.340	54
					caudate nucleus	0.150	0.132	54
		inhalation	Immediat.	Rat	brain	0.331	0.294	50
		inhalation (for 20 min)			brain	0.332	0.611	1
Ethyl alcohol, see Ethanol								
Fear (stress by fear)	3 days			Hare	brain	0.075	0.096	47
	7 days						0.102	47
	10 days						0.040	47
	14 days						0.059	47

Substance	Dose mg/kg	Route of administration	Time between drug adm. and killing	Animal	tissue	5-HT µg/g or ml content controls	5-HT µg/g or ml content treated	Reference
Fludrocortisone	10 (daily for 4 days)	I.M.		Rat	abdom. skin	1.9	0.70	80
					pyloric stomach	2.1	3.31	80
					jejunum	3.2	2.33	80
Frenquel	75 (fractionated doses)	I.V.	4 h	Rabbit	brain	0.57	0.53	11
							0.59	11
Gramine	40	I.P.	2–5 h	Rat	brain	0.66	0.90	23b
Guanethidine	20	I.V.	24 h	Human	$\mu g/10^9$ platelets	0.715	0.652	21a
Harmaline	10	I.P.	30 min	Rat	brain	0.43	0.69	6a
	10		2–5 h			0.66	0.92	23b
Heparin	15,000 U. (fractionated doses)	I.P. and S.C.	4 h	Rat	brain	0.331	0.307	50
Hexobarbital	80	I.P.	35 min	Rat	hemispheres	0.37	0.49	6
					cerebellum	0.117	0.124	6
					diencephalic area	0.81	0.99	6
					mesencephalic area	0.56	0.74	6
					total brain	0.47	0.70	6

Hydrocortisone	50 (fractionated doses)	S.C.	4 h	Rat	brain	0.331	0.303	50
Hydroxypregnanedione (Viadril)	100		30 min	Mouse	brain	0.471	0.380	86
5-Hydroxytryptamine (Serotonin; 5-HT)	25	S.C.	24 h	Rat	serum	0.68	1.23	24
					spleen	1.42	2.65	24
	10	I.V.	10 min		brain	0.49	1.2	40
			5 min		blood	0.205	2.7	40
	10	I.P.	1 h		brain	0.38	0.39	30
					spleen	3.16	5.22	30
					lung	0.59	2.03	30
					kidney	0.20	1.65	30
		Incubation (37° in vitro)		Man	μg/10^8 platelets	0.06	0.245	9
	0.2	I.P.	5 min	Guinea pig	lung	0.15	0.27	42
		I.V.	3 min			0.15	0.41	42
	0.25	Incubation (37°)	120 min	Rabbit	brain	0.675	0.986	17
5-Hydroxytryptophan (5-HTP)	100	I.P.		Rat	brain	0.33	0.879	8
	75		3 h		brain	0.63	1.3	83
					liver	0.3	3.7	83
	150				brain	0.4	2.4	83
					liver	<0.1	6	83

Substance	Dose mg/kg	Route of administration	Time between drug adm. and killing	Animal	tissue	5-HT μg/g or ml content		Reference
						controls	treated	
5-Hydroxytryptophan (5-HTP)	150			Rat	heart	<0.1	3.1	83
					uterus	<0.2	2.5	83
	200	I.P.	2 h		brain	0.4	1	22
	(daily for				liver	<0.1	15.65	22
	12 weeks)				kidney	<0.1	18.7	22
	(for 16		4–6 h		brain	0.4	0.54	22
	weeks)				liver	<0.1	1.95	22
					kidney	<0.1	3.35	22
	(for 8	I.P.	12 h		brain	0.4	0.35	22
	weeks)				liver	0.1	1	22
					kidney	0.1	1.37	22
	50		2 h		heart	0.1	2.5	22
	200				foetus	0.1	5	22
					uterus	0.1	2.5	22
	100		1 h		brain	0.33	0.879	8
	60	I.V.	90 min	Dog	hypothalamus	1.5	13.3	83
					cortex	0.2	2.4	83
					liver	0.1	6.2	83
					blood	0.43	1.9	83
					plasma		1.5	83
	80		30 min	Mouse	brain	0.85	1	84
			3 h		carcasse	0.7	3.9	84
	200				in toto	2.4	8.8	89
				Chicken	brain	0.8	5	89
					liver	1.6	10.8	89

75	I.V.	1 h	Rabbit	telencephalon	0.282	0.741	21
				hippocampus	0.272	0.942	21
				mesencephalon	0.647	1.836	21
				medulla + pons	0.582	1.332	21
				cerebellum	0.112	0.284	21
		2 h		telencephalon	0.282	0.628	21
				hippocampus	0.272	1.226	21
				mesencephalon	0.647	1.807	21
				medulla + pons	0.582	1.668	21
				cerebellum	0.112	0.293	21
		3 h		telencephalon	0.282	0.445	21
				hippocampus	0.272	0.559	21
				mesencephalon	0.647	1.654	21
				medulla + pons	0.582	0.902	21
				cerebellum	0.112	0.206	21
1	I.P.	30 min	Mouse	brain			
				(cerebellum-free)	0.82	0.87	32a
10						0.99	32a
20						1.02	32a
50						1.12	32a
100						1.63	32a
200						2.2	32a
600						4.25	32a
50		1 h	Rat	brain	0.241	0.867	17
				spleen	2.48	6.36	17
				intestine	2.559	7.114	17
Hyperoxia			Rat	hemispheres	0.37	0.49	27
				cerebellum	0.117	0.161	27
				diencephalic area	0.81	0.85	27

Substance	Dose mg/kg	Route of administration	Time between drug adm. and killing	Animal	tissue	5-HT µg/g or ml content		Reference
						controls	treated	
Hyperoxia				Rat	mesen-cephalic			
					area	0.56	0.69	27
					total brain	0.47	0.56	27
Hypophysectomy			24 h	Rat	hemispheres	0.85	0.87	71
					base	0.97	1.54	71
					medulla oblongata	0.37	0.81	71
					serum	0.50	0.20	71
Imipramine (Tofranil)	7.5	I.P.	5 h	Rat	brain	0.47	0.55	19
					small and large intestine	4.76	4.06	19
					kidney	0.34	0.33	19
					lung	2.86	2.79	19
					spleen	3.51	3.57	19
	15				brain	0.47	0.70	19
					small and large intestine	4.76	2.43	19
					kidney	0.34	0.32	19
					lung	2.86	1.77	19
					spleen	3.51	2.57	19
	10				brain	0.241	0.516	19
	20	I.P.	1 h	Rat	brain	0.60	0.63	61a
			4 h				0.60	61a

Iproniazid (Marsilid)	75			Rat	brain	0.63	1.1	83
					liver	0.3	0.5	83
	100			Rabbit	brain	0.5	0.8	83
	240			Rat	brain	0.35	0.8	84
	200			Rabbit	brain	0.5	1	84
	100		16 h	Rat	brain	0.55	0.9	37
							1.1	37
	100	I.P.	7 h	Rat	brain	0.331	0.880	50
	70		4–16 h	Dog	hippocampus	0.260	0.370	52
					hypothalamus	0.375	0.537	52
					olfactory bulb	0.078	0.160	52
					caudate nucleus	0.229	0.342	52
					cerebral white matter	0.012	0.23	52
	25	I.P.	48 h	Rabbit	midbrain	0.378	0.756	20
	25		6 h	Rat	brain	0.44	0.65	15
							0.94	15
	100	I.V.	16 h	Rabbit	brain	0.49	1.05	61
	100	S.C.	24 h		brain stem	0.72	1.45	77
			17 h			0.50	0.86	77
			24 h			0.50	0.80	77
	25 (daily for 4 days)					0.52	0.98	77
	(for 6 days)					0.52	0.88	77
	240	I.P.	6 h	Rat	brain	0.35	0.9	84
	25 (daily for 8 days)	S.C.		Rabbit	blood	5.5	14.2	72
	100	I.P.	12 h	Rat	brain	0.47	1.01	29

Substance	Dose mg/kg	Route of administration	Time between drug adm. and killing	Animal	tissue	5-HT µg/g or ml content		Reference
						controls	treated	
Iproniazid (Marsilid)	150	I.P.	12 h	Mouse	brain (cerebellum-free)	0.82	1.25	29
	100	I.V.	1 h	Rabbit	brain	0.55(+)	0.63(+)	70
	125 (fractionated doses)	I.P.	2 h	Rat		0.241	1.311	17
	25 (daily for 3 days)	I.V.	2 h	Rabbit	telencephalon	0.265	0.407	17
					mesencephalon	0.663	0.816	17
	100	S.C.	18 h	Rat	brain	0.32	0.81	51
					small intestine	3.95	5.8	51
			17 h		brain	0.49	1.11	40
					blood	0.205	0.33	40
	125	I.P.	2 h		brain	0.241	1.311	17
					spleen	2.480	4.128	17
	100	I.V.	24 h	Rabbit	brain	0.55	1.2	57
			16 h	Rat	brain	0.52	1.2	57
					intestine	3.5	4	57
	10 (daily for 5 days)	S.C.		Rabbit	brain	0.58	0.92	5
	5 (daily for 5 days)					0.58	0.60	5
	25 (daily for 6 days)	I.V.			brain stem	0.7	1.6	71
	100		24 h			0.7	1.4	71

	100		15 h		brain	0.78	1.47	68
Isobutylserpentinate	20	I.P.	4 h	Rat	brain	0.38	0.32	29
					spleen	2.52	1.72	29
	10		12 h		brain	0.38	0.38	29
Isoniazid	100	I.V.	48 h	Rabbit	brain	0.55	0.58	57
	50 (daily for 5 days)	S.C.				0.58	0.58	5
	100		16 h	Rat	brain	0.55	0.6	58
Isopropylhydrazine hydrochloride	66	I.P.	6 h	Rat	brain	0.47	1.86	16
Isopropylhydrazine Na-methanesulphonate	130	I.P.	6 h	Rat	brain	0.47	1.40	16
Isoraunescine	2	I.V.	4 h	Rabbit	brain	0.57	0.47	10 / 74
	5	I.P.	6 h	Rat	brain	0.366	0.369	49
			16 h			0.366	0.363	49
Isoreserpidine	2	I.V.	4 h	Rabbit	brain	0.57	0.48	10 / 74
Isoreserpine	2	I.V.	4 h	Rabbit	brain	0.57	0.54	11
							0.42	11

JB 516, see Phenylhydrazinopropane

Substance	Dose mg/kg	Route of administration	Time between drug adm. and killing	Animal	tissue	5-HT µg/g or ml content		Reference
						controls	treated	
JB 835, see Phenylhydrazinobutane								
LSD, see Lysergic diethylamide								
Luminal, see Phenobarbitone								
Lysergic diethylamide (LSD)	0.12 (divided doses)	I.V.	4 h	Rabbit	brain	0.57	0.47	11
							0.49	11
	0.300 (fractionated doses)	S.C.		Rat	brain	0.331	0.299	50
	0.52–1.3			Rat	brain	0.525	0.628	24a
Malathion	1600	S.C.	4 h	Rat	brain	0.78	0.85	68
Marsilid, see Iproniazid								
Meperidine	25	I.P.	25 min	Rat	brain	0.36	0.63	6a
Meprobamate	450 (fractionated doses)	I.P. S.C.	4 h	Rat	brain	0.331	0.350	50
	250	I.P.	2½ h	Rat	brain	0.331	0.331	50
Mescaline	75	I.V.	4 h	Rabbit	brain	0.57	0.46	11
							0.50	11

Compound	Dose	Route	Time	Animal	Tissue			Ref.
...thiazole	18	oral	24 h	Rat	thyroid gland	4.52	4.62	55
Methoxypropylamino-methylbenzodioxane HCl (S 6011)	20	I.P.	1 h	Rat	brain	0.38	0.47	29
					spleen	2.52	4.78	29
			4 h		brain	0.38	0.50	29
					spleen	2.52	5.00	29
			12 h		brain	0.38	0.38	29
					spleen	2.52	2.66	29
	40		1 h		brain	0.38	0.62	29
					spleen	2.52	4.21	29
	60				brain	0.38	0.56	29
					spleen	2.52	5.46	29
N-Methyl-N-benzyl-2-propynyl-amine (MO-911)	100	I.P.	48 h	Mouse	brain	0.82	1.16	90a
α-Methyl-3,4-dihydroxy-DL-phenylalanine (α-Methyl-DOPA)	200	I.V.	3 h	Dog	caudate nuclei	1.03	0.43	34
			24 h				1.13	34
	500	I.V.	24 h				1.00	34
	100	S.C.	1 h	Mouse	brain	0.88(+)	0.64(+)	76
					intestine	2.83	3.16	76
	400	S.C.			brain	0.99(+)	0.43(+)	76
	(4-hourly for 24 h)					3.67	2.31	76
	200	I.V.	3 h	Rabbit	brain stem	0.80	0.35	65a
				Guinea pig	brain	0.29(+)	0.025(+)	76
	400	S.C.	12 h	Mouse	brain*	21.2	10.2	75
					intestine*	204	170	75
			24 h		brain*	21.2	8.7	75
					intestine*	204	127	75

* Values are expressed as mg/kg body weight.

References p. 317

Substance	Dose mg/kg	Route of administration	Time between drug adm. and killing	Animal	tissue	5-HT µg/g or ml content		Reference
						controls	treated	
Methyl-18-O-(3-N,N-dimethylaminobenzoyl)-reserpate (SU 5171)	2.5	I.P.	4 h	Rat	brain	0.47	0.48	31
	5						0.25	31
	7.5						0.12	31
	10						0.14	31
	5				small intestine			
					+ colon	4.76	3.98	28
					kidney	0.34	0.49	28
					lung	2.86	1.84	28
					spleen	3.51	2.23	28
	7.5				small intestine			
					+ colon	4.76	2.73	28
					kidney	0.34	0.28	28
					lung	2.86	1.21	28
					spleen	3.51	0.59	28
	5		12 h		brain	0.581	0.409	17
					spleen	2.412	1.832	17
2-Methyl-fluorocortisone	10 (daily for 4 days)	I.M.		Rat	abdom. skin	1.9	1.16	88
					pyloric stomach	2.1	1.89	88
					jejunum	3.2	3.17	88
α-Methyl-3-hydroxy-DL-phenylalanine	200	I.V.	5 h	Dog	caudate nuclei	1.03	0.41	76
	400	I.V.					0.50	76

Drug	Dose	Route	Time	Animal	Region			Ref.
5-Methyltryptophan	44	I.A.		Rabbit	mesencephalon+ diencephalon	0.375	0.250	77
MO 911, see N-Methyl-N-benzyl-2-propynylamine								
Morphine	20	I.V.	4 h	Rabbit	brain	0.57	0.43	11
							0.57	11
	80	S.C.		Rat		0.331	0.414	50
	100 (fract.dos.)		12 h			0.331	0.278	50
	10	I.P.	2 h	Rat	brain	0.36	0.44	6a
	150–200 (chronic)	I.P.		Rat	brain (minus cerebellum)	0.371	0.358	44a
	60–124 (chronic)	I.P.		Dog	brain stem	0.430	0.436	44a
	200 (chronic)	I.P.		Rabbit	brain stem	0.425	0.441	44a
Nembutal	35	I.P.	35 min	Rat	hemispheres	0.37	0.54	6
					cerebellum	0.117	0.108	6
					diencephalic area	0.81	1.18	6
					mesencephalic area	0.56	0.78	6
					total brain	0.47	0.67	6
	300 (fract.dos.)	I.P.	4 h		brain	0.331	0.339	50
Neomycin	215 (daily for 20–21 days)	O.		Mouse	small intestine	1.96	2.52	79
					spleen	0.95	1.44	79
					brain	0.16	0.15	79

Substance	Dose mg/kg	Route of administration	Time between drug adm. and killing	Animal	tissue	5-HT μg/g or ml content		Reference
						controls	treated	
Nicotinylisopropyl-sulphonylhydrazide (K 577)	25	I.P.	6 h	Rat	brain	0.44	0.73	15
							1.13	15
Oestradiol	10 (on alternate days for 21 days)	I.M.		Rat	abdom. skin	1.9	0.57	80
					skin of the feet	1.5	0.57	80
					ears	1.4	0.41	80
					pyloric stomach	2.1	2.04	80
					jejunum	3.2	1.82	80
Ozone				Rat	brain	0.63	0.39	74a
Paramethadione	400	I.P.	2–5 h	Rat	brain	0.392	0.52	7
					spleen	5	4.9	7
					small intestine	4.6	2.9	7
					caval blood	0.5	0.4	7
	200 (4 doses)		12 h		brain	0.341	0.501	7
Paraoxon	1.72	I.P.	2 h	Rat	brain	0.78	0.88	68
			18 h			0.78	0.90	68
Parathion	7	I.P.	1 h	Rat	brain	0.78	0.86	68
Pentobarbital	25	I.P.	5 min	Rat	brain	0.332	0.632	1

	450					0.392	0.740	7
					spleen	5	4.6	7
					small intestine	4.6	2.4	7
					caval blood	0.5	0.42	7
Phenobarbitone	100	I.P.	30 min			0.48	0.50	23a
(Luminal)			60 min				0.54	23a
			150 min				0.54	23a
			2 h	Rat	brain	0.332	0.572	1
		I.V.	4 h	Rabbit		0.57	0.51	11
	125		2–5 h	Rat		0.366(+)	0.621(+)	7
					spleen	5	5.5	7
					small intestine	4.6	4.9	7
					caval blood	0.5	0.52	7
	120	I.P.	35 min		hemispheres	0.37	0.47	6
					cerebellum	0.117	0.112	6
					diencephalic area	0.81	0.98	6
					mesencephalic area	0.56	0.76	6
					total brain	0.47	0.65	6
Phenylacetic acid	4% (for 2 weeks)	Diet		Rat	brain	0.47	0.49	21b
Phenylalanine	5000 (daily for 71–74 days)	Diet		Rat	brain	0.584	0.468	93
	7% (for 3 weeks)	Diet		Rat	brain	0.70	0.16	88a
					liver	0.48	0.17	88a

Substance	Dose mg/kg	Route of administration	Time between drug adm. and killing	Animal	tissue	5-HT µg/g or ml content		Reference
						controls	treated	
Trans-2-Phenylcyclopropylamine (SKF 385)	2	I.V.	4 h	Rabbit	midbrain	0.378	1.10	20
Phenylethylhydrazine (W 1544)	40	I.P.	30 min	Rat	hemispheres	0.37	0.52	29
					cerebellum	0.117	0.128	29
					diencephalic area	0.81	1.36	29
					mesencephalic area	0.56	1.04	29
					total brain	0.47	0.88	29
			12 h		hemispheres	0.37	1.16	29
					cerebellum	0.117	0.245	29
					diencephalic area	0.81	2.09	29
					mesencephalic area	0.56	1.78	29
					total brain	0.47	1.23	29
	5 (daily for 6 days)		16 h		brain	0.38	0.92	29
					spleen	2.52	5.83	29
					lung	0.51	2.31	29
					kidney	0.20	0.46	29
Phenylhydrazinobutane (JB 835)	10 (daily for 5 days)	S.C.		Rabbit	brain stem	0.74	2.43	78

Drug	Dose (mg/kg)	Route	Time	Species	Tissue			Ref.
	2 (daily for 5 days)	S.C.	60 min			0.72	1.46	78
	3					0.64	1.25	78
	2 (daily for 12 days)			Dog		0.65(+)	5.05	78
	2 (daily for 5 days)			Cat		0.75(+)	1.92	78
1-Phenylisopropyl-2-iso- propylhydrazine	10	I.V.	4 h 24 h	Rabbit	brain stem	0.70 0.70	1.2 1.25	14 14
Phenylpropyl-1,1-diphenyl- propylamine (Segontin)	50	S.C.	12 h 4 h	Rat	brain spleen brain spleen	0.48 3.15 0.48 3.15	0.49 3.21 0.50 2.33	32 32 32 32
Phenylpropyl-1,1-diphenyl- propylamine (Segontin)	100	S.C.	12 h 4 h		brain spleen brain spleen	0.48 3.15 0.48 3.15	0.43 3.13 0.43 2.84	32 32 32
	25 (daily for 8 days)	S.C.			brain spleen	0.48 3.15	0.41 2.84	32 32
Phenylpyruvic acid	4.5% (for 2 weeks)	Diet		Rat	brain	0.47	0.40	21b

Substance	Dose mg/kg	Route of administration	Time between drug adm. and killing	Animal	tissue	5-HT μg/g or ml content		Reference
						controls	treated	
Phenytoin, see Diphenylhydantoin								
Piradol	150 (fract.dos.)	S.C.	4 h	Rat	brain	0.331	0.300	50
Pirogallol	40	I.P.	4 h	Rat	brain	0.38	0.38	32
Prednisolone	10 (daily for 9 days)	I.M.		Rat	abdom. skin	1.9	0.47	80
					skin of the feet	1.5	0.48	80
					ears	1.4	0.32	80
					fundic stomach	1.8	2.32	80
					pyloric stomach	2.1	3.63	80
					duodenum	5.6	3.19	80
					jejunum	3.2	2.08	80
					ileum	5.5	3.57	80
					colon	3.8	2.43	80
					lung	3.9	2.57	80
					whole rat	24.3	10.45	80
Primidone	500 (4 doses)		12 h	Rat	brain	0.341	0.546	7
Prominal (Mebaral, Gemonil)	125		2–5 h	Rat	brain	0.341	0.524	7
Propylthiouracil	sol. 0.075%	O. ad		Rat	brain	0.98(±)	1.75	81

Drug	Dose	Route	Time	Species	Tissue			Ref.
Pyridoxine (deficiency)				Chicken	brain	1.1	0.4	89
					intestine	5.3	1.2	89
					liver	1.6	0.4	89
					blood	2.9	0.5	89
					brain	1.13	0.39	85
					intestine	5	1.19	85
					blood	5.5	0.4	85
			6 weeks	Rat	spleen	2.14	1.14	92
Raunescine	5	I.P.	6 h	Rat	brain	0.366	0.174	49
			16 h				0.227	49
	5	I.V.	4 h	Rabbit		0.57	0.07	74
	0.5	I.P.		Rat		0.21	0.25	39
					intestine	2.79	2.68	39
	2.5				brain	0.21	0.189	39
					intestine	2.79	2.39	39
	5				brain	0.21	0.169	39
					intestine	2.79	2.850	39
	2.5				brain	0.45	0.19	10a
	5		1 h		plasma	0.004	0.034	41
					platelets	2.8	1.7	41
			6 h		plasma	0.004	0.002	41
					platelets	2.8	0.30	41
Rescinnamine	2	I.V.	4 h	Rabbit	brain	0.57	0.1	74
	2					0.57	0.09–0.12	11

Substance	Dose mg/kg	Route of administration	Time between drug adm. and killing	Animal	tissue	5-HT μg/g or ml content		Reference
						controls	treated	
Resection (small bowel, caecum and ascending colon res.)				Human	serum	0.21	0.01	66
(entire gastrointestinal tract res.)			3 days	Rat	brain	0.33	0.26	4
					serum	0.82	0.14	4
					spleen	2.40	1.30	4
					lung	3.00	0.76	4
					paws	0.25	0.13	4
(large intestine res.)			30 days		brain	0.29(+)	0.25(+)	4
					stomach and small intestine	2.51	1.87	4
					serum	0.59	0.62	4
					spleen	2.25	2.05	4
					lungs	2.14	1.56	4
					kidneys	0.10	0.05	4
					ears	0.32	0.63	4
					paws	0.15	0.15	4
Reserpic acid	2	I.V.	4 h	Rabbit	brain	0.57	0.52	74
						0.57	0.48	11
							0.55	11
Reserpine	2	I.V.	4 h	Rabbit	brain	0.57	0.05	11
	2.5	S.C.	6 h	Rat		0.33	0.1	8
	1	I.P.	5 h			0.241	0.160	17

10	S.C.	24 h		brain	0.4	0.2	38
2.5	I.P.	1 h			0.063 (free)	0.032	33
					0.38 (bound)	0.061	33
1	I.V.	4 h	Rabbit		0.57	0.09	57
5					0.57	0.06	57
5		1 h			0.55(+)	0.10(+)	70
0.5	I.P.	22 h	Puppy	hypothalamus	0.443	0.031	54
		11 h				0.039	54
		$4\frac{1}{2}$ h				0.247	54
0.25		3 h				0.105	54
0.5		22 h		caudate nucleus	0.150	0.016	54
		11 h				0.020	54
		$4\frac{1}{2}$ h				0.097	54
0.25		13 h				0.063	54
0.5		22 h		area postrema	0.858	0.107	54
		11 h				0.063	54
		$4\frac{1}{2}$ h				0.141	54
5		16 h	Rabbit	intestine	19	3	62
1		1 day	Man	μg/10⁸ platelets	0.036	0.014	37
(fract.dos.)		3 days				0.001	37
		9 days				0.007	37
10	S.C.		Rat	abdominal skin	1.68	0.29	55
2	I.V.	4 h	Rabbit	brain	0.57	0.06	74
0.25					0.7	0.2	63
0.5						0.18	63
1						0.15	63
2.5						0.1	63
5						0.04	63

Substance	Dose mg/kg	Route of administration	Time between drug adm. and killing	Animal	tissue	5-HT μg/g or ml content		Reference
						controls	treated	
Reserpine	0.5	I.P.		Rat	brain	0.21	0.196	39
					intestine	2.79	2.3	39
	1.25				brain		0.125	39
					intestine		2.14	39
	2.5				brain		0.102	39
					intestine		1.65	39
	12.5				brain		0.019	39
					intestine		2.1	39
	1		19 h	Rabbit	total blood	3–5	0.7	88
					intestine	6.5–16	6.5–12	88
	0.1	I.P.		Rat	brain	0.23	0.4	24
					gastrointest. tract	2.07	1.45	24
					spleen	1.61	1.03	24
					serum	0.6	0.38	24
					brain	0.23	0.1	24
					gastrointest.tract	2.07	1.3	24
					spleen	1.61	0.09	24
					serum	0.6	0.02	24
	2	I.V.	4 h	Rabbit	brain	0.57	0.05	11
							0.07	11
	2.5	S.C.		Rat		0.33	0.245	8
	1	I.V.		Rabbit		0.55	0.10	69
	5						0.05	69
	5		16 h		whole blood	3.75	0.16	73
	5		24 h		whole blood	4.6	0.5	87

			(schizophrenic)				
1		4 h	Rabbit	blood	4.5	0.45	64
				intestine	6	3.3	64
2.5		4–6 h	Rat	brain	0.331	0.082	50
	I.P.	4 h		hemispheres	0.37	0.102	29
				cerebellum	0.127	0.048	29
				diencephalic area	0.82	0.172	29
				mesencephalic area	0.53	0.178	29
				total brain	0.47	0.17	29
				small intestine + colon	4.76	2.95	29
				kidney	0.34	0.19	29
				lung	2.86	1.6	
				spleen	3.51	0.99	29
5				total brain	0.47	0.12	29
				small intestine + colon	4.76	2.04	29
0.6				total brain	0.47	0.6	29
10 (daily for 6 days)		24 h	*Bufo bufo bufo*	intestine	0.87(+)	0.90	56
20 (daily for 4 days)		10 days			0.87(+)	0.25	56
5 (daily for 3 days)		24 h	*Scylliorhinus canicula*	intestine	1.5	1.3	56
10 (daily for 2 days)					1.5	1.11	56

Substance	Dose mg/kg	Route of administration	Time between drug adm. and killing	Animal	tissue	5-HT µg/g or ml content		Reference
						controls	treated	
Reserpine	5	S.C.	32 h	*Octopus vulgaris*	optic ganglia	4	0.22	56
					post saliv. glands	125	162	56
	10		3 days		optic ganglia	4	0.08	56
					post saliv. glands	125	135	56
	5		48 h	*Eledone moschata*	optic ganlia	141	0.15	56
					post saliv. glands	490	650	56
	1			Rabbit	stomach (fundus)	13.5	16	60
					stomach (prepyloric part)	10	3.5	60
	2				stomach (fundus)	13.5	8.5	60
					stomach (prepyloric part)	10	2.5	60
	5	I.P.	18 h	Rat	brain	0.78	0.18	68
Reserpinine	2	I.V.	4 h	Rabbit	brain	0.57	0.49	11
						0.57	0.49	74

RO 1–9569, see Benzoquinolizine derivative

Secobarbital	40	I.P.	45 min	Rat	brain	0.36	0.62	6a
Segontin, see Phenylpropyl-1,1-diphenylpropylamine								
Semicarbazide	240	I.P.		Mouse	*in toto*	2.4	2.2	89
Sensibilization	Horse serum		12 days	Rat	spleen	4	4.2	67
					lung	2	2.2	67
					jejunum	4.6	3.8	67
					subcutaneous	1	0.9	67
Serotonin, see 5-Hydroxytryptamine								
Serpentine	2	I.V.	4 h	Rabbit	brain	0.57	0.43	31
							0.46	31
	2					0.57	0.45	74
Sodium bromide	1000 (in 5 doses)	I.P.	4 h	Rat	brain	0.341	0.531	7
Sodium salicylate	1500 (for 4 days)	I.P.		Rat	skin	1.087	0.504	45
		I.P.			ileum	1.020	1.10	45
Stilboestrol	10 (on alternate days for 21 days)	I.M.		Rat	abdom. skin	1.9	0.76	80
					skin of the feet	1.5	0.72	80
					ears	1.4	0.29	80
					pyloric stomach	2.1	2.08	80
					jejunum	3.2	2.30	80
Stress (sham operated) see also Fear				Rat	brain	0.98(+)	1.04	81

Substance	Dose mg/kg	Route of administration	Time between drug adm. and killing	Animal	tissue	5-HT µg/g or ml content		Reference
						controls	treated	
Strychnine	1.5	I.P.	15 min	Rat	hemispheres	0.37	0.49	27
					cerebellum	0.117	0.013	27
					diencephalic area	0.81	0.92	27
					mesencephalic area	0.56	0.70	27
					total brain	0.47	0.57	27
Substance P	100 U./kg (for 3 days)	I.P.	24 h	Rat	ileum	0.964	1.773	2
					stomach	1.249	1.933	2
					spleen	1.121	1.489	2
Succinylcholine	1	I.P.	35 min	Rat	hemispheres	0.37	0.50	6
					cerebellum	0.117	0.123	6
					diencephalic area	0.81	0.95	6
					mesencephalic area	0.56	0.80	6

SU 3118, see Carbethoxysyringoylmethylreserpate

SU 5171, see Methyl-18-*O*-(3-*N*,*N*-dimethylaminobenzoyl)reserpate

Syrosingopine, see Carbethoxysyringoylmethylreserpate

Substance	Dose mg/kg	Route of administration	Time between drug adm. and killing	Animal	tissue	5-HT µg/g or ml content		Reference
						controls	treated	
Testosterone	10 (on alternate days for	I.M.		Rat	abdom. skin	1.9	1.06	80
					skin of the feet	1.5	0.93	80
					ears	1.4	0.64	80

β-Tetrahydronaphthylamine carbonate	32	S.C.	$4\frac{1}{2}$ h	Puppy	hypothalamus	0.342	0.398	54
					caudate nucleus	0.145	0.151	54
Thiocyanate (Potassium th.)	65	O.	20 h	Rat	thyroid gland	4.32	5.16	53
Thyroidectomy				Rat	brain	0.98(+)	0.98	81
L-Thyroxin	250–500 μg (daily)			Rat	brain	0.98(+)	1.47	81
Tofranil, see Imipramine								
p-Toluenesulphonylmethyl-reserpate	2	I.V.	4 h	Rabbit	brain	0.57	0.49	11
3,5,5-Trimethyloxazolidine-2,4-dione (Troxidione; Tridione)	150		12 h	Rat	brain	0.341	0.481	7
L-Tryptophan	1000	I.P.	4 h	Rat	brain	0.241	0.496	17
					spleen	2.48	2.275	17
					intestine	2.559	2.01	17
			5 h		brain	0.331	0.430	50
	5% (for 3 weeks)	Diet		Rat	brain	0.77	1.17	88a
					liver	0.64	0.77	88a
Vagotomy (cervical subdiaphragmatic)			3–14 days	Rat	brain	0.331	0.295	50
						0.331	0.364	50
Viadril, see Hydroxypregnanedione								
W 1544, see Phenylethylhydrazine								

Substance	Dose mg/kg	Route of administration	Time between drug adm. and killing	Animal	tissue	5-HT μg/g or ml content		Reference
						controls	treated	
Vincamin	50	I.P.	4 h	Rat	brain	0.286	0.166	34a
X-radiation	1000 R		2 h	Rat	hypothalamus	2.75(+)	1.28(+)	65
	4500 R		18 min		brain	0.53	0.44	76a
	9000 R		0 min		brain	0.53	0.42	76a
Yohimbine	2	I.V.	4 h	Rabbit	brain	0.57	0.48	11
	10	I.P.		Rat	brain	0.21	0.163	39
	(divided doses)				intestine	2.79	1.52	39
	20				brain	0.21	0.28	39
					intestine	2.79	3.95	39
	30				brain	0.21	0.225	39
					intestine	2.79	3.10	39
	25	S.C.			brain	0.331	0.378	50

REFERENCES

1   ANDERSON, E. G. AND BONNYCASTLE, D. D., *J. Pharmacol. Exptl. Therap.*, 130 (1960) 138.
1a   ANDERSON, E. G., MARKOWITZ, S. D. AND BONNYCASTLE, D. D., *J. Pharmacol. Exptl. Therap.*, 136 (1962) 179.
2   BELESLIN, D. AND VARAGIC, V., *Experientia*, 15 (1959) 186.
3   BERTACCINI, G., *J. Neurochem.*, 4 (1959) 217.
4   BERTACCINI, G., *J. Physiol. (London)*, 153 (1960) 239.
5   BIEL, J., DRUKKER, E. A., SHORE, P. A., SPECTOR, S. AND BRODIE, B. B., *J. Am. Chem. Soc.*, 80 (1958) 1519.
6   BISIANI, M., GARATTINI, S., KATO, R. AND VALZELLI, L., *Atti soc. lombarda sci. med. biol.*, 13 (1958) 345.
6a   BONNYCASTLE, D. D., BONNYCASTLE, M. F. AND ANDERSON, E. G., *J. Pharmacol. Exptl. Therap.*, 135 (1962) 17.
7   BONNYCASTLE, D. D., GIARMAN, N. J. AND PAASONEN, M. K., *Brit. J. Pharmacol.*, 12 (1957) 228.
8   BONNYCASTLE, D. D., PAASONEN, M. K. AND GIARMAN, N. J., *Nature*, 178 (1956) 990.
9   BORN, G. V. R., INGRAM, G. J. C. AND STACEY, R. S., *Brit. J. Pharmacol.*, 13 (1958) 62.
9a   BREITNER, C., PICCHIONI, A., CHIN, L. AND BURTON, L. E., *Diseases Nervous System (Monograph Suppl.)*, 22 (1961) 1.
10   BRODIE, B. B., in G. P. LEWIS (Ed.) *5-Hydroxytryptamine*, Pergamon, London, 1957, p. 64.
10a BRODIE, B. B., FINGER, K. F., ORLANS, F. B., QUINN, G. P. AND SULSER, F., *J. Pharmacol. Exptl. Therap.*, 129 (1960) 250.
11   BRODIE, B. B., SHORE, P. A. AND PLETSCHER, A., *Science*, 123 (1956) 992.
12   BRODIE, B. B., SPECTOR, S., KUNTZMAN, R. G. AND SHORE, P. A., *Naturwissenschaften*, 45 (1958) 243.
13   BRODIE, B. B., SPECTOR, S., KUNTZMAN, R. G. AND SHORE, P. A., *Naturwissenschaften*, 45 (1958) 243.
14   BRODIE, B. B., SPECTOR, S. AND SHORE, A., *Ann. N. Y. Acad. Sci.*, 80 (1959) 609.
15   CANAL, N., MAFFEI-FACCIOLI, A. AND MANZINI, B., *Boll. soc. ital. biol. sper.*, 34 (1958) 1525.
16   CANAL, N., MAFFEI-FACCIOLI, A. AND MANZINI, B., *Atti soc. lombarda sci. med. biol.*, 14 (1959) 210.
16a CASS, R. AND MARSHALL, P. B., *Arch. intern. pharmacodyn.*, 136 (1962) 311.
17   COSTA, E., personal communication.
18   COSTA, E. AND GARATTINI, S., unpublished observations.
19   COSTA, E., GARATTINI, S. AND VALZELLI, L., *Experientia*, 16 (1961) 461.
20   COSTA, E., PSCHEIDT, G. R., VAN METER, W. G. AND HIMWICH, H. E., *J. Pharmacol. Exptl. Therap.*, 130 (1960) 81.
21   COSTA, E. AND RINALDI, F., *Am. J. Physiol.*, 194 (1958) 214.
21a CROSTI, P. F. AND LUCCHELLI, P. E., *Atti soc. lombarda sci. med. biol.*, 16 (1961) 1.
21b CULLEY, W. J., SAUNDERS, R. N., MERTZ, E. T. AND JOLLY, D. H., *Proc. Soc. Exptl. Biol. Med.*, 111 (1962) 444.
22   DAVIDSON, J., SJOERDSMA, A., LOUIS, N. AND UDENFRIEND, S., *J. Clin. Invest.*, 36 (1957) 1594.
23   DE MAIO, D., *Science*, 129 (1959) 1678.
23a DUBNICK, B., LEESON, G. A. AND PHILLIPS, G. E., *J. Neurochem.*, 9 (1962) 299.
23b EFRON, D. H. AND GESSA, G. L., *Arch. intern. pharmacodyn.*, 142 (1963) 111.
24   ERSPAMER, V., *Experientia*, 12 (1956) 63.
24a FREEDMAN, D. X., *Am. J. Psychiatry*, 119 (1963) 843.
25   FRESIA, P., GENOVESE, E., KATO, R. AND VALZELLI, L., *Boll. soc. ital. biol. sper.*, 34 (1958) 1397.
26   FRETER, K., WEISSBACH, H. AND REDFIELD, B. G., *J. Am. Chem. Soc.*, 80 (1958) 983.
27   GARATTINI, S., KATO, R. AND VALZELLI, L., *Atti soc. lombarda sci. med. biol.*, 13 (1958) 1228.
28   GARATTINI, S., KATO, R. AND VALZELLI, L., *Experientia*, 16 (1960) 120.
29   GARATTINI, S., LAMESTA, L., MORTARI, A. AND VALZELLI, L., *J. Pharm. and Pharmacol.*, 13 (1961) 548
30   GARATTINI, S., LAMESTA, L., MORTARI, A., PALMA, V. AND VALZELLI, L., *J. Pharm. and Pharmacol.*, 13 (1961) 385.
31   GARATTINI, S., MORTARI, A., VALSECCHI, A. AND VALZELLI, L., *Nature*, 183 (1959) 1273.
32   GARATTINI, S. AND VALZELLI, L., unpublished observations.

[32a] GARATTINI, S. AND VALZELLI, L., *Acta Symp. "Le sindromi depressive"*, Rapallo, 1960, p. 7.

[33] GLUCKMAN, M. I., *Federation Proc.*, 19 (1960) 265.

[34] GOLDBERG, L. I., DACOSTA, F. M. AND OZAKI, M., *Nature*, 188 (1960) 502.

[34a] GÖRÖG, P. AND SZPORNY, L., *Biochem. Pharmacol.*, 11 (1961) 165.

[35] GREEN, J. P., PAASONEN, M. K. AND GIARMAN, N., *Proc. Soc. Exptl. Biol. Med.*, 94 (1957) 428.

[35a] GURSEY, D. AND OLSON, R. E., *Proc. Soc. Exptl. Biol. Med.*, 104 (1960) 280.

[36] HALPERN, B. M., LIACOPOULOS, P. AND LIACOPOULOS BRIOT, M., *Compt. rend. soc. biol.*, 151 (1957) 1692.

[37] HARDISTY, R. M., INGRAM, G. I. C. AND STACEY, R. S., *Experientia*, 12 (1956) 424.

[38] HOLTZ, P., BALZER, H. AND WESTERMANN, E., *Arch. exptl. Pathol. Pharmakol. Naunyn-Schmiede-berg's*, 231 (1957) 361.

[39] KÄRKI, N. T. AND PAASONEN, M. K., *J. Neurochem*, 3 (1959) 352.

[40] KÄRKI, N. T. AND PAASONEN, M. K., *Acta Pharmacol. Toxicol.*, 16 (1959) 20.

[41] KÄRKI, N. T. AND PAASONEN, M. K., *Nature*, 185 (1960) 4706.

[42] KATO, R., MARIANI, L. AND VALZELLI, L., *Atti soc. lombarda sci. med. biol.*, 13 (1958) 297.

[43] KATO, R. AND VALZELLI, L., *Boll. soc. ital. biol. sper.*, 34 (1958) 1402.

[43a] KINDWALL, E. P., BORÉUS, L. O. AND WESTERHOLM, B., *Am. J. Physiol.*, 203 (1962) 389.

[44] KRUEGER, A. P., SMITH, R. F. AND REED, E., *J. Gen. Physiol.*, 44 (1960) 269.

[44a] MAYNERT, E. W., KLINGMAN, G. I. AND KAJI, H. K., *J. Pharmacol. Exptl. Therap.*, 135 (1962) 296.

[45] MEDAKOVIĆ, M. AND RADMANOVIĆ, B., *J. Pharm. and Pharmacol.*, 12 (1960) 695.

[46] MEDAKOVIĆ, M. AND SPUŽIĆ, I., *Nature*, 183 (1959) 1685.

[47] MILINE, R., ŠTERN, P. AND HUKOVIĆ, S., *Experientia*, 14 (1958) 415.

[48] MUSSINI, E. AND VALZELLI, L., *Atti soc. lombarda sci. med. biol.*, 13 (1958) 310.

[49] PAASONEN, M. K. AND DEWS, P. B., *Brit. J. Pharmacol.*, 13 (1958) 84.

[50] PAASONEN, M. K. AND GIARMAN, N. J., *Arch. intern. pharmacodyn.*, 114 (1958) 189.

[51] PAASONEN, M. K. AND KÄRKI, N. T., *Brit. J. Pharmacol.*, 14 (1959) 164.

[52] PAASONEN, M. K., MCLEAN, P. D. AND GIARMAN, N. J., *J. Neurochem.*, 1 (1957) 326.

[53] PAASONEN, M. K. AND PELTOLA, P., *Ann. Med. Exptl. et Biol. Fenniae (Helsinki)*, 38 (1960) 227.

[54] PAASONEN, M. K. AND VOGT, M., *J. Physiol. (London)*, 131 (1956) 617.

[55] PARRATT, J. R. AND WEST, G. B., *J. Physiol. (London)*, 137 (1956) 232.

[56] PICCINELLI, D., *Arch. intern. pharmacodyn.*, 117 (1958) 452.

[57] PLETSCHER, A., *Helv. Physiol. et Pharmacol. Acta*, 14 (1956) C 76.

[58] PLETSCHER, A., *Experientia*, 12 (1956) 479.

[59] PLETSCHER, A., *Science*, 126 (1957) 507.

[60] PLETSCHER, A., in G. B. LEWIS (Ed.), *5-Hydroxytryptamine*, Pergamon, London, 1958, p. 84.

[61] PLETSCHER, A., BESENDORF, H. AND BÄTCHOLD, H., *Arch. exptl. Pathol. Pharmakol. Naunyn-Schmiedeberg's*, 232 (1957) 499.

[61a] PLETSCHER, A. AND GEY, K. F., *Med. Exptl.*, 6 (1962) 165.

[62] PLETSCHER, A., SHORE, P. A. AND BRODIE, B. B., *Science*, 122 (1955) 374.

[63] PLETSCHER, A., SHORE, P. A. AND BRODIE, B. B., *J. Pharmacol. Exptl. Therap.*, 116 (1956) 84.

[64] QUINN, G. P., SHORE, P. A. AND BRODIE, B. B., *J. Pharmacol. Exptl. Therap.*, 127 (1959) 103.

[65] RENSON, J. AND FISHER, P., *Arch. intern. physiol. et biochim.*, 67 (1959) 142.

[65a] ROOS, B. E. AND WERDINIUS, B., *Life Sci.*, 2 (1963) 92.

[66] ROSEMBERG, J. C., DAVIS, R., MORAU, W. H. AND ZIMMERMANN, B., *Federation Proc.*, 18 (1959) 503.

[67] SANYAL, R. K. AND WEST, G. B., *J. Physiol. (London)*, 142 (1958) 571.

[68] SCHWABE, U., HEYE, D., HILDEBRANDT, H. J. AND HUKUHARA, T., *Arch. exptl. Pathol. Pharmakol. Naunyn-Schmiedeberg's*, 241 (1961) 254.

[69] SHORE, P. A. AND BRODIE, B. B., in S. GARATTINI AND V. GHETTI (Eds.), *Psychotropic Drugs*, Elsevier, Amsterdam, 1957, p. 421.

[70] SHORE, P. A. AND BRODIE, B. B., *Proc. Soc. Exptl. Biol. Med.*, 94 (1957) 433.

[71] SHORE, P. A. AND BRODIE, B. B., *J. Clin. Exptl. Psychopathol. and Quart. Rev. Psychiat. Neurol.*, 19 (1958) 56.

[72] SHORE, P. A., GILLESPIE JR., L., SPECTOR, S. AND PROCKOP, D., *Naturwissenschaften*, 45 (1958) 340.

[73] SHORE, P. A., PLETSCHER, A., TOMICH, E. G., KUNTZMAN, R. AND BRODIE, B. B., *J. Pharmacol. Exptl. Therap.*, 117 (1957) 232.

74 SHORE, P. A., PLETSCHER, A., TOMICH, E. G., CARLSSON, A., KUNTZMAN, R. AND BRODIE, B. B., *N. Y. Acad. Sci.*, 66 (1957) 609.

74a SKILLEN, R. G., THIENES, C. H., CANGELOSI, J. AND STRAIN, L., *Proc. Soc. Exptl. Biol. Med.*, 108 (1961) 340.

75 SMITH, S. E., *J. Physiol. (London)*, 148 (1959) 18P.

76 SMITH, S. E.. *Brit. J. Pharmacol.*, 15 (1960) 319.

76a SPECK, L. B., *J. Neurochem.*, 9 (1962) 573.

77 SPECTOR, S., PROCKOP, D., SHORE, P. A. AND BRODIE, B. B., *Science*, 127 (1958) 704.

78 SPECTOR, S., SHORE, P. A. AND BRODIE, B. B., *J. Pharmacol. Exptl. Therap.*, 128 (1960) 15.

79 SULLIVAN, T. J., *Brit. J. Pharmacol.*, 16 (1961) 90.

80 TELFORD, J. M. AND WEST, G. B., *Brit. J. Pharmacol.*, 15 (1960) 532.

81 TOWNE, J., PUT, T. AND SCHWARTZ, M., *Federation Proc.*, 18 (1959) 452.

82 TOWNE, J. C. AND SHERMAN, J. O., *Proc. Soc. Exptl. Biol. Med.*, 103 (1960) 721.

83 UDENFRIEND, S., WEISSBACH, H. AND BOGDANSKI, D. F., *J. Biol. Chem.*, 224 (1957) 803.

84 UDENFRIEND, S., WEISSBACH, H. AND BOGDANSKI, D. F., *J. Pharmacol. Exptl. Therap.*, 120 (1957) 255.

85 UDENFRIEND, S., WEISSBACH, H. AND BOGDANSKI, D. F., *N. Y. Acad. Sci.*, 66 (1957) 603.

86 VACEK, L., *Scripta Med.*, 32 (1959) 271.

87 WAALKES, T. P. AND WEISSBACH, H., *Proc. Soc. Exptl. Biol. Med.*, 93 (1956) 394.

88 WAALKES, T. P., WEISSBACH, H., BORICEVICH, J. AND UDENFRIEND, S., *Proc. Soc. Exptl. Biol. Med.*, 95 (1957) 479.

88a WANG, H. L., HARWALKAR, V. H. AND WAISMAN, H. A., *Arch. Biochem. Biophys.*, 97, (1962) 181.

89 WEISSBACH, H., BOGDANSKI, D. F., REDFIELD, B. G. AND UDENFRIEND, S., *J. Biol. Chem.*, 227 (1957) 617.

90 WEST, G. B., *J. Pharm. and Pharmacol.*, 10 (1958) 92 T.

90a WIEGAND, R. G. AND PERRY, J. E., *Biochem. Pharmacol.*, 7 (1961) 181.

91 WOOLLEY, D. W. AND EDELMAN, P. M., *Science*, 127 (1958) 281.

92 YEH, S. D. J., SOLOMON, J. D. AND CHOW, B. W., *Federation Proc.*, 18 (1959) 357.

93 YUWILER, A. AND LOUTTIT, R. T., *Science*, 134 (1961) 831.

ACTION OF SOME COMBINATIONS OF DRUGS AND TREATMENTS ON 5-HT CONTENT OF VARIOUS TISSUES OF DIFFERENT ANIMAL SPECIES

LEGEND

I.V. = intravenously
I.P. = intraperitoneally
S.C. = subcutaneously
(+) = mean of authors' data
U.   = units

Combination	Dose mg/kg	Route of administration	Time between last admin. and determination	Animal	Tissue	5-HT μg/g or ml content		Reference
						controls	treated	
Adrenalectomy +			1 h	Rat	brain	0.37	0.30	7
(71 h after)					spleen	3.10	3.50	7
5-Hydroxytryptamine	10	S.C.			lung	0.58	1.85	7
					kidney	0.21	1.02	7
Adrenalectomy +			4 h	Rat	brain	0.37	0.18	8
(68 h after)					spleen	3.10	0.82	8
Reserpine	0.62	I.P.			lung	0.58	0.61	8
					kidney	0.21	0.17	8
Adrenalectomy + Traumatic shock			45 min	Rat	blood	0.276	0.371	11
L-Amphetamine +	100	I.P.	5 min	Mouse	brain	0.68	1.05	4a
(10 min after)								
5-Hydroxytryptophan	25	I.V.						
Chlorpromazine +	10	I.V.	1 h	Rabbit	telencephalon	0.282	1.342	4
(2 h after)					hippocampus	0.272	1.725	4
5-Hydroxytryptophan	75				midbrain	0.647	3.324	4
					medulla			
					+ pons	0.582	2.563	4
					cerebellum	0.112	0.381	4
Chlorpromazine + (1 h after)	15	I.P.	4 h	Rat	brain	0.47	0.34	8

(8 h after) Reserpine	(daily for 15 days) 2.5				cerebellum	0.127	0.103	8
					diencephalic area	0.82	0.28	8
					mesencephalic area	0.53	0.22	8
					brain	0.43	0.18	8
Diphenylhydantoin +	100	I.P.	15 min	Rat	hemispheres	0.37	0.60	1
(20 min after)					cerebellum	0.117	0.128	1
Electroshock	110 V — 0.2 sec				diencephalic area	0.81	1.45	1
					mesencephalic area	0.56	1.11	1
					brain	0.47	0.73	1
Evipal +	80	I.P.	15 min	Rat	hemispheres	0.37	0.52	1
(20 min after)					cerebellum	0.117	0.160	1
Electroshock	110 V — 0.2 sec				diencephalic area	0.81	1.31	1
					mesencephalic area	0.56	0.92	1
					brain	0.47	0.79	1
Gramine + (1 h after)	40	I.V.	2 h	Mouse	brain	0.67	0.95	4b
Phenylethylhydrazine	10	I.V.						
Gramine + (1 h after)	40	I.V.	2 h	Mouse	brain	0.67	0.62	4b
Reserpine	3	I.P.						
Harmaline + (10 min after)	10	I.P.	5 min	Mouse	brain	0.67	2.22	4a
5-Hydroxytryptophan	25	I.V.						

Combination	Dose mg/kg	Route of administration	Time between last admin. and determination	Animal	Tissue	5-HT μg/g or ml content		Reference
						controls	treated	
Harmaline + (1 h after)	10	I.P.	2 h	Mouse	brain	0.66	1.06	4b
Phenylethylhydrazine	10	I.V.						
Harmaline + (30 min after)	10	I.P.	2 h	Mouse	brain	0.66	0.69	4b
Reserpine	3	I.P.						
Hexobarbital, see Nembutal								
5-HT, see 5-Hydroxytryptamine								
5-HTP, see 5-Hydroxytryptophan								
5-Hydroxytryptamine +	10	O.		Rat	brain	0.39	0.42	8
Calcium aspartate	200 (daily for 6 days)				spleen	2.96	5.70	8
idem	10	S.C.			brain	0.39	0.42	8
	400	I.P.			spleen	2.96	7.93	8
5-Hydroxytryptamine +	10	I.P.	1 h	Rat	brain	0.39	0.54	8
Calcium chloride	259.4				spleen	2.96	4.5	8
idem			4 h		brain	0.39	0.59	8
					spleen	2.96	8.03	8
5-Hydroxytryptophan +	240	I.V.		Chicken	brain	0.4	2	21
Pyridoxine deficiency					liver	0.5	3.2	21
5-Hydroxytryptophan + (30 min after)	200	I.P.		Mouse	in toto	2.4	6.5	21

Phenylethylhydrazine	2.5 (daily for 6 days)				diencephalic area	0.72	1.68	8
					mesencephalic area	0.79	1.61	8
Imipramine + (1 h after)	10	I.P.	4 h	Rat	brain	0.47	0.30	8
Reserpine	1.25							
idem	15					0.47	0.32	8
	2.5							
Iproniazid + (17 h after)	100	S.C.	20 min	Rat	brain	0.49	2.65	10
					blood	0.205	3.19	10
5-Hydroxytryptamine	10	I.V.						
Iproniazid + (30 min after)	300	I.V.	30 min	Mouse	brain	0.8	3.2	19
					carcass	0.7	4	19
5-Hydroxytryptophan	80							
Iproniazid + (12 h after)	150	I.P.	30 min	Mouse	brain (cere-bellum-free)	0.82	1.28	8
5-Hydroxytryptophan	1							
	10						1.45	8
	20						1.66	8
	50						2.57	8
	100						3.74	8
	200						7.42	8
	600						11.59	8

Combination	Dose mg/kg	Route of administration	Time between last admin. and determination	Animal	Tissue	5-HT μg/g or ml content		Reference
						controls	treated	
Iproniazid +	75	I.P.	4 h	Mouse	brain	0.8	3.2	19
(1 h after)					liver	0.9	2.1	19
5-Hydroxytryptophan	100							
idem	100		6 h		in toto	1.2	8	19
	150							
idem	75		3 h	Rat	brain	0.63	4.1	19
	75				liver	0.3	6.3	19
Iproniazid +	100	I.V.	$1\frac{1}{2}$ h	Rabbit	brain	0.50	3.8	19
(16 h after)								
5-Hydroxytryptophan	70							
idem	100	S.C.	3 h	Rat	thyroid gland	6.0	29.3	12a
	100	I.P.			stomach fundus	1.0	1.8	12a
					pylorus	6.1	10.2	12a
					duodenum	3.3	14.5	12a
Iproniazid +	300	I.P.	$1\frac{1}{2}$ h	Mouse	brain	0.85	3.2	20
(30 min after)					carcass	0.7	5	20
5-Hydroxytryptophan	80							
Iproniazid +	100	I.P.	6 h	Rat	brain	0.470	1.865	3
(12 h after)								
Isopropylhydrazine	100							
Iproniazid +	100	I.P.	2 h	Rat	brain	0.78	1.23	16

Phenylethylhydrazine (W 1544)	40								
Iproniazid + (18 h after)	100	S.C.	1 h	Rat	brain	0.32	0.79	12
					small intestine	3.95	5.07	12
Raunescine	5	I.P.						
idem			3–4 h		brain	0.32	1.103	12
					small intestine	3.95	2.97	12
Iproniazid + (4 h after)	100	I.M.	20 h	Rat	brain	0.4	0.6	9
Raunescine	10	S.C.						
Iproniazid + (16 h after)	100	S.C.	4 h	Rat	brain	0.55	1	13
Reserpine	3							
Iproniazid + (16 h after)	100	I.V.	3 h	Rabbit	brain	0.49	0.88	14
Tetrabenazine (RO 1-9569)	40							
Isoniazid + (16 h after)	100	S.C.	4 h	Rat	brain	0.55	0.2	13
Reserpine	3							

JB 516, see Phenylhydrazinepropane

Luminal, see Phenobarbitone

Marsilid, see Iproniazid

References p. 333

Combination	Dose mg/kg	Route of administration	Time between last admin. and determination	Animal	Tissue	5-HT µg/g or ml content		Reference
						controls	treated	
Nembutal +	35	I.P.	15 min	Rat	hemispheres	0.37	0.69	1
(20 min after)					cerebellum	0.117	0.145	1
Electroshock	110 V —				diencephalic			
	0.2 sec				area	0.81	1.17	1
					mesencephalic			
					area	0.56	0.97	1
					brain	0.47	0.78	1
Phenobarbital, see Phenobarbitone								
Phenobarbitone +	120	I.P.	15 min	Rat	hemispheres	0.37	0.63	1
(20 min after)					cerebellum	0.117	0.116	1
Electroshock	110 V —				diencephalic			
	0.2 sec				area	0.81	1.24	1
					mesencephalic			
					area	0.56	1.09	1
					brain	0.47	0.74	1
L-Phenylalanine +	7% ⎱	diet (for 3 weeks)		Rat	brain	0.77	0.98	20a
L-Tryptophan	5% ⎰				liver	0.64	0.74	20a
Phenylethylhydrazine +	10	I.P.	2 h	Rat	brain	0.47	0.72	8
(16 h after)					spleen	2.84	4.01	8
Carbethoxysyringoylmethyl-								
reserpate								
(SU 3118)	2.5	I.V.						
Phenylethylhydrazine +	10	I.P.	2 h	Rat	brain	0.47	0.82	8
(16 h after)					spleen	2.84	3.45	8

10-Methoxydeserpidine (10 MD)	5	I.V.							
Phenylethylhydrazine + (16 h after) Methyl-18-O-(3-N,N-dimethylaminobenzoyl)-reserpate (SU 5171)	10 2.5	I.P. I.V.	2 h	Rat	brain spleen	0.47 2.84	0.65 3.65	8 8	
Phenylethylhydrazine + (4 h after) 5-Hydroxytryptophan (5-HTP)	10 200	I.P. I.V.	1 h	Rat	brain spleen	0.47 2.84	5.3 26.33	8 8	
Phenylethylhydrazine + (3½ h after) Cyproheptadine + (30 min after) 5-Hydroxytryptophan	10 10 10	I.P.		Rat	brain spleen	0.47 2.84	5.69 18.24	8 8	
Phenylethylhydrazine + (16 h after) Reserpine	5 2.5	I.P.	4 h	Rat	brain	0.38	0.36	8	
Phenylethylhydrazine + (30 min after) Reserpine	10 3	I.V. I.P.	2 h	Mouse	brain	0.66	0.73	4b	
Phenylethylhydrazine + Reserpine	10 0.3 (daily for 15 days)	I.P.				0.38	0.76	8	

Combination	Dose mg/kg	Route of administration	Time between last admin. and determination	Animal	Tissue	5-HT µg/g or ml content		Reference
						controls	treated	
Reserpine + Catechol	2.5 5	I.P.	4 h	Rat	brain	0.38	0.17	8
Reserpine + 5-Hydroxytryptamine	1 25	I.P. S.C.	24 h	Rat	serum	0.68	0.57	5
					spleen	1.42	2.58	5
					gastrointestinal tract	2.3	1.96	5
					brain	0.44	0.2	5
idem	0.5 25	I.P. S.C.	24 h		serum	0.68	0.74	5
					spleen	1.42	3.62	5
					gastrointestinal tract	2.3	2.25	5
					brain	0.44	0.4	5
Reserpine + (4 h after) 5-Hydroxytryptophan	2.5 100	I.P.	45 min	Mouse	brain	0.82	1.30	6
Reserpine + (1 h after) Imipramine	2.5 15	I.P.	3 h	Rat	brain	0.47	0.51	8
Reserpine + (3 h after) Imipramine	2.5 7.5 15	I.P.	1 h	Rat	brain	0.47	0.49 0.52	8 8
Reserpine +	2.5	I.P.	30 min	Rat	brain	0.47	0.31	8

Iproniazid	110								
Reserpine + (1 h after) Iproniazid	2.5 110	I.P.	3 h	Rat	brain	0.42	0.33	8	
Reserpine + (12 h after) Isopropylhydrazine	5 100	I.P.	6 h	Rat	brain	0.470	0.453	3	
Reserpine + (18 h after) Paraoxon	5 1.72	I.P.	2 h	Rat	brain	0.78	0.15	16	
Reserpine + (18 h after) 1-Phenyl-2-hydrazinopropane (JB 516)	2.5 3	I.V.	2 h	Rabbit	brain stem	0.64(+)	0.46	2	
Reserpine + Pyrogallol	2.5 40	I.P.	4 h	Rat	brain	0.38	0.18	8	
Reserpine + (1 h after) Phenylethylhydrazine (W 1544)	2.5 40	I.P.	3 h	Rat	brain	0.47	0.51	8	

RO 1–9569, see Tetrabenazine

Serotonin, see 5-Hydroxytryptamine

Combination	Dose mg/kg	Route of administration	Time between last admin. and determination	Animal	Tissue	5-HT μg/g or ml content		Reference
						controls	treated	
Streptomycin +	70	O.		Mouse	small intestine	1.31	1.72	17
Chlortetracycline	70				spleen	0.69	0.69	
	(daily for 9 days)							
SU 3118, see Phenylethylhydrazine + Carbethoxysyringoylmethylreserpate								
SU 5171, see Phenylethylhydrazine + Methyl-18-O-(3-N,N-dimethylaminobenzoyl)reserpate								
Succinylcholine +	1	I.P.	15 min	Rat	hemispheres	0.37	0.58	1
(20 min after)					cerebellum	0.117	0.123	1
Electroshock	110 V —				diencephalic			
	0.2 sec				area	0.81	0.99	1
					mesencephalic			
					area	0.56	0.84	1
					brain	0.47	0.77	1
Tetrabenazine (RO 1–9569) +	50		24 h	Rabbit	brain stem	0.75	0.35	15
(15 min after)								
Reserpine	1							
Thyroidectomy +				Rat	brain	0.98(+)	0.92	18
Propylthiouracil	4 μg							
	(daily)							
Tofranil, see Imipramine								
Tranylcypromine + (10 min after)	7.2	I.P.	5 min	Mouse	brain	0.67	1.83	4a
5-Hydroxytryptophan	25	I.V.						

REFERENCES

[1] BISIANI, M., GARATTINI, S., KATO, R. AND VALZELLI, L., *Atti soc. lombarda sci. med. biol.*, 13 (1958) 345.

[2] BRODIE, B. B., SPECTOR, S., KUNTZMAN, R. G. AND SHORE, P. A., *Naturwissenschaften*, 45 (1958) 243.

[3] CANAL, N., MAFFEI-FACCIOLI, A. AND SCAMAZZO, T., *Boll. soc. ital. biol. sper.*, 35 (1959) 436.

[4] COSTA, E. AND RINALDI, F., *Am. J. Physiol.*, 194 (1958) 214.

[4a] DUBNICK, B., LEESON, G. A. AND PHILLIPS, G. E., *Biochem. Pharmacol.*, 13 (1962) 45.

[4b] DUBNICK, B., LEESON, G. A. AND PHILLIPS, G. E., *J. Neurochem.*, 9 (1962) 299.

[5] ERSPAMER, V., *Experientia*, 12 (1956) 63.

[6] FRESIA, P., GARATTINI, S. AND VALZELLI, L., unpublished observations.

[7] GARATTINI, S., LAMESTA, L., MORTARI, A., PALMA, V. AND VALZELLI, L., *J. Pharm. and Pharmacol.*, 13 (1961) 385.

[8] GARATTINI, S. AND VALZELLI, L., unpublished observations.

[9] HOLTZ, P., BALZER, H. AND WESTERMANN, E., *Arch. exptl. Pathol. Pharmakol. Naunyn-Schmiedeberg's*, 231 (1957) 362.

[10] KÄRKI, N. T. AND PAASONEN, M. K., *Acta Pharmacol. Toxicol.*, 16 (1959) 20.

[11] MEDAKOVIC, M., *Nature*, 183 (1959) 1685.

[12] PAASONEN, M. K. AND KÄRKI, N. T., *Brit. J. Pharmacol.*, 14 (1959) 164.

[12a] PAASONEN, M. K., KÄRKI, N. T. AND MOLKKA, S., *Ann. Med. Exptl. et Biol. Fenniae (Helsinki)*, 39 (1961) 405.

[13] PLETSCHER, A., *Experientia*, 12 (1956) 479.

[14] PLETSCHER, A., BESENDORF, H. AND BÄTCHOLD, H. P., *Arch. exptl. Pathol. Pharmakol. Naunyn-Schmiedeberg's*, 232 (1958) 499.

[15] QUINN, G. P., SHORE, P. A. AND BRODIE, B. B., *J. Pharmacol. Exptl. Therap.*, 127 (1959) 103.

[16] SCHWABE, U., HEYE, D., HILDEBRANDT, H. J. AND HUKUHARA, T., *Arch. Pathol. Pharmakol.*, 241 (1961) 254.

[17] SULLIVAN, T. J., *Brit. J. Pharmacol.*, 16 (1961) 90.

[18] TOWNE, J., PUT, T. AND SCHWARTZ, M., *Federation Proc.*, 18 (1959) 452.

[19] UDENFRIEND, S., WEISSBACH, H. AND BOGDANSKI, D. F., *J. Pharmacol. Exptl. Therap.*, 120 (1957) 255.

[20] UDENFRIEND, S., WEISSBACH, H. AND BOGDANSKI, D. F., *J. Biol. Chem.*, 224 (1957) 803.

[20a] WANG, H. L., HARWALKAR, V. H. AND WAISMAN, H. A., *Arch. Biochem. Biophys.*, 97 (1962) 181.

[21] WEISSBACH, H., BOGDANSKI, D. F., REDFIELD., B. G. AND UDENFRIEND, S., *J. Biol. Chem.*, 227 (1957) 617.

DRUGS AFFECTING 5-HT IN SEVERAL BIOLOGICAL TESTS

LEGEND

$++$ = strong antagonism
$+$ = antagonism
$\pm$ = weak antagonism
$-$ = no antagonism
x = stimulant
f = facilitating
$\neq$ = other derivatives are listed in the quoted paper

CAT

Drug	Tests		
	Arterial pressure	Respiration	Other tests
1-Acetyllysergic diethylamide		++ (76)	
Atropine	+ (78, 97) — (39, 42)	± (76)	
Bromolysergic diethylamide	+ (81) — (102)	++ (76)	
Chlorpromazine	+ (58, 59)		
Dibenamine	± (31, 39)		
Diethazine	± (58, 59)		
Diphenhydramine (Benadryl)	± (92)		
Hexamethonium	— (56)	— (56)	
Lysergic acid diethylamide (LSD)	— (102)	++ (76)	+ (83)(A), (83)(B), (46, 55)(C), (46, 55)(D)
Phenergan	± (58, 59, 92)		
Phenylguanide	f (35)		
Procaine	+ (97)		
Thenalidine tartrate (Sandosten)		— (76)	
Thenyldiamine (Thenfadil)	± (92)		
Tolazoline (Priscol)	— (92)		
Tryptamine	++ (56)	+ + (56)	
Yohimbine	+ (92) ± (95, 96)		

(A) = Heart auricle
(B) = Heart
(C) = Perfused lungs
(D) = Perfused hind limb

DOG

	Tests	
Drug	*Arterial pressure*	*Other tests*
Atropine	± (91, 111)	
Benadryl	± (92)	
1-Benzyl-2,5-dimethylbufotenine	+ (134)	
1-Benzyl-2-methyl-5-hydroxytryptamine (BAS-phenol)	+ + (115)	
1-Benzyl-2-methyl-5-methoxytryptamine (BAS)	+ + (115)	
Bromolysergic acid diethylamide (BOL)	± (102)	
Dibenamine	± (92)	+ (33)(A)
Dibenzyline	+ (27)	
2,5-Dimethylbufotenine	+ (114)	
2,5-Dimethylserotonin	+ (131)	
Ergotamine	+ (92)	
Lysergic acid diethylamide (LSD)	± (102)	
2-Methyl-3-ethyl-5-dimethylaminoindole (Medmain)	+ (32, 111)	
1-Methyllysergic acid butanolamide (UML 491)	— (122)	
	+ (33a)	
Noradrenaline	f (92)	
Phenergan	± (92)	
Phentolamine	+ (92)	
	± (108)	
Piperoxan	± (28, 92)	
Procaine	± (111)	
Reserpine	± (109, 110)	+ (109, 110) (B)
Tetrahydrocarbazole	+ (117) ≠	
Thenyldiamine (Thenfadil)	±(92)	
Tolazoline (Priscol)	—(92)	
Yohimbine	± (104, 105)	

(A) = Isolated bladder
(B) = Respiration

GUINEA PIG

Drug	Tests			
	Isolated ileum	Isolated colon	Bronchospasm	Other tests
Acetylcholine			— (25)	
Acetylpromazine	±(101a)			
Adrenaline	+ + (67)		± (135)	
			+ + (67)	
3, β-Aminoethyl-5-hydroxyindazole	x (1)			
Aminophylline			+ (64)	
Amphetamine	± (67)		— (67)	
Antazoline	+ (85)		+ (85, 106)	
Antergan (RP 2339)			+ + (67)	
Atropine	+ (78), (M receptors) (48) ± (15, 36, 67, 94, 97, 99) — (42, 79)		+ + (67) + (65) ± (23, 61)	± (80, 81)(A) — (83)
Azamethonium (Pendiomid)			— (67)	
Benzylbufotenine (BAB)				± (121)(C)
1-Benzyl-2-methyl-5-methoxytryptamine (BAS)				± (121)(C) — (89)(A)
5-Benzyloxygramine	+ (D receptors) (7, 43, 48)			
5-Benzyloxy-N,N'-dimethyltryptamine	+ (7)			
5-Benzyloxytryptamine			— (25)	
6-Benzyloxytryptamine			— (25)	
Benzylphenoxyisopropyl-chloroethylamine	+ (D receptors) (43, 48) ± (77)			
Botulinum toxin	± (2)			
Bromolysergic acid diethylamide (BOL)	+ (120)			
5-Bromotryptamine		f (93)		
Bufotenine			+ (25)	
Butyrylperazine	+ +(101a)			

GUINEA PIG

	Tests			
Drug	Isolated ileum	Isolated colon	Bronchospasm	Other tests
Butyrylpromazine	+(101a)			
Caffeine			± (67)	
Chlorcyclizine			+ (71)	
Chlorisondamine			+ (67)	
5-Chlorotryptamine		f (93)		
5-Chlorotryptophan		+ (82)		
Chlorpromazine	+ + (29, 85)	+ + (67, 85)	+ + (9)	+ (17)(B)
	+ (67)	+ (69)	+ (25, 64, 67, 106, 135)	
Chlorpyramine (Synopen)			+ (106)	
Chlorpyribenzamine			+ (106)	
Cocaine	+ (M receptors) (48)		+ (64)	
	± (43, 72, 99, 117)		— (67)	
Dibenamine			+ (12)	— (83)(A)
			— (83)	
Dibucaine			— (67)	
Diethazine (Diparcol)	± (29)			
N,N'-Diethyltryptamine	+ + (11a) ≠			
Dihydroergotamine	+ (D receptors) (43, 48)		± (12, 65)	— (83)(A)
			— (83)	
1,4-Dimethyl-7-isopropylazulene (Guaiazulen)			+ (54)	
N,N'-Dimethyltryptamine	+ + (7)			
Diphenhydramine (Benadryl)	+ (85, 94)		+ (85)	
Dock leaf extract	+ + (22)			
D-Tubocurarine				— (81, 88)(A)
Ephedrine	+ (67)		+ (67)	± (80, 81)(A)
Ethopropazine	— (29)			
5-Fluorotryptamine		f (93)		
5-Fluorotryptophan		+ (93)		
Hexamethonium	+ (81, 100)		+ (64)	
			— (83)	
Histamine			— (25)	

GUINEA PIG

Drug	Tests			
	Isolated ileum	*Isolated colon*	*Bronchospasm*	*Other tests*
Homochlorcyclizine (SA-97)			+ (71)	
Hydralazine			— (67)	
DL-5-Hydroxyacetyl-tryptophan			— (25)	
5-Hydroxyindoleacetic acid			± (25)	
Hydroxyzine	+ + (118)			
Imipramine (Tofranil)	+ (29, 85)		+ (85)	
Isopropylnoradrenaline	+ + (67)		+ + (62, 64, 67)	(17)(B)
Isothipendyl (Andantol)			+ (106)	
Levomepromazine			+ + (9) + (135)	
Lysergic acid diethylamide (LSD)	+ (D receptors (48) ± (43, 44, 67, 81, 118)		+ + (9, 67) + (135) ± (59)	+ (17)(B), (83)(A)
Meprobamate			— (9, 135)	
Mepyramine	± (15)		— (62, 63, 65)	
Methamphetamine (Pervitin)			+ (67)	
Methoxyharmalan	+ + (85a)			
5-Methoxytryptamine		f (93)		
2-Methyl-5-chlorogramine		+ + (93)		
5-Methyltryptamine		f (93)		
5-Methyltryptophan		+ (93)		
Morphine	+ (M receptors) (43, 48) ± (69)			
Noradrenaline	+ (67)		± (67)	
Oxyphenonium (Antrenyl)			+ (67)	
Papaverine			— (67)	
Perphenazine (Trilafon)	+ (85) ± (78)		+ (85)	

GUINEA PIG

	Tests			
Drug	Isolated ileum	Isolated colon	Bronchospasm	Other tests
Phenindamine (Thephorin)			+ (106)	
Phentolamine (Regitin)	± (67)		± (12) — (67)	
Procaine	+ (117) ± (81)		— (13, 67)	— (80, 81)(A)
Prochlorperazine	+ (29)			
Prochlorpromazine			+ (25)	
Promazine	+ + (29, 85)		+ + (85)	
Promethazine (Phenergan)	+ (85) — (29, 81)		+ (85) ± (65)	+ (17) (B) — (80, 81)(A)
Propantheline			+ (64)	
Prothipendyl (Dominal)			+ (106)	
Reserpine	+ (67)		+ + (67) — (9, 135)	+ (17)(B)
Substance P	f (3)			
Thenalidine tartrate (Sandosten)	+(73)		+(72, 73)	
Theophylline			± (67)	
Thiopropazate (Dartal)	± (85)		± (85)	
Trasentin	+ (67)		— (67)	
Triflupromazine (Vesprin)	+ (85)		+ (85)	
Tripelennamine (Pyribenzamine)	+ (85, 94) ± (61)		+ + (67, 85) + (85, 106)	
Tryptamine	+ (41)			
Tryptophan			— (25)	
Urethane			— (84)	
Valeroylperazine	+ +(101a)			
Yohimbine	+ (67)		± (67)	

(A) = Heart auricle

(B) = Gall bladder

(C) = Perfused lung

HORSE

Drug	Tests
	Carotid strips
Hydralazine	$++$ (74, 75)
Lysergic acid diethylamide (LSD)	$++$ (74)
Reserpine	$++$ (74)

HUMAN

Drug	Tests	
	Umbilical vessels	Forearm blood flow
Bromolysergic acid diethylamide (BOL)		$++$ (57)
Chlorpromazine	$++$ (4)	
Heparin	$-$ (4)	
Hexamethonium	$-$ (4)	
Lysergic acid diethylamide (LSD)	$++$ (4)	
Mescaline	f (4)	
Phenoxybenzamine (Dibenzyline)	$\pm$ (4)	
Phentolamine	$+$ (4)	
Reserpine	$-$ (4)	
Sodium salicylate		$+$ (57)
Tetraethylammonium	$+$ (4)	
Trimetaphan (Camphor sulphonate)	$\pm$ (4)	
Tryptamine	$+$ (4)	
Yohimbine	$+$ (4)	

MOLLUSCAN HEART

	Tests					
Drug	Venus mer- cenaria	Spisula solida	Helix aspersa	Mya are- naria	Cardium edule	Anodonta cignea
Acetylcholine					+ (47)	— (34)
Adrenaline	± (129)	xx (47)	— (47)	x (47)	+ (47)	— (34)
Benzoquinonium					— (47)	— (34)
5-Benzyloxygramine		± (47)	± (47)		± (47)	
Bromolysergic acid diethylamide (BOL)	+ + (126)					
Cocaine				— (47)		x (34)
Decamethonium				— (47)		
Dihydroergotamine		x (47)	x (47)	x (47)		x (34)
Ephedrine				— (47)		
Ergometrine		x (47)	x (47)	x (47)		
Hexamethonium				— (47)		
Histamine	— (129)	— (47)	— (47)	x (47)	— (47)	— (34)
Hydergin		x (47)	x (47)	x (47)		
Isopropylnoradrenaline (Isoprenaline)				± (47)		
Lysergic acid diethylamide (LSD)	x (47, 116, 128)	x (47)	x (47)	x (47)	± (47)	
6-Methylgramine		± (47)	± (47)		± (47)	
Noradrenaline		xx (47)	— (47)	x (47)	+ (47)	— (34)
Substance P		— (47)				— (34)
Tryptamine	x(124)					
Tubocurarine				— (47)		

MOUSE

	Test
	Uterus
Drug	
Lysergic acid diethylamide (LSD)	+ + (38)
Reserpine	+ + (38)

PIG

Drug	Test
	Ureter
1-Benzyl-2-methyl-5-methoxytryptamine (BAS)	— (21)
Chlorpromazine	+ (21)
Isopropylnoradrenaline	+ (21)
Lysergic acid diethylamide (LSD)	+ (21)
Reserpine	± (21)

RABBIT

Drug	Tests				
	Perfused ear	Intestine	Heart auricle	Perfused hind limbs	Other tests
Acetylcholine				+ + (86, 87)	
Adenosine				+ (86)	
Adrenaline				— (86)	
Amphetamine	+ (123)				
Atropine	— (44, 123)		+ (80, 81)	+ (86)	± (92)
(high doses)		+ (63)			
Azamethonium (Pendiomid)				+(86)	
Benactyzine	— (123)				
Botulinum toxin		± (81)			
Bromolysergic acid diethylamide (BOL)	+ (103)	— (120)			+ + (14) (C)
Caffeine	+ (123)				
Caramiphen				+ (87, 127) ≠	
Chlorpromazine	+ + (123)			+ + (86)	
Cocaine	— (44, 123)				
Deoxyephedrine	+ (123)				
Dibenamine					+ (40) (B) ± (33) (A)
Dihydroergotamine	+ (44)			+ + (86, 87) ≠	
Ephedrine			± (80, 81)		
Ergometrine	+ (103) — (44)				
Ergotamine				+ (135)	

RABBIT

Drug	Tests				
	Perfused ear	Intestine	Heart auricle	Perfused hind limbs	Other tests
Eserine			± (80, 81)		
Ferrous salts	± (88)				
Hexamethonium	— (101)				
Histamine				f (135)	
Hydergin				++ (86)	
Hydralazine				++ (86)	
Lysergic acid diethylamide (LSD)	+ (42, 44, 45) ++ (103, 107, 123)	± (120) + (108)		++ (86, 135)	++ (14) (C)
(small doses)	f (26)				
Meprobamate	— (123)				
Mescaline	— (123)			— (135)	
1-Methyllysergic acid butanolamide (UML 491)					++ (33a) (D)
Methylphenidate (Ritalin)	+ (123)				
Morphine	+ (123) — (123)				
Noradrenaline				f (86)	
Papaverine				++ (86)	
Phentolamine (Regitin)				+ (86)	
Pipradol	— (123)				
Procaine			± (80, 81, 119)	+ (86)	
Promazine	++ (123)				
Promethazine			± (80, 81)	++ (86)	
Reserpine	± (123)			+ (86, 135)	
Scopolamine	± (123)				
Substance P	± (3)				
Transentin				++ (86)	
Tripelennamine (Pyribenzamine)				++ (86)	
D-Tubocurarine			± (80, 81)		
Vasopressin (Pitressin)				f (86)	
Veratrine				f (135)	
Yohimbine				++ (86)	
284 C 51 (Cholinesterase inhibitor)		f (98)			

(A) = Arterial pressure; (B) = Aorta strips; (C) = Intestinal peristalsis; (D) = Uterus *in situ*

RAT

Drug	Tests			
	Isolated uterus	Isolated colon	Diuresis	Other tests
Acetylcholine	x (3)			
Adrenaline	+ (3, 33, 52)	+ (50, 67)		+ (126)(C)
Adrenoglomerulotropin	+ (34a)			
Aminoether J.L. 1276				+ + (82)(D)
1279				+ + (82)(D)
1280				+ + (82)(D)
1281				± (82)(D)
1285				+ (82)(D)
1287				+ (82)(D)
1289				± (82)(D)
1297				± (82)(D)
3, β-Aminoethyl-5-hydroxyindazole	x (1)			
Amphetamine (high doses)	± (66) f (27, 66)	— (67)		
Angiotonin	x (33)			
Antazoline	— (30)			
Atropine	± (78)	— (67, 81)	± (31)	— (79)(D)
Azacyclonol	+ (23, 24)			
Azamethonium (Pendiomid)		— (67)		
Benactzine		+ (11)		
4, N-Benzylanilino-1-methylpiperidine	± (30)			
1-Benzyl-2-methyl-5-(β-aminoethoxy)-tryptamine	+ (130)			
1-Benzyl-2-methyl-5-benzyloxytryptamine	± (130)			
1-Benzyl-2-methyl-1′,10′-decamethylenebis-(5-oxy)tryptamine	± (130)			
1-Benzyl-2-methyl-N-isoindoline-5-(β-aminoethoxy)-tryptamine	± (130)			

RAT

Drug	Tests			
	Isolated uterus	Isolated colon	Diuresis	Other tests
1-Benzyl-2-methyl-5-methoxytryptamine (BAS)	+ + (114) + (129, 130)	— (89)	— (89)	+ + (90)(A) + (5)(C)
1-Benzyl-2-methyl-5-(p-methoxybenzyloxy)-tryptamine	+ + (130)			
1-Benzyl-2-methyl-5-oxyacetohydrazido-tryptamine	+ + (130)			
5-Benzyloxy-3-(2-dimethylaminoethyl)-indole	+ (5)			+ (5, 6)(C)
5-Benzyloxydimethyl-tryptamine	f (5)			+ (5, 6)(C)
5- or 6-Benzyloxygramine	+ (45)			+ (5, 6)(C)
Caffeine		— (67)		
Carbazole derivatives	— (45) ≠			
4-Carbomethoxygramine	+ + (60) ≠			
4-Carboxygramine	± (60) ≠			
Chlorpromazine	+ + (59) + (23, 24, 30, 58, 75)	+ + (49) + (10, 11, 67)	+ (51)	± (16)(B)
Chlorpyramine	— (30)			
Cocaine	+ (119)	— (67)		
4-Cyanogramine	+ (60) ≠			
Deserpidine	+ (23)			
Dibenamine	+ (43, 45) — (32)	— (67)	+ (33) — (125)	+ (31)(D), (90)(A) ± (33)(A)
Dibenzyline	+ (33, 44)		— (33)	
1-(3,4-Dichlorophenyl)-2-isopropylaminoethanol				± (90)(A)
4,4'-Dicyano-3,3'-diindolylmethane	+ + (60) ≠			

RAT

Drug	Tests			
	Isolated uterus	Isolated colon	Diuresis	Other tests
Diethazine (Diparcol)	+ (58, 59) — (30)	± (49)		
3-(2-Diethylaminoethyl)-indole	± (5)			+ (5)(C)
N,N'-Diethyltryptamine	+ (11a)≠			
Dihydroergotamine	+ (3, 32, 37, 133), (19)≠		++ (33)	++ (90)(A)
3,3'-Diindolylmethane	+ (60) ≠			
5,6-Dimethoxygramine	— (33)			
3-(2-Dimethylaminoethyl)-indole	+ (5)			+ (5, 6)(C)
3-(2-Dimethylaminoethyl)-2-methylindole	± (5)			+ (5)(C)
N,N-Dimethyl-5-hydroxytryptamine (Bufotenine)	xx (33)			
N,N-Dimethyltryptamine	f (5) ± (32, 33)		+ (32, 33)	+ (5)(C)
Diphenhydramine (Benadryl)	+ (33) — (30)			
Dock leaf extracts	++ (22)	+ (22)		
Ephedrine (high doses)	± (66) f (66)	± (67)		
Ergobasine	± (19a)			
Ergocornine	— (19) ≠			
Ergometrine	+ (41)			
Ergosine	— (19) ≠			
Ergotamine			± (31)	
Ergotoxine			± (31)	
Gramine	+ (32, 33)		+ (32, 33)	
Harmaline	+ (85a) — (33)		— (33)	+ (85a) (A)
Harmalole	— (33)		— (33)	
Harmine	— (33)		— (33)	
Harmole	— (33)		— (33)	
Heparin	+ (70)	+ (70)		
Histamine	± (3, 33)		± (31)	
Hydroxyzine	++ (118)			

RAT

Drug	Tests			
	Isolated uterus	*Isolated colon*	*Diuresis*	*Other tests*
Hyoscine				— (126)(C)
4-Indolecarboxylic acid diethylamide	+ + (60) ≠			
Iproniazid				— (127)(C) f (8)(A)
Isopropylnoradrenaline		+ + (50, 67)	+ + (51)	+ + (16)(B), (100)(C)
Lysergic acid 1-acetyldiethylamide (ALD 52)	+ + (18, 19, 20) + (101)			+ + (20)(E)
Lysergic acid amide	— (19)			
Lysergic acid amylamide	+ (18, 19)			
Lysergic acid 2-bromodiethylamide (BOL)	+ + (18, 19, 20, 45) + (79, 102)			+ + (5, 6)(C), (20)(E),(53, 102) (A), (126) (C)
Lysergic acid butylamide	± (18, 19)			
Lysergic acid dibutylamide	+ (18, 19)			
Lysergic acid diethylamide (LSD)	+ + (3, 30, 42, 45, 46, 52, 101, 120)	± (67)	+ + (33) + (28)	+ + (16)(B), (79)(D) + (90, 102) (A)
(small doses)	± (44)			
(high doses)	x (44)			
Lysergic acid diisopropylamide	± (18, 19)			
Lysergic acid dipropylamide	+ (18, 19)			
Lysergic acid ethylamide	— (18, 19)			
Lysergic acid isopropylamide	± (18, 19)			
Lysergic acid methylamide	— (18, 19)			
Lysergic acid morpholide	— (19)			
Lysergic acid piperidide	— (19)			
Lysergic acid propylamide	± (18, 19)			

RAT

Drug	Isolated uterus	Isolated colon	Diuresis	Other tests
Lysergic acid pyrrolidide	— (19)			
Lysergic acid pyrrolimide	— (19)			
Medmain	+ (113)			
Mepyramine	— (30)			
Mescaline	— (107) f (23, 24)			
Methapyrilene	± (33)			
Methergin	+ (19a)			
Methoxyharmalan	+ + (85a)			+ + (85a) (A)
3-(2-Methylaminoethyl)-indole	± (5)			
1-Methyl-2-bromo-lysergic acid diethylamide (MOB 61)	+ + (19)			
2-Methyl-3-ethyl-5-aminoindole			+ (33)	
2-Methylgramine	+ (32, 33)		+ (32, 33)	
N-Methyl-5-hydroxytrypt-amine	— (45)			
1-Methyllysergic acid butanolamide (UML 491)	+ + (10a)			
1-Methyllysergic acid diethylamide (MLD 41)	+ + (19)			
1-Methyl-6-methoxy-1,2,3,4-tetrahydro-2-carboline	+ (34a)			
N-Methyltryptamine	± (32, 33)		+ (32, 33)	
Morphine				— (126)(C)
Neohetramine	± (3, 33, 52)	+ (50, 67)		+ (108)(C)
Noradrenaline				+ + (8)(A)
Octopamine	+ (33)			
Oxyphenonium (Antrenyl)		— (67)		
Papaverine		± (67)	± (31)	+ (79)(D)
Pervitin	± (78)	— (67)		
Phenindamine (Thephorin)	— (30)			
Phentolamine (Regitin)				+ (8)(A)
1-Phenylethyl–N,N-dimethyl-5-methoxytrypt-amine	+ + (68) ≠			

RAT

| Drug | Tests | | | |
	Isolated uterus	Isolated colon	Diuresis	Other tests
Phenylisopropylhydrazine				— (127)(C)
Piperoxan	— (31, 44)			
Procaine	+ (119)	— (67)		
Promethazine	+ + (30) + (58, 59) — (23)	+ (49)		± 16(B), 79(D)
Reserpine	+ (11, 23, 24)	+ (49) ± (67)	+ (51)	+ (16)(B)
Substance P	f (33)	f (31) — (97)		+ (3)(D)
1,2,3,4-Tetrahydro-6-(N,N-dimethylaminoethyl)-9-benzylcarbazole	+ + (117)			
1,2,3,4-Tetrahydro-6-(N-phenylcarboxamidino)-carbazole	± (117)			
Thenalidine tartrate (Sandosten)	± (30)			
Theophylline			± (31)	
Thiazinamium methyl-sulphate	± (59)			
Tripelennamine (Pyribenzamine)	+ (33) — (30)			
Tryptamine	x (33)			± (126)(C)
Tyramine	+ (33)			
Yohimbine	± (31)	± (67)		

(A) = Arterial pressure
(B) = Isolated stomach (*in toto*)
(C) = Isolated stomach (strips)
(D) = Isolated duodenum
(E) = Perfused kidneys

SHEEP

Drug	Tests
	Carotid segments
Aminoharmine	+ (112, 132)
Dibenamine	+ (112)
2,3-Dimethyl-4-aminoindole	± (132)
2,3-Dimethyl-5-aminoindole	+ (132)
2,3-Dimethyl-6-aminoindole	+ + (132)
Ergotamine	+ (112, 132)
Ergotoxine	± (112, 132)
3-Ethyl-5-aminoindole	± (132)
Harmane	+ (112, 132)
2-Methyl-3-butyl-5-aminoindole	+ (132)
2-Methyl-3-ethyl-5-dimethylaminoindole (Medmain)	+ (112, 132)
1-Methylmedmain	+ (112, 132)
2-Methyl-3-(β-piperidinoethyl)5-aminoindole	+ (132)
1,2,3,4-Tetrahydro-6-aminocarbazole	+ (132)
1,2,3,4-Tetrahydrocarbazole	— (132)
1,2,3,4-Tetrahydro-6-(N-phenylcarboxamidino)carbazole	+ + (112)
Yohimbine	+ (112, 132)

REFERENCES

1   AINSWORTH, C., *J. Am. Chem. Soc.*, 79 (1957) 5245.
2   AMBACHE, N., in G. P. LEWIS (Ed.), *5-Hydroxytryptamine*, Pergamon, London, 1957, p. 203.
3   AMIN, A. H., CRAWFORD, T. B. B. AND GADDUM, I. H., *J. Physiol. (London)*, 126 (1954) 596.
4   ASTRÖM, A. AND SAMELIUS, U., *Brit. J. Pharmacol.*, 12 (1957) 410.
5   BARLOW, R. B. AND KAHN, J., *Brit. J. Pharmacol.*, 14 (1959) 99.
6   BARLOW, R. B. AND KHAN, J., *Brit. J. Pharmacol.*, 14 (1959) 265.
7   BARLOW, R. B. AND KHAN, I., *Brit. J. Pharmacol.*, 14 (1959) 553.
8   BELESLIN, D. AND VARAGIĆ, V., *Arch. intern. pharmacodyn.*, 128 (1960) 100.
9   BENDA, P. AND MIRAVET, L. F., *Compt. rend. soc. biol.*, 151 (1957) 2064.
10   BENDITT, E. P. AND ROWLEY, D. A., *Science*, 123 (1956) 24.
10a BERDE, B., DÖPFNER, W. AND CERLETTI, A., *Helv. Physiol. Acta*, 18 (1960) 537.
11   BERGER, F. M., CAMPBELL, G. L., HENDLEY, C. D., LUDWIG, B. J. AND LYNES, T. E., *Ann. N. Y. Acad. Sci.*, 66 (1957) 686.
11a BERTACCINI, G. AND ZAMBONI, P., *Arch. intern. pharmacodyn.*, 133 (1961) 138.
11b BUÑAG, R. D. AND WALASZEK, E. J., *Experientia*, 17 (1961) 503.
12   BHATTACHARYA, B. K., *Arch. intern. pharmacodyn.*, 103 (1955) 357.
13   BHATTACHARYA, B. K. AND ATANACKOVIC, D., *Arch. intern. pharmacodyn.*, 104 (1956) 275.
14   BÜLBRING, E. AND LIN, R. C. Y., *J. Physiol. (London)*, 140 (1958) 381.
15   CAMBRIDGE, G. W. AND HOLGATE, J. A., *Brit. J. Pharmacol.*, 10 (1955) 326.
16   CASENTINI, S. AND GALLI, G., *Boll. soc. ital. biol. sper.*, 32 (1956) 1640.
17   CASENTINI, S. AND LANZETTA, A., *Atti soc. lombarda sci. med. biol.*, 12 (1957) 1640.
18   CERLETTI, A. AND DÖPFNER, W., *XX Congr. Physiol.*, Brussels, 1956, Abstr. of papers, p. 165.
19   CERLETTI, A. AND DÖPFNER, W., *J. Pharmacol. Exptl. Therap.*, 122 (1958) 124.
19a CERLETTI, A. AND DÖPFNER, W., *Helv. Physiol. Acta*, 16 (1958) 55.
20   CERLETTI, A. AND KONZETT, H., *Arch. exptl. Pathol. Pharmakol. Naunyn-Schmiedeberg's*, 228 (1956) 146.
21   CIMA, G. AND FRESCHI, C., *Boll. soc. ital. biol. sper.*, 33 (1957) 867.
22   COLLIER, H. D. J., in G. P. LEWIS (Ed.), *5-Hydroxytryptamine*, Pergamon, London, 1957, p. 204.
23   COSTA, E., *Proc. Soc. Exptl. Biol. Med.*, 91 (1956) 39.
24   COSTA, E., *Psychiat. Research Repts.*, 4 (1956) 11.
25   COURVOISIER, S. AND LEAN, O., in P. B. BRADLEY, P. DENIKER AND C. RADOUCO-THOMAS (Eds.), *Neuropsychopharmacology*, Elsevier, Amsterdam, 1959, p. 303.
26   DELAY, J. AND THUILLIER, J., *Compt. rend. soc. biol.*, 150 (1956) 1335.
27   DELAY, J. AND THUILLIER, J., *Compt. rend.*, 242 (1956) 3138.
28   DEL GRECO, F., MASSON, G. M. C. AND CORCORAN, A. C., *Am. J. Physiol.*, 187 (1956) 509.
29   DOMENJOZ, R. AND THEOBALD, W., *Arch. intern. pharmacodyn.*, 120 (1959) 450.
30   DÖPFNER, W. AND CERLETTI, A., *Intern. Arch. Allergy Appl. Immunol.*, 10 (1957) 348.
31   ERSPAMER, V., *Arch. intern. pharmacodyn.*, 93 (1953) 293.
32   ERSPAMER, V., *Ricerca sci.*, 22 (1953) 1203.
33   ERSPAMER, V., *Rend. sci. Farmitalia*, 1 (1954) 1.
33a FANCHAMPS, A., DÖPFNER, W., WEIDMANN, H. AND CERLETTI, A., *Schweiz. med. Wochschr.*, 90 (1960) 1040.
34   FÄNGE, R., *Experientia*, 11 (1955) 156.
34a FARRELL, G. AND MCISAAC, W. M., *Arch. Biochem. Biophys.*, 94 (1961) 543.
35   FASTIER, F. N., MCDOWALL, M. A. AND WAAL, H., *Brit. J. Pharmacol.*, 14 (1959) 527.
36   FELDBERG, W. AND TOH, C. C., *J. Physiol. (London)*, 119 (1953) 352.
37   FINGL, E. AND GADDUM, J. H., *Federation Proc.*, 12 (1953) 320.
38   FINK, M. A., *Proc. Soc. Exptl. Biol. Med.*, 96 (1956) 673.
39   FREYBURGER, W. A., GRAHAM, B. E., RAPPORT, M. M., SEAY, P. H., GOVIER, W. M., SWOAP, O. F. AND VAN DER BROOK, M. J., *J. Pharmacol. Exptl. Therap.*, 105 (1952) 80.
40   FURCHFOTT, R. F., *J. Pharmacol. Exptl. Therap.*, 111 (1955) 265.
41   GADDUM, J. H., *J. Physiol. (London)*, 119 (1952) 363.
42   GADDUM, J. H., *J. Physiol. (London)*, 121 (1953) 15 P.
43   GADDUM, J. H., in G. P. LEWIS (Ed.), *5-Hydroxytryptamine*, Pergamon, London, 1957, p. 195.
44   GADDUM, J. H. AND HAMEED, K. A., *Brit. J. Pharmacol.*, 9 (1954) 240.

45 GADDUM, J. H., HAMEED, K. A., HATAWAY, D. E. AND STEPHENS, F. F., *Quart. J. Exptl. Physiol.*, 40 (1955) 49 L.

46 GADDUM, J. H., HEBB, C. O., SILVER, A. AND SWAN, A. A. B., *Quart. J. Exptl. Physiol.*, 38 (1953) 255.

47 GADDUM, J. H. AND PAASONEN, M. K., *Brit. J. Pharmacol.*, 10 (1955) 474.

48 GADDUM, J. H. AND PICARELLI, Z. P., *Brit. J. Pharmacol.*, 12 (1957) 323.

49 GARATTINI, S. AND VALZELLI, L. *Boll. soc. ital. biol. sper.*, 31 (1955) 1648.

50 GARATTINI, S, AND VALZELLI, L., *Boll. soc. ital. biol. sper.*, 32 (1956) 288.

51 GARATTINI, S. AND VALZELLI, L., *Boll. soc. ital. biol. sper.*, 32 (1956) 295.

52 GARVEN, J. D., *Brit. J. Pharmacol.*, 11 (1955) 66.

53 GESSNER, P. K., KHAIRALLAH, P. A., McISAAC, W. M. AND PAGE, I. H., *J. Pharmacol. Exptl. Therap.*, 130 (1960) 126.

54 VON GIERTZ, H. AND HAN, F., *Arzneimittel-Forsch.*, 9 (1959) 553.

55 GINZEL, K. H. AND KOTTEGODA, S. E., *Quart. J. Exptl. Physiol.*, 38 (1953) 225.

56 GINZEL, K. H. AND KOTTEGODA, S. E., *J. Physiol. (London)*, 123 (1954) 277.

57 GLOVER, W. E., MARSHALL, R. J. AND WHELAN, R. F., *Brit. J. Pharmacol.*, 12 (1957) 498.

58 GYERMEK, L., *Lancet*, 269 (1955) 724.

59 GYERMEK, L., LAZAR, I. AND CSÁK, A. Z., *Arch. intern. pharmacodyn.*, 107 (1956) 62.

60 HARRIS, L. S. AND UHLE, F. C., *J. Pharmacol. Exptl. Therap.*, 128 (1960) 358.

61 HERXHEIMER, H., *J. Physiol. (London)*, 120 (1953) 65 P.

62 HERXHEIMER, H., *J. Physiol. (London)*, 122 (1953) 49 P.

63 HERXHEIMER, H., *J. Physiol. (London)*, 128 (1955) 435.

64 HERXHEIMER, H., *Arch. intern. pharmacodyn.*, 106 (1956) 371.

65 HERXHEIMER, H., in G. P. LEWIS (Ed.), *5-Hydroxytryptamine*, Pergamon, London, 1957, p. 163.

66 HOWIG, T. AND NAESS, K., *Acta Pharmacol. Toxicol.*, 11 (1955) 336.

67 JAQUES, R., BEIN, H. J. AND MEIER, R., *Helv. Physiol. Acta*, 14 (1956) 269.

68 JULIA, M., IGOLEN, J., FELIX, M. AND JACOB, J., *Compt. rend.*, 250 (1960) 1741.

69 KATO, R., MARIANI, L. AND VALZELLI, L., *Atti soc. lombarda sci. med. biol.*, 13 (1958) 297.

70 KELLER, R., *Experientia*, 13 (1957) 112.

71 KIMURA, E. T., YOUNG, P. R. AND RICHARDS, R. K., *J. Allergy*, in press (1960).

72 KING, T. O., *XX Congr. Intern. Physiol.*, Bruxelles, 1956, Abstracts of papers, p. 499.

73 KING, T. O., *Arch. intern. pharmacodyn.*, 110 (1957) 71.

74 KIRKPEKAR, S. M. AND LEWIS, J. J., *J. Pharm. and Pharmacol.*, 10 (1958) 255.

75 KIRKPEKAR, S. M. AND LEWIS, J. J., *J. Pharm. and Pharmacol.*, 10 (1958) 307.

76 KONZETT, H., *Brit. J. Pharmacol.*, 11 (1955) 289.

77 KOSTERLITZ, H. W. AND ROBINSON, J. A., *J. Physiol. (London)*, 129 (1955) 18P.

78 LEITCH, J. L., MARYN, D., DEBLEY, V. G. AND HALEY, T. J., *J. Pharmacol. Exptl. Therap.*, 120 (1957) 408.

79 LÉVY, J., *J. physiol. (Paris)*, 49 (1957) 879.

80 LÉVY, J. AND MICHEL BER, E., *Compt. rend.*, 243 (1956) 326.

81 LÉVY, J. AND MICHEL BER, E., *J. physiol. (Paris)*, 48 (1956) 1051.

82 LÉVY, J. AND MICHEL BER, E., *Compt. rend. soc. biol.*, 152 (1958) 750.

83 MAGISTRETTI, M. AND VALZELLI, L., *Boll. soc. ital. biol. sper.*, 31 (1955) 1035.

84 MARIANI, L., personal communication.

85 MARIANI, L. AND VILLANI, R., *Boll. soc. ital. biol. sper.*, 35 (1959) 1598.

85a McISAAC, W., KHAIRALLAH, P. A. AND PAGE, I. H., *Science*, 134 (1961) 674.

86 MEIER, R., TRIPOD, J. AND STUDER, A., *Arch. intern. pharmacodyn.*, 117 (1958) 185.

87 MEIER, R., TRIPOD, J. AND WIRZ, E., *Arch. intern. pharmacodyn.*, 109 (1957) 55.

88 MILKOVIC, S., SUPEK, Z. AND TABORSKY, J., *Arch. exptl. Pathol. Pharmakol. Naunyn-Schmiedeberg's*, 227 (1955) 221.

89 MURELLI, B., VALSECCHI, A. AND VALZELLI, L., *Boll. soc. ital. biol. sper.*, 33 (1957) 859.

90 OUTSCHOORN, A. S. AND JACOB, J., *Brit. J. Pharmacol.*, 15 (1960) 131.

91 PAGE, I. H., *J. Pharmacol. Exptl. Therap.*, 105 (1952) 58.

92 PAGE, I. H. AND McCUBBIN, J. W., *Am. J. Physiol.*, 174 (1953) 436.

93 QUABECK, G. AND RÖHM, E., *Hoppe-Seyler's Z. physiol. Chem.*, 297 (1954) 229.

94 RAPPORT, M. AND KOELLE, G. B., *Arch. intern. pharmacodyn.*, 92 (1953) 464.

95 REID, G., *J. Physiol. (London)*, 118 (1952) 435.

[96] REID, G. AND RAND, M., *Nature*, 169 (1952) 801.
[97] ROBERTSON, P. A., *J. Physiol. (London)*, 121 (1953) 54 P.
[98] ROBERTSON, P. A., *J. Physiol. (London)*, 125 (1954) 37 P.
[99] ROCHA Y SILVA, M. AND DO VALLE, J. R., *XIX Intern. Physiol. Congress, Montreal*, 1953, Abstr. of papers, p. 708.
[100] ROCHA Y SILVA, M., DO VALLE, J. R. AND PICARELLI, Z. P., *Brit. J. Pharmacol.*, 8 (1953) 378.
[101] ROTHLIN, E., *Ann. N. Y. Acad. Sci.*, 66 (1957) 668.
[101a] RUMMEL, W., *Med. Exptl.*, 4 (1961) 126.
[102] SALMOIRAGHI, G. C., McCUBBIN, J. W. AND PAGE, I. H., *J. Pharmacol. Exptl. Therap.*, 119 (1957) 240
[103] SAVINI, E. C., *Brit. J. Pharmacol.*, 11 (1956) 313.
[104] SCARINCI, V., *Boll. soc. ital. biol. sper.*, 31 (1955) 777.
[105] SCARINCI, V., *Farmaco (Pavia)*, *Ed. sci.*, 10 (1955) 783.
[106] SCHMID, E., ZICHA, L., SHEIFFARTH, F. AND BÜTTNER, O., *Arzneimittel-Forsch.*, 9 (1959) 474.
[107] SCHNECKLOTH, R., PAGE, I. H., DEL GRECO, F. AND CORCORAN, A. C., *Circulation*, 16 (1957) 523.
[108] SCHNEIDER, J. A., GAUNT, T. AND EARL, A. E., *J. Pharmacol. Exptl. Therap.*, 110 (1954) 45.
[109] SCHNEIDER, J. A. AND RINEHART, R. K., *J. Pharmacol. Exptl. Therap.*, 116 (1956) 51.
[110] SCHNEIDER, J. A. AND RINEHART, R. K., *Arch. intern. pharmacodyn.*, 105 (1956) 253.
[111] SCHNEIDER, J. A. AND YONKMAN, R., *J. Pharmacol. Exptl. Therap.*, 84 (1954) 11.
[112] SHAW, E. AND WOOLLEY, D. W., *J. Biol. Chem.*, 203 (1953) 979.
[113] SHAW, E. AND WOOLLEY, D. W., *J. Pharmacol. Exptl. Therap.*, 111 (1954) 43.
[114] SHAW, E. AND WOOLLEY, D. W., *J. Pharmacol. Exptl. Therap.*, 116 (1956) 164.
[115] SHAW, E. AND WOOLLEY, D. W., *Proc. Soc. Exptl. Biol. Med.*, 93 (1956) 217.
[116] SHAW, E. AND WOOLLEY, D. W., *Science*, 124 (1956) 121.
[117] SHAW, E. AND WOOLLEY, D. W., *J. Am. Chem. Soc.*, 79 (1957) 3561.
[118] SHERROD, T. R. AND BOBB, C. A., *Federation Proc.*, 16 (1957) 335.
[119] SINHA, Y. K. AND WEST, G. B., *J. Pharm. and Pharmacol.*, 5 (1953) 370.
[120] SOLLERO, L., PAGE, I. H. AND SALMOIRAGHI, G. C., *J. Pharmacol. Exptl. Therap.*, 117 (1956) 10.
[121] STORMORKEN, H., *Arch. intern. pharmacodyn.*, 119 (1959) 232.
[122] TRAPOLD, J. H., MULFORD, R. AND GROHAM, D., *Federation Proc.*, 19 (1960) 87.
[123] TRIPOD, J., in S. GARATTINI AND V. GHETTI (Eds.), *Psychotropic Drugs*, Elsevier, Amsterdam, 1957, p. 437.
[124] TWAROG, B. M. AND PAGE, I. H., *Am. J. Physiol.*, 175 (1953) 157.
[125] VAN ARMAN, C. G., *Arch. intern. pharmacodyn.*, 108 (1956) 356.
[126] VANE, J. R., *Brit. J. Pharmacol.*, 12 (1957) 344.
[127] VANE, J. R., *Brit. J. Pharmacol.*, 14 (1959) 87.
[128] WELSH, J. H. AND McCOY, A. C., *Science*, 125 (1957) 348.
[129] WOOLLEY, D. W., *Science*, 128 (1958) 1277.
[130] WOOLLEY, D. W., *Biochem. Pharmacol.*, 3 (1959) 51.
[131] WOOLLEY, D. W. AND SHAW, E., *J. Am. Chem. Soc.*, 74 (1952) 2948.
[132] WOOLLEY, D. W. AND SHAW, E., *J. Biol. Chem.*, 203 (1953) 69.
[133] WOOLLEY, D. W. AND SHAW, E., *Federation Proc.*, 12 (1953) 293.
[134] WOOLLEY, D. W., VAN WINKLE, E. AND SHAW, E., *Proc. Nat. Acad. Sci. U. S.*, 43 (1956) 128.
[135] ZAMBONI, P., *Arch. intern. pharmacodyn.*, 113 (1957) 203.

Appendix V

DISTRIBUTION OF SOME BIOLOGICAL COMPOUNDS IN VARIOUS PARTS OF THE BRAIN OF DIFFERENT ANIMAL SPECIES

LEGEND

Ach.	= Acetylcholine
Ach. synth.	= Acetylcholine synthesis
Adr.	= Adrenaline
Ch.est.	= Choline esterase
COMT	= Catechol-O-methyltransferase
DOPA	= Dihydroxyphenylalanine
DOPA dec.	= DOPA decarboxylase
DPA	= Dopamine
DPO	= Dopamine-β-oxidase
Enc.	= Encephaline
Est.	= Esterase
GA	= Glutamic acid
GABA	= γ-Aminobutyric acid
GA dec.	= Glutamic acid decarboxylase
GAm.	= Glutamine
GOT	= Glutamic oxalacetic transaminase
GTh.	= Glutathione
Hist.	= Histamine
Histam.	= Histaminase
Histid.dec.	= Histidine decarboxylase
5-HT	= Serotonin
5-HTP dec.	= 5-Hydroxytryptophan decarboxylase
MO	= Monoamine oxidase
Nadr.	= Noradrenaline
Subst.P	= Substance P
Th.	= Thiamine
Trypt.	= Tryptophan
Trypt. hydrox.	= Tryptophan hydroxylase
(+)	= Means of authors' data
g.m.	= Grey matter
w.m.	= White matter
Figures in brackets	= References

VALUES ARE EXPRESSED AS

μg/acetylcholine formed/g dried tissue/h	for Acetylcholine synthesis
μl CO_2 evolved/g wet weight of tissue/10 min (dog) or	
g Ach. hydrolysed/h/g tissue (human) or	
mequiv. Ach. hydrolysed/min/mg wet weight of tissue (rabbit) or	
moles Ach. $\times$ 10^{10} hydrolysed/min/mg tissue (rat)	for Cholinesterase activity
μmoles of Metanephrine formed/90 min/g tissue	for COMT activity
μmoles of PhAc. hydrolysed/h/g tissue	for Esterase activity
μg of DPA formed/3 h/g tissue	for DOPA decarboxylase activity
μg of 5-HT formed/h/g tissue	for 5-Hydroxytryptophan decarboxylase activity
μg of Hist. formed/24 h/g tissue	for Histidine decarboxylase activity
μmoles of GABA formed/h/kg wet tissue	for Glutamic decarboxylase activity
μmoles of DPNH oxidized/g wet tissue/h	for Glutamic oxalacetic transaminase activity
μg of 5-HT destroyed/h/g tissue or	
μl O_2 consumed/mg mitochondrial protein/h (human[53]) or	
μl O_2 consumed/g tissue/h (dog[53])	for Monoamine oxidase activity
μg of Nadr. formed/3 h/g tissue	for Dopamine-β-oxidase activity
units/g tissue	for Substance P
μequiv./g tissue	for Histamine[30] and Encephaline
μmoles/g tissue	for Glutamic acid, Glutamine, γ-Aminobutyric acid[5], Glutathione
mg %	for γ-Aminobutyric acid[56]
μg/g tissue	for Acetylcholine, DOPA, Dopamine, Adrenaline, Noradrenaline, Histamine, 5-Hydroxytryptamine, Thiamine, Tryptophan

CAT

Tissue	Ach.	5-HTP dec.	5-HT	DOPA dec.	Nadr.
Album et griseum cerebri				51(29)	
Amygdala		35(55)	1.6 (29)	291(29)	0.30(29)
Area acoustica			0.41(29)	69(29)	0.35(29)
visiva				51(29)	
Area ovalis et					
elliptica medullae	1.95(33)				
Bulbus olfactorius	1.3 (33)			69(29)	
Capsula interna	1.29(33)				
Cerebellum	0.18(33)			21(29)	
Colliculus inferior		40(29)	0.76(29)	201(29)	0.26(29)
superior	1.65(33)			201(29)	
Columnae dorsales	0.47(33)				
Corpora quadrigemina,					
see Colliculi					
Corpus callosum	2.1 (33)			120(29)	
Corpus geniculatum					
laterale				99(29)	
mediale		30(29)		129(29)	0.50(29)
Corpus striatum	2.7 (33)				
Cortex cinguli			0.41(29)	51(29)	0.35(29)
cruciata				99(29)	
piriformis		20(29)	1.4 (29)	159(29)	0.35(29)
Formatio reticularis					
medullae				351(29)	
mesencephali		185(29)	2.6 (29)	750(29)	1.4 (29)
pontis		85(29)	0.89(29)	383(29)	0.97(29)
Fornix				129(29)	
Ganglia basalia	7 (33)				
Griseum centrale			1.6 (29)	519(29)	1.2 (29)
Hippocampus				81(29)	
Hypothalamus	1.85(33)			750(29)	
anterior		185(29)	2.4 (29)	750(29)	3.9 (29)
dorsalis				870(29)	
medialis				792(29)	
posterior		175(29)	2.5 (29)	896(29)	2.0 (29)
ventralis				600(29)	
Lobus frontalis (g.m.)	4.5 (33)				
occipitalis (g.m.)	2.2 (33)				
(w.m.)	0.4 (33)				
parietalis (g.m.)	2.8 (33)				
(w.m.)	0.75(33)				

CAT

Tissue	Ach.	5-HTP dec.	5-HT	DOPA dec.	Nadr.
Lobus temporalis (g.m.)	3.2 (33)				
(w.m.)	0.7 (33)				
Medulla				219(29)	
(w.m.)				120(29)	
Mesencephalon	1.6 (33)			648(29)	
Neocortex			0.69(29)	39(29)	0.49(29)
Nervus opticus				51(29)	
Nucleus caudatus		300(29)	1.6 (29)	1260(29)	0.36(29)
ruber				540(29)	
supraopticus				780(29)	
Nuclei cochleares		20(29)		90(29)	0.32(29)
intralaminares et					
mediales thalami		80(29)	0.56(29)	400(29)	0.40(29)
Pituitary				159(29)	
Pons		40(29)	0.33(29)	300(29)	0.25(29)
pars superficialis					
lateralis	1.4 (54)				
medialis	4.1 (54)				
Pulvinar				159(29)	
Pyramis	0.34(54)				
Septum			2.0 (29)	420(29)	1.5 (29)
Thalamus	3.00(54)	50(29)	0.43(29)	231(29)	0.17(29)
basis		135(29)	0.88(29)	660(29)	1.00(29)
Tuberculus olfactorius		264(29)	2.00(29)	930(29)	0.95(29)
Zona hypnogena (Hess)		45(29)	1.8 (29)	261(29)	0.45(29)

DOG

Tissue	Ach. synth.	Ach.	Ch.est.	Trypt.	5-HTP dec.	5-HT	MO	DPA	DPO	Nadr.	Adr.	Hist.	Enc.	Subst. P
Album centrale			53							0.02				3.6
			(12)							(52)				(1)
Area acoustica														
(medullae)										0.39				
										(52)				
Area acoustica														
(23 et 36a)										0.05				
										(52)				
Area frontalis (6)										0.08				
										(52)				
Area motoria (4)	229.33					0.078				0.18				19
	(21)					(22)				(52)				(1)
	71.00													
	(12)													
Area olfactoria	261.63					0.016				0.12				29
(51)	(21)					(49)				(52)				(1)
	8.1													
	(12)													
Area postrema						0.215				1.04				
						(49)				(52)				
Area somaesthetica														
(3)	184.11		282							0.19				290
	(21)		(12)							(52)				(1)
	57.00													457
	(12)													(40)
Area septalis					109	1.5	1217							

	(21) 58.00 (12)				(22)				(52)		(1)
Brachium conjunctivum cerebelli									0.09 (52)		
Bulbus olfactorius	177.65 (21)				0.39 (51) 0.078 (39)	573 (51)			0.05 (52)	0.35(+) (23)	
Capsula interna											
pars anterior	122.74 (21)										
pars centralis	335.92 (21)										
pars posterior	87.21 (21)										
Cerebellum		1 (32)	1–4 (43)	9 (51)	0.09 (51) 0.04 (49) 0.02 (43)	930 (51)	0.03 (13)	0.08 (50) 0.06 (13)		1.5 (23)	1.6 (1) 15.0 (41)
(g.m.)				48 (53)							
Colliculus											
inferior	142.12 (21)								0.11 (52)		20 (1)
superior	335.92 (21)								0.16 (52)		

DOG

Tissue	Ach. synth.	Ach.	Ch.est.	Trypt.	5-HTP dec.	5-HT	MO	DPA	DPO	Nadr.	Adr.	Hist.	Enc.	Subst. P
Cornua anteriora														
(g.m.)	494.19									0.18				
	(21)									(52)				
lateralia														
(g.m.)										0.19				
										(52)				
posteriora														
(g.m.+w.m.)	332.69									0.12				
	(21)									(52)				
Cornu Ammonis														
(g.m.)	348.84									0.04				
	(21)									(52)				
(w.m.)										0.03				
										(52)				
Corpora mammillaria	148.58									0.41				
	(21)									(52)				
Corpora quadrigemina														
inferiora			811											
			(12)											
superiora			1619											
			(12)											
Corpus callosum	83.98		62	53						0.08			1.1	5.9
	(21)		(12)	(53)						(52)			(44)	(1)
	26.00													8.0

	(21)	(12)	(51)	(51)		(52)		
mediale	193.8 (21)	938 (12)	7.6 (51)	865 (51)		0.13 (52)		
Cortex cerebellaris	27.45 (21)					0.07 (52)		
Cortex cerebralis			0.17 (51)	844 (51) 53 (53)	0.004 (50)	0.11 (52)	0.05 (30)	1.7 (44)	
Fibrae cerebellares						0.05 (52)			
Formatio reticularis pars anterior						0.36 (52)			
pars posterior						0.34 (52)			
Funiculus gracilis et cuneatus	32.30 (21)								27 (1)
Griseum centrale						0.42 (52)			130 (41)
Gyrus cingulatus									76 (1)
anterior (24)			0.029 (39)			0.08 (52)			
medialis (23, 24)		377 (12)				0.09 (52)			
posterior (26, 29, 30)			0.019 (39)			0.09 (52)			
Gyrus postcentralis (4a)		330 (12)							

DOG

Tissue	Ach. synth.	Ach.	Ch.est.	Trypt.	5-HTP dec.	5-HT	MO	DPA	DPO	Nadr.	Adr.	Hist.	Enc.	Subst. P
Gyrus praecentralis (6a)			371 (12)											
prorealis (8)			331 (12)											
suprasplenialis (18, 19)			231 (12)											
sylvianus (52)			407 (12)											
uncinatus (51)			897 (12)											
Hemispheria cerebelli	8.5 (12)		1850 (12)											
cerebri				0.1 (43)		0.01 (43)								20 (41)
Hippocampus	3.06 (21)	3.06 (32)			15.5 (51)	0.64 (51) 0.26 (39)	1176 (51)	0.13 (13)		0.14 (13)				15 (1)
Hypothalamus	115 (12)	3.39 (32)	866 (12)	5–40 (43)	117 (51) 74 (53)	1.65 (+) (51) 0.28 (1) 0.99 (22) 0.40 (43)	1624 (51)	0.26 (13)	2.18 (50)	1.03 (52) 0.76 (13)	0.17 (52)			70 (1) 154 (40) 180 (41)

dorsalis							0.43 (52)	
lateralis							0.71 (52)	22 (1)
medialis							1.12 (52)	22 (1)
posterior				0.225 (19)			0.86 (52)	70 (1)
ventralis							0.95 (52)	
Infundibulum	87.21 (21)							
Liquor				0.004 (19)			0.05 (52)	
Lobus frontalis	1.69 (32)							
temporalis	2.43 (32)							
Mesencephalon				0.72 (22) 0.205 (19)	0.20 (13)		0.33 (13)	68 (1)
Midbrain			97.6 (51)	1.00 (51)	842 (51)		0.37 (52)	44 (1)
Nervus facialis	300.39 (21)							
opticus	18.09 (21) 5.6 (12)	283 (12)					0.02 (52)	6 (1)

References p. 378

DOG

Tissue	Ach. synth.	Ach. Ch.est.	Trypt.	5-HTP dec.	5-HT	MO	DPA	DPO	Nadr.	Adr.	Hist.	Enc.	Subst. P
Nervus vestibularis	13.67 (21)												
Nuclei cerebellares	67.83 (21)												
Nuclei thalamici laterales									0.08 (52)				
mediales									0.24 (52)				
Nucleus amygdalae				17.6 (51)	2.1 (51) 0.489 (39)	968 (51)			0.06 (52)				
caudatus	410.21 (21)	7450 (12)		306 (51)	0.76 (+) (51)	935 (51)	6.5 (8)	1.60 (50)	0.06 (52)			3.2 (44)	46 (1)
	127.00 (12)				0.229 (39)		5.9 (13)		0.10 (53)				201 (40)
cuneatus et gracilis	277.78 (21) 86.00 (12)	933 (12)							0.11 (52)				110 (1)
hypoglossi et vagi motor	574.94 (21)												
lenticularis		4002				1.62			0.08				

Region									
(continued)	(21) 87.00 (12)	(12)							
ruber et fossa interpeduncularis								0.26 (52)	
supraopticus	371.45 (21)								
vestibularis	138.89 (21)								
Pedunculi cerebellares									
pes	71.06 (21)	237 (12)							41 (1)
inferior	116.28 (21)								
medianus	193.8 (21)	582 (12)							
superior	190.57 (21)	645 (12)							
Pituitary									
lobus anterior	19.26 (21)								
lobus posterior	22.61 (21)	252 (12)							
Pons			1–20 (43)	28 (51)	0.38 (51) 0.05 (43)	936 (51)	0.10 (13)	0.41 (13)	45 (41)
pars ventromedialis								0.20 (52)	
Pyramis	39.73 (21)	177 (12)						0.06 (52)	6 (1)

References p. 378

Tissue	Ach. synth.	Ach.	Ch.est.	Trypt. 5-HTP dec.	5-HT	MO	DPA	DPO	Nadr.	Adr.	Hist.	Enc.	Subst. P
Regio nuclei X et XII									0.31 (52)				
Regio praeoptica									0.28 (52)				
Substantia nigra subcorticalis	164.73 (21)					37 (53)							
Thalamus				38 (51)	0.57 (+) (51) 0.47 (22) 0.018 (1)	940 (51) 81 (53)							12.5 (1) 85.0 (41)
dorsolateralis			808 (12)										
massa intermedia			1437 (12)		0.00 (1)				0.079 (52)				8.4 (1)
medialis					0.067 (1)				0.234 (52)				11.0 (1)
Tractus olfactorius	151.81 (21)										<0.1 (30)		
Tractus opticus	113.05 (21)		237 (12)	7.5 (51)		701 (51)							50 (41)
pars lateroventralis	38.76 (21)												
pars mediodorsalis	132.43												

Tissue	Hist.	DOPA	DPA	Nadr.	5-HT	MO	Ach.	Ch.est.	Est.	Subst. P	Enc.
				(1)							
pars centralis								0.27 (52)			(1) 77 (40)
Whole brain						0.19 (8) 0.22 (13)		0.14 (13)	0.3 (54)	2 (26)	15 (15)

HUMAN

Tissue	Hist.	DOPA	DPA	Nadr.	5-HT	MO	Ach.	Ch.est.	Est.	Subst. P	Enc.
Area postrema					0.57(16)						
Bulbus olfactorius	0.6(+) (30)										
Capsula interna		0.03(47)	0.38(47) 0.42 (7)	0.04(47) 0.00 (7)					409(6)		
Cerebellum		0.02(47)	0.00(47)	0.01(47)	0.03(16)	6.8(53)			305(6)		1.5(44)
Claustrum			0.44 (7)	0.02 (7)							
Colliculi, see Corpora quadrigemina											
Corpora mammillaria					0.39(16)		124(56)			34.3(56)	
Corpora quadrigemina		0.04(47)	0.07(47)	0.15(47) 0.12(24)	0.635(+) (16) 0.76(24)						
anteriora			0.13 (7)	0.12 (7)	0.81(16)		340(56)			55(56)	
posteriora			0.10 (7)	0.15 (7)	0.46(16)		147(56)			140.7(56)	

HUMAN

Tissue	Hist.	DOPA	DPA	Nadr.	5-HT	MO	Ach.	Ch.est.	Est.	Subst. P	Enc.
Corpus callosum		0.01(47)	0.05(47)	0.00(47)	0.25(16)	9.6(53)	89(56)	0.4(4)	283(6)	1.6(56)	0.7(44)
Corpus geniculatum											
laterale							324(56)			4.9(56)	
mediale							204(56)			36.8(56)	
Cortex cerebellaris			0.02 (7)	0.02 (7)					710(6)		
cerebralis					0.048(16)	9.8(53)		0.8(4)			1.2(44)
Crura cerebri										2.9(56)	
Fimbria hippocampi		0.02(47)	0.13(47)	0.04(47)							
Fornix anterior pars					0.36(16)						
Globus pallidus			0.32 (7)	0.05 (7)	0.08 (7)				719(6)	103 (56)	
Gyrus											
cinguli		0.01(47)	0.00(47)	0.05(47)	0.042(16)				730(6)	84.5(56)	
dentatus		0.01(47)	0.12(47)	0.03(47)							
frontalis											
inferior		0.00(47)	0.02(47)	0.02(47)							
medianus		0.00(47)	0.03(47)	0.02(47)							
superior		0.00(47)	0.11(47)	0.01(47)							
hippocampi		0.01(47)	0.04(47)	0.02(47)	0.16(16)				478(6)		
occipitalis lateralis		0.03(47)	0.05(47)	0.01(47)							
parietalis inferior		0.03(47)	0.02(47)	0.01(47)							
parietalis superior		0.00(47)	0.07(47)	0.00(47)							
postcentralis		0.03(47)	0.17(47)	0.03(47)			190(56)		755(6)	38.5(56)	
			0.06 (7)	0.03 (7)							
praecentralis		0.00(47)	0.13(47)	0.06(47)			186(56)		803(6)	43.3(56)	
			0.01 (7)	0.02 (7)							

medialis	0.03(47)	0.06(47)	0.01(47)							
superior	0.01(47)	0.08(47)	0.02(47)							
Hippocampus				0.088(16)						
Hypothalamus	0.06(47)	1.12(47)	1.11(47)	0.81(16)	16.6(53)			518(6)		
				0.52(24)						
pars anterior		0.18 (7)	0.96 (7)	0.35 (7)						
pars intermedia		0.14 (7)	1.19 (7)	0.48 (7)						
pars posterior		0.22 (7)	0.31 (7)	0.30 (7)						
Insula	0.01(47)	0.08(47)	0.04(47)					768(6)		
Liquor cisternalis										
lumbaris				0.00(24)						0.4(44)
Medulla oblongata	0.02(47)	0.17(47)	0.14(47)	0.34(16)						0.2(44)
Monticulus	0.02(47)	0.03(47)	0.01(47)							
Nervus opticus	0.003(+)									
	(54)									
vagus	3(54)									
Nucleus amygdalae	0.01(47)	0.13(47)	0.06(47)	0.32(16)						
caudatus	0.02(47)	5.74(47)	0.04(47)	0.44(16)		435(56)	31.8(4)		84.3(56)	1.9(44)
		3.12 (7)	0.04 (7)	0.30(24)						
				0.27 (7)						
dentatus cerebelli	0.04(47)	0.03(47)	0.01(47)					920(6)		
		0.08 (7)	0.01 (7)							
pallidus	0.02(47)	1.01(47)	0.02(47)					719(6)		
ruber	0.08(47)	1.17(47)	0.23(47)			216(56)		442(6)	29.8(56)	
		0.19 (7)	0.22 (7)							
Olivae	0.03(47)	0.05(47)	0.04(47)							
Operculum								844(6)		
Pinealis	0.04(47)	0.5 (47)	0.1 (47)							
		0.02 (7)	0.00 (7)							
Pituitary										3.6(44)
Plexus chorioidaeus	0.07(47)	0.11(47)	0.04(47)							1.3(44)

HUMAN

Tissue	Hist.	DOPA	DPA	Nadr.	5-HT	MO	Ach.	Ch.est.	Est.	Subst. P	Enc.
Polus frontalis			0.00 (7)	0.01 (7)	0.014(16)				786(6)		
occipitalis			0.06 (7)	0.00 (7)	0.02 (16)				698(6)		
									796(6)		
temporalis					0.037(16)				829(6)		
Pons		0.013(+)	0.11(+)	0.06(+)	0.7 (16)				470(6)	2.6(56)	
		(47)	(47)	(47)	0.19 (7)						
			0.00 (7)	0.04 (7)							
dorsalis		0.01(47)	0.07(47)	0.02(47)							
ventralis		0.03(47)	0.08(47)	0.15(47)							
brachium pontis			0.17(47)	0.01(47)							
Putamen		0.03(47)	8.25(47)	0.07(47)	0.55(16)					63.7(56)	
			5.27 (7)	0.02 (7)	0.23 (7)						
					0.38(24)						
+ globus pallidus						16.3(53)					
+ nucleus caudatus					0.51(16)						
Pyramis		0.03(47)	0.17(47)	0.01(47)							
Septum pellucidum			0.04 (7)	0.03 (7)	0.03 (7)						
+ fornix		0.02(47)	0.07(47)	0.17(47)							
Substantia										118.6(56)	
grisea centralis		0.04(47)	0.38(47)	0.07(47)	1.42(16)		126(56)		594(6)	698.8(56)	
nigra			0.40 (7)	0.04 (7)	1.03(24)						
perforata anterior					0.98(16)						
subcorticalis						9.1(53)		0.5(4)			
subcorticalis frontalis									280(6)		
subcorticalis occipitalis									352(6)		
Thalamus		0.023(47)	0.29(47)	0.043(47)	0.62(16)	19(53)	323(56)		656(6)	12.2(56)	1.2(44)
lateralis		0.03(47)	0.3 (47)	0.04(47)	0.22 (7)						

Tissue						
		0.03 (7)	0.09 (7)			
rostralis	0.02(47)	0.11(47)	0.00(47)			
		0.07 (7)	0.02 (7)			
Trigonum hypoglossi					425(56)	37.3(56)
Uncus				0.22(16)		

MONKEY

Tissue	5-HT	NAD	COMT
Amygdala	0.29(43a)	0.11(43a)	0.37(4a)
Area postrema			0.94(4a)
Corpus callosum			0.82(4a)
Cortex cerebellaris (lobus ant.)			0.34(4a)
Gyrus praecentralis			0.58(4a)
Hippocampus (ventralis)			0.46(4a)
Hypothalamus	0.77(43a)	1.25(43a)	0.73(4a)
Nucleus caudatus	0.36(43a)	0.27(43a)	0.78(4a)
lenticularis thalami	0.32(+)(43a)	0.38(43a)	
Pons et medulla	0.49(43a)	0.52(43a)	
Various cortical structures	0.17(43a)	0.13(43a)	

OX

Tissue	Ach.	Histam.	Hist.	Histid. dec.	Trypt. hydrox.	5-HTP dec.	5-HT	MO	Nadr.
Pineal body	0.6(27)	Absent(26)	10(26)	Absent(26)	Absent(26)	Present(26)	0.4(26)	Present(26)	0.4(26)

RABBIT

Tissue	Histid. dec.	Hist.	Ach.	Ch.est.	5-HT	DPA	Nadr.	GA dec.	Enc.
Brain stem					0.87 (42)		0.50(48)		
Bulbus olfactorius		0.1(30)			0.17 (25)				
Cerebellum					0.112(17)			11.8(31)	
Colliculus caudalis	9.36(37)								
rostralis	10.47(37)								
superior								36.9(31)	
Cornu Ammonis								16.6(31)	
Corpora quadrigemina									
inferiora				9.12(+) (3)					
superiora				26.3(+) (3)					
Corpus callosum	5.94(37)								
Corpus geniculatum									
laterale	7.83(37)								
mediale	11.3 (37)								
Corpus mammillare	7.83(37)								
Cortex frontalis			0.88(+) (35)	6.13(+) (3)				16.2(31)	
Gyrus postcentralis	12 (37)								
praecentralis	11.5 (37)								
Hippocampus					0.272(17)				
Hypothalamus	14.4 (37)				0.39 (25)			26.7(31)	
Medulla								11.3(31)	
Medulla + pons					0.582(17)				
Mesencephalon					0.647(17)				
Nervus opticus	9.9 (37)	0.03(+)							

		(2)	(3)					
Nucleus lenticularis	14.4 (37)							
Substantia nigra	8.64(37)							
Substantia rubra	6.48(37)							
Telencephalon			0.17(17)					
Thalamus	13.44(37)		13.30(+) (3)					
Tractus olfactorius		<0.1(30)						
Vermis cerebelli	16.32(37)							
Whole brain		0.6(55)		0.65 (+) (42) 0.46 (9)	0.31 (8) 0.4 (14) 0.28(13)	0.99(48) 0.22(13)	19.7(31)	1.2(44)

References p. 378

R A T

Tissue	GOT	GA	GAm.	GABA	GTh.	Ach.	Ch.est.	Subst. P	DPO	DPA	Nadr.	Adr.	5-HT	Enc.	Th.
Area diencephalica													0.81 (23)		
mesencephalica													0.56 (23)		
motoria							66.1(+) (28)								
spatialis							62.1(+) (28)								
visiva							52.6(+) (28)								
Basis pontis															13.1 (19)
Bulbus olfactorius		9.8 (5)	5.1 (5)	3.2 (5)	2 (5)										
Cerebellum	4640 (34)	10.2 (5)	5.8 (5)	1.6 (5) 28 (46)	1.9 (5)								0.117 (23)		
Colliculi				57 (46)											
Corpus callosum		8.9 (+) (5)	4.2 (+) (5)	0.55 (+) (5)	2.15 (+) (5)										
Cortex (g.m.)	4420 (34)	11.6 (5)	5 (5)	2 (5)	2.3 (5)										13.0 (19)
frontalis													0.162 (27a)		
Diencephalon				58											

	C1	C2	C3	C4	C5	C6	C7	C8	C9	C10	C11	C12	C13	C14
Hemispheria												0.57 (23)		
Hypothalamus	3980 (34)											0.315 (27a)		13.0 (19)
Hypothalamus + Thalamus		10.7 (5)	4.8 (5)	3.5 (5)	2.1 (5)									
Medulla	3960 (34)	5.7 (5)	3 (5)	1.5 (5)	1.2 (5)									13.4 (19)
Nucleus caudatus														17.4 (19)
Pons	5130 (34)	6.4 (5)	3.2 (5)	1.7 (5)	1.4 (5)									
Regio mammillaris														12.4 (19)
Regio periaqueductalis														13.2 (19)
Regio subcorticalis (w.m.)	4410 (34)													
Tegmentum pontinum laterale														11.5 (19)
Thalamus	4600 (34)													9.2 (19)
Vermis														21.1 (19)
Whole brain		1.53 (45)		0.203 (45)		3.18 (11) 2 (18) 3.45 (20) 3.06 (10)	20 (15)	0.26 (50)	0.60 (8) 0.60 (13)	0.44 (26) 0.22 (38) 0.49 (13)	0.21 (36)	0.47 (23) 0.40 (9)	2.8 (44)	

REFERENCES

1 AMIN, A. H., CRAWFORD, T. B. B. AND GADDUM, J. H., *J. Physiol. (London)*, 126 (1954) 596.
2 APRISON, M. H. AND NATHAN, P., *Arch. Biochem. Biophys.*, 66 (1957) 388.
3 APRISON, M. H., NATHAN, P. AND HIMWICH, H. E., *Am. J. Physiol.*, 177 (1954) 175.
4 ASHBY, W., GARZOLI, R. F. AND SCHUSTER, E. M., *Am. J. Physiol.*, 170 (1952) 116.
4a AXELROD, J., ALBERS, W. AND CLEMENTE, C. D., *J. Neurochem.*, 5 (1959) 68.
5 BERL, S. AND WAELSCH, H., *J. Neurochem.*, 3 (1958) 161.
6 BERNSOHN, J., POSSLEY, L. AND LIEBERT, E., *J. Neurochem.*, 4 (1959) 191.
7 BERTLER, A., *Acta Physiol. Scand.*, 51 (1961) 97.
8 BERTLER, A. AND ROSENGREN, E., *Experientia*, 15 (1959) 10.
9 BOGDANSKI, D. F., PLETSCHER, A., BRODIE, B. B. AND UDENFRIEND, S., *J. Pharmacol. Exptl. Therap.*, 117 (1956) 82.
10 BOSE, B. C., GUPTA, S. S. AND SHARMA, S., *Arch. intern. pharmacodyn.*, 117 (1958) 248.
11 BOSE, B. C., GUPTA, S. S. AND SHARMA, S., *Arch. intern. pharmacodyn.*, 117 (1958) 254.
12 BURGEN, A. S. V. AND CHIPMAN, L. M., *J. Physiol. (London)*, 114 (1951) 296.
13 CARLSSON, A., *Pharmacol. Revs.*, 11 (1959) 490.
14 CARLSSON, A., LINDQVIST, M., MAGNUSSON, T. AND WALDECK, B., *Science*, 127 (1958) 471.
15 CORREALE, P., *Arch. intern. pharmacodyn.*, 119 (1959) 435.
16 COSTA, E. AND APRISON, M. H., *J. Nervous Mental Disease*, 126 (1958) 289.
17 COSTA, E. AND RINALDI, F., *Am. J. Physiol.*, 194 (1958) 214.
18 CROSSLAND, J., ELLIOT, K. A. AND PAPPIUS, H. M., *Am. J. Physiol.*, 183 (1955) 32.
19 DREYFUS, P. M., *J. Neurochem.*, 4 (1959) 183.
20 ELLIOT, K. A. C., SWANK, R. L. AND HENDERSON, N., *Am. J. Physiol.*, 162 (1950) 469.
21 FELDBERG, W. AND VOGT, M., *J. Physiol. (London)*, 107 (1948) 372.
22 FRESIA, P., GENOVESE, E., KATO, R. AND VALZELLI, L., *Boll. soc. ital. biol. sper.*, 34 (1958) 1397.
GARATTINI, S., KATO, R. AND VALZELLI, L., *Atti soc. lombarda sci. med. biol.*, 13 (1958) 300.
24 GARATTINI, S., LODI, F. AND VALZELLI, L., Unpublished data (1960).
25 GARVEN, J. D., *Brit. J. Pharmacol.*, 11 (1956) 66.
26 GIARMAN, N. J. AND DAY, M., *Biochem. Pharmacol.*, 1 (1958) 235.
27 GIARMAN, N. J., DAY, M. AND PEPEU, G., *Federation Proc.*, 18 (1959) 1555.
27a JOYCE, D., *Brit. J. Pharmacol.*, 18 (1962) 370.
28 KECH, D., ROSENZWEIG, H. R., BENNETT, E. L. AND KRUECKEL, B., *Science*, 120 (1954) 994.
29 KUNTZMAN, R., SHORE, P. A., BOGDANSKI, D. AND BRODIE, B. B., *J. Neurochem.*, 6 (1961) 226.
30 KWIATOWSKI, H., *J. Physiol. (London)*, 102 (1943) 32.
31 LOWE, I. P., ROBINS, E. AND EYERMAN, G. S., *J. Neurochem.*, 3 (1958) 8.
32 MALHOTRA, C. I. AND PUNDLIK, P. C., *Brit. J. Pharmacol.*, 14 (1959) 46.
33 McINTOSH, F. C., *J. Physiol. (London)*, 99 (1941) 436.
34 MAY, L., MIYAZAKI, M. AND GRENELL, R. G., *J. Neurochem.*, 4 (1959) 269.
35 MICHAELIS, M., FINESINGER, J. E., DE BALBIAN-VERSTER, F. AND ERICKSON, R. W., *Brit. J. Pharmacol.*, 111 (1954) 169.
36 MONTAGU, K. A., *Biochem. J.*, 63 (1956) 559.
37 NAITO, T. AND KURIAKI, K., *Arch. exptl. Pathol. Pharmakol., Naunyn-Schmiedeberg's*, 232 (1956) 481.
38 PAASONEN, M. K. AND DEWS, P. B., *Brit. J. Pharmacol.*, 13 (1958) 84.
39 PAASONEN, M. K., MACLEAN, P. D. AND GIARMAN, N. J., *J. Neurochem.*, 1 (1957) 326.
40 PAASONEN, M. K. AND VOGT, M., *J. Physiol. (London)*, 131 (1956) 617.
41 PERNOW, B., *Nature*, 171 (1953) 746.
42 PLETSCHER, A., SHORE, P. A. AND BRODIE, B. B., *J. Pharmacol. Exptl. Therap.*, 116 (1956) 84.
43 PRICE, S. A. P. AND WEST, G. B., *Nature*, 185 (1960) 470.
43a PSCHEIDT, G. R. AND HIMWICH, H. E., *Biochem. Pharmacol.*, 12 (1963) 65
44 RAAB, W. AND GIGEE, W., *Proc. Soc. Exptl. Biol. Med.*, 76 (1951) 97.
45 RINDI, G. AND FERRARI, G., *Nature*, 183 (1959) 608.
46 ROBERTS, E., BAXTER, C. F. AND EIDELBERG, E., in D. B. TOWER AND J. P. SCHADÉ (Eds.), *Structure and Function of the Cerebral Cortex, Proceedings of the 2nd International Meeting of Neurobiologists, Amsterdam*, Elsevier, Amsterdam, 1959, p. 392.
47 SANO, I., GAMO, T., KAKIMOTO, Y., TANIGUCHI, K., TAKESADA, M. AND NISHINUMA, K., *Biochim. Biophys. Acta*, 32 (1959) 586.

48 SHORE, P. A. AND OLIN, J. S., *J. Pharmacol. Exptl. Therap.*, 122 (1958) 295.
49 UDENFRIEND, S., in G. P. LEWIS (Ed.), *5-Hydroxytryptamine*, Pergamon, London, 1957, p. 43.
50 UDENFRIEND, S. AND CREVELING, C. R., *J. Neurochem.*, 4 (1959) 350.
51 UDENFRIEND, S., WEISSBACH, H. AND BOGDANSKI, D. F., *Hormones, Brain Function and Behaviour*, Academic Press, New York, 1957, p. 147.
52 VOGT, M., *J. Physiol. (London)*, 123 (1954) 451.
53 WEINER, N., *J. Neurochem.*, 6 (1960) 79.
54 WERLE, E. AND PALM, D., *Biochem. Z.*, 323 (1952) 255.
55 WEST, G. B., *J. Pharm. and Pharmacol.*, 11 (1959) 17.
56 ZETLER, G. AND SCHLOSSER, L., *Naturwissenschaften*, 41 (1953) 46.

SUBJECT INDEX

The reader is referred to the Appendices (I–V) for the information concerning:

– distribution of 5-HT in given tissues of various animal and vegetal species (pages 241–276);
– effect of given drugs and treatments alone or in combination on the level of tissue 5-HT of different animal species (pages 277–333);
– effect of given drugs on the activity of 5-HT in various biological tests (pages 335–355);
– distribution of 5-HT and other biological compounds in various parts of the brain of different animal species (pages 356–379).